Space Storms and Space Weather Hazards

NATO Science Series

A Series presenting the results of scientific meetings supported under the NATO Science Programme.

The Series is published by IOS Press, Amsterdam, and Kluwer Academic Publishers in conjunction with the NATO Scientific Affairs Division

Sub-Series

I. Life and Behavioural Sciences	IOS Press
II. Mathematics, Physics and Chemistry	Kluwer Academic Publishers
III. Computer and Systems Science	IOS Press
IV. Earth and Environmental Sciences	Kluwer Academic Publishers

The NATO Science Series continues the series of books published formerly as the NATO ASI Series.

The NATO Science Programme offers support for collaboration in civil science between scientists of countries of the Euro-Atlantic Partnership Council. The types of scientific meeting generally supported are "Advanced Study Institutes" and "Advanced Research Workshops", and the NATO Science Series collects together the results of these meetings. The meetings are co-organized bij scientists from NATO countries and scientists from NATO's Partner countries – countries of the CIS and Central and Eastern Europe.

Advanced Study Institutes are high-level tutorial courses offering in-depth study of latest advances in a field.
Advanced Research Workshops are expert meetings aimed at critical assessment of a field, and identification of directions for future action.

As a consequence of the restructuring of the NATO Science Programme in 1999, the NATO Science Series was re-organized to the four sub-series noted above. Please consult the following web sites for information on previous volumes published in the Series.

http://www.nato.int/science
http://www.wkap.nl
http://www.iospress.nl
http://www.wtv-books.de/nato-pco.htm

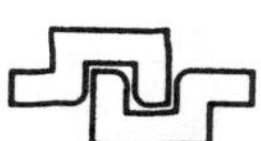

Series II: Mathematics, Physics and Chemistry – Vol. 38

Space Storms and Space Weather Hazards

edited by

I.A. Daglis

Institute for Space Applications and Remote Sensing,
National Observatory of Athens,
Athens, Greece

Kluwer Academic Publishers

Dordrecht / Boston / London

Published in cooperation with NATO Scientific Affairs Division

Proceedings of the NATO Advanced Study Institute on
Space Storms and Space Weather Hazards
Hersonissos, Crete, Greece
19–29 June, 2000

A C.I.P. Catalogue record for this book is available from the Library of Congress.

ISBN 1-4020-0030-8 (HB)
ISBN 1-4020-0031-6 (PB)

Published by Kluwer Academic Publishers,
P.O. Box 17, 3300 AA Dordrecht, The Netherlands.

Sold and distributed in North, Central and South America
by Kluwer Academic Publishers,
101 Philip Drive, Norwell, MA 02061, U.S.A.

In all other countries, sold and distributed
by Kluwer Academic Publishers,
P.O. Box 322, 3300 AH Dordrecht, The Netherlands.

Printed on acid-free paper

Printed in the Netherlands.

TABLE OF CONTENTS

PREFACE

Following the advent of spaceflight in 1957, space has become the new frontier of human exploration and exploitation. Successive spacecraft missions led us to realise that space is far from the absolute void that we once thought of. "Clouds" of plasma and magnetic field emanating from the Sun traverse the interplanetary space and create weather disturbances in geospace, much like the motion of cloud masses in the troposphere are responsible for weather on the Earth's surface. Space storms, the manifestation of bad weather in space, have a number of physical effects in near-Earth space environment: Acceleration of charged particles in space, intensification of electric currents in space and on the ground, impressive aurora displays, and global magnetic disturbances on the Earth surface – a defining storm feature and the origin of the classical name "magnetic storm".

Aurora, the glorious visible revelation of solar-terrestrial electromagnetic coupling, has impressed humans since antiquity. Records of aurora sightings exist in Chinese myths and the Old Testament. However, the first non-mythical, non-mystical descriptions of aurora are credited to the Greek philosophers of nature Anaximenes and Xenophanes (6th century BC). Space weather, the invisible but nevertheless effective collective result of solar activity, geomagnetism and the resulting solar-terrestrial coupling, has affected the operations of communication systems since the appearance of telegraph in the 19th century, though as an unknown trouble factor. Nowadays, the effects of space weather affect in various ways all communication modes, from cable to wireless to space-based systems. Moreover, space weather hazards include malfunction or even permanent damage of power distribution grids and of telecommunication, navigation and surveillance satellites, disturbances of over-the-horizon (OTH) radar, HF, VHF and UHF communications, surveying and navigation systems that use Global Positioning System (GPS) satellites, surveillance (optical and radar), and satellite tracking. Space weather influences on the Earth's weather and climate is still a developing topic. Little is known concerning the biological effects of space weather, especially regarding astronauts and airline crew staff.

This Institute could not be timelier. Space Weather, which has been defined as "conditions on the sun and in the solar wind, magnetosphere, ionosphere, and thermosphere that can influence the performance and reliability of space- and ground-based technological systems and can endanger human life", encompasses a broad range of phenomena impacting both space science and technology. The effects of Space Weather, while present throughout the solar cycle, gain particular prominence around the peak of the 11-year sunspot activity; the maximum of cycle 23 occurred in April 2000. A large number of new data sets are becoming available to help us in our efforts to successfully reveal the complex physics of space storms and solar-terrestrial coupling in general, and to attain the capability of accurate and reliable space weather forecasting. In order to achieve these goals it is imperative to maintain appropriate instrumental capabilities and assure training of young scientists who will lead the effort tomorrow.

As space science missions become more complex and demanding, the need to design tolerance to Space Weather effects into spacecraft systems and scientific payloads is apparent. Examples include sensitivity to radiation, leading to increased backgrounds and even detector damage, as well as the complete failure of key components. Critical conclusions and recommendations made during the meeting were following: Continued improvements in space weather forecasting will absolutely require the application of space-based monitoring. Important observational infrastructures need to be recognised from an operational forecast perspective. The most important and expensive assets are the space-based systems such as SOHO for solar observations and ACE for continuous real-time solar wind monitoring. These systems have been funded under investigative science programs, but the long-term operational forecast role and the commitment to funding a continuing series of such observation platforms has not been adequately addressed or committed to by any nation. Key issues regarding the relevance of space weather, and especially its hazards, to society should be brought to the appropriate policy makers.

The 19 chapters of this book have been developed from the tutorial lectures given by leading world-class experts at the NATO Advanced Study Institute on "Space Storms and Space Weather Hazards" held in Hersonissos, Crete, June 19-29, 2000. All manuscripts were refereed and subsequently meticulously edited by the editor to ensure the highest quality for this monograph. I owe particular thanks to the lecturers of the Advanced Study Institute for producing these excellent tutorial reviews, which convey the essential knowledge and the latest advances in our field. Due to the breadth, extensive literature citations and quality of the reviews we expect this publication to serve well as a reference book. Multimedia material referring to individual chapters of the book is accessible on the WWW site of Kluwer Academic Publishers (www.wkap.nl), through the catalogue page for this book.

The fully met objectives of the Advanced Study Institute were:

- to provide a systematic overview and rigorous introduction to the physics of space storms;
- to review recent spacecraft measurements that have provided new insight into the dynamics and effects of space storms;
- to review space weather hazards associated with space storms and pertinent to the operation of technological systems;

- to discuss and assess methods of space weather forecasting, as well as national and international initiatives towards an efficient development of space climatology.

The Advanced Study Institute was held in Crete - the southernmost corner of Europe - the country that inherited Europe its name. In Greek mythology, Europa was the daughter of Agenor, the Phoenician king of Tyre, and sister of Cadmus, the legendary founder of Thebes. One morning, when Europa was gathering flowers by the seashore, the god Zeus saw her and fell in love with her. Assuming the guise of a beautiful chestnut-coloured bull, he appeared before her and enticed her to climb onto his back. He then sped away with her across the ocean to Crete. Among the sons she bore him was Minos, the legendary king of Minoan Crete and founder of the Labyrinth, a building made up of intricate, maze-like chambers or passages so designed that a person entering one would find it difficult to find a way out.

The fact that this Institute on Space Storms and Space Weather Hazards was held in Europe is particularly appropriate. European countries host a considerable repository of knowledge and world-class facilities in this field. Nevertheless, while Space Weather activities in countries such as the U.S.A. and Japan are well established, Europe has yet to undertake a co-ordinated programme in this area. An important step towards an autonomous European Space Weather programme has recently been taken with the initiation of a broad-based study in the context of the European Space Agency General Studies Programme.

I am grateful to the NATO Scientific and Environmental Affairs Division for the generous support that made this Advanced Study Institute possible. Furthermore, I have the pleasure to acknowledge the following co-sponsors for their contribution to the success of this Institute:

Belgian Institute for Space Aeronomy
Committee on Space Research (COSPAR)
European Office of Aerospace Research and Development, London (UK),
     Air Force Office of Scientific Research, US Air Force Research
     Laboratory (EOARD/AFOSR/AFRL)
European Space Agency (ESA)
Greek General Secretariat of Research and Technology, Ministry of
     Development
National Observatory of Athens
Scientific Committee on Solar-Terrestrial Physics (SCOSTEP)
United States National Aeronautics and Space Administration (NASA)
United States National Science Foundation (NSF)
United States Office of Naval Research (ONR)
University of Crete.
AMPTEK

I am indebted to the many individuals who contributed to the success of the meeting. In particular, I would like to personally thank Joe Allen, Tasos Anastasiadis, Dan Baker, Paul Bellaire, Roger Bonnet, Eamonn Daly, Gerhard Haerendel, Dimitri Lalas, Tim Lawrence, Joseph Lemaire, Bob McCoy, John Pantazis, Jerry Sellers, and Kanaris Tsinganos for their personal contribution to the success of this meeting. Thanks are due to Aleka Marangoudaki (National Observatory of Athens) and Maria Papadaki (Triaena Tours & Congress), who were instrumental in solving the many different small problems that make life and conferences difficult.

Most of all, I wish to thank my wife Anna and my sons Alexandros, Athanasios and Dimitrios for their patience during the long period of the Institute preparation, when I was "unavailable" for my family.

Athens, May 2001

Ioannis A. Daglis

Chapter 1

Space Storms, Ring Current and Space-Atmosphere Coupling

Critical elements of space weather

Ioannis A. Daglis
*Institute for Space Applications and Remote Sensing, National Observatory of Athens
Penteli, 15236 Athens, Greece*

Abstract

Space storms are the prime complex processes of space weather. They interconnect, in a uniquely global manner, the Sun, the interplanetary space, the terrestrial magnetosphere and atmosphere, and occasionally the surface of the Earth. Energy from the Sun drives a continuous interaction of these distinct but coupled regions. The essential element of space storms in the near-Earth space environment is the ring current, which is an electric current flowing toroidally around the Earth, centred at the equatorial plane and at altitudes of ~10,000 to 60,000 km. The trinity of ring current "life" includes its sources, its buildup processes, and its decay mechanisms. The ring current is formed by the injection into the inner magnetosphere of ions originating in the solar wind and the terrestrial ionosphere. The injection process involves electric fields, associated with enhanced magnetospheric convection and/or magnetospheric substorms. The main carriers of the storm-time ring current are positive ions, with energies from ~1 keV to a few hundred keV, which are trapped by the geomagnetic field and undergo an azimuthal drift around the Earth. The usually dominant ion species is H^+, while the abundance of O^+ ions – originating in the terrestrial atmosphere, increases with storm intensity. During the main phase of great storms, O^+ ions, which originate in the terrestrial ionosphere, dominate the ring current. Intensity enhancements of the ring current decrease the horizontal component of the magnetic field in the vicinity of the Earth. Ground magnetograms recording this decrease are used for the construction of the *Dst* index, which is the main measure of space storm intensity and therefore attracts special attention. A large part of ongoing disputes on storm dynamics actually relates to characteristics of *Dst* variations rather than ring current dynamics. Space-atmosphere coupling during storms is an important part of space weather: space disturbances are communicated to

1

I.A. Daglis (ed.), Space Storms and Space Weather Hazards, 1–42.
© 2001 *Kluwer Academic Publishers. Printed in the Netherlands.*

the atmosphere and the atmosphere can in turn drastically influence storm dynamics through the massive outflow of oxygen ions.

Keywords   Space storm, magnetic storm, ring current, radiation belts, geospace, magnetosphere-ionosphere coupling, particle acceleration, space-atmosphere coupling, space weather, space hazards.

1.   INTRODUCTION

The classical term "magnetic storms" is much older than the modern terms "space weather" (NSWC, 1995) and "space storms" (Daglis, 1997b; 1999a; 1999b): it was coined in 1808 by Alexander von Humboldt, the German naturalist and explorer who gained attention by his expedition of South America in 1799-1804. Space storms, the key phenomenon of space weather, have a number of effects in near-Earth space environment: Acceleration of charged particles in space, intensification of electric currents in space and on the ground, impressive aurora displays, and global magnetic disturbances on the Earth surface – a defining storm feature and the origin of the denomination "magnetic storms".

After a six-month journey through Russia in 1829, von Humboldt convinced the Czar to set up a network of magnetic observatories across Russia and Alaska. Later, more observatories were established by the Royal Society in the outer reaches of the British Empire (Canada, Africa and Australia). This network clearly showed that magnetic storms were essentially identical in morphology all over the world: a steep decrease of the horizontal component of the geomagnetic field over many hours, followed by a gradual recovery which lasted several days. The change in the magnetic field was small, about 50-300 nT out of a total intensity of 30,000-60,000 nT, but its world-wide appearance suggested that a large-scale disturbance in space was the physical reason: a huge "ring current" in space circling the Earth emerged as a good candidate (Figure 1).

Chapman (1919) had proposed that the magnetic field depression during magnetic storms was due to electrical currents flowing near the Earth, which were fed by singly-charged particle streams originating at the Sun. Lindemann (1919) pointed out that mutual electrostatic repulsion would destroy a singly-charged stream, and proposed an electrically neutral solar stream containing charged particles of both signs in equal numbers. This critically important suggestion was adopted by Sydney Chapman and led to his later seminal work with Vincent Ferraro. Chapman and Ferraro (1930, 1931) proposed a transient stream of outflowing solar ions and electrons responsible for terrestrial magnetic storms; once the solar stream had reached the Earth, charged particles would leak into the magnetosphere and drift around the Earth, creating a current whose field would oppose the main geomagnetic field. This is surprisingly close to what we believe today. The only major element of Chapman's theory that changed is the existence of a continuous (instead of transient) stream of ionised gas from the Sun. This stream was named solar wind by Parker (1958) and its existence was later confirmed from observations made by the Venus-heading Mariner 2 spacecraft (Neugebauer and Snyder, 1962). Just prior to the discovery of the radiation belts by Explorer 3 (Van Allen et al., 1958), Singer (1956) proposed that particles arriving from the Sun could, by collective motion, perturb the dipolar magnetic field of the Earth sufficiently to allow entry of the particles into the trapping regions identified by Størmer (1955). A more detailed theory followed (Singer, 1957), suggesting that the gradient drift of the energetic particles trapped in the geomagnetic field carries a westward electric current, which effectively decreases the horizontal component of the magnetic field in the vicinity of the Earth. Thus the existence of an "extraterrestrial ring current" (Figure 1) was inferred before the dawn of the space era and the discovery of the radiation belts by James Van Allen.

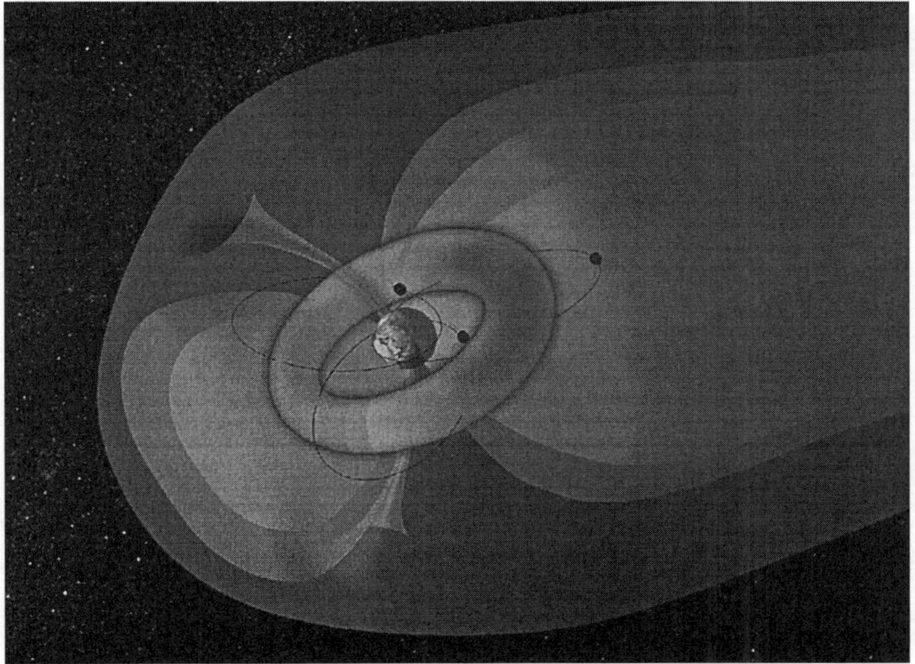

Figure 1. Ideally symmetric ring current visualised as a toroid encircling the earth (courtesy of H. Koskinen, Finnish Meteorological Institute).

However, the fact that radiation belt particles contribute little to the ground effects of space storms, was recognised rather early (Akasofu et al., 1961) – actually before any direct observations of the ring current itself, which were first published by Davis and Williamson (1963). Now we know that the abundance of the radiation belt particles is relatively low, so that they have no appreciable effect on currents and on the ground geomagnetic field. In contrast, particles of comparatively lower energies but higher concentrations are the ones that make up the ring current, responsible for the global geomagnetic disturbances on the Earth's surface – they are the ring current particles.

After four decades of space exploration, we know quite a few things about weather in space. Space storms are often preceded by a well-defined mark, which is the arrival of an interplanetary shock. We know that the interplanetary condition for a storm to develop is a prolonged, southward-directed IMF and that the solar antecedents of intense storms are coronal mass ejections rather than solar flares (i.e., Gosling, 1993; Bothmer and Schwenn, 1995; Gonzalez and Tsurutani, 1997). The main effect of space storms on the terrestrial magnetosphere is the injection of energetic ions and electrons from the near-Earth magnetotail (Figure 2) into the inner magnetosphere, causing the westward-flowing ring current to grow significantly.

The "traditional" graphical representation of a space storm, or more correctly of its effects on the Earth, is the time profile of the *Dst* index (Figure 3). The westward-flowing ring current decreases the horizontal component of the geomagnetic field. Low-latitude ground magnetometers that record this decrease, provide the data for the construction of the *Dst* index, which is a geomagnetic index commonly used as a measure of magnetic storm intensity. The *Dst* index was conceived as a ring current measure, based on the assumption of Sydney Chapman that the global decrease of the geomagnetic *H*-component is solely due to an external westward electric current system around the earth, i.e. the ring current (Akasofu and Chapman, 1961). Sugiura (1964) defined *Dst* as the average global variation of the low-latitude *H*-component. The general morphology of a storm-*Dst* can be seen in Figure 3: a relatively sharp and large decrease of *Dst* signifies the "main phase" of the storm, and the subsequent slow increase of *Dst* marks the storm recovery. Some storms, especially the largest ones, begin with a sudden impulse (positive excursion of *Dst*), which marks the arrival of an interplanetary shock.

The *Dst* index is widely used to monitor and predict magnetic storm activity and therefore attracts special attention. The original assumption was that *Dst* is influenced only by the ring current fluctuations. Today the prevailing perception is that there are other magnetospheric currents (cross-tail current, substorm current wedge, magnetopause current, Birkeland field-aligned currents), which also fluctuate during space storms and influence the ground magnetic field and, consequently, the *Dst* index.

The precise definition of the causes of *Dst* variations is among the open issues of storm dynamics. Actually, the "fidelity" of *Dst* in representing the intensity of the ring current and of the storm itself, is among a number of open issues. The list further includes:

- the role of substorms in storm development – particularly the ring current growth.
- the actual large-scale morphology (symmetry) of the ring current.
- the main mechanism of ring current decay (storm recovery).

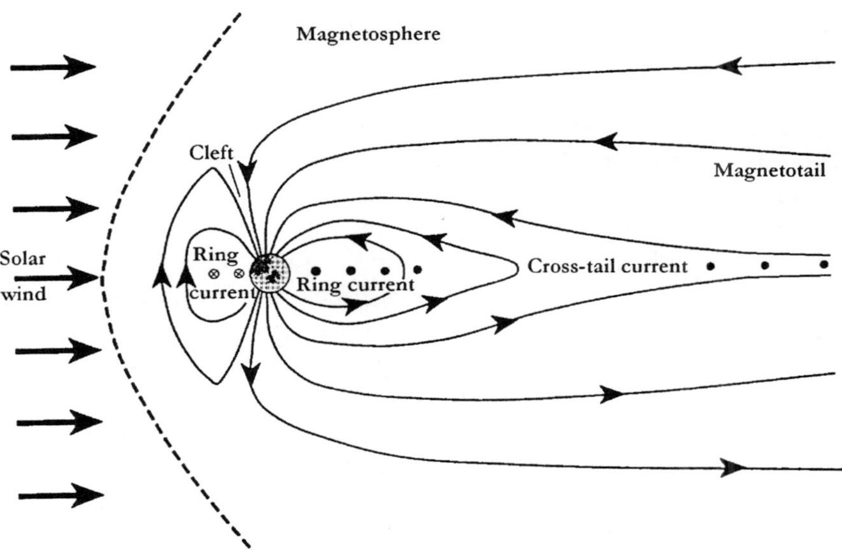

Figure 2. A schematic picture of the terrestrial magnetosphere (Daglis et al., 1999c).

This list refers to prominent old-standing paradigms currently under dispute. Actually these paradigms have seemed to hold well for many years. Perhaps the oldest paradigm de-commissioned was the solar origin paradigm, which prevailed in the 1960's and 1970's. According to that, solar wind ions penetrate the terrestrial magnetosphere and dominate it; consequently, the storm-time ring current has also solar origin. This misconception was utterly dismissed after the detailed full-compositional ring current measurements provided by the AMPTE and CRRES missions (see Section 2.2).

6

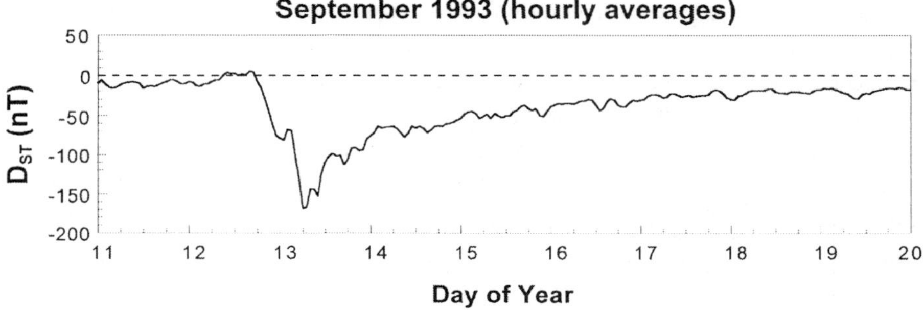

Figure 3. Example of storm-time *Dst* index (courtesy of J. K. Arballo, Jet Propulsion Laboratory).

Now we are in the course of questioning other classical paradigms, such as the "substorm-added-value" paradigm, the paradigm of the symmetric ring current, the charge-exchange decay paradigm. Substorms are questioned as ring current feeders, charge exchange is questioned as ring current killer, and the classical symmetric ring current is probably becoming a myth.

These issues represent the scientific challenge of space storms, and will be treated in the following sections. However, in an era of augmenting human presence and activity in space, one should not underestimate the technological challenge of space storms. Our society has become increasingly dependent on electricity and electronics, which are vulnerable to space storms. Repeated incidents have shown that space weather can have severe impacts on space-borne and ground technological systems, including malfunction or even permanent damage of telecommunication, navigation and surveillance satellites. Intense electric currents flowing in the near-Earth space and closing through the upper atmosphere, lead to geomagnetically induced currents (GICs) that can damage power transmission lines, long-distance telephone cables and oil pipelines. These issues are discussed in detail by Baker (2001), Ginet (2001), Kappenman (2001) and Lanzerotti (2001). Finally, the case of solar and space weather influence on tropospheric weather and climate should is addressed by Friis-Christensen (2001).

2. SPACE STORM EFFECTS IN THE TERRESTRIAL MAGNETOSPHERE: RADIATION BELTS AND RING CURRENT

The most distinct result of space storms in the terrestrial magnetosphere is the intensification of the radiation belts and the ring current. As already described in the Introduction, energetic charged particles can be trapped by the geomagnetic field and thereafter perform a drift motion around the Earth. The most energetic of these trapped particles comprise the "radiation belts". The first diagnostic instrumentation in space was a simple Geiger Mueller tube and the first observations were interpreted as the result of very intense radiation. Van Allen correctly suggested that the radiation was corpuscular rather than electromagnetic (Van Allen et al., 1958; Van Allen, 1959). The ability of the geomagnetic field to trap relativistic electrons was experimentally verified by the Argus experiment, which was proposed by Nicholas Christofilos in 1957 and carried out in 1959 (Christofilos, 1959). Christofilos had actually notified the US Army in the early 1950s that many charged particles, due to the dipole magnetic field, could be trapped around the Earth. He further proposed that an artificial radiation belt, due to beta decay, could be created by exploding a nuclear bomb at high altitude (Papadopoulos, 2000, personal communication). This proposal evolved into Argus - the first active experiment in space.

Today we know that the Van Allen radiation belts have a fairly stable basic structure (Figure 4), and a highly dynamic nature that is influenced by intense space storms (e.g., Blake et al., 1992; Baker et al., 1997). The radiation belts population includes high-energy (>1 MeV) ions and electrons. The abundance of these particles is relatively low, so that they have no appreciable effect on currents. However, their impacts on space technological systems are appreciable, and at times severe (Baker, 2001). The term ring current refers to those particles in the inner magnetosphere, which contribute substantially to the total current density and to the global geomagnetic disturbances on the Earth surface. These particles are mainly ions in the medium-energy range of ~10 keV to a few hundreds of keV (Daglis et al., 1999c).

2.1   Basic properties and structure of the ring current

The ring current can be envisioned as a toroidal-shaped electric current that flows westward around the Earth, with variable density at geocentric distances of ~2 to 9 R_E (Daglis et al., 1999c). As originally suggested by Singer (1957), charged particles that are trapped in the inner magnetosphere undergo an azimuthal drift and form the ring current.

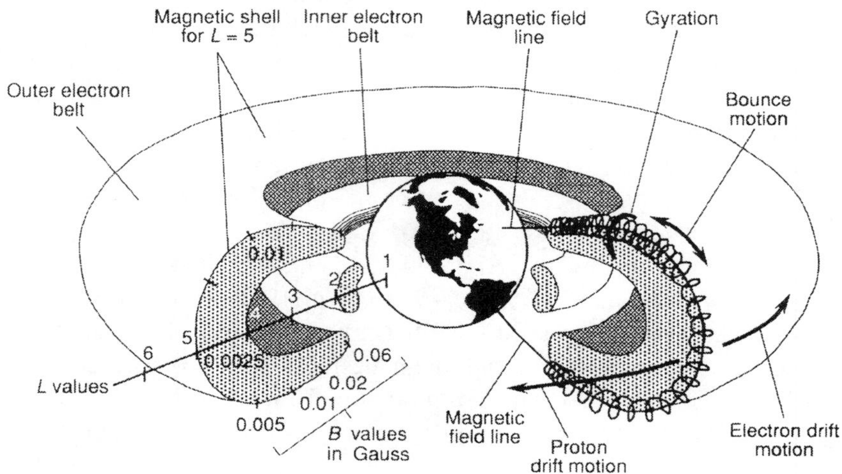

Figure 4. Sketch of a three-dimensional representation of the inner and outer radiation belts forming a ring current around Earth (after Mitchell, 1994).

Geomagnetically trapped charged particles execute two elementary motions: the cyclotron motion (or gyration) around the ambient field due to the Lorentz force and the bounce motion of the particle along field lines and between mirror points (Figure 5). Additionally, the particles are subject to drift motions due to the gradient and curvature of the magnetic field (e.g., Baumjohann and Treumann, 1996). As the particles bounce, they also drift around the Earth, moving on a closed three-dimensional drift-shell around the magnetic field axis (c.f. Figure 3). The total effect is a collective azimuthal drift, which is oppositely directed for ions and electrons: electrons move eastward and most ions (with energies above a relatively low threshold; c.f. De Michelis et al., 1997) move westward. This drift constitutes a net charge transport; the current associated with the charge transport is the ring current. As first worked out by Parker (1957), the total current density perpendicular to the magnetic field $j_\perp$ for a plasma under equilibrium with the magnetic stress balancing the particle pressure is:

$$j_\perp = \frac{\vec{B}}{B^2} \times \left[\nabla P_\perp + (P_{//} - P_\perp)\frac{(\vec{B}\cdot\nabla)}{B^2}\vec{B} \right]$$

where $\vec{B}$ is the local magnetic field vector, while $P_\parallel$ and $P_\perp$ are the pressure tensor components perpendicular and parallel to the magnetic field. Among the three pressure terms on the right-hand side of the above equation, the first term is due to a particle pressure gradient, the second term is due to field line curvature driven drift, and the third one is due to the crowding of gyro-orbits inside a curved field line. The second and the third term vanish when magnetic field lines are straight or when the particle distribution is isotropic ($P_\parallel = P_\perp$). The current system is then established only by particle pressure gradients. It is noteworthy that the electron contribution to the plasma pressure is usually neglected, because the electron temperature is generally very low as compared to the proton temperature ($T_p \approx 7T_e$, Baumjohann, 1993). The horizontal component H of the surface terrestrial magnetic field decreases as a consequence of the magnetic field induced by the ring current during magnetic storms. Dessler and Parker (1959) and Sckopke (1966) showed that the disturbance ΔB of the equatorial surface geomagnetic field, which is the key diagnostic of space storms on Earth, is proportional to the total energy of the ring current particles:

$$\frac{\Delta B}{B_0} = -\frac{2}{3}\frac{E}{E_m}$$

where B_0 is the average surface geomagnetic field intensity at the magnetic equator (~0.3 gauss), E is the total energy of the ring current particles and $E_m = B_0^2 R_E^3 / 3 = 10^{18}$ J, is the energy of the Earth's dipole field at the Earth's surface.

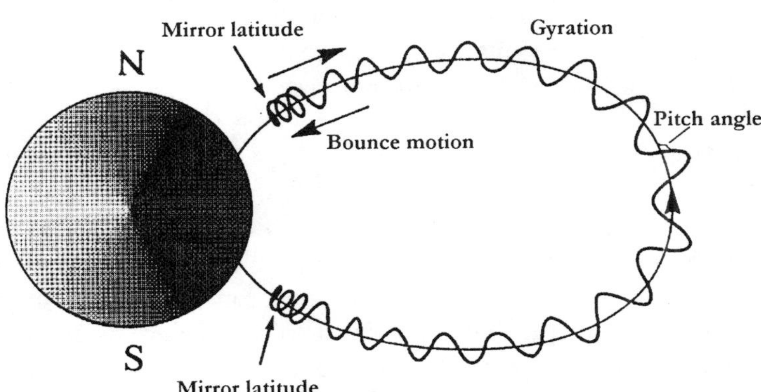

Figure 5. Cyclotron motion (gyration) and bounce motion of a charged particle along a geomagnetic field line.

During the past decades, theoretical and observational work established the general location and the driving forces of the principal current system of space storms. Spacecraft measurements confirmed the existence of the ring current and showed that it is a permanent yet dynamic feature. The development of mass-resolving space instrumentation in the 1970s made it possible to distinguish between ionic species in space. However, the detailed composition and energy of the ring current were not clarified until the Active Magnetospheric Particle Tracer Explorer (AMPTE) mission of the late 1980s. In particular, AMPTE observations enabled us to outline the position of the ring current population (Lui et al., 1987), to establish the particle energies and intensities involved (Stüdemann et al., 1985), to measure the ion angular distributions on a given magnetic field line, and to determine the composition of the ions responsible for the ring current (Krimigis et al., 1985; Hamilton et al., 1988; Daglis et al., 1993). However, little is known about the global dynamic variability of the ring current. This is simply due to the inherently undersampling measurements of single spacecraft in a fundamentally variable and huge system.

2.2 Sources and growth of the storm-time ring current

Among the critical issues of space storms we consider the origin and the build-up of the storm-time ring current. They are both part of the "trinity" of ring current "life", which may be summarised as: sources – growth – decay. Several prominent old-standing paradigms regarding the ring current trinity have come under disfavour, despite the fact that they seemed to hold well for many years. The most prominent of these paradigms refer to the ring current sources and the process of the storm-time ring current growth.

2.2.1 Ring current sources

There was considerable uncertainty on the sources of ring current ions until the end of the 1980's. In the early years of space exploration it was generally assumed that solar wind protons penetrate and dominate the magnetosphere. Accordingly, the ring current was considered a solar-origin proton current (e.g., Frank, 1967). The contribution of the terrestrial atmosphere/ionosphere was originally considered negligible. This assumption was mainly due to the low initial energy of ionospheric ions, which is of the order of several eV to several tens of eV. In comparison, solar origin ions, arriving as solar wind to the Earth, have much higher initial energies of the order of several keV.

This solar-oriented attitude was further nurtured by the inherent inability of the first spacecraft to conduct compositional measurements. The first generation of space instrumentation was unable to provide full identification (mass and charge state) of the detected ions. The lack of compositional information led to the erroneous conclusion that essentially all ions in geospace are protons of solar origin. Opposing theories (Dessler and Hanson, 1961; Axford, 1970) were discarded due to the lack of supporting observations.

In the early 1970s the solar-origin paradigm was challenged by the discovery of energetic heavy ions (i.e., ions with M/q=16) at energies up to 17 keV (Shelley et al., 1972) by the first mass spectrometer in space, on board the polar-orbiting satellite 1971-089A. These ions were presumably O^+ ions originating in the terrestrial ionosphere. In the following years, a new type of ion composition instrumentation made it possible to obtain charge state information in addition to M/q analysis of energetic ions. Observations by the missions GEOS 1 and 2, PROGNOZ 7 and SCATHA confirmed the existence of O^+ in the magnetosphere, and the significant contribution of the ionospheric source during magnetic storms (e.g., Balsiger et al., 1980). However, early composition observations existed only in the low ($\leq$15 keV) and high ($\geq$600 keV) energy portion of the ring current energy density distribution (Williams, 1980). The lack of composition information at intermediate energies was critical, since the ring current energy distribution has its maximum value within this energy range. Williams (1980) had estimated that the bulk (~90%) of the ring current is contained in the 15- to 250-keV energy range and that the mean energy of the ring current is several tens of keV. He further noted that no information at all (either direct measurements or inferences) existed concerning the peak (50-100 keV) of the ring current energy density distribution. Shortly after the launch of AMPTE, Williams (1985) was remarking that the source of the ring current, its generation, and its composition were totally unknown and were questions that the data from AMPTE should go a long way toward answering.

The estimates of Williams (1980) were later confirmed by the AMPTE mission (Williams, 1987), which was the first mission to adequately measure the composition of the main part of the ring current. Case studies (Krimigis, 1985; Hamilton, 1988), and statistical studies of the ion population in the inner magnetosphere (Daglis, 1993) clarified the contribution of the individual ion species to the quiet-time and storm-time ring current. Figure 6 is adopted from Daglis et al. (1993) and shows the energy distribution of the ion energy density for the 4 main ion species, as well as for the total energy density. The curves represent averages over 2.5 years of measurements by the CHEM instrument onboard AMPTE/CCE (Gloeckler et al., 1985). They show the accumulated percentage of the ion energy density at geosynchronous altitude (i.e., the outer ring current) as a function of energy. Plotted are curves for the total energy density as well as the energy density

12

of H^+, O^+, He^{++}, and He^+. The left panel shows the average energy density distribution in the outer ring current at geomagnetically quiet times, while the right panel shows it for active times. There are two outstanding features: a. the bulk of the total measured ion energy density is contained in the energy range ~10-100 keV; b. H^+ is the dominant ion species, with the O^+ contribution increasing drastically (from 6% to 21%) during active times. It should be stressed that these numbers are averages over all local times and over all kinds of events (storms/substorms) with AE<30 nT (quiet times) and AE>700 nT (active times). This means that the characteristics of intense events (for example, the February 1986 storm, which is mentioned in the next paragraph) are smoothed out. We should also note that according to previous ring current measurements covering energies above 300 keV, a non-negligible percentage (~5%) of the energy density resides at energies >300 keV (Williams, 1987, Figure 1).

In February 1986 an intense magnetic storm was observed by AMPTE/CCE. Hamilton et al. (1988) showed that the abundance of O^+ not only rose continuously during the storm, but it eventually became the dominant ion species near the storm's maximum phase, contributing 47% of the total energy density in the inner ring current ($L = 3$-5), compared with 36% in H^+. The situation in the outer ring current ($L = 5$-7) was less dramatic: the maximum O^+ contribution there was 31%, with H^+ contributing 51% at that time.

The next space mission suitable for ring current observations was CRRES (Combined Release and Radiation Effects Satellite) which operated during 1990-1991, i.e. around solar maximum. CRRES observations of several moderate and large magnetic storms showed that the ring current characteristics observed by AMPTE in the February 1986 storm (Hamilton et al., 1988) were not exceptional. Daglis (1997a) showed that O^+ dominates not only in the inner ring current, but also in the outer ring current. We should note here that there is an earthward gradient for the O^+, H^+ and He^{++} density (i.e. of those ion species that are partly or exclusively of terrestrial origin), while there is an anti-earthward gradient for the solar-origin He^{++} ions. Daglis (1997a) concluded that the O^+ contribution to the ring current rises with the storm intensity. Moreover, Daglis (1997a) showed that the *Dst* magnitude and the O^+ contribution to the ring current increase concurrently. This feature was present in all moderate to large storms observed by CRRES in 1991.

A particularly intense storm was the storm of March 24-26, 1991, which reached a minimum Dst of −320 nT. Figure 7 shows the time profiles of compositional changes and of the *Dst* index. Here we have used a special high-resolution (5-min) *Dst* index, which is produced by the Solar-Terrestrial Environment Laboratory, Nagoya University (courtesy Y.

Kamide). The first two panels show the contribution of the two main ion species H^+ and O^+ to the total energy density of the ring current population in the L-range 5 to 6. The storm exhibits a two-phase profile in Dst, which is known from previous studies of intense storms (e.g., Tsurutani and Gonzalez, 1997). An additional feature to notice here, is that the two-phase profile exists both in the level of Dst and of O^+ contribution.

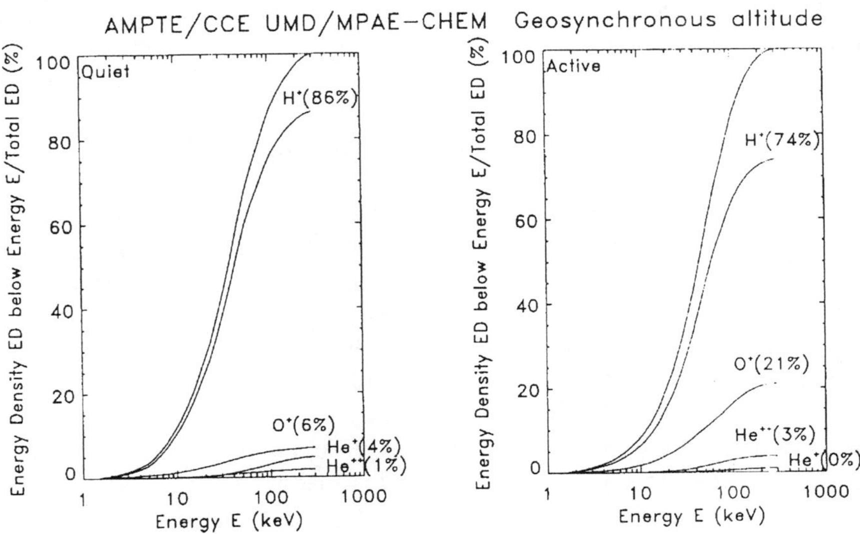

Figure 6. Accumulated percentage of the ion energy density at geosynchronous altitude (i.e., outer ring current) as a function of energy (Daglis et al., 1993). Plotted are curves for the total energy density as well as for the energy density of the 4 main ion species (H^+, O^+, He^{++}, He^+).

Kamide et al. (1998b) recently suggested that such two-phase storms actually are a superposition of two consecutive storms with two consecutive main phases (instead of one storm with a two-step main phase). Following the SSC at 0341 UT, on March 24, and during the first main phase of the storm, when Dst reaches –130 nT, we notice the O^+ curve gradually rising from the pre-storm level of ~15% to above 50%. During the second main phase of the storm (which is the maximum phase), the O^+ contribution reaches ~78%, while Dst reaches its minimum (–320 nT). After the storm maximum, there is a decrease of O^+ contribution to ~45%, concurrently with a quick recovery of Dst to –250 nT. However, neither the change in O^+ abundance, nor the recovery of Dst continue with the same rate: the O^+ abundance remains extraordinarily high for many hours, and the Dst recovery is much slower than the initial recovery following storm maximum.

In summary, the immediate particle sources of the ring current are the plasma sheet and the ionosphere. Since the plasma sheet population

14

originates in the ionosphere and the solar wind, the two main sources of the ring current are the terrestrial ionosphere and the solar wind. We still lack knowledge of the exact access paths of the two source populations, although significant progress has been achieved through numerous observational and numerical studies (e.g., Fujimoto et al., 1998; Winglee, 1998). The relative contribution of the two sources is not completely clarified. Generally the contribution of the ionosphere to the ring current (mainly O^+ ions), increases with the magnitude of the storm. Large storms have ring currents that are dominated by ions of terrestrial origin.

2.2.2   Ring current growth and substorm influence

The second member in the "trinity" of ring current dynamics is the build-up of the storm-time ring current. The source question has been answered satisfactorily, but the questions of acceleration and transport of the seed population are still open. Ring current ions enter the inner magnetosphere from the plasma sheet by drifting earthward across field lines, according to the basic rules described in Section 2.1. Outflowing ionospheric ions move along magnetic field lines from the ionosphere to the magnetosphere, being accelerated upward by various small-scale plasma processes (e.g., Hultqvist, 1999). The demonstration of ionospheric dominance in large storms (Daglis, 1997a) made the issue of ionospheric ion supply to the ring current more appealing. What can be the cause of explosive O^+ enhancements during intense storms? Is it mainly due to changes in the source strength, in the acceleration efficiency or in the transport paths? Since the ionospheric ions are rather cold, the question of efficient acceleration of the ionospheric ions and associated extraction into the magnetosphere is of central interest.

The basic transport and acceleration process for ions moving from the plasma sheet to the inner magnetotail is the ExB drift imposed by the large-scale electric field in the nightside magnetosphere. The particles gain energy while they move from regions of weaker to stronger magnetic field, conserving their first adiabatic invariant – this is the elementary mechanism underlying the build-up of the ring current. While approaching the inner magnetosphere, the particles are transported across magnetic field lines primarily by gradient and curvature drift, as well as by ExB drift in a complicated combination of potential and induction electric fields.

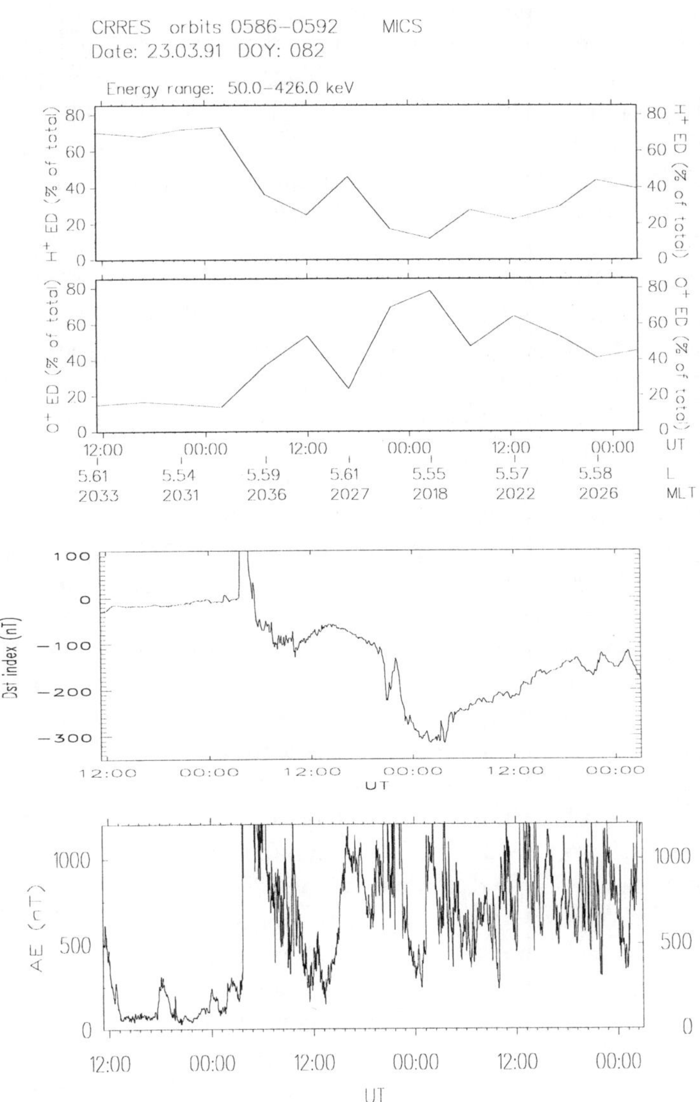

Figure 7. The diagram shows the time profile of the H$^+$ and O$^+$ contribution to the total ion energy density in the ring current region at *L*=5-6 during the great magnetic storm of March 24, 1991. Data come from CRRES-MICS. The prominent features to be observed are the dominance of O$^+$ during storm maximum, as well as the concurrent increase of the |*Dst*| level and of the O$^+$ contribution to the total ion energy density (adopted from Daglis et al., 1999a).

16

There is an ongoing dispute on the relative importance of the large-scale convection electric field and the substorm-associated impulsive electric fields in the energisation and transport of ions into the ring current. The dispute usually appears as a "substorm effects" issue, and refers to the storm-substorm paradigm according to which storms are the result of a superposition of successive "sub-storms" (Akasofu, 1968). Actually the substorm dispute relates to two coupled but distinct issues, which are often confused: the effects of substorms on the ring current growth and the effects of substorms on *Dst* profile variations. The first issue shall be treated here, the latter one in Section 4.

Is a space storm and its ring current the result of a series of intense substorms? That was the view of Sydney Chapman, who introduced the term "substorm" to suggest this idea (Chapman, 1962; Akasofu et al., 1965). Chapman noted that the same storms, which at near-equator magnetic observatories (e.g. in Hawaii) followed simple curves of growth and decay, in Alaska seemed to consist of a number of distinct "sub-storms". However, substorms can exist at other times as well. Substorms do not need much of a stimulus: during times of southward interplanetary field, the magnetotail seems to quickly reach the rim of instability, and small changes in the solar wind can then trigger a substorm. Although magnetic storms seem to come from more powerful sources, such as the arrival of interplanetary shocks, they too require a southward IMF. A strong shock arriving with a northward IMF may shake the magnetosphere, but not to the point of creating a storm.

The storm-substorm relation paradigm according to which storms are the result of a superposition of successive "sub-storms", has been questioned in recent years. Several studies have addressed the issue, without however achieving conclusive evidence (e.g., Kamide, 1992; Kamide and Allen, 1997; Daglis et al., 1998; Kamide et al., 1998a; Daglis, 2000; Daglis et al., 2000). Studies opposing the "Chapman-Akasofu paradigm" have claimed that substorm occurrence is incidental to the main phase of storms, and that ion transport into the ring current is accomplished solely by enhanced large-scale convection electric fields, with no contribution from substorms at all. One argument of such studies is that ion energisation and earthward penetration is much larger during storms than during substorms. However this is not a real argument – it is simply the defining difference between storms and substorms. Hence it does not prove anything regarding substorm "modular" functionality during storms. Actually the anti-substorm polemic is based on the (usually untold) *a priori* assumption that storm-time substorms do not differ from non-storm substorms, hence the "inability" of non-storm substorms to produce storms condemns all substorms to "storm-impotence". However, there are no sound research results that could justify this assumption.

The fact that there are no storms without substorms suggests a synergy between substorms and other processes during storms, which leads to the ring current growth. "Other processes" certainly include the enhanced large-scale convection electric field in the nightside magnetosphere and the enhanced ionospheric outflow, both resulting from increased solar wind – magnetosphere coupling (which is the fuel of it all). The occurrence of fast and bulky ionospheric ion feeding of the inner magnetosphere during substorms was shown by Daglis et al. (1991, 1994), however it is still unclear if the ionospheric dominance during intense storms is again a substorm product, or the result of other conditions (Daglis et al., 1998, 2000).

The establishment of an intense ring current – a key feature of space storms – is a matter of balance between incoming and outgoing particles, i.e. between particle supply from the plasma sheet and particle loss by any means. Any process contributing to the particle supply of the inner magnetosphere automatically contributes to ring current growth and is important for storm development. If substorms contribute to the acceleration and transport of particles into the inner magnetosphere, allowing ring current enhancement, they are important for storms.

During convection of plasma from the magnetotail to the inner magnetosphere, plasma and magnetic field are frozen together as in a superconductor and move together with a bundle of field lines forming a flux tube filled with plasma (because the magnetospheric plasma is collision-free and, hence, resistance-free). For adiabatic transport, the temporal change of the flux tube volume and pressure are related as $dpV^\gamma/dt=0$, where $\gamma=5/3$ is the adiabatic index and the differential flux tube volume is related to the magnetic field strength, B, by the path integral along the field line, $V=\int ds/B$. Due to the 1/B dependence, a flux tube volume in the tail is very large, while it is comparatively small in the inner magnetosphere. Hence, on its way from the tail to the inner magnetosphere, the volume of a flux tube decreases rather strongly. Since pV^γ has to be kept constant, the pressure increases even more dramatically. The massive build-up of plasma pressure strangles the flow and stops further convection. This pressure catastrophe problem, first described by Erickson and Wolf (1980), can be solved by substorms. Substorm-associated reconnection at a near-earth neutral line will cut off the tailward part of the flux tube and thus strongly reduce the flux tube volume. Consequently, the flow will be unclogged and can proceed into the inner magnetosphere. This simple "storm-functionality" of substorms was originally suggested by Michael Hesse at a small Storms Workshop a few years ago (c.f., Daglis et al., 1999c).

Resuming the discussion on the convection/inductive electric fields, I shall note that the short-lived impulsive electric fields reported in the literature, are induced by magnetic field reconfigurations ("dipolarizations" from stretched tail-like to dipole-like configuration) at substorm onset (Aggson and Heppner, 1977; Moore et al., 1981; Aggson et al., 1983). Wygant et al. (1998) showed that during the large March 1991 storm, the large-scale electric field repeatedly penetrated earthward, maximizing between $L=2$ and $L=4$ with magnitudes of 6 mV/m. Such magnitudes are 60 times larger than quiet-time values. However, Wygant et al. (1998) also noted that strong impulsive electric fields with amplitudes of up to 20 mV/m were observed during magnetic field dipolarizations in the inner magnetosphere, i.e. during substorm expansions or intensifications.

Analysis of single-particle dynamics in simulations of magnetic field dipolarizations reveals prominent short-lived accelerations of plasma sheet ions during the expansion phase of substorms (Delcourt, 2001). Delcourt (2001) has shown that at low latitudes, these ions are centrifugally accelerated toward the mid-plane and locally experience parallel energisation up to several tens of keV. Furthermore, under the effect of the electric field induced by relaxation of the magnetic field lines, ions with gyroperiods comparable to the field variation time scale can experience dramatic nonadiabatic heating. For instance, when considering a smooth 1-min magnetic transition, low-energy O^+ originating from the terrestrial ionosphere are found to be accelerated up to a few hundreds of keV during earthward injection. These newly accelerated ions, which subsequently drift rapidly around Earth, evidently can provide a significant – or even major – part of the ring current. By performing such short-lived intense energisation, inductive electric fields accordingly are of considerable importance for the storm-time dynamics of the inner magnetosphere.

Delcourt's results are consistent with repeated evidence from observational studies that acceleration mechanisms for ions in the near-Earth plasma sheet are mass dependent (e.g., Daglis, 1997a; Daglis et al., 1999a; Nosé et al., 2001). Such studies have shown that the energy density of O^+ increases much more drastically than the energy density of H^+ during substorms, and even more so during storms.

Having realised that *in situ* measurements in a vast medium such as the magnetosphere, although significant and indispensable, are not sufficient for drawing definite conclusions, several groups have endeavoured in the development of comprehensive computer simulations. Simulations not only complement the observational studies, but are in fact essential to the understanding of ring current and storm dynamics. Among the efforts regarding the "substorm-induced electric field" question, we should mention Chen et al. (1994), who used spike-like enhancements of the convection electric field to simulate the effect of individual substorms, and Fok et al. (1996), who used an iductive localised electric field, tied to successive cycles of stretching and diplarisation of the Tsyganenko model magnetic field. The results of both efforts suggested that the substorm contribution was subtle, and possibly negative to the development of a ring current.

More recently however, Fok et al. (1999) found that global convection and substorm dipolarizations do cooperate to inject plasma energy more deeply into the magnetosphere than either would individually. Fok et al. made some substantial modifications to their model. First, the range of the ring current model was extended out to 12 R_E, in the nightside magnetosphere, setting the boundary condition well outside of geosynchronous orbit, at the outer limits of the region of validity of the adiabatic bounce-averaged ring current code. This provides a plasma input that is realistically influenced by substorm-dipolarization electric fields in the inner plasma sheet. Second, a 3D test particle code was used to construct the ion velocity distribution by backtracking particles from a velocity space grid to source regions assumed to have constant properties independent of the storm/substorm process. Fok et al. concluded that the substorm-associated induced electric fields significantly enhance the ring current by redistributing plasma pressure earthward.

In summary, here is what we know today about the sources and the build-up of the storm-time ring current. Although all trapped particles in the inner magnetosphere contribute to the ring current, it is the ions in the medium-energy range of ~10 keV to a few hundreds of keV that contribute substantially to the total current density. The immediate particle sources of the ring current are the magnetospheric plasma sheet and the terrestrial ionosphere. The plasma sheet is fed by the ionosphere and the solar wind. Hence, the ultimate main sources of ring current particles are the solar wind and the terrestrial ionosphere. H^+ and O^+ are the main ring current ions. H^+ ions originate both in the terrestrial and in the solar atmosphere, reaching the Earth as solar wind; this complicates the identification of the dominant source. In contrast, the vast majority of magnetospheric O^+ originates in the upper atmosphere of the Earth. To further complicate matters, solar-origin oxygen with higher charge states is transformed to ionosphere-like lower charge state oxygen, and solar He^{++} into ionospheric-like He^+. However, only a negligible percentage of magnetospheric O^+ ions originate through charge exchange from solar wind oxygen ions with high charge states (O^{6+}). Therefore O^+ ions are considered tracer ions of ionospheric outflow associated with space-atmosphere coupling.

A current hot issue is the predominant mechanism of ring current buildup. There is an ongoing dispute on the relative importance of the large-scale convection electric field and the substorm induced electric fields in the energisation and transport of ions into the ring current. The debate usually appears as the «storm-substorm relation» question, because it was initiated by the questioning of the Chapman-Akasofu paradigm, according to which storms are the result of a superposition of successive "sub-storms". Today it is generally accepted that the basic transport and acceleration process for ions moving from the plasma sheet to the inner magnetotail is the ExB drift imposed by the large-scale electric field in the nightside magnetosphere. However, there is no sound evidence yet that ion transport into the ring current is accomplished solely by the large-scale convection electric field without any contribution from substorms at all. In contrary, there are studies showing that substorms do contribute to the ring current build-up.

3. STORM RECOVERY AND RING CURRENT DECAY

According to the traditional *Dst* storm representation, storm recovery manifests itself in the *Dst*-index profile as the increase of *Dst* from its low, negative values achieved during the storm main phase, to its pre-storm level. As already mentioned, Sugiura (1964) designed *Dst* as a ring current measure, according to the original storm / ring current paradigm of Chapman and Singer. Therefore, storm recovery as seen in the *Dst* profile has been traditionally and implicitly linked to ring current decay. However, we now know that *Dst* change does not necessarily reflect ring current change. This is currently a hot, disputed topic. Keeping this limitation in mind, we shall first discuss ring current decay per se.

The loss mechanisms of ring current ions involve charge exchange, Coulomb collisions with thermal plasma, and wave-particle interactions that cause pitch angle scattering into the atmospheric loss cone. The main mechanism of ring current decay is charge exchange of the ring current ions with cold hydrogen atoms of the geocorona. The geocorona is an exospheric extension of relatively cold (~1000 K) neutral atoms, which resonantly scatter solar Lyman-α radiation, thus optically resembling the solar corona. Since oxygen atoms must have an energy of about 10 eV to overcome the Earth's gravitational field, while lighter atoms and molecules need much less energy to escape, the geocorona is essentially a hydrogen gas. The geocoronal density falls off quickly with radial distance, so that at altitudes $\geq 10R_{E}$, collisions between ions and geocoronal hydrogen are rare. At ring current altitudes however, such collisions are frequent enough to account for significant loss of ring current ions.

All ring current ions are subject to charge-exchange decay. However the decay rate depends on the ion mass and energy. While the O^{+}-H charge exchange cross section hardly depends on ion energy, the H^{+}-H charge exchange cross section changes reduces dramatically with increasing energy, resulting in relatively long charge-exchange lifetimes for higher energy H^+ (Smith and Bewtra, 1978). This is illustrated in Table 1: while at 50 keV H^+ and O^+ lifetimes are comparable, at 100 keV they already differ by an order of magnitude. Furthermore, the charge exchange decay rate grows with exospheric hydrogen density, i.e. at lower altitudes. Therefore, ions with mirror points at lower altitudes (i.e., ions with smaller equatorial pitch angles) will charge-exchange easier.

Accordingly, high-energy O^+ will be lost much faster than H^+, and field-aligned pitch-angle distributions will experience larger losses than pancake pitch-angle distributions. It is noteworthy that storm-time O^+ distributions tend to be more field-aligned than H^+ ones (Daglis et al., 1993); consequently, the storm-time O^+ population will additionally experience a faster charge exchange decay because of their smaller pitch angles.

Ring current decay models have shown that change-exchange is the most important loss process, but losses due to Coulomb collisions are not negligible at lower energies (below 10 keV), especially for heavier ions. Coulomb collisions are collisional interactions between charged particles due to their electric fields. Coulomb pitch angle diffusion scatters particles into the loss cone, thus increasing precipitation at low energies and at low L shells. The Coulomb energy degradation process builds up a low-energy heavier ion population; when all loss mechanisms are included, the heavier ion flux increases at low energies (~1-5 keV), while it decreases at higher energies. It follows that during the storm recovery phase, the high-energy part (above 10 keV) of the ring current distribution is dominated by H^+ ions, while conversely the heavier ions dominate the lower energy range. Similar results have been presented by Noël and Prölss (1997), where the decay of the magnetospheric ring current is modelled by means of a Monte Carlo simulation.

Energy Ion species	50 keV	100 keV
H^+	86.4	467.0
O^+	61.2	45.7

Table 1. Charge-exchange lifetimes (in hours) at $L=5$ for H^+ and O^+ ions for a two different energy levels. The difference between the two ion species increases dramatically towards higher energies.

22

Large-scale models incorporating the effects of charge exchange and Coulomb scattering (e.g., Fok et al., 1995) tend to overestimate the flux of protons above tens of keV. Furthermore, they yield a pitch-angle distribution which is too flat for energies >100 keV. Additional pitch-angle scattering by plasma waves has been suggested as a mechanism to account for these discrepancies (Fok et al., 1996). Intense plasma waves also provide an efficient process for energy transfer between different components of the plasma. Waves are particularly important as a heating mechanism for thermal heavy ions (Gendrin and Roux, 1980) and they can also transfer energy from ring current H^+ to O^+ during magnetic storms (Thorne and Horne, 1994).

The most widely studied interactions between plasma waves and the ring current involve electromagnetic ion cyclotron (EMIC) waves. The time scales for scattering of ions into the loss cone during resonant interactions with EMIC waves can be rapid. EMIC waves, which are excited during storms, can cause rapid scattering loss into the atmosphere. The propagation characteristics of EMIC waves and the resulting growth rates are strongly dependent on the relative ion abundance at different phases of a storm. During great magnetic storms, when the injected O^+ concentration in the ring current may exceed that of H^+, the growth of EMIC waves is likely to be suppressed or confined to frequencies below the O^+ gyrofrequency (e.g., Thorne and Horne, 1997). The modulation of EMIC instability by O^+ injection should therefore also change the ability of waves to provide a rapid loss process for ring current H^+ during the main phase of a storm.

The short time scales of EMIC-destruction of the ring current is attractive, because studies of the ring current energy balance (e.g., Prigancová and Feldstein, 1992) suggest that energy loss time scales during the main phases of intense-to-great geomagnetic storms may reach values as low as 0.5-1.0 hrs. Such time scales are far too short to be the result of charge exchange or Coulomb collision processes, but reasonable for the action of EMIC waves. Feldstein et al. (1994) report decay times for the asymmetric component of the ring current with values of the order of an hour in the dusk to noon MLT sector. In addition, distributions unstable to the amplification of ion cyclotron waves are produced naturally in the inner magnetosphere through the betatron acceleration of ions moving along adiabatic drift paths. The enhanced charge exchange loss of ring current ions with small pitch angles deepens the loss cone and increases the anisotropy of the drifting ion distributions making them even more unstable to the generation of plasma waves.

Jordanova et al. (1996) investigated the modification of the ring current ion distributions caused by Coulomb collisions energy degradation, pitch angle scattering and charge exchange. The changes in the distribution functions of the ring current H^+ and O^+ ions were analyzed considering each loss effect. Jordanova et al. found that the loss of O^+ ions proceeds at a much slower rate, so that the modification of their distribution functions becomes significant only after 32 hours. Change-exchange is the most important loss process (especially for H^+), but Coulomb collisions are not negligible at lower energies (below 10 keV), especially for heavier ions. The Coulomb energy degradation process builds up a low-energy heavier ion population (first shown by Fok et al., 1993); when all loss mechanisms are included, the heavier ion flux increases at low energies (~1-5 keV), while it decreases at higher energies. It was also shown that Coulomb pitch angle diffusion scatters ions into the loss cone, thus increasing precipitation at low energies and at low L shells. From the previous considerations it follows that during the storm recovery phase the high-energy part (above 10 keV) of the ring current distribution is eventually dominated by H^+ ions, while heavier ions dominate the lower energy range.

The integrated effect of losses due to the scattering of ions by plasma waves on the dissipation of the storm-time ring current is one of the major unresolved questions in ring current dynamics. Isolated single-point observations of plasma waves and/or changes in ion pitch angle distributions attributed to plasma waves have been made on spacecraft which document the presence and impact of these wave modes on the ring current distribution in restricted local time intervals. On the other hand, observations and recent theoretical studies of wave excitation indicate that the occurrence of particular plasma wave modes may be limited in time and/or confined to localised regions (such as the plasmapause density gradient) in the inner magnetosphere, bringing into question the ability of wave scattering processes to affect the global energy balance of the ring current.

A clarification of the ring current decay problem is expected from global imaging of the entire inner magnetospheric plasma. Through unfolding procedures, the detection of ENA coming from different magnetospheric regions allows imaging of the global magnetospheric dynamics, so that the spatial and temporal evolution of the hot plasma components could be separately inspected. Image unfolding can be done on the basis of specific mathematical procedures (iterative forward modelling) that rely on iterative comparison of the actual images with model images until the differences are minimised (e.g., Mitchell et al., 1998).

The convection of energetic ions on open drift paths past the dusk magnetosphere and into the magnetopause and the subsequent loss out from the magnetosphere is yet another mechanism of ring current "decay", which has caught the attention of recent studies (Liemohn et al., 1999; Kozyra et al., 2001). The "convective drift loss" out the dayside magnetopause has been suggested as the dominant (and fast) process in removing particles from

24

the inner magnetosphere (and therefore from the ring current) during the initial rapid recovery of storms.

At this point I would like to stress that it is conceivable that we misidentify the start of the *Dst* recovery as the start of the ring current decay, at a time when the ring current is actually intensifying due to freshly injected particles in association with substorm occurrence. Recently Ohtani et al. (2001) suggested that if the onset of a major substorm expansion takes place near storm maximum, the positive *Dst* variation due to the substorm current wedge may be larger than the negative *Dst* variation due to the substorm-injected ring current ions. The result is the familiar quickly rising *Dst* profile, usually interpreted as a fast ring current decay. The reality may be a quickly recovering *Dst*, with a still growing ring current. The critical message is that *Dst* does not accurately represent ring current reality. Substorms may contribute to ring current growth, and simultaneously to *Dst* growth - instead of (expected) *Dst* decrease.

In summary: Although classical collisional processes provide the dominant loss process for ring current ions, there is evidence that pitch-angle diffusion by plasma waves also contributes to ion loss, especially during the main phase of a storm. Furthermore, convective loss of energetic ions through the dayside magnetopause is suggested as the main process of fast initial storm recovery, consistent with the short decay times reported by Feldstein et al. (1994). Finally, the initial fast *Dst* recovery may also be the effect on *Dst* of a major substorm current wedge, rather than the effect of a fast ring current loss process.

4. STORM-SUBSTORM RELATION AND DST VARIATION

As mentioned in section 2.2.2, the "storm-substorm-dispute" refers to two distinct, though related issues. These are the effects of substorms on the ring current growth and the effects of substorms on *Dst* variations. Scientists often mix the two effects, although there is ground for doubt that they are equivalent. The first issue was already addressed in section 2.2.2. Here I shall address the relation of substorm occurrence and the *Dst* profile during storms.

Two most frequent aspects of the storm-substorm-*Dst* problem used by the substorm opponents are:

a. the correlation between the westward auroral electrojet index *AL* (i.e., substorm expansion activity) and the global storm disturbance index *Dst* (i.e., storm activity) on one side, and between the rectified solar wind electric field vB_s (i.e., rate of dayside reconnection) and *Dst* on the other side.

b. the variation of *Dst* following substorm expansion onsets.

A major argument of the substorm-opponents is that vB_s is a better predictor of *Dst* than *AL* is (aspect a). McPherron (1997), who is often cited in this context, found that the best solar wind-magnetosphere coupling function (very nearly the rectified solar wind electric field) predicted 76% of the *Dst* variance during 1979, while *AL* predicted 71% of the *Dst* variance. The second major argument, put forth by McPherron (1997) and other ivestigations, is the absence of apparent *Dst* response to substorm expansion onsets (aspect b). Based on these two arguments, the substorm opponents suggest that storms are a direct consequence of the solar wind electric field, without the need for intermediate substorm action.

However, most relevant studies neglect the existence of substorm associated currents other than the (symmetric) ring current, which also contribute to *Dst*. These are the partial ring current, the magnetopause current, the cross-tail current, the substorm expansion current wedge (e.g., Kaufmann, 1987; Alexeev *et al.*, 1996; Turner et al., 2000) and the field-aligned currents (e.g., Sun and Akasofu, 2000). The cross-tail current flows westward across the centre of the tail, and is responsible for the highly stretched magnetic field geometry of the magnetotail. The partial ring current and the region 2 current systems flow in the nightside magnetospheric equatorial plane and close through the ionosphere, to which they are linked by field-aligned currents. All these current systems undergo changes with geomagnetic activity, substorms included. The clear assessment of the location, the intensity and the changes of these current systems as a function of storm/substorm phase, is necessary for a concluding comprehensive theory on the storm-substorm relation and its role in storm dynamics.

Substorms have the potential to influence storm development through their efficiency in altering plasma composition: the abundance of terrestrial plasma (particularly O^+) in the inner magnetosphere increases quickly, as a fast response of the ionosphere to enhanced activity during substorms (i.e., Daglis and Axford, 1996; Daglis et al., 1996). Storm studies, on the other hand, have demonstrated that the O^+ abundance in the ring current increases with storm size, $|Dst|$ and O^+ increase concurrently during storms, and O^+ becomes the dominant ion species during the main phase of large storms (Daglis, 1997a; Daglis, 1997b; Daglis et al., 1999a).

Based on these facts, I have suggested that series of successive storm-time substorms sustain an enhanced ionospheric feeding of the inner plasma sheet, leading to a rapid ultimate enhancement of the ring current at storm maximum, as clearly observed through the O^+ dominance during large storms (Daglis, 1997b). The explosively enhanced ring current at storm maximum, could in large be a partial ring current that would "decay" relatively quickly through drift loss at the dayside magnetopause (Figure 8).

This scenario is consistent with a number of other observations and theoretical model predictions as presented below.

Baker et al. (1982) had suggested an O^+ influence on the localisation of substorm onset due to the potential role of O^+ in ion tearing instability growth. Actually, a substorm case study showed that the suggested O^+ influence was consistent with an earthward and duskward displacement of substorm onset associated with loading of the inner magnetosphere with O^+ (Baker et al., 1985). Later, a model by Rothwell et al. (1988) predicted that a higher concentration of O^+ in the nightside magnetosphere would permit substorm onset at lower L-values. Interestingly, a relatively old storm study by Konradi et al. (1976) had shown that the substorm injection boundary was displaced earthward with each successive substorm during the storm. Furthermore, there are numerous indications from DMSP satellites that the substorm process occurs progressively closer to Earth during the main phase of magnetic storms (e.g., Shiokawa et al., 1996).

Combining model predictions with observations, one comes up with a scenario of a feedback between enhanced (in quantity and spatial extension) O^+-feeding of the plasma sheet and/or the inner magnetosphere and series of intense substorms occurring at progressively lower L-shells. Such a combination of successive substorms and continuous O^+ supply would facilitate successive inward penetration of substorm ion injections, according to the Rothwell et al. (1988) model and consistent with the observations reported by Konradi et al. (1976) and Daglis (1997a). The result of successive inward penetration of substorm injections would be the transport of increasingly more energetic ions into the inner magnetosphere, resulting in the intensification of the storm-time ring current. This scenario can explain why some substorms seem to influence the storm-time ring current growth, while others don't. An experimental verification would be possible through global imaging of storms.

Finally, with regard to the role of the substorm current wedge in *Dst* morphology, we had mentioned in the previous section the possibility of an apparent fast decay of the ring current, as suggested by a quick *Dst* recovery that actually is due to the effect of the current wedge.

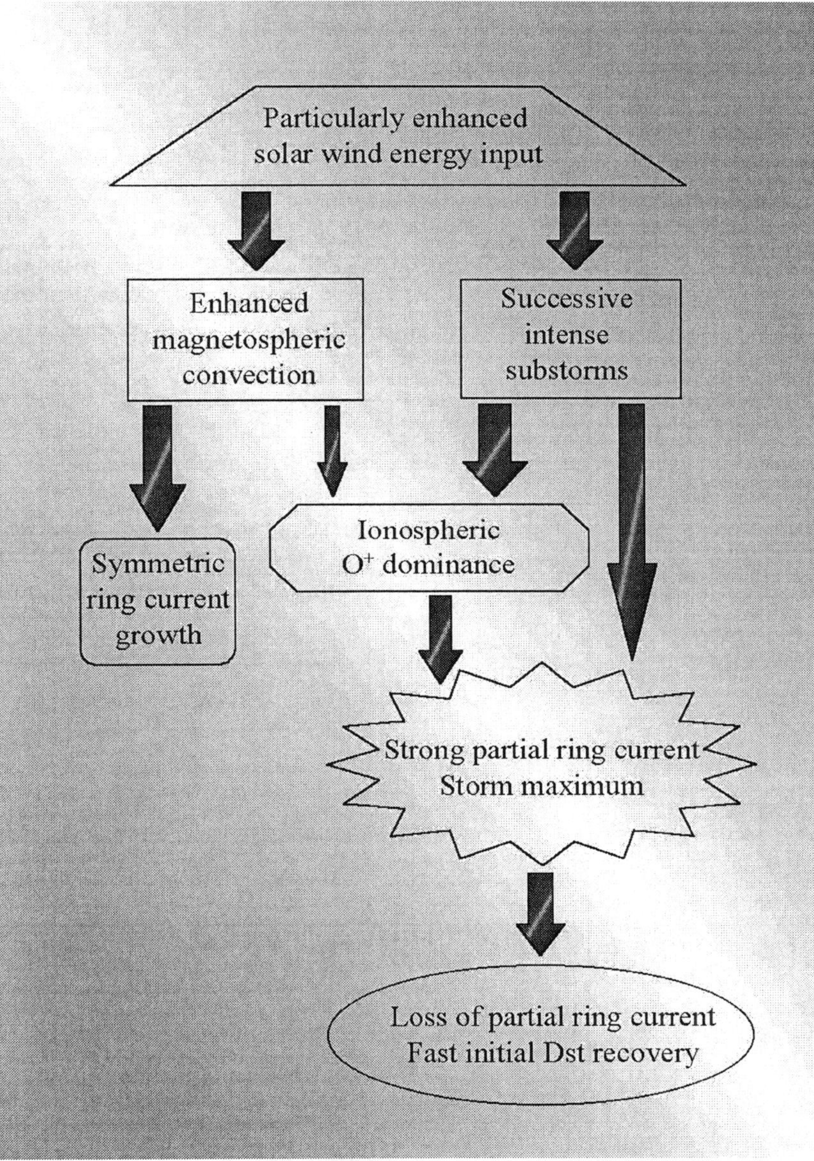

Figure 8. Suggested substorm role in storm development.

5. SPACE STORMS AND SPACE-ATMOSPHERE COUPLING

The thermosphere is the upper layer of the earth's atmosphere, which lies above the mesosphere at altitudes between ~80 and 300 km, and is defined by an exponential increase in temperature as a function of altitude. The temperature peaks at about 1,500 degrees Kelvin at about 300 km. The thermosphere is mainly composed of nitrogen and oxygen molecules. The region is very sensible to changes in energy input, which can cause large changes in temperature. Therefore, the thermosphere temperature is very sensitive to solar activity (i.e., solar ultraviolet radiation). The high temperatures in the thermosphere can cause molecules to ionise, creating the ionosphere, which is defined as the region of the atmosphere mostly composed of gasses that have been ionised. The ionosphere has a large abundance of free electrons and consequently can reflect radio transmissions, making it possible to bounce radio waves off of it and establishing communication links between distant sites of the globe.

The thermosphere-ionosphere system is known to vary substantially with altitude, latitude, longitude, universal time, season, solar cycle and geomagnetic activity, as a result of mechanisms inherent to the system, as well as a result of space weather. The primary driving mechanism is solar radiation (EUV and UV), but precipitation of charged magnetospheric particles and magnetospheric electric fields also have significant effects on the ionosphere-thermosphere system. The driving processes determine the density, composition, and temperature of the ionised and neutral constituents of the upper atmosphere.

5.1 Influence of the terrestrial atmosphere on space storm dynamics

Space-atmosphere coupling was revealed by the first oxygen measurements of the polar-orbiting satellite 1971-089A (Shelley et al., 1972). The importance of terrestrial matter in dynamical space processes has been progressively recognized, through both observational and theoretical studies (e.g., Baker et al., 1982; Moore and Delcourt, 1985; Daglis et al., 1991; Cladis and Francis, 1992; Rothwell et al., 1995; Lakhina and Tsurutani, 1997; Daglis, 1997a,b; Silevitch et al., 2000).

In particular, Daglis (1997a) and Daglis et al. (1999a) demonstrated that atmospheric-origin O^+ ions dominate during the main phase of large magnetic storms throughout the ring current. Daglis (1997a) further suggested that the cause of the intense ring current during large storms is terrestrial.

Recently, Moore et al. (2001) generalized this idea, suggesting that ring current growth is not a natural response of a magnetosphere to solar wind driving, unless there is an internal (atmospheric) particle supply within the magnetosphere, absorbing energy and producing a ring current. Moore et al. have also suggested testing this hypothesis not only on geospace, but also in the magnetosphere of Mercury.

The cause of explosive O^+ enhancements during large storms is still unclear. Although Daglis and Axford (1996) showed that the response of the ionospheric outflow to enhanced solar wind - magnetosphere coupling is rather fast, it seems that a prolonged action of ionospheric ion acceleration mechanisms is a prerequisite for ionospheric dominance in the magnetosphere. The basic mechanisms of ionospheric ion extraction and acceleration are more or less known and have been observed by many polar orbiting spacecraft. A variety of internal magnetospheric processes influence the acceleration of ionospheric O^+ and its extraction to the magnetosphere.

Daglis et al. (1997, 1999b) suggested that the extent of ionospheric outflow is additionally influenced, or even controlled, by solar and/or interplanetary factors. Recent POLAR spacecraft observations of intense direct ionospheric outflow after the passage of an interplanetary shock and associated coronal mass ejection confirmed these suggestions (Moore et al., 1999). A persistent southward interplanetary magnetic field (which is the pre-requisite of intense storms) leads to sustained ionospheric outflows that can reach the magnetotail and eventually dominate the ring current at high energies.

However, the influence of atmospheric material is not limited to the growth of the storm-time ring current. Daglis (1997b) nominated O^+ ions the "first nail in the coffin" of the storm-time ring current, since O^+ induces a rapid initial recovery of *Dst* after storm maximum, following the decline of O^+ supply from the upper atmosphere. The physical reason for this property of O^+ is that its charge exchange lifetime is considerably smaller than that of H^+ at typical ring current energies (≥ 40 keV).

Figure 9. Artistic presentation of ionospheric outflow from the northern polar ionosphere (courtesy of T. E. Moore, D. L. Gallagher, and NASA Marshall Space Flight Center). It has been suggested that the enhanced ionospheric outflow is a pre-requisite of intense space storms (Daglis, 1997a; Moore et al., 2001).

5.2 Space effects on the terrestrial atmosphere

There are many different effects of space storm and generally space and solar disturbances on the terrestrial atmosphere. In the following we list a variety of selected effects of space disturbances on atmospheric dynamics, to illustrate the extent and significance of space-atmosphere coupling.

A direct electrodynamic coupling between space and the upper atmosphere exists due to the magnetosphere-ionosphere coupling through precipitation of magnetospheric particles, field aligned currents and the electric coupling along the magnetic field lines connecting the magnetosphere to the ionosphere.

Ionospheric currents that are enhanced during space storms can induce voltages on ground, large enough to disrupt transformer operation and cause power blackouts (more in next section). Furthermore, space weather disturbances lead to electron density fluctuations in the ionosphere. Even small fluctuations in electron density scintillate radio signals, causing amplitude and phase fluctuations in ground-to-space links. These affect HF communications and over-the-horizon (OTH) radars, as well as links to communication, navigation, and reconnaissance satellites.

Another space-atmosphere link is due to atmospheric heating from space: The total energy of the ring current particles ($\sim 10^{16}$joule), which precipitate to the upper atmosphere, is sufficiently large to cause significant atmospheric heating ionisation, if released in a short time. A number of effects such as stable auroral red arcs, precipitation of energetic ions and neutrals and subauroral electron temperature enhancements, are observed at midlatitudes as consequences of this energy release.

Co-ordinated observations of magnetospheric electron fluxes by the Solar Anomalous and Magnetospheric Particle Explorer (SAMPEX) satellite and NO in the mesosphere by the Upper Atmospheric Research Satellite (UARS), have provided evidence of mesospheric and lower thermospheric NO formation due to precipitating electrons (Callis et al., 1996).

An analysis of concurrent observations of precipitating magnetospheric electron fluxes made by the Television and Infrared Observation Satellite (TIROS) and nighttime NO_2 observed by the Improved Stratsopheric and Mesospheric Sounder (ISAMS) also confirms the linkage (Callis and Lambeth, 1998). These observations demonstrate the formation of NO_y constituents during these electron events as well as their subsequent transport into the stratosphere during periods of advective descent.

Simulations further confirmed the effects of energetic electron precipitation on stratospheric NO_y, NO_2, and O_3 (Callis et al., 1998). Odd nitrogen participates in a catalytic cycle that controls the abundance of stratospheric ozone (Crutzen, 1970), so that its distribution plays an important role in calculations of the atmospheric ozone content.

A space influence on the protective ozone layer has also been studied extensively by K. Labitzke and H. van Loon, who have found an apparent signal of the 11-year sunspot cycle in the lower stratosphere-upper troposphere (e.g., Labitzke and van Loon, 1997a,b). The authors have reported correlations between the total column ozone observed by the Earth Probe TOMS (Total Ozone Mapping Spectrometer) and the 11-yr sunspot cycle. The correlations are lowest in the equatorial region, where ozone is produced, and in the subpolar regions, where the largest amounts are found. In the annual mean the highest, statistically significant, correlations lie between the 5° and 30° parallels of latitude in either hemisphere – between the area of production and the areas of plenty. This position of the largest correlations suggests that the association between the Sun and the ozone is not a direct, radiative one, but that it is due to solar induced changes in the transport of ozone, that is, to changes in the atmospheric circulation.

Independent data sets have shown that there is a correlation of measured changes in atmospheric dynamics with measured changes in vertical atmospheric air-earth current density, which are due to external modulation of the global electric circuit by the solar wind (e.g., Tinsley, 1996). The link has been explored semi-quantitavely by a study of the effects on the microphysics of clouds of these changes in current density (Tinsley, 2000). For storm cloud systems the effect can be to enhance precipitation rates and latent heat release, intensifying the storm. This may be a link between storms in space and storms in the troposphere.

There have been many reports of apparent tropospheric responses to space storms (e.g., Roberts and Olsen, 1973; Stolov and Shapiro, 1974; Olsen et al., 1975). Effects were found beginning on the day of storm onset, and lasting 4 or more days. The onset of space storms is coincident with a sudden decrease of galactic cosmicray flux known as a Forbush decrease. Thus apparent tropospheric responses to space storm activity may also be regarded as responses to MeV-GeV particle flux changes (Tinsley and Deen, 1991).

The MeV particles also include fluxes of relativistic electrons precipitating from the radiation belts, and these affect conductivity at stratospheric heights. During periods when the stratospheric conductivity is abnormally low due to the presence of volcanic aerosols, the conductivity changes due to the relativistic electron precipitation appear to be sufficiently large to modulate the ionosphere-earth current density in the global electric circuit. This has been shown to be likely to affect cloud microphysics and can explain correlations of tropospheric dynamics with the relativistic flux changes (Tinsley, 2000).

The above listed studies have been motivated by the realisation that global change is not confined only to the lower atmosphere, but that the consequences of Man's activities may be more global and extend into the upper atmosphere and even affect plasma processes that couple the solar wind into the Earth's upper atmosphere. Indeed, the application of current, well tested models of the upper atmosphere, suggest that significant changes to the present day structure of the atmosphere above about 50 km could occur by the middle of the of the 21st century. The studies conducted thus far are only suggestive of possible changes and there is an important need for further investigations, not only to establish present day structure in a region of the atmosphere designated as the ``ignorosphere" because of the lack of modern measurements, but also to detect natural and anthropogenic trends into the 21st century.

A few critical open issues are:

Will an altered upper atmosphere structure have an important feedback on troposphere climate and weather systems?

Will space technological systems have to be redesigned to adapt to a changed upper atmosphere environment?

Will present day solar-terrestrial coupling be altered by changed upper atmosphere and ionosphere structures?

6. SPACE STORM EFFECTS ON TECHNOLOGICAL SYSTEMS IN SPACE AND ON THE GROUND

Perhaps the most prominent space storm effect on humanity's technological achievements is the effect on telecommunications, beginning with the earliest electric telegraph systems and continuing to today's wireless communications using satellites and land links. Notable examples of space storm impacts on communication systems include the failures of Telesat Canada Anik E1 in 1994, of the AT&T communication satellite Telstar 401 in 1997, and of the Galaxy 4 satellite in 1998. The loss of Telstar 401 cost AT&T several hundred million dollars, while the failure of Galaxy 4 caused widespread loss of pager service to 45 million customers and numerous other communication outages. A typical communication spacecraft has an estimated value of $250 million. Communication and other service satellites in geosynchronous orbit and low earth orbit amount to more than $50 billion in hardware. A comprehensive review on this topic is presented by Lanzerotti (this volume).

As technology evolves, so do space effects on it. Space enterprise is also affected by bad space weather. The simplest known effect is an electric

34

charge on a satellite, usually negative, raising its voltage to hundreds or even thousands of volts. Charging by itself has little effect on the satellite's operation, although on a scientific satellite it would seriously distort observations (if the satellite is charged to, say, -500 volts, electrons with less energy than 500 electron-volts are repelled and cannot be detected). However, if different parts of the satellite are charged to different voltages, the current between them can cause damage. The relevant chapter in this volume has been prepared by Baker (2001).

Unfortunately, space effects are not confined in space. We have witnessed dramatic demonstrations of space storm effects on ground power distribution grids. While currents of a few amperes can disrupt transformer operation, space storm-induced currents of over 300 amperes have been measured in the grounding connections of transformers in affected areas. Different than tropospheric storms, space storms can create large-scale problems because the footprint of a space storm can extend across a continent. The collapse of the Hydro-Quebec power system during the intense storm of March 1989 led to complete blackout of the second largest utility grid in North America for 9 hours. It is now recognized that North America did not experience the largest and most intense geomagnetic disturbances associated with this storm. This same storm produced dB/dt fluctuations twice as intense over the lower Baltic, than any that were experienced in North America. The topic is presented in detail by Kappenman (2001).

7. SUMMARY

The space storm is – as one would expect – the prime complex phenomenon of space weather. Space storms interconnect, in a uniquely global manner, the Sun, the interplanetary space, the terrestrial magnetosphere and atmosphere, and occasionally the surface of the Earth. The essential element of space storms in the near-Earth space environment is the ring current. The trinity of ring current "life" includes the origin of ring current particles, the build-up processes, and the decay mechanisms. A couple of very successful magnetospheric missions (AMPTE and CRRES) led to a satisfactory clarification of the source problem. The immediate particle sources of the ring current are the plasma sheet and the ionosphere. Since the plasma sheet population originates in the ionosphere and the solar wind, the two main sources of the ring current are the solar wind and the upper atmosphere of the earth. Although the exact relative contribution of the two sources remains unclear, we know that the contribution of the ionosphere to the ring current (mainly O^+ ions) increases with storm magnitude. Large storms have ring currents that are dominated by terrestrial-origin ions. It has been suggested that a significant source of ionospheric outflow is a pre-requisite of a large ring current (Daglis, 1997a; Moore et al., 2001).

The late 1990's have brought into focus the issues of storm-time growth and configuration of the ring current, as well as the efficient destruction agents of it. The "rivals" in the storm-time build-up of the ring current are large-scale convection versus substorms, or − more correctly − large-scale electric field versus impulsive, substorm-induced localized electric fields. The build-up issue relates to the access of particle sources to the inner magnetosphere, transport paths and efficiency of acceleration mechanisms, and the maximum storm intensity. It has proven to be a rather complicated issue, partly because a lot of relevant studies tend to use *Dst* as the main diagnostic tool, thereby increasing the problem instead of decreasing it. The reason is that it is not clear yet to what extent *Dst* is mainly a result of the ring current or mainly a collective result of all current systems in geospace. Therefore "tracking" ring current intensity through *Dst* variations (and correlating it with, e.g., solar wind parameters) is a "weak" method.

Actually, the fact that there are no storms without substorms suggests a synergy between substorms and other processes during storms, leading to the ring current growth. "Other processes" certainly include the action of the large-scale convection electric field and the enhanced ionospheric outflow. Substorms have been shown to play a key role in the extraction and acceleration of ionospheric ions, which dominate the ring current during large storms. However, we do not know yet how the ionosphere plays a decisive role in the build-up of large ring currents. We do not know if and how solar/interplanetary causes and ionospheric causes "co-operate" (possibly through substorms) to create great storms. There is probably a substorm "modular" functionality during storms. A more definitive assessment of the relative importance of induced and convective electric fields is needed to solve this issue.

The other hot issue of ring current trinity is the question of the most efficient loss processes, especially during the early recovery phase of storms, when the decay rate is highest. Again there is an "overlap" of ring current and *Dst* properties here, because what is actually firmly documented is not a fast ring current loss, but a fast *Dst* recovery. This decisive point may easily lead to wrong conclusions if overlooked. To bring an example, if the ground effect of a very intense substorm current wedge at storm maximum is "large enough", there will be a quick *Dst* recovery despite the fact that the ring current will actually continue growing for some time after the onset of the intense substorm, because of substorm-injected ions (Ohtani et al., 2001).

Charge exchange, the classical ring current killer (i.e., Daglis et al., 1999c) is now being disputed as the "suspect" for the rapid storm recovery just after storm maximum, the suggested alternatives still have to prove themselves. The main "rival" of charge exchange is the "convective drift loss" of the partial ring current out the dayside magnetopause. This is actually a double paradigm refutation, because it postulates an overwhlemingly asymmetric ring current during the main phase of storms – while the classical picture of space storms involves a strong, symmetric ring current at storm maximum. In the new proposed paradigm, charge exchange is left with the "glory" of terminating the residual ring current after the initial fast loss.

Besides "drift loss", pitch-angle diffusion by plasma waves is also suggested as an alternative for charge exchange loss. The strong point of both alternatives, with respect to charge exchange, is their relatively shorter time scale, which is consistent with the fast Dst recovery times observed after the maximum of especially large storms. However, there is still the chance that initial fast Dst recovery of large storms may also be the effect on Dst of a major substorm current wedge, rather than the effect of a fast ring current loss process.

In situ spacecraft measurements of the ring current have serious limitations, while the various indices used as proxies of storm and substorm intensity also have their inherent weaknesses. It is obvious that multi-point measurements and co-ordination with ground observations (e.g., Lockwood, 1997; Opgenoorth and Lockwood, 1997), empirical models of the ring current (e.g., Milillo et al., 2001), and comprehensive simulations of ring current growth and decay (e.g. Fok et al., 1995, 1999) not only complement the observational studies, but moreover are essential to the full understanding of ring current and storm dynamics.

Furthermore, there is an important need to continue model development using coupled models of the Earth's atmosphere, ionosphere and magnetosphere to evaluate the entire atmospheric readjustment to global change induced by human activity and the implications of natural space-atmosphere interaction.

Finally, the societal importance of modern technological applications in space imposes an increasingly comprehensive understanding of space storms and of their impact on the near-Earth space environment. Currently more than 300 operational commercial satellites, mostly for communications, provide services circling the Earth. Because of the occasionally detrimental effects of bad space weather on these technological systems, there is a need for an ever more sophisticated understanding of the physical phenomena that will ultimately lead to reliable forecasting capabilities.

8.   REFERENCES

Aggson, T. L., and J. P. Heppner, Observations of large transient magnetospheric electric fields, *J. Geophys. Res.*, *82*, 5155-5164, 1977.

Aggson, T. L., J. P. Heppner, and N. C. Maynard, Observations of large magnetospheric electric fields during the onset phase of a substorm, *J. Geophys. Res.*, *88*, 3981-3990, 1983.

Akasofu, S.-I., *Polar and magnetospheric substorms*, D. Reidel, Dordrecht, Holland, 1968.

Akasofu, S.-I., and S. Chapman, The ring current, geomagnetic disturbance, and the Van Allen radiation belts, *J. Geophys. Res.*, *66*, 1321-1350, 1961.

Akasofu, S.-I., J. C. Cain, and S. Chapman, The magnetic field of a model radiation belt, numerically computed, *J. Geophys. Res.*, *66*, 4013-4026, 1961.

Akasofu, S.-I., S. Chapman, and C.-I. Meng, The polar electrojet, *J. Atmos. Terr. Phys.*, *27*, 1275-1305, 1965.

Axford, W. I., On the origin of radiation belt and auroral primary ions, in *Particles and fields in the magnetosphere*, edited by B. M. McCormac, pp. 46-59, D. Reidel, Dordrecht, 1970.

Alexeev, I. I., E. S. Belenkaya, V. V. Kalegaev, Y. I. Feldstein, and A. Grafe, Magnetic storms and magnetotail currents, *J. Geophys. Res.*, *101*, 7737-7747, 1996.

Baker, D. N., Satellite anomalies due to space storms - The effects of space weather on spacecraft systems and subsystems, this volume, 2001.

Baker, D. N., E. W. Hones Jr., D. T. Young, and J. Birn, The possible role of ionospheric oxygen in the initiation and development of plasma sheet instabilities, *Geophys. Res. Lett.*, *9*, 1337-1340, 1982.

Baker, D. N., T. A. Fritz, W. Lennartsson, B. Wilken, H. W. Kroehl, and J. Birn, The role of heavy ions in the localization of substorm disturbances on March 22, 1979: CDAW 6, *J. Geophys. Res.*, *90*, 1273-1281, 1985.

Baker, D. N., et al., Recurrent geomagnetic storms and relativistic electron enhancements in the outer magnetosphere: ISTP coordinated measurements, *J. Geophys. Res.*, *102*, 14,141-14,148, 1997.

Balsiger, H., P. Eberhardt, J. Geiss, and D. T. Young, Magnetic storm injection of 0.9- to 16-keV/e solar and terrestrial ions into the high-altitude magnetosphere, *J. Geophys. Res.*, *85*, 1645-1662, 1980.

Baumjohann, W., The near-Earth plasma sheet: An AMPTE/IRM perspective, *Space Sci. Rev.*, *64*, 141-163, 1993.

Baumjohann, W., and R. A. Treumann, *Basic space plasma physics*, Imperial College Press, London, 1996.

Blake, J. B., W. A. Kolanski, R. W. Fillius, and E. G. Mullen, Injection of electrons and protons with energies of tens of MeV into $L<3$ on 24 March 1991, *Geophys. Res. Lett.*, *19*, 821-824, 1992.

Bothmer, V., and R. Schwenn, The interplanetary and solar causes of major geomagnetic storms, *J. Geomagn. Geoelectr.*, *47*, 1127-1132, 1995.

Callis, L. B., and J. D. Lambeth, NO_y formed by precipitating electron events in 1991 and 1992: Descent into the stratosphere as observed by ISAMS, *Geophys. Res. Lett.*, *25*, 1875-1878, 1998.

Callis, L. B., R. E. Boughner, D. N. Baker, R. A. Mewaldt, J. B. Blake, R. S. Selesnick, J. R. Cummings, M. Natarajan, G. M. Mason, and J. E. Mazur, Precipitating electrons: Evidence for effects on mesospheric odd nitrogen, *Geophys. Res. Lett.*, *23*, 1901-1904, 1996.

Callis, L. B., M. Natarajan, J. D. Lambeth, and D. N. Baker, Solar atmospheric coupling by electrons (SOLACE) 2. Calculated stratospheric effects of precipitating electrons, 1979-1988, *J. Geophys. Res.*, *99*, 28,421-28,438, 1998.

Chapman, S., An outline of a theory of magnetic storms, *Proc. Roy. Soc. London, A95*, 61, 1919.

Chapman, S., and V. C. A. Ferraro, A new theory of magnetic storms, *Nature, 126*, 129-130, 1930.

Chapman, S., and V. C. A. Ferraro, A new theory of magnetic storms, I. The initial phase, *Terrest. Magn. Atmosph. Elec., 36*, 77-97, 1931.

Chapman, S., Earth storms: Retrospect and prospect, *J. Phys. Soc. Japan, 17* (Suppl. A-I), 6-16, 1962.

Chen, M. W., L. Lyons, and M. Schultz, Simulations of phase space distributions of storm time proton ring current, *J. Geophys. Res., 99*, 5745-5759, 1994.

Christofilos, N. C., The Argus Experiment, *J. Geophys. Res., 64*, 869-875, 1959.

Cladis, J. B., and W. E. Francis, Distribution in magnetotail of O^+ ions from cusp/cleft ionosphere: a possible substorm trigger, *J. Geophys. Res., 97,* 123-130, 1992.

Crutzen, P. J., The influence of nitrogen oxides on the atmospheric ozone content, *Q. J. R. Meteorol. Soc., 96*, 320, 1970.

Daglis, I. A., The role of magnetosphere-ionosphere coupling in magnetic storm dynamics, in *Magnetic Storms, Geophys. Monogr. Ser.*, vol. 98, edited by B. T. Tsurutani, W. D. Gonzalez, Y. Kamide, and J. K. Arballo, pp. 107-116, American Geophysical Union, Washington, DC, 1997a.

Daglis, I. A., Terrestrial agents in the realm of space storms: Missions study oxygen ions, *Eos Trans. AGU, 78* (24), 245-251, 1997b.

Daglis, I. A., Space storms and space weather hazards, *Proposal for an Advanced Study Institute to the NATO Scientific and Environmental Division*, Athens, February 1999a.

Daglis, I. A., Space Storms, *Human Potential Research Training Network Proposal RTN1-1999-00285,* Athens, May 1999b.

Daglis, I. A., Intense magnetic storms: The topic of intense scientific discussion, *Eos Trans. AGU, 81* (6), 56, 2000.

Daglis, I. A., and W. I. Axford, Fast ionospheric response to enhanced activity in geospace: Ion feeding of the inner magnetotail, *J. Geophys. Res., 101,* 5047-5065, 1996.

Daglis, I. A., E. T. Sarris, and G. Kremser, Ionospheric contribution to the cross-tail current enhancement during the substorm growth phase, *J. Atmos. Terr. Phys., 53*, 1091-1098, 1991.

Daglis, I. A., E. T. Sarris, and B. Wilken, AMPTE/CCE observations of the ion population at geosynchronous altitudes, *Ann. Geophys., 11*, 685-696, 1993.

Daglis, I. A., S. Livi, E. T. Sarris, and B. Wilken, Energy density of ionospheric and solar wind origin ions in the near-Earth magnetotail during substorms, *J. Geophys. Res., 99,* 5691-5703, 1994.

Daglis, I. A., W. I. Axford, S. Livi, B. Wilken, M. Grande, and F. Søraas, Auroral ionospheric ion feeding of the inner plasma sheet during substorms, *J. Geomagn. Geoelectr., 48*, 729-739, 1996.

Daglis, I. A., W. I. Axford, E. T. Sarris, S. Livi, and B. Wilken, Particle acceleration in geospace and its association with solar events, *Sol. Phys., 172*, 287-296, 1997.

Daglis, I. A., Y. Kamide, G. Kasotakis, C. Mouikis, B. Wilken, E. T. Sarris, and R. Nakamura, Ion composition in the inner magnetosphere: Its importance and its potential role as a discriminator between storm-time substorms and non-storm substorms, in *Fourth International Conference on Substorms (ICS-4)*, edited by S. Kokubun and Y. Kamide, Terra/Kluwer Publications, Tokyo, pp. 767-772, 1998.

Daglis, I. A., G. Kasotakis, E. T. Sarris, Y. Kamide, S. Livi, and B. Wilken, Variations of the ion composition during a large magnetic storm and their consequences, *Phys. Chem. Earth, 24*, 229-232, 1999a.

Daglis, I. A., E. T. Sarris, W. I. Axford, G. Karagevrekis, G. Kasotakis, S. Livi, and B. Wilken, Influence of interplanetary disturbances on the terrestrial ionospheric outflowmagnetic storm and their consequences, *Phys. Chem. Earth, 24*, 61-65, 1999b.

Daglis, I. A., R. M. Thorne, W. Baumjohann, and S. Orsini, The terrestrial ring current: Origin, formation, and decay, *Rev. Geophys*, *37*, 407-438, 1999c.

Daglis, I. A., Y. Kamide, C. Mouikis, G. D. Reeves, E. T. Sarris, K. Shiokawa, and B. Wilken, "Fine structure" of the storm-substorm relationship, *Adv. Space Res.*, *25* (12), 2369-2372, 2000.

Davis, L. R., and J. M. Williamson, Low-energy trapped protons, *Space Res.*, *3*, 365-375, 1963.

De Michelis, P., I. A. Daglis, and G. Consolini, Average terrestrial ring current derived from AMPTE/CCE-CHEM measurements, *J. Geophys. Res.*, *102*, 14103-14111, 1997.

Dessler, A. J., and W. B. Hanson, Possible energy source for the aurora, *Astrophys. J.*, *134*, 1024-1025, 1961.

Dessler, A. J., and E. N. Parker, Hydromagnetic theory of geomagnetic storms, *J. Geophys. Res., 64,* 2239-2252, 1959.

Erickson, G. M., and R. A. Wolf, Is steady convection possible in the Earth's magnetotail?, *Geophys. Res. Lett.*, *7*, 897-900, 1980.

Feldstein, Y. I., A. E. Levitin, S. A. Golyshev, L. A. Dremukhina, U. B. Vestchezerova, T. E. Valchuk, and A. Grafe, Ring current and auroral electrojets in connection with interplanetary medium parameters during magnetic storm, *Ann. Geophys.*, *12*, 602-611, 1994.

Fok, M.-C., J. U. Kozyra, A. F. Nagy, C. E. Ramussen and V. Khazanov, Decay of equatorial ring current ions and associated aeronomical consequences, *J. Geophys. Res.*, *98*, 19,381-19,393, 1993.

Fok, M.-C., T. E. Moore, J. U. Kozyra, G. C. Ho, and D. C. Hamilton, Three-dimensional ring current decay model, *J. Geophys. Res.*, *100*, 9619-9632, 1995.

Fok, M.-C., T. E. Moore, and M. E. Greenspan, Ring current development during storm main phase, *J. Geophys. Res.*, *101*, 15,311-15,322, 1996.

Fok, M.-C., T. E. Moore, and D. C. Delcourt, Modeling of inner plasma sheet and ring current during substorms, J. Geophys. Res., *104*, 14,557-14,569, 1999.

Frank, L. A., On the extraterrestrial ring current during geomagnetic storms, *J. Geophys. Res.*, *72*, 3753-3767, 1967.

Friis-Christensen, E., Solar activity variations and possible effects on climate, this volume, 2001.

Fujimoto, M., T. Terasawa, T. Mukai, Y. Saito, T. Yamamoto, and S. Kokubun, Plasma entry from the flanks of the near-Earth magnetotail: Geotail observations, *J. Geophys. Res.*, *103*, 4391-4408, 1998.

Gendrin, R., and A. Roux, Energization of helium ions by proton-induced hydromagnetic waves, *J. Geophys. Res.*, *85*, 4577-4586, 1980.

Ginet, G., Space weather: An Air Force Research Laboratory perspective, this volume, 2001.

Gloeckler, G., et al., The charge-energy-mass (CHEM) spectrometer for 0.3 to 300 keV/e ions on the AMPTE/CCE, *IEEE Trans. Geosci. Remote Sens.*, *GE-23*, 234-240, 1985.

Gonzalez, W. D., and B. T. Tsurutani, Criteria of interplanetary parameters causing intense magnetic storms (*Dst* < -100nT), *Planet. Space Sci.*, *35*, 1101-1109, 1987.

Gosling, J. T., The solar flare myth, *J. Geophys. Res.*, *98*, 18,937-18,949, 1993.

Hamilton, D. C., G. Gloeckler, F. M. Ipavich, W. Stüdemann, B. Wilken, and G. Kremser, Ring current development during the great geomagnetic storm of February 1986, *J. Geophys. Res.*, *93*, 14,343-14,355, 1988.

Hultqvist, B., Acceleration of ionospheric outflowing ions, *Phys. Chem. Earth*, *24*, 247-257, 1999.

Jordanova, V. K., L. M. Kistler, J. U. Kozyra, G. V. Khazanov, and A. F. Nagy, Collisional losses of ring current ions, *J. Geophys. Res.*, *101*, 111-126, 1996.

Kamide, Y., Is substorm occurrence a necessary condition for a magnetic storm?, *J. Geomagn. Geoelectr.*, *44*, 109-117, 1992.

Kamide, Y., and J. H. Allen, Some outstanding problems of the storm/substorm relationship, in *Solar-Terrestrial Predictions – V*, edited by G. Heckman, K. Marubashi, M. A. Shea, D. F. Smart and R. Thompson, pp. 207-216, Communications Research Laboratory, Tokyo, Japan, 1997.

Kamide, Y., W. Baumjohann, I. A. Daglis, W. D. Gonzalez, M. Grande, J. A. Joselyn, R. L. McPherron, J. L. Phillips, G. D. Reeves, G. Rostoker, A. S. Sharma, H. J. Singer, B. T. Tsurutani, and V. M. Vasyliunas, Current understanding of magnetic storms: Storm/ substorm relationships, *J. Geophys. Res.*, *103*, 17,705-17,728, 1998a.

Kamide, Y., Yokoyama, N., Gonzalez, W. D., Tsurutani, B. T., Daglis, I. A., Brekke, A., and Masuda, S., Two-step develement of geomagnetic storms, *J. Geophys. Res.*, *103*, 6917-6921, 1998b.

Kappenman, J. G., An introduction to power grid impacts and vulnerabilities from space weather, this volume, 2001.

Kaufmann, R. L., Substorm currents: growth phase and onset, *J. Geophys. Res.*, *92*, 7471-7486, 1987.

Konradi, A., C. L. Semar, and T. A Fritz, Injection boundary dynamics during a geomagnetic storm, *J. Geophys. Res.*, *81*, 3851-3865, 1976.

Kozyra, J. U., M. W. Liemohn, C. R. Clauer, A. J. Ridley, M. F. Thomsen, J. E. Borovsky, J. L. Roeder, V. K. Jordanova and W. D. Gonzalez, Multi-step *Dst* development and ring current composition changes during the 4-6 June 1991 magnetic storm, *J. Geophys. Res.*, submitted, 2001.

Krimigis, S. M., G. Gloeckler, R. W. McEntire, T. A. Potemra, F. L. Scarf, and E. G. Shelley, Magnetic storm of 4 September 1985: A synthesis of ring current spectra and energy densities measured with AMPTE-CCE, *Geophys. Res. Lett.*, *12*, 329-332, 1985.

Labitzke, K., and H. van Loon, Total ozone and the 11-yr sunspot cycle, *J. Atmos. Terr. Phys.*, *59*, 9-19, 1997a.

Labitzke, K., and H. van Loon, The signal of the 11-year sunspot cycle in the upper troposphere-lower stratosphere, *Space Sci. Rev.*, *80*, 393-410, 1997b.

Lakhina, G. S., and B. T. Tsurutani, Helicon modes driven by ionospheric O^+ ions in the plasma sheet region, *Geophys. Res. Lett.*, *24,* 1463-1466, 1997.

Lanzerotti, L. J., Space weather effects on communications, this volume, 2001.

Liemohn, M. W., J. U. Kozyra, V. K. Jordanova, G. V. Khazanov, M. F. Thomsen, and T. E. Cayton, Analysis of early phase ring current recovery mechanisms during geomagnetic storms, *Geophys. Res. Lett.*, *26*, 2845-2849, 1999.

Lindemann, F. A., Note on the theory of magnetic storms, *Phil. Mag.*, *38*, 669, 1919.

Lockwood, M., Testing substorm theories: The need for multipoint observations, *Advances in Space Research*, *20* (4/5), *883-894*, 1997.

Lui, A. T. Y., R. W. McEntire, and S. M. Krimigis, Evolution of the ring current during two geomagnetic storms, *J. Geophys. Res.*, *92*, 7459-7470, 1987.

McPherron, R. L., The role of substorms in the generation of magnetic storms, in *Magnetic Storms, Geophys. Monogr. Ser.*, vol. 98, edited by B. T. Tsurutani, W. D. Gonzalez, Y. Kamide, and J. K. Arballo, pp. 131-147, American Geophysical Union, Washington, DC, 1997.

Milillo, A., S. Orsini, and I. A. Daglis, Empirical model of proton fluxes in the equatorial inner magnetosphere. 1. Development, *J. Geophys. Res.*, *105*, in press, 2001.

Mitchell, D. G., The space environment, in *Fundamentals of space systems*, edited by V. L. Pisacane and R. C. Moore, pp. 45-98, Oxford Univ. Press, Oxford, 1994.

Mitchell, D. G., H. O. Funsten, M. Gruntman, M. Hesse, B. H. Mauk, R. R. Meier, D. J. McComas, E. C. Roelof, and E. E. Scime, Multi-point magnetospheric reconnaisance imaging: Visualization of ion dynamics, evolution, origins, and structure, in *Science closure and enabling technologies for constellation class missions*, edited by V. Angelopoulos and P. V. Panetta, pp. 44-50, Univ. of Berkeley, Berkeley, 1998.

Moore, T. E., and D. C. Delcourt, The geopause, *Rev. Geophys.*, *33*, 175-209, 1995.

Moore, T. E., R. L. Arnoldy, J. Feynman, and D. A. Hardy, Propagating substorm injection fronts, *J. Geophys. Res.*, *86*, 6713-6726, 1981.

Moore, T. E., M. O. Chandler, Mr. R. Collier, H. L. Collin, R. Fitzenreiter, B. L. Giles, W. K. Peterson, C. J. Pollock, and C. T. Russell, Ionospheric mass ejection in response to a coronal mass ejection, *Geophys. Res. Lett.*, *26*, 2339-2342, 1999.

Moore, T. E., M.O. Chandler, M.-C. Fok, B.L. Giles, D.C. Delcourt, J. L. Horwitz, and C.J. Pollock, Ring currents and internal plasma sources, *Space Sci. Rev.*, *95*, 555-568, 2001.

Neugebauer, M., and C. Snyder, The mission of Mariner II: Preliminary observations: Solar plasma experiments, *Science*, *138*, 1095-1096, 1962.

Noël, S., and G. W. Prölss, A Monte Carlo model of the ring current decay, *Advances in Space Research*, *20* (3), 335-338, 1997.

Nosé, M., S. Ohtani, K. Takahashi, A. T. Y. Lui, R. W. McEntire, and D. J. Williams, S. P. Christon, and K. Yumoto, Ion composition of the near-Earth plasma sheet in storm and quiet intervals: Geotail/EPIC measurements, *J. Geophys. Res.*, in press, 2001.

NSWC (National Space Weather Committee), The National Space Weather Program: The Strategic Plan, Office of the Federal Coordinator for Meteorological Services and Supporting Research, *FCM-P30-1995*, Washington, DC, 1995.

Ohtani, S.-I., M. Nosé, G. Rostoker, H. J. Singer, A. T. Y. Lui, and M. Nakamura, Storm-Substorm Relationships: Near-Earth Dipolarization and the Recovery of *Dst*, *J. Geophys. Res.*, in press, 2001.

Olsen, R. H., W. O. Roberts, and C. S. Zerefos, Short term relationships between solar flares, geomagnetic storms, and tropospheric vorticity patterns, *Nature*, *257*, 113, 1975.

Opgenoorth, H. J., and M. Lockwood, Opportunities for magnetospheric research with coordinated Cluster and Ground-Based Observations, *Space Sci. Rev.*, *79*, 599-637, 1997.

Parker, E. N., Newtonian development of the dynamical properties of the ionised gases at low density, *Phys. Rev.*, *107*, 924-933, 1957.

Parker, E. N., Interaction of the solar wind with the geomagnetic field, *Phys. Fluids*, *1*, 171, 1958.

Prigancová, A., and Y. I. Feldstein, Magnetospheric storm dynamics in terms of energy output rate, *Planet. Space Sci.*, *40*, 581—588, 1992.

Roberts, W. O., and R. H. Olsen, Geomagnetic storms and wintertime 300 mb trough development in the North Pacific –North America area, *J. Atmos. Sci.*, *30*, 135, 1973.

Rothwell, P. L., L. P. Block, M. B. Silevitch, and C.-G. Fälthammar, A new model for substorm onsets: The pre-breakup and triggering regimes, *Geophys. Res. Lett.*, *15*, 1279-1282, 1988.

Rothwell, P. L., M. B. Silevitch, L. P. Block, and C.-G. Fälthammar, Particle dynamics in a spatially varying electric field, *J. Geophys. Res.*, *100*, 14,875-14,885, 1995.

Sckopke, N., A general relation between the energy of trapped particles and the disturbance field near the Earth, *J. Geophys. Res.*, *71*, 3125-3130, 1966.

Shelley, E. G., R. G. Johnson, and R. D. Sharp, Satellite observations of energetic heavy ions during a geomagnetic storm, *J. Geophys. Res.*, *77*, 6104-6110, 1972.

Silevitch, M. B., P. L. Rothwell, L. P. Block, and C.-G. Fälthammar, O^+ phase bunching as a source for stable auroral arcs, *J. Geophys. Res.*, *105*, 10739-10749, 2000.

Singer, S. F., Trapped orbits in the Earth's dipole field, *Bull. Am. Phys. Soc. Series II*, *1*, 229 (A), 1956.

Singer, S. F., A new model of magnetic storms and aurorae, *Trans. Am. Geophys. Union*, *38*, 175-190, 1957.

Smith, P. H., and N. K. Bewtra, Charge exchange lifetimes for ring current ions, *Space Sci. Rev.*, *22*, 301-318, 1978.

Stolov, H. L., and R. Shapiro, Investigation of the responses of the general circulation at 700 mb to solar geomagnetic disturbance, *J. Geophys. Res.*, *79*, 2161, 1974.

Størmer, C., The Polar Aurora, Oxford Univ. Press, Oxford, 1955.

Stüdemann, W., G. Gloeckler, B. Wilken, F. M. Ipavich, G. Kremser, D. C. Hamilton, and D. Hovestadt, Ion composition of the bulk ring current during a magnetic storm: Observations with the CHEM-instrument on AMPTE/CCE, in *Solar Wind-Magnetosphere Coupling*, edited by Y. Kamide and J. A. Slavin, pp. 697-705, Terra Scientific Publishing Company, Tokyo, 1986.

Sugiura, M., Hourly values of the equatorial *Dst* for IGY, in *Ann. Int. Geophys. Year, Vol. 35*, pp. 945-948, Pergamon Press, Oxford, 1964.

Sun, W., and S.-I. Akasofu, On the formation of the storm-time ring current belt, *J. Geophys. Res.*, *105*, 5411-5418, 2000.

Thorne, R. M., and R. B. Horne, Energy transfer between energetic ring current H^+ and O^+ by electromagnetic ion cyclotron waves, *J. Geophys. Res.*, *99*, 17,275-17,282, 1994.

Thorne, R. M., and R. B. Horne, Modulation of electromagnetic ion cyclotron instability due to interaction with ring current O^+ during magnetic storms, *J. Geophys. Res.*, *102*, 14,155-14,163, 1997.

Tinsley, B. A., Correlations of atmospheric dynamics with solar wind-induced changes of air-earth current density into cloud tops, *J. Geophys. Res.*, *101*, 29,701-29,714, 1996.

Tinsley, B. A., Influence of solar wind on the global electric circuit, and inferred effects on cloud microphysics, temperature and dynamics of the troposphere, *Space Sci. Rev.*, *94*, 231-258, 2000.

Tinsley, B. A., and G. W. Deen, Apparent tropospheric responses to MeV-GeV particle flux variations: A connection via electrofreezing of supercooled water in high-level clouds?, *J. Geophys. Res.*, *96*, 22,283-22,296, 1991.

Tsurutani, B. T., and W. D. Gonzalez, The interplanetary causes of magnetic storms: A review, in *Magnetic Storms, Geophys. Monogr. Ser.*, vol. 98, edited by B. T. Tsurutani, W. D. Gonzalez, Y. Kamide, and J. K. Arballo, pp. 77-89, American Geophysical Union, Washington, DC, 1997.

Turner, N. E., D. N. Baker, T. I. Pulkkinen, R. L. McPherron, Evaluation of the tail current contribution to *Dst*, *J. Geophys. Res.*, *105*, 5431-5439, 2000.

Van Allen, J. A., The geomagnetically trapped corpuscular radiation, *J. Geophys. Res.*, *64*, 1683-1689, 1959.

Van Allen, J. A., G. H. Ludwig, E. C. Ray, and C. E. McIlwain, Observations of high intensity radiation by satellites 1958 Alpha and Gamma, *Jet Propul.*, *28*, 588-592, 1958.

Williams, D. J., Ring current composition and sources, in *Dynamics of the magnetosphere*, ed. by S.-I. Akasofu, pp. 407-424, D. Reidel, Dordrecht, Holland, 1980.

Williams, D. J., Exploration and understanding in space physics, *Geophys. Res. Lett.*, *12*, 303, 1985.

Williams, D. J., Ring current and radiation belts, *Rev. Geophys.*, *25*, 570-578, 1987.

Winglee, R. M., Multifluid simulations of the magnetosphere: The identification of the geopause and its variation with IMF, *Geophys. Res. Lett.*, *25*, 4441-4444, 1998.

Wygant, J., D. Rowland, H. J. Singer, M. Temerin, F. Mozer, and M. K. Hudson, Experimental evidence on the role of the large spatial electric field in creating the ring current, *J. Geophys. Res.*, *103*, 29,527-29,544, 1998.

Chapter 2

Geomagnetic Storms as a Dominant Component of Space Weather: Classic Picture and Recent Issues

Reviewing the progress in studies of geomagnetic storms to present-day space physics

Yohsuke Kamide
Solar-Terrestrial Environment Laboratory, Nagoya University
Toyokawa 442-8507, Japan

Abstract   It was in the mid-1800s that extraordinary, worldwide disturbances in the Earth's magnetic field were coined "geomagnetic storms." It would not be too much of an exaggeration to state that space weather predictions originated in early studies of geomagnetic storms. The effects of geomagnetic storms in space surrounding Earth result from a chain of processes involving flow/transformation of solar wind energy, and electrodynamic coupling among the interplanetary medium, magnetosphere, ionosphere, and upper atmosphere. The importance of predicting geomagnetic storms lies not only in its "academic" purposes in understanding physical processes in the solar-terrestrial environment, but also in its practical aspects, influencing societal problems such as the effects on communications and satellite anomalies. This chapter discusses the characteristic signatures of geomagnetic storms, obtained from a number of statistical studies, and addresses recent major issues which impact directly our fundamental understanding of solar wind effects on magnetospheric and ionospheric processes, i.e., solar wind control of geomagnetic storms, storm/substorm relationships, ring current constituents, and solar cycle and seasonal dependence of geomagnetic storms. The following are the main points of the discussion: (1) Most of the Dst variance during intense geomagnetic storms can be reproduced by knowledge about changes in large-scale electric fields in the solar wind. A continuing controversy exists, however, as to whether the successive occurrence of substorms plays a direct role in the energization of storm-time ring current particles. (2) The increase in the ring current of about 50% of the largest geomagnetic storms goes through two steps at the main phase. The solar wind causes of this double enhancement in the ring current must be identified. (3) CMEs (coronal mass ejections) and CIRs (corotating interaction regions)

I.A. Daglis (ed.), Space Storms and Space Weather Hazards, 43–77.

44

appear to be the primary sources leading to major geomagnetic storms. These are dominant near the maximum phase and during the declining phase of the solar cycle, respectively. The 22-year solar cycle dependence of geomagnetic activity must also be quantitatively evaluated in terms of CMEs and CIRs. (4) Recent satellite observations in the inner magnetosphere have shown that the abundance of ionospheric origin ions is high and is correlated well with substorm activity during the main phase of geomagnetic storms. The relative importance of solar wind-origin and ionosphere-origin ions in constituting ring current particles is currently a critical unsolved question.

Keywords   Geomagnetic storms, geomagnetism, *Dst* index, substorms, auroral electrojets, solar wind, solar cycle, seasonal geomagnetic variation, magnetospheric processes, magnetosphere-ionosphere coupling, ring current.

1.   INTRODUCTION: FROM GEOMAGNETISM TO SPACE PHYSICS

As reviewed by Kamide (2000), Solar-Terrestrial Physics, or Space Physics, has its roots in Geomagnetism in the 19th century, whose aim was to locate and estimate the intensities of electric currents that generate world geomagnetic disturbances, such as geomagnetic storms and substorms in the present terminologies. Since the end of the 19th century, it became clear that great magnetic perturbations are generated by currents flowing in the magnetosphere and in the upper atmosphere, i.e., in the region of auroras. It was about fifty years ago when the existence of the solar wind and the magnetosphere was predicted and subsequently discovered by means of satellite measurements. The average configuration of the magnetosphere was modeled. The presence of solar wind particles inside the magnetosphere was confirmed, and a number of plasma regions within the magnetosphere were characterized. The role of each plasma region in geomagnetic storm processes was identified. In the phase of comprehensive understanding of the solar-terrestrial environment, an integration of the data from satellite and ground-based observations became essential. More and more the importance of interactions between different plasma regions in the magnetosphere and the ionosphere was realized. Computer simulations have become a powerful tool in quantitative understanding of such complicated processes that occur in space between the Sun and the Earth.

The goal of space physics or space weather is now to become able to predict the chain of processes that occur in the entire solar-terrestrial system. Our operational purpose is to input data of the current solar wind status into a computer system which, in turn, will produce output about how, when, and where geomagnetic storms will begin, what magnitude these storms will reach, and even how and where they will develop and subside. Research of geomagnetic storms is the result of a convergence of multi-disciplinary sciences, which developed from several traditional fields of research, such as solar physics, geomagnetism, auroral physics, and aeronomy. In a sense geomagnetic disturbance research transformed its focus from geomagnetism to space weather.

Geomagnetic storms are known to be multi-faceted phenomena that originate at the solar corona and occur in the solar wind, the magnetosphere, the ionosphere, and the thermosphere: see Kamide et al. (1997). What in the solar wind causes geomagnetic storms; how is the ring current intensified during geomagnetic storms; how do changes in the magnetosphere-ionosphere system affect the Earth's upper atmosphere throughout the chain of processes; and how can we predict storm occurrence and intensity? These are some of the major questions relating to magnetic storms. Because of the nature of geomagnetic storms, taking place in a wide range of plasma regions, geomagnetic storms must be understood as a chain of processes from the Sun to the Earth by combining the many different aspects of storms and by testing different scenarios on the cause-and-effect relationship of a variety of storm phenomena.

This article attempts to discuss, starting with some of the historical accounts of geomagnetic storm studies, the primary signatures of geomagnetic storms that resulted from a number of statistical studies, and to identify recent issues which remain unsolved but fundamental to our understanding of solar wind coupling with the magnetosphere and ionosphere.

2.   WHAT IS A GEOMAGNETIC STORM?

2.1   Who introduced this terminology into the scientific community?

As Stern (1989) summarized, studies of global geomagnetic data with some help from solar and auroral observations were conducted in parallel with discoveries of the important laws of electromagnetism, on which we presently rely. Some of them are:

(1) The discovery of geomagnetic storms (later term) by Graham in 1724

(2) The discovery by Oersted in 1820 that electric currents produce magnetic forces

46

(3) The law of Ampère's force in 1821

(4) Electromagnetic induction by Faraday in 1831

which were to lead subsequently to Maxwell's equations of electromagnetism.

It is well known that Graham in London and Hiorter and Celsius in Uppsala noticed, communicating notes by letters, that large irregular disturbances, i.e., the D component (according to the present-day definition), in the geomagnetic field occurred simultaneously at the two places, indicating that magnetic disturbances are not of local nature. They were observing small motions of a compass needle by a microscope. Gauss and Weber later showed these magnetic disturbances to be a worldwide phenomenon such that magnetic "weather" is much less local than ordinary weather. Although Gauss was well aware of the necessity of measuring the complete vector of disturbances rather than only the D component, it was not practical at that time. It was Sir Edward Sabine who introduced complete vector measurements.

Who, and when, introduced the terminology geomagnetic storm into the scientific community? According to Schröder (1997), Alexander von Humboldt (1769-1859) used "magnetisches Ungewitter" (magnetic thunderstorms) to describe the variability of geomagnetic needles, which were associated with the occurrence of "light meteor," i.e., auroras. Humboldt thought that magnetic disturbances on the ground and auroras in the polar sky are two manifestations of the same phenomenon. He maintained a lifelong interest in geomagnetic disturbances, being instrumental in the establishment of a number of magnetic stations around the world through his diplomatic contacts, notably in Britain, Russia, and in lands then under British and Russian rule, e.g., at Bombay, Toronto, and Sitka (see, for example, Chapman, 1967; Malin, 1987). It was found by Humboldt that the storm-time disturbance generally reduces the daily mean value of the horizontal intensity. During the First Polar Year (1882-1883), scientists defined "geomagnetic storms" as intense, irregular variabilities of geomagnetic field which occur as a consequence of solar disturbances.

It was Chapman (1919) who thought that worldwide geomagnetic disturbances during geomagnetic storms are a result of electric currents in space, which are enhanced by streams of particles from the Sun. Chapman and Ferraro (1932) proposed that charged particles originated from the Sun would drift around the Earth and cause the main phase decrease in the geomagnetic horizontal component at the Earth surface. The stream of plasmas was the solar wind.

2.2   The phases of geomagnetic storms

As seen in the history above, every disturbance in the Earth's magnetic field was thought to be a geomagnetic storm until the early 1960s, when a substorm was defined as an elementary disturbance that occurs during a magnetic storm. At present, a geomagnetic storm is best defined by the existence of a main phase during which the horizontal component of the magnetic field on the Earth's surface is significantly decreased. This depression is caused by an enhancement of the trapped particle population in the magnetosphere, and thus by the ring current encircling the Earth. In the early 1900s it was believed that there is a ring current of electrons and positive ions encircling the Earth in the opposite directions. Störmer, and subsequently Schmidt, suggested that the ring current must be the cause of the main phase of geomagnetic storms.

As shown in Figure 1, it is now commonly assumed that the magnitude of geomagnetic storms can be monitored by the *Dst* index. The magnetic disturbance field was analyzed geometrically into a part *Dst* that is symmetrical about the Earth's axis and the remaining part *DS*. The characteristic signature of a geomagnetic storm is a depression in the *H* component of the magnetic field lasting over some tens of hours. This depression is caused by the ring current in the magnetosphere flowing westward. Note that *Dst* is practically the average of magnetic perturbations at mid- and low-latitude stations distributed in longitude.

A geomagnetic storm customarily involves three phases, beginning with a sudden increase in the *H* component (storm sudden commencement or SSC), followed by a period of arbitrary length in which the elevated field does not change very much. This period is called the initial phase. The sudden increase in the geomagnetic field is caused typically by an interplanetary shock. The initial phase is followed by the main phase where the development of a depressed *H* component occurs, enduring over a period of a few to several hours. The storm concludes with a slow recovery toward the pre-storm level over hours to tens of hours, i.e., the recovery phase. The depression in *H* has a strong latitudinal dependence, being maximum at the equator and minimum at the poles. It is noted that an SSC is not a necessary condition for a geomagnetic storm to occur and hence the initial phase is not an essential feature (see Akasofu, 1965; Tsurutani et al., 1988; Joselyn and Tsurutani, 1990).

During the main phase of a geomagnetic storm, a number of intense substorms occur successively, accompanied by dynamic displays of auroras and by dissipation of Joule heating from auroral electrojet currents. A substorm is a transient phenomenon in which a significant amount of energy is released from the magnetotail and is deposited in the polar ionosphere

48

(e.g., Akasofu, 1968; Rostoker et al., 1980). Each substorm has a lifetime on the order of a few hours, much shorter than the lifetime of a magnetic storm. Thus, magnetic storms and substorms are, in a sense, low-latitude and high-latitude phenomena, respectively. How are these two phenomena coupled is a matter of debate, as will be discussed in Section 4.

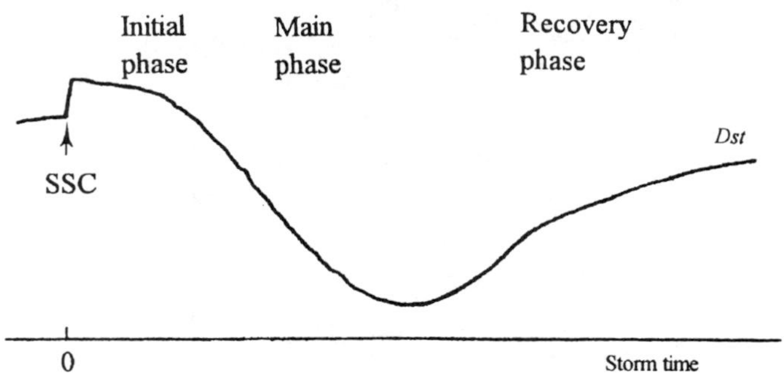

Figure 1. Schematic illustration of the *Dst* variation for a typical geomagnetic storm. For the definition of the initial, main, and recovery phases, see text.

What is the minimum level in *Dst* is required for a particular *H* disturbance to be assigned as a geomagnetic storm? In fact, assigning a threshold serves only an operational purpose in arranging "storm-time" data and has no physical basis. It is quite possible that small storms can be substorms. It is understood that intense storms are those with the peak *Dst* value of -100 nT or less (Joselyn and Tsurutani, 1990).

3.   PARTICLE MEASUREMENTS OF THE RING CURRENT

3.1   The intensity of geomagnetic storms

The intensity of a geomagnetic storm is commonly defined by the minimum *Dst* value, or the maximum depressed *Dst* magnitude, at the main phase: see Figure 1. The depression of the magnetic field during the main phase is explained as the effect of the ring current in the magnetosphere. The ring current is carried primarily by energetic (10 - 200 keV) ions: see Daglis et al. (1999). During the recovery phase, the ring current in the $L \sim 2 - 7$ region decays due to charge-exchange, Coulomb interaction, and wave-particle interaction processes in the volume of space occupied by the ring current particles.

The concept of charged particles trapped in the magnetic field and encircling the Earth, e.g., Störmer (1955), Chapman and Ferraro (1932), Alfvén (1950), and Singer (1957), was understood well before the discovery of trapped radiation by Van Allen et al. (1959). See textbooks such as ones by Roederer, (1970), Schulz and Lanzerotti (1974), Nishida (1978), Lyons and Williams (1984), and Baumjohann and Treumann (1996) for charged-particle motions under the influence of magnetic and electric fields in the magnetosphere.

The principal property of a geomagnetic storm is the creation of an enhanced ring current, located usually between 2 to 7 R_E and producing a magnetic field perturbation, which is opposite to the Earth's dipole field. The strength of this perturbation on the Earth's surface is approximately given by the so-called Dessler-Parker-Sckopke relationship (Dessler and Parker, 1959; Sckopke, 1966):

$$\Delta B / B_0 = 2E/3E_\mathrm{m}$$

where ΔB is the field decrease at the center of the Earth caused by the ring current, B_0 (~ 0.3 gauss) is the average equatorial surface field, E is the total energy of the ring current particles, and E_m (= 8×10^{24} ergs) is the total magnetic energy of the geomagnetic field outside the Earth: see Olbert et al. (1968) and Carovillano and Siscoe (1973) for a complete derivation. According to the above relationship, the *Dst* value is, in a first approximation, linearly proportional to the total energy of the ring current particles. This is the reason the *Dst* index is being used practically as a measure of the magnitude of geomagnetic storms.

3.2   Observations of ring current particles

Parker (1957) established a hydromagnetic formalism, relating the magnetospheric currents to particle pressures both parallel and perpendicular to the magnetic field. The total current j summing over motions of individual particles is

$$j = j_\mathrm{D} + j_\mathrm{c}$$

where j_D is the drift current caused by the magnetic field gradient and field line curvature, and j_c is the current driven by gyration effects within the particle distribution: see Lyons and Williams (1984) for more details.

50

Following the discovery by IMP-1 that the Earth's magnetic field is consistently confined and distorted by the solar wind (Ness, 1965; Beard, 1965; Smith et al., 1965; Coleman, 1966; Schatten et al., 1968), various plasma regions in the magnetosphere, including the ring current and the plasma sheet, were identified. Figure 2 shows one of the early measurements of the differential energy density of the ring current both

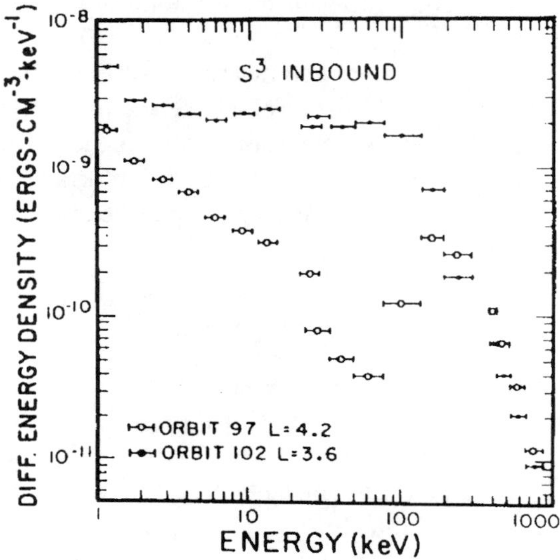

Figure 2. Satellite-measured spectra of ion energy density (proton assumed) during the pre-storm quiet (Orbit 97 of the Explorer 45 satellite, or S^3) and during a large geomagnetic storm (Orbit 102). After Smith and Hoffman (1973).

during a quiet period and during an intense geomagnetic storm (Smith and Hoffman, 1973).

As energetic particles are injected into the inner magnetosphere on the night side, they are influenced by forces due to curvature and gradient of the magnetic field. Because of these forces, protons drift westward from midnight toward dusk and electrons drift eastward from midnight toward dawn, comprising the net effects as a ring current encircling the Earth westward. A geomagnetic storm is nothing but an enhancement of this ring current.

Conducting an extensive particle measurement, Frank (1967; 1970) was one of the first who discovered the asymmetric nature of the ring current. Measurements of the differential energy spectrums of protons and electrons over the energy range extending from 200 eV to 50 keV were used. The total energy of these low-energy particles was found to be sufficient to account for the depression of the geomagnetic field in terms of the *Dst* index.

Figure 3 shows intensities of protons as functions of L during the different phases of a magnetic storm. It is evident that a severe increase in proton intensities over $3 < L < 5.5$ is apparent in the main phase observations, with a maximum located at $L = 3.6$, and that by the recovery phase, this distribution has substantially decreased in intensities with a peak positioned at $L = 4.5$.

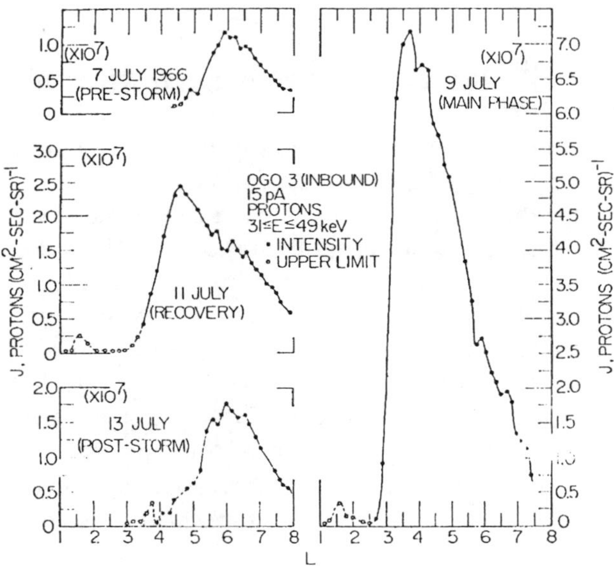

Figure 3. Directional intensities of protons ($31 < E < 49$ keV) as functions of L during the pre-storm, main phase, recovery phase, and post-storm periods of geomagnetic storms. After Frank (1967).

It should be noted that *Dst* includes the magnetic effects not only of the symmetric ring current but of other currents, such as ionospheric, field-aligned, and tail currents (e.g., Campbell, 1973, 1996; Alexeev et al., 1996). In particular, using a numerical modelling of various current systems, Alexeev et al. (1996) have demonstrated that the ground effect of the tail current during the main phase of geomagnetic storms can be of the same order as the ring current. Dremukhina et al. (1999) have applied this model to major geomagnetic storms, showing that during the main phase, the contribution of the tail current to *Dst* is roughly equal to that of the

52

symmetric ring current, although the ring current becomes dominant during the recovery phase. It is interesting to note, however, that the ring current intensity estimated from data of the CRRES magnetometer in the inner magnetosphere is consistent with what *Dst* values require (Jorgensen et al., 2001).

4. IS "THE FREQUENT OCCURRENCE OF SUBSTORMS" ENERGY SOURCE OF THE RING CURRENT?

4.1 "Mini" ring currents associated with substorms

Intense substorms occur frequently during the main phase of geomagnetic storms (Akasofu and Chapman, 1961; Akasofu et al., 1974). Many researchers believe, or tacitly assume, that a geomagnetic storm develops as a result of the frequent occurrence of substorms. There are certainly good reasons to believe that a geomagnetic storm consists of intense substorms. During episodes of substorm activity, energy can be deposited into the inner magnetosphere, leading to the formation of the so-called partial ring current, which is connected to the substorm electrojet through field-aligned currents. If intense substorms occur successively while the effects of previous substorms still remain in terms of the partial ring current, the local time extent of that partial ring current could increase and evolve into a complete ring. In other words, an individual substorm may cause only a "mini" ring current, but if substorms occur frequently enough, the injected ring current particles accumulate in the trapping region, forming the symmetrical ring current as a geomagnetic storm.

It is not clear, however, whether the occurrence of substorms is a necessary condition for a magnetic storm, or merely coincidental (Kamide, 1992; Gonzalez et al., 1994; Siscoe, 1997). Thus, a basic question remains unanswered involving whether a magnetic storm is a superposition of intense substorms, each of which constitutes an elementary storm. In fact, the storm-substorm relationship is only morphologically and qualitatively seen, and therefore basic questions remain unanswered involving the hypothesis whether a magnetic storm is a non-linear (or linear) superposition of intense substorms, each of which constitutes an elementary storm, or if the main phase of magnetic storms occurs as a result of large southward IMF (interplanetary magnetic field) which enhances magnetospheric convection and increases the occurrence probability of substorms.

Although not all substorms occur during magnetic storms – that is, most substorms occur without being associated with magnetic storms, we all are aware that substorms do occur during the main phase of magnetic storms. Can the main phase commence without these substorms? According to Chapman (1962), who first defined polar substorms in terms of magnetic storms, a magnetic storm consists of sporadic and intermittent polar disturbances, with a life time of usually one to two hours. Chapman referred to these disturbances as polar substorms.

Can we reproduce *Dst* only through knowledge of the *AE* indices which are a measure of substorm activity? According to Chapman, the simplest solution is to suppose schematically that

$$\text{STORM} = \sum_i (\text{SUBSTORM})_i$$

In the energy-balance equation for the ring current:

$$dE/dT = Q - E/T$$

where E is the ring current energy, Q is the rate of energy supply into the ring current, and T is the decay rate. To test whether *Dst*, which is proportional to E, can be reproduced exclusively by the occurrence and the intensity of substorms, it is assumed that Q is simply proportional to substorm activity. In other words, the storm main phase is assumed to be described by a linear superposition of substorm activity. The logic or physics behind this assumption is not, of course, that the auroral electrojets in the ionosphere directly generate the ring current in the magnetosphere, but that there is an energy reservoir in the magnetosphere from which some energy channels into the ring current and some other energy touches the polar ionosphere, generating polar substorms.

During the last two decades, IMF data from satellites in the interplanetary medium became available. Many independent studies on storms and substorms showed that there is one-to-one correspondence between southward turnings of the IMF and substorm occurrence, and also that the main phase of magnetic storms is associated with a large, sustained southward IMF. In particular, Russell et al. (1974) found that southward B_z must exceed a threshold level in order to trigger a main phase of major magnetic storms, and further that a weak southward B_z does not necessarily lead to an increase in the ring current, even though such southward fields persist. In fact, Siscoe and Crooker (1974) developed a linear relation between the time rate of change in *Dst*, representing the energy transfer to the inner magnetosphere, and the "merging" electric field.

4.2 Two scenarios for storm-substorm relationship

The question regarding the storm/substorm relationship can therefore be summarized as follows: Is the main phase of magnetic storms a result of (a) the impact of the southward IMF which also relates to substorm activity, or (b) the successive occurrence of substorms which also have a direct relationship with the southward IMF? In this argument it is assumed that there is no intrinsic difference between storm-time substorms and non-storm substorms: see Baumjohann et al. (1996).

The main question, which then arises regarding the storm/substorm relationship, involves the nature of the physical processes, which lead to the growth of the ring current during magnetic storms. Substorms are in some way responsible for the growth of the ring current. In fact, early studies of "injections" of energetic particles into the inner magnetosphere suggested that the occurrence of substorms led to the acceleration of particles to energies which allowed them to be effective current carriers in the ring current. However, it later became clear that the storm-time ring current was carried by energetic ions with energies typically in excess of several tens of keV. The question of how ring current particles attain their energies and whether substorm disturbances play an integral role in that process is still open.

As is shown in Figure 4, the present question can be illustrated in terms of whether Scenario 1 or Scenario 2 is more realistic: see Kamide (2001) for details. Scenario 1 is close to the original definition of substorms by Chapman (1962) except that in the "modern" view, the important role of the southward IMF in generating substorms has been taken into account. This is qualitatively consistent with the working model of a magnetic storm, which is viewed as the superposition of individual substorms. It is assumed that each substorm involves a partial ring current, and that before this partial ring current has died away, a second substorm produces a second partial ring current, and so forth. If substorms occur frequently enough, particles of the partial ring current accumulate in the trapping region, forming a complete ring and causing a significant decrease in the intensity of low-latitude geomagnetic field over the entire local-time range. In other words, an intense ring current can be created only when intense particle injection occurs successively. The essence of this scenario is that the IMF and the solar wind are important in generating substorms, which in turn create the storm-time ring current.

Scenario 1

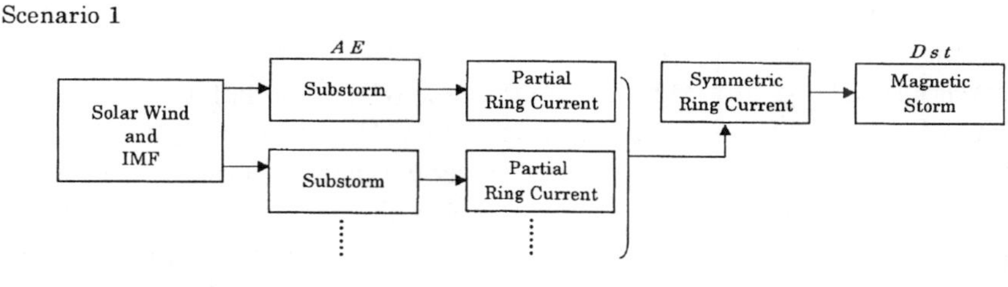

Scenario 2

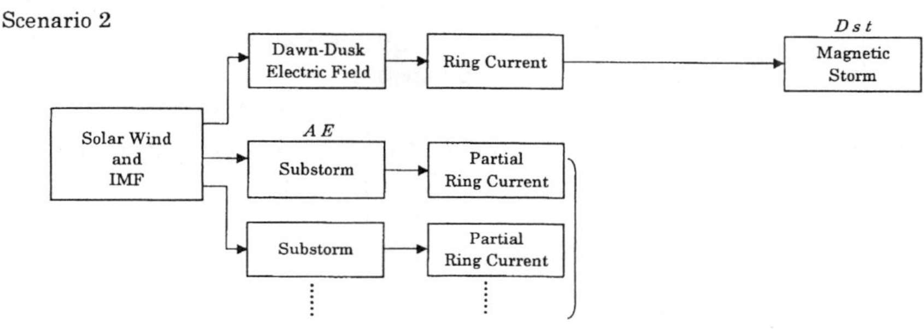

Figure 4. Block diagrams outlining two contrasting scenarios for the problem of the storm/substorm relationship.

On the other hand, in Scenario 2, the IMF and the substorm have separate roles in geomagnetic storms. Burton et al. (1975) presented an algorithm for predicting the storm *Dst* signature only from information on the solar wind and the southward component of the IMF. Assuming the injection rate to be linearly proportional to the dawn-to-dusk component of the interplanetary electric field, the algorithm resulting from the energy budget equation seems to reproduce observed *Dst* quite successfully. In Scenario 2, substorms occur coincidentally during the main phase of magnetic storms associated with southward turnings of the IMF. Contrary to Scenario 1, the main phase of magnetic storms, namely the development of a symmetrical ring current, is not a direct result of the accumulation of substorm-induced partial ring currents.

One may think that it is not a very difficult task to evaluate these scenarios since solar wind data as well as information on substorms and storms are all available. We will immediately notice, however, that simple assumptions, on which data interpretation is based, do not work well when explaining individual geomagnetic storms. It may be correct to state that actual observations are so complicated that it is difficult to test whether either scenario accurately captures the true essence of magnetospheric processes.

Figure 5 is a "good" illustration of such complications, representing a difficulty with Scenario 1 (Gonzalez et al., 1994). Consider the first half and the second half of the interval, separately. The first half is rather quiet in terms of Dst, yet quite disturbed in terms of AE activity. Accordingly, this interval cannot be identified as a major geomagnetic storm. The main phase of an intense magnetic storm commenced during the second half of the interval. With nearly the same intensity as that in the auroral electrojet measured in AE, the ring current developed more efficiently during the main phase of magnetic storms (the second half of the time period) than during the non-main-phase period (the first half), indicating that the energy injection rate into the ring current is not simply proportional to substorm activity. This observation contradicts what Scenario 1 implies. It is clear that the difference between the first and second halves lies in IMF B_z behavior. That is, the extreme value of the southward component of the IMF is reached at –25 nT in the second half, signaling the arrival of a large-scale magnetic cloud in the interplanetary medium. One may argue, however that AE does not accurately register the substorm intensity, particularly during the main phase of magnetic storms when the region of the most intense auroral electrojet tends to expand equatorward, beyond the field of view of the AE observatories. For this argument to be valid, however, a systematic "correction" factor for AE as a function of the electrojet latitude must be presented.

Scenario 2 does not always work perfectly, either. Kamide (2001) has shown such examples in which during major magnetic storms associated with magnetic clouds, the Dst variations calculated by the Burton et al. formula are much larger than the observed Dst variations, while in other storm cases, the Burton et al. algorithm underestimates the Dst variations significantly. It was thus argued that although the Burton et al. formula can reproduce the general trend of storm-time Dst variations, the reproduction efficiency varies considerably from storm to storm.

What are the causes of such high variability in the reproduction efficiency of the Burton et al. formula? One likely suspect is the inefficiency of solar wind energy entering into the magnetosphere for a given value of the interplanetary electric field. Since the decay rate of the ring current is determined primarily by charge exchange, Coulomb collisions, and wave-particle interactions in ring current particles, it is essential to identify what the primary ion species of the ring current are.

In fact, the decay rate, which is often called the decay constant, is not constant at all, but varies throughout the recovery phase during a single magnetic storm. The decay rate must be a complex function of several loss processes of the ring current particles. Wrenn (1989) suggested a solar cycle dependence of the decay rate. Feldstein et al. (1984, 1994) proffered that the decay rate is a strong function of local time, implying that loss processes differ depending on local time.

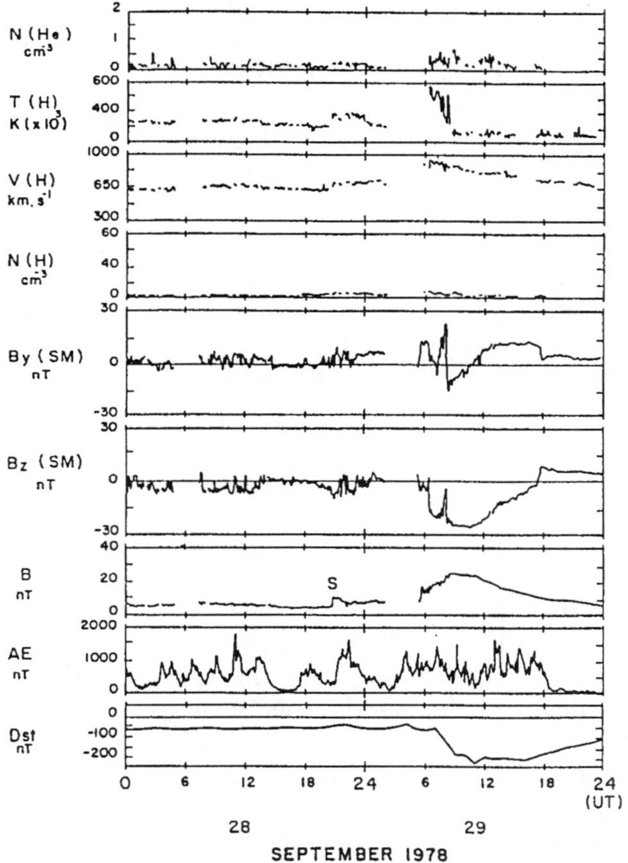

Figure 5. Interplanetary field and plasma data for a two-day interval, in which an intense magnetic storm with a peak *Dst* value of −215 nT occurred. After Gonzalez et al. (1994).

To unify the concept of the long-standing issue on the storm/substorm relationship and to account consistently for a vast number of complicated observations during magnetic storms, the following four points must be taken into account:

(1) A major magnetic storm occurs when the IMF experiences (more than three hours) an intense (more than 10 nT) southward component (Gonzalez and Tsurutani, 1987).

58

(2) No magnetic storms have occurred in the absence of intense substorms. This finding implies that storms and substorms have a common cause; it does not necessarily suggest, however, that one results from the other.

(3) Geomagnetic activity at high latitudes is always high during a magnetic storm. When substorm activity is of a lesser magnitude, however, a magnetic storm may or may not be underway.

(4) Some large substorms are associated with a significant main phase, while other equally large substorms have little effect. Some storms recover very quickly as soon as the IMF turns northward, while others take a long time to recover.

These observations imply that magnetic storms and substorms occur as rather independent processes. This also implies that the occurrence of substorms is not a necessary condition for a magnetic storm. In this view, substorms occur during a magnetic storm simply because the IMF condition for substorms is included in the IMF condition for magnetic storms.

4.3   A "thought" experiment

Each of the two scenarios shown in Figure 4 has its own problems in providing solid evidence that it can account for every detail of storm development and decay. As in a "thought" experiment, Kamide (2001) proposed that only the quasi-steady electric field in the solar wind is responsible for Dst changes during geomagnetic storms. In this model, it is fluctuations in the solar wind electric field that generate substorms, which may or may not enhance the ring current. If we were to control the solar wind, generating a purely southward IMF lasting for a long time, we would be able to create a geomagnetic storm during which no substorm expansions occur. This scenario is presented in Figure 6.

Figure 6(a) shows an "imaginary" plot of IMF B_z, Dst, and AE (AU and AL) variations, representing solar wind conditions, magnetic storms, and substorms, respectively, in which the IMF is steadily southward-directed for approximately ten hours. It is imaginary simply because the IMF is in reality almost always fluctuating. Even when the IMF appears to be strongly steady, there are often significant changes in the interplanetary dynamic pressure (e.g., Farrugia et al., 1993). If the conditions shown in this figure were to occur in nature, the Burton et al. algorithm predicts that Dst would develop significantly but there would be no substorm expansions.

As the so-called directly-driven auroral electrojet would contribute to the AU and AL indices even without substorm expansions (Akasofu, 1981), the Dst index could grow considerably, part of which results from the development of the magnetotail current (Alexeev et al., 1996). Note that this viewpoint is inconsistent with what nonlinear dynamical models predict (e.g., Baker et al., 1990; Klimas et al., 1992; Vassiliadis et al., 1995). Figure 6(b), on the other hand, shows a usual, realistic geomagnetic storm in which the solar wind is unsteady, demonstrating that fluctuations in the IMF are associated with individual substorm expansions.

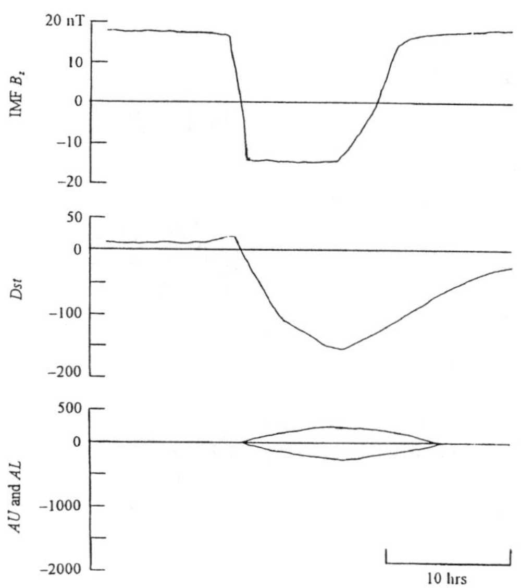

Figure 6(a). Schematic illustration of the proposed model for an extreme condition of the IMF, magnetic storms (in terms of *Dst*), and substorms (in terms of *AE*) relationship. This is an idealized case where the IMF is steadily southward for approximately 10 hours or more without any short-term changes or fluctuations. This model predicts that under such a circumstance, no substorm expansions would result.

The model proposed here accords with a number of observations that have thus far been reported:

(1) McPherron (1997) has shown that when the solar wind electric field is used to predict both *Dst* and *AL*, the prediction residuals for these two geomagnetic indices are uncorrelated. This means that the effect of substorm expansions is undetectable in storm-time *Dst* variations.

(2) A northward turning of, or becoming less southwardly directed, IMF has been found to trigger most substorm expansions (e.g., Rostoker, 1983; Lyons, 1995; Lyons et al., 1997). McPherron et al. (1986) showed earlier that more than 50% of substorms are triggered by changes in the IMF.

(3) Ballif et al. (1967) indicated that geomagnetic activity is proportional to the variance of IMF. It is quite possible that this variance is identical to the fluctuations shown in Figure 6(b).

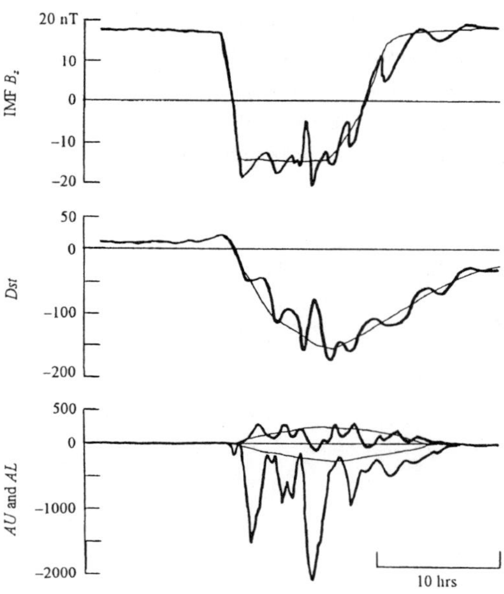

Figure 6(b). Schematic illustration of the proposed model for the IMF, magnetic storms, and substorms relationship. The IMF is almost always fluctuating. Thus, even when the IMF is directed southward for approximately 10 hours, it is the short-term changes or fluctuations that trigger substorm expansions.

5. DOUBLE GEOMAGNETIC STORMS

5.1 Two-step growth in the ring current

What is the best indicator of the intensity of geomagnetic storms? Because of the close theoretical relationship between the total energy of ring current particles and the geomagnetic *Dst* index (Dessler and Parker, 1959; Sckopke, 1966; Siscoe, 1970), the minimum *Dst* value at the main phase of magnetic storms has been used extensively in the literature as a measure of the storm intensity. Sugiura and Chapman (1960) divided magnetic storms into three categories based on peak *Dst* values: weak, moderate, and intense storms. In that "classic" statistical study, they identified magnetic storms on the basis of the existence of SSCs, thus excluding the so-called gradual storms. In more recent studies, Taylor et al. (1994), Loewe and Prölss (1997), and Yokoyama and Kamide (1997) have conducted statistical studies of geomagnetic storms in which *Dst* variations were compared with auroral electrojet activity, as well as with their interplanetary causes. These studies followed essentially the same approach as Sugiura and Chapman, where the variability in duration for different storms was obscured in their averaging process.

However, we often find that intense magnetic storms develop in two-steps during the main phase (e.g., Burlaga et al., 1987; Tsurutani et al., 1988; Cliver and Crooker, 1993). In particular, Cliver and Crooker (1993) noted that about two thirds of intense storms have two or more minima in *Dst*. It is of great interest to examine how often the ring current develops in such a two-step fashion during magnetic storms. We then ask ourselves the following major questions: What magnetospheric parameter represents quantitatively the intensity of magnetic storms? How can one define the magnetic storm strength when geomagnetic storms develop in two steps?

Using the *Dst* index, Yokoyama and Kamide (1997) and Kamide et al. (1998) examined statistically more than 1200 geomagnetic storms, from weak to intense, spanning over three solar cycles. It is surprisingly found that for more than 50% of intense magnetic storms, the main phase undergoes a two-step growth. That is, before the ring current has decayed significantly to the pre-storm level, a new major particle injection occurs, leading to a further development of the ring current, and making *Dst* decrease a second time. Thus intense magnetic storms may often be the result of two closely-spaced moderate storms.

The entire data set was grouped into three classes: weak, moderate, and intense, according to the magnitude of the storms, according to the commonly-used peak *Dst* values. Each of the three classes of geomagnetic storms is further classified into two types: Type 1 and Type 2, according to how *Dst* reaches the peak through the main phase. Figure 7 shows schematically these two types of geomagnetic storms. Type 1 represents a "normal" magnetic storm that consists of a main phase and a subsequent recovery phase. On the other hand, Type 2 magnetic storms are those which have a two-step growth in the ring current, i.e., a two-step decrease in *Dst*. To differentiate properly Type 2 from Type 1, several parameters for classification criteria were introduced. The corresponding *AE* indices and the IMF/solar wind data were also examined whenever they were available.

Table 1 summarizes the statistics. It is seen that the percentage of Type 2 occurrence increases statistically as the peak intensity in *Dst* increases. About 2/3 of intense storms have a two-step growth, whereas a relatively simple growth in *Dst* can is found in less than 1/3 of the magnetic storms.

Figure 8 shows the average *Dst* profile of Type 1 and Type 2 intense magnetic storms. As expected, the average diagrams in Loewe and Prölss are a mixture of the two diagrams. It should be noted that there is no obvious difference among weak, moderate, and intense storms in terms of the overall difference between Type 1 and 2 storms, except for their durations and the peak intensities.

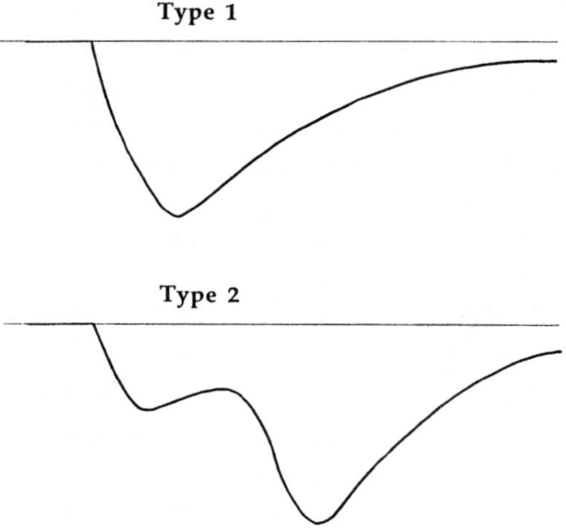

Figure 7. Schematic representation of *Dst* for Type 1 and Type 2 geomagnetic storms. See Kamide et al. (1998) for details of parameters that differentiate Type 1 and Type 2 storms.

Table 1. Classification of geomagnetic storms into two types

	Type 1	Type 2	Uncertain	All
Weak	47%	47%	6%	100%
Medium	35	56	9	100
Intense	29	67	4	100
All	40	53	61	100

Figure 9 shows the corresponding variations in auroral electrojet activity in AL and in the B_z component of the IMF. Both quantities consist of two peaks in Type 2. This effect is particularly pronounced in the AL plot, where the second peak is more intense that the first. For both Type 1 and 2 storms, peaks in AL and IMF B_z occur well before the corresponding peaks in *Dst*. The two peaks in B_z are almost equal, whereas the second AL peak seems to be more intense than the first. In individual cases, AL often returns to a very quiet state close to zero between the two peaks.

5.2   What is intense in intense geomagnetic storms?

When one observes an intense magnetic storm, it is natural to assume that some single solar event occurred and something intense propagated through the interplanetary medium to the Earth. Figures 8 and 9 clearly demonstrate, however, that this picture may be oversimplified. What actually happens in many cases is that before a *Dst* decrease has fully recovered to the pre-storm level, a second decrease often follows. In fact, auroral electrojet activity at high latitudes is found to go through two steps as well. The IMF also has a structure of two southward field regions. This means that some of the "largest" geomagnetic storms consist of two or more superposed medium-size storms. Thus, an "intense" magnetic storm in terms of the peak *Dst* value may result from the superposition effect, rather than a single, intense disturbance in the interplanetary field.

It is suggested that it is not physically very meaningful to rely on the minimum *Dst* value to define storm intensity, particularly for intense magnetic storms. It is interesting to speculate as to why earlier studies did not reveal that intense magnetic storms often go through two steps during the main phase. Figure 8 indicates that having a single, large disturbance in the solar wind is neither necessary nor sufficient to generate an intense geomagnetic storm. Future efforts should then be directed toward identifying the cause for a two-stage structure in the southward IMF, not one large southward turning. This structure has in fact been observed in some of the intense magnetic storms (see Gonzalez and Tsurutani, 1987; Tsurutani et al., 1988, 1992; Gonzalez et al., 1989).

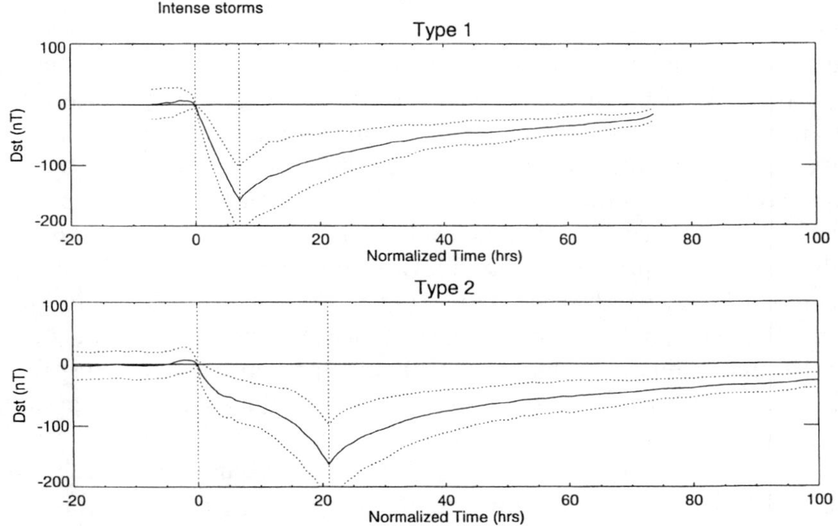

Figure 8. *Dst* variations for Type 1 and Type 2 intense geomagnetic storms. Dotted lines above and below the solid lines show the standard deviations. Two vertical dotted lines in each diagram indicates the start time and the end time of the main phase.

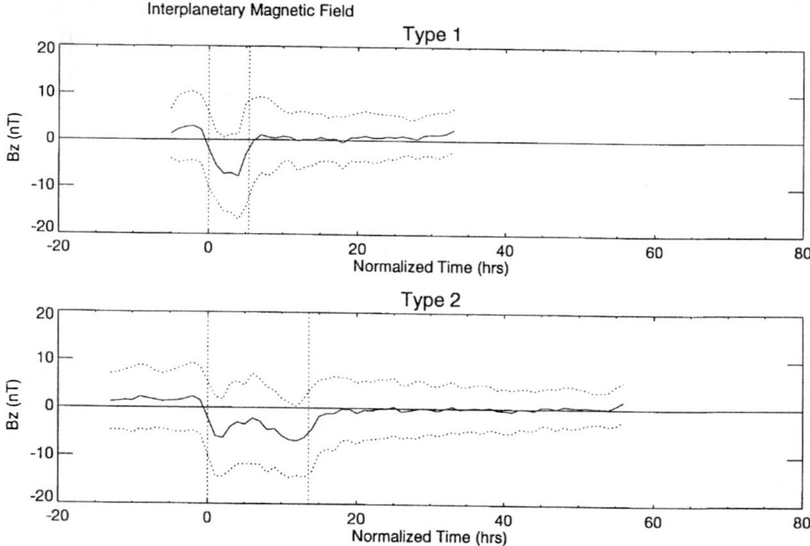

Figure 9. Variations in (a) the *AL* index, and (b) the B_z component of the interplanetary magnetic field for Type 1 and Type 2 geomagnetic storms, shown in solid lines. Dotted lines above and below the solid lines show the standard deviations. Two vertical dotted lines in each diagram indicates the start time and the end time of the main phase.

The importance of both sheath (or draped) fields and driver gas fields, carrying southward IMFs, was pointed out by Tsurutani et al. (1988) for the generation of major geomagnetic storms, displaying two-stage development characteristics. Grande et al. (1996), following this suggestion, have recently shown that CRRES heavy ion charge states were distinctly different during the two particle injections (which led to the two main phases) of the March 1991 great storm. Their interpretation was that these represent ion populations from two different coronal regions, corresponding to sheath and driver gas plasmas.

In connection with finding the double IMF B_z structure, one important candidate is a shocked B_s (negative B_z) field followed by a magnetic cloud field in the interplanetary extension of a CME. When the solar ejecta propagates at a speed greater than the upstream slow solar wind such that the speed differential is greater than the magnetosonic speed, a fast forward shock develops. The greater the speed differential, the stronger (in Mach number) the shock: the shock will compress the upstream magnetic fields and create a high intensity field sheath region downstream from the shock. If the upstream field is originally southward, shock compression will lead to intense B_s in the sheath (there are also other mechanisms to create B_s sheath fields). Following the sheath, the internal field of the ejecta itself, often called a magnetic cloud, can take on a helical structure with a cross sectional rotation in the Z-X plane, showing a rotation from south to north (or vice versa) in the IMF. The southern part of that field can become the second large B_s structure, responsible for the second stage of a Type 2 storm.

6. IONOSPHERIC ORIGIN OF RING CURRENT PARTICLES

Questions have continued as to the formation and composition of the storm-time ring current, as well as the loss mechanism of ring current particles during the storm recovery phase that sometimes needs only several hours, but occasionally several weeks. The relative importance of the solar wind and the ionosphere as the origin of ring current ions in the 10-200 keV energy range has been identified by recent satellite observations (Daglis, 2001). The ionospheric component has been shown to be most abundant during the main phase of intense magnetic storms. This implies that ionospheric ions are accelerated upward by parallel electric fields or by perpendicular heating (involving plasma waves) leading to ion conic distributions, both of which are associated with substorm expansion onsets.

66

Although some of the early measurements by mass spectrometers in space discovered energetic heavy ions of ionospheric origin (e.g., Shelley et al., 1972; Balsiger et al., 1980), it was relatively recently that the role of ionospheric particles in the ring current evolution during storms became evident after the AMPTE mission (Krimigis et al., 1982). The AMPTE lifetime coincided with the solar minimum, and thus only one great storm was observed. The great storm of February 1986 was studied in detail by Hamilton et al. (1988), finding that the ionospheric-origin ions dominated the ring current near the storm's maximum phase. O^+ alone contributed 47% of the total ion energy density compared with 36% contributed by H^+. It was estimated that 67-80% of the ring current density near the maximum of the storm was of ionospheric origin (since also a fraction of H^+ and He^+ is of ionospheric origin).

The next opportunity for multi species measurements was provided by the CRRES mission, which coincided with solar maximum. Observations regarding the ring current composition of the great storm in March 1991 were first presented by Wilken et al. (1992). The spectra and pitch angle distributions showed that new particles of predominantly ionospheric origin entered the inner magnetosphere during the storm main phase. O^+ was the dominant ion species near the storm maximum phase. Its contribution to the total energy density in the L-range 5 to 6 reached the extraordinary level of 75%. Daglis (1997) studied the importance of the ionospheric ion component in the ring current during several storms observed by CRRES, showing that during the main phase of great storms, the abundance of ionospheric-origin ions (O^+ in particular) in the inner magnetosphere is extraordinarily high. The outstanding storm feature was the concurrent increase of Dst magnitude and of O^+ contribution to the total particle energy density. Considering the domination of O^+ and taking into account that a fraction of H^+ is also of ionospheric origin, Daglis (1997) suggested that the cause of the intense ring current during large storms is terrestrial, although the energy source is unambiguously of solar origin. A very intense ring current is only created when the ionospheric response to the solar wind-magnetosphere coupling is of sufficient strength.

The increased relative abundance of ionospheric O^+ ions in the inner magnetosphere during storms, besides influencing the ring current enhancement, influences the decay rate of the ring current, since the charge-exchange lifetime of O^+ is considerably shorter than the H^+ lifetime for ring current energies (> 40 keV): see, for example, Kozyra et al. (1997). This implies that O^+-dominated ring current will decay faster, at least initially. Such a fast initial ring current decay, associated with a large O^+ component during the storm main phase, has indeed been observed (Hamilton et al.,

1988; Daglis, 1997). The initial fast recovery of *Dst* is concurrent with an initially fast drop of the O^+ contribution to the total energy density.

It is also quite possible that two major particle sources for the ring current (i.e., the solar wind and the ionosphere) play the leading role in the two successive enhancements in the ring current, shown in Section 5. The first enhancement in the ring current may be due to the magnetospheric convection driven by the southward IMF (e.g., Burton et al., 1974; McPherron, 1997), while the second ring current enhancement may be due to the substorm-associated accumulation of a new O^+ population. This second growth of the ring current must be driven by "highly fluctuating" electric fields (e.g., Chen et al., 1994), resulting from substorm expansions. It is well known that toward the end of the main phase of magnetic storms, the occurrence of intense substorms is very frequent and that the O^+ energy density in the inner magnetosphere is strongly correlated with these substorm activities (Daglis et al., 1994). Thus, the first development phase seems to prime the ring current, setting up a precondition for the second phase, which is dominated by injections of ionospheric ions.

The issue of the importance or non-importance of substorm occurrence for the storm-time ring current growth relates to the connection of substorms with ionospheric outflow. On the basis of a large set of substorm observations by AMPTE/CCE, Daglis et al. (1994) showed the association of strong substorms (as observed during storms) and enhanced ionospheric ion abundance in the inner plasma sheet. A recent study of substorms observed by CRRES confirmed the AMPTE/CCE results (see Daglis et al, 1994, 1996). Further clues to this issue could be provided by studies of the processes of ionospheric ion extraction. Viking observations of ionospheric outflow and associated electric fields (Lundin et al., 1987, 1990; Hultqvist et al., 1988) along with simulation studies showed that outflowing ionospheric ions are accelerated very efficiently by low-frequency large-amplitude electric field fluctuations (Lundin and Hultqvist, 1989; Hultqvist, 1996). Since such electric fields occur during intense substorm activity, it is expected that this type of acceleration of ionospheric ions at low altitudes operates during substorm expansion.

Observations of field-aligned ionospheric ions indicate that injection occurs over a broad range of L. These ions are trapped through an efficient magnetospheric process, becoming the main contributors to the storm-time ring current at $L < 4$. Therefore, the successive occurrence of intense substorms appears to be a necessary condition for the main phase of magnetic storms. This in fact supports the classical notion that a magnetic storm consists of continual substorm activity.

7. SOLAR ACTIVITY AND GEOMAGNETIC ACTIVITY

There is no doubt that geomagnetic storms begin when disturbances in the solar wind reach the Earth's magnetosphere. CMEs and solar flares are the most energetic phenomena among various types of coronal disturbances that occur near the maximum sunspot phase of the solar cycle, although how these two elements of solar activity are interrelated is not well understood (Kosugi and Shibata, 1997). These events occur on a wide variety of spatial scales and have a variety of speeds ranging from < 100 to nearly 2000 km/s, but the ones that are most effective in creating space storms are shown to be rapid, with speeds exceeding the ambient wind speed, so that a forward shock is formed. Considerable uncertainties exist also regarding their magnetic configuration which must play a key role in unveiling the physics for different sizes of Interplanetary CMEs (ICMEs) and such changes as the dimming of coronal features, which is known to be crucial for generating intense geomagnetic storms. "Halo" CMEs, originating near disk center and therefore directed at Earth, have shown promise as an early warning indicator of geomagnetic storms (e.g., Webb et al., 2000), but key questions concerning the origins, evolution, and basic structure of CMEs remain unanswered. One of the major problems lies in difficulties at present of visualizing the three-dimensional structure of these solar ejecta and the corresponding magnetic field lines extending into interplanetary space.

During the declining phase of the solar cycle, another type of solar event is shown to dominate, in which the coronal holes emerge from polar regions and extend into the equatorial regions. Coronal holes represent low-temperature regions and the areas of open magnetic field lines. Fast-speed plasma is continuously emitted from these solar regions. Because these hole regions are long-lived, they appear to corotate with the Sun as observed from Earth, leading to the so-called recurrent geomagnetic storms. Crooker and Cliver (1994) contended that it is the CIRs (corotating interaction regions), rather than the high-speed streams from cororal holes alone, that are responsible for peak recurrent activity.

As discussed in the previous sections, the causes of intense geomagnetic storms are strong dawn-to-dusk electric fields in the solar wind, driving plasma convection in the magnetosphere. These electric fields are caused by a combination of two factors in the solar wind: velocity and southward IMF. Of the two, the southward field is probably more important for individual storms because of its greater variability, while the solar wind velocity is empirically found to be an important factor in controlling long-term geomagnetic activity.

Within a solar cycle, there are two peaks in geomagnetic activity, one somewhat before and the other following solar maximum (Cliver et al., 1996; Richardson et al., 2000). For two solar cycles, Figure 10 shows the yearly-averaged sunspot number and the hours when $Dst < -50$ and -100 nT. It is clear that the number of intense geomagnetic storms follow the sunspot cycle faithfully at the beginning and toward the ending of the solar cycle, but have pronounced dips during the years of solar maximum. The occurrence rate of ICME is shown to follow the sunspot cycle, while strong CIRs are most predominant during the declining phase due to the presence of large polar coronal holes.

Another statistical trend in geomagnetic activity is its modulation in accordance with the 22-year cycle (Svalgaard, 1972; Cliver et al., 1996). As shown in Figure 11, which is the result of an extensive statistical study using the *aa* index over 150 years (for the years 1844-1994), the average levels in geomagnetic activity appear to be highest during the rising phase of odd-numbered sunspot cycles and during the declining phase of even cycles. These "double-solar-cycle" variations indicate that an intrinsic difference exists in large-scale solar magnetic fields, manifested as changes in the solar wind between odd and even cycles.

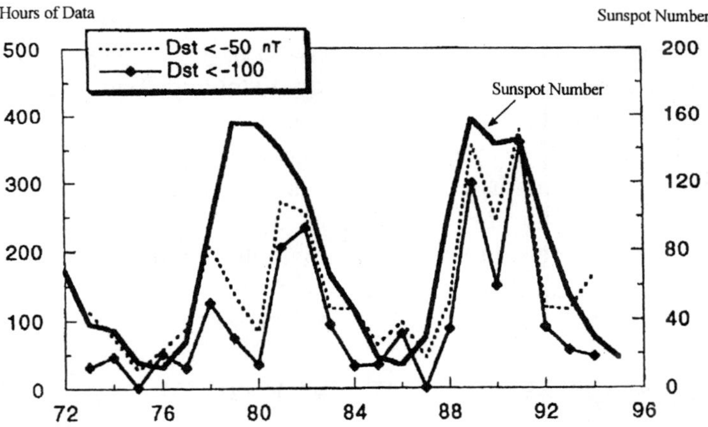

Figure 10. Yearly averaged number of hours with *Dst* less than −100 nT (solid trace with diamonds), and with *Dst* less than −50 nT, divided by 5 (dashed trace). Yearly averaged sunspot number (heavy trace) for the interval 1972-1996 is also shown. After Kamide et al. (1998).

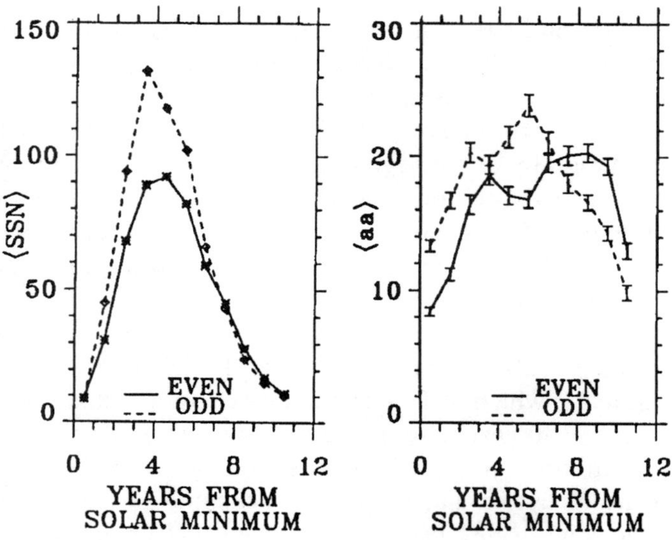

Figure 11. Composite averages of sunspot number for even- and odd-numbered solar cycles that formed complete Hale cycles, 1844-1994, and corresponding composite annual averages of the geomagnetic *aa* index for the six even-numbered and six odd-numbered solar cycles since 1844. After Cliver et al. (1996).

The odd- and even-numbered cycles have some differences in average solar wind speed profiles. This 22-year pattern in geomagnetic activity may be a reflection of the solar dynamo coupling of "poloidal" magnetic fields of one cycle to the "toroidal" fields at the maximum of the next cycle and the low-high alternation of even-odd sunspot maxima since ~1850. It must be noted, however, that studies to date have used only the *aa* index due to its availability for over 100 years. Evidence for a 22-year period in average values of this index does not necessarily guarantee similar modulation in the occurrence of geomagnetic storms.

8.   SEASONAL DEPENDENCE OF GEOMAGNETIC ACTIVITY

Another subject, which attracts intensive re-examination, is the semiannual variation of geomagnetic activity. The semiannual variation of geomagnetic disturbances – the tendency for geomagnetic storms to occur more frequently and to be larger at the equinoxes (March and September) than at the solstices (June and December) – was first noticed in the 19th century (see, for example, Sabine, 1856). Among the three proposed hypotheses to account for this seasonal modulation, i.e., the axial hypothesis, the equinoctial hypothesis, and the R-M (Russell and McPherron) effect, the R-M mechanism is generally regarded as the principal cause (e.g., Orlando et al., 1995; Siscoe and Crooker, 1996). The R-M effect is based on the variation of the tilt angle between the Earth's rotational axis and the solar equatorial plane (Russell and McPherron, 1973). In essence, it works by "converting" the IMF in the Sun's equatorial plane to southward IMF in the geocentric solar magnetospheric coordinate (GSM) system. In this way, observers on Earth can see more southward IMF at the equinoxes via coordinate transformation.

Recently, Cliver et al. (2000, 2001) argued, contrary to current popular thinking, that the R-M effect makes a relatively small contribution to the seasonal modulation of geomagnetic activity, while the equinoctial effect is most dominant. It was suggested that the efficiency of reconnection between the IMF and the geomagnetic field must be modulated by the angle between the flow direction of the solar wind and Earth's dipole axis. It is still uncertain how this seasonal variation in the average aa index relates quantitatively to the occurrence of individual major geomagnetic storms. It is interesting to see that between 1932 and 1989, no great (A_p greater than 100) storms occurred during December and only five were observed in June, which can be compared with a total 46 such major storms in March and September (Crooker et al., 1992; Phillips et al., 1993).

9.   TOWARD PREDICTIONS OF SPACE WEATHER

Throughout human history, people have stood in awe of the wildly changing colors and forms of auroras. This fascinating auroral display is a manifestation of substorm expansions, which begin sporadically in the polar ionosphere as a result of a chain of processes occurring in space between the Sun and the Earth. "Space weather" refers to conditions in the solar wind, magnetosphere, ionosphere, and atmosphere that can influence the reliability of space and ground-based technological systems. In other words, the aurora is one of the end results of space weather (National Space Weather Program Strategic Plan, 1995).

Space weather plays an ever-increasing role in our society as human beings develop more and more delicate technologies in space and on Earth. In view of the crucial effects of geomagnetic storms on such human-societal systems as radio communications and satellite drag, there is a strong need for prediction schemes continually to be upgraded. In fact, the risks for disruption of satellite operations, communications, and power grids on the ground could be minimized if reliable forecasts of space weather were possible. Studies have already begun aimed at predicting the sequence of extreme conditions that occur in the solar-terrestrial coupling system. It is strongly required to integrate our observation and modelling efforts in the space weather project.

Geomagnetic storms are the essential component of space weather, and the ring current is the dominant element of geomagnetic storms. Although since the discovery of the close connection between solar disturbances and near-Earth disturbances in the 19th century, we have learned a lot about geomagnetic storms and the ring current growth/decay, we face a long way to go before a complete understanding the structure and dynamics of the solar-terrestrial environment is achieved. Until that day comes, the role of empirical formulas cannot be insignificant.

10.　　ACKNOWLEDGEMENTS

I am indebted to Dr. Wilfried Schröder for his informative and illuminating account for the history of geomagnetic storm studies in Europe. I would also like to thank E. W. Cliver, I. A. Daglis, W. D. Gonzalez, N. S. Ness, and B. T. Tsurutani for their useful discussions throughout the various phases of the writing of this chapter. I acknowledge a number of questions and remarks made by the attendants, including students, on my presentation at the NATO Advanced Study Institute on Space Storms and Space Weather Hazards.

11.　　REFERENCES

Akasofu, S.-I., The development of geomagnetic storms without a preceding enhancement of the solar plasma pressure, *Planet. Space Sci.*, *13*, 297, 1965.

Akasofu, S.-I., *Polar and Magnetospheric Substorms*, D. Reidel Publ. Co., Dordrecht, Holland, 1968.

Akasofu, S.-I., Energy coupling between the solar wind and the magnetosphere, *Space Sci. Rev.*, *28*, 121, 1981.

Akasofu, S.-I., and S. Chapman, The ring current, geomagnetic disturbance and the Van Allen radiation belts, *J. Geophys. Res.*, *66*, 1321, 1961.

Akasofu, S.-I., S. DeForest, and C. E. McIlwain, Auroral displays near the "foot" of the field line of the ATS-5 satellite, *Planet. Space Sci.*, *22*, 25, 1974.

Alexeev, I. I., E. S. Belenkaya, V. V. Kalegaev, Y. I. Feldstein, and A. Grafe, Magnetic storms and magnetotail currents. *J. Geophys. Res.*, *101*, 7737, 1996.

Alfvén, H., *Cosmical Electrodynamics*, 1st edition, Oxford University Press, London, 1950.

Baker, D. N., A. J. Klimas, R. L. McPherron, and J. Buechner, The evolution from weak to strong geomagnetic activity: An interpretation in terms of deterministic Chaos, *Geophys. Res. Lett.*, *17*, 41, 1990.

Ballif, J. R., D. E., Jones, P. J. Coleman, L. Davis, and E. J. Smith, Transverse fluctuations in the interplanetary magnetic field: A requisite for geomagnetic variability., *J. Geophys. Res.*, *72*, 4357, 1967.

Balsinger, H., P. Eberhardt, J. Geiss, and D. T. Young, Magnetic storm injection of 0.9- to 16-keV/e solar and terrestrial ions into the high-altitude magnetosphere, *J. Geophys. Res.*, *85*, 1645, 1980.

Baumjohann, W., and R. A. Treumann, *Basic Space Plasma Physics*, Imperial College Press, London, 1996.

Baumjohann, W., Y. Kamide, and R. Nakamura, Substorms, storms, and the near-Earth tail, *J. Geomag. Geoelectr.*, *48*, 177, 1996.

Beard, D. B., The solar wind geomagnetic field boundary, *Rev. Geophys.*, *2*, 335, 1965.

Behannon, K. W., Heliocentric distance dependence of the interplanetary magnetic field, *Rev. Geophys. Space Phys.*, *16*, 125, 1978.

Burlaga et al., Compound streams, magnetic clouds, and major geomagnetic storms, *J. Geophys. Res.*, *92*, 5275, 1987.

Burton, R. K., R. L. McPherron, and C. T. Russell, An empirical relationship between interplanetary conditions and *Dst*, *J. Geophys. Res.*, *80*, 4204, 1975.

Campbell, W. H., The field levels near midnight at low and equatorial geomagnetic stations, *J. Atmos. Terr. Phys.*, *35*, 1127, 1973.

Campbell, W. H., Geomagnetic storms, the *Dst* ring-current myth, and lognormal distribution, *J. Atmos. Terr. Phys.*, *58*, 1171, 1996.

Carovillano, R. L., and G. L. Siscoe, Energy and momentum theorems in magnetospheric processes, *Rev. Geophys.*, *11*, 289, 1973.

Chapman, S., An outline of a theory of magnetic storms, *Proc. Roy. Soc. London*, *A95*, 61, 1919.

Chapman, S., and V. C. A. Ferraro, A new theory of magnetic storms, *Terr. Mag. Atmos. Electr.*, *37*, 147, 1932.

Chapman, S., Earth storms: Restropect and prospect, *J. Phys. Soc. Japan*, *7*, 6, 1962.

Chapman, S., Perspective, in *Physics of Geomagnetic Phenomena*, edited by S. Matsushita and W. H. Campbell, pp. 3-28, Academic Press, New York, 1967.

Chen, M. W., M. Schulz, and L. R. Lyons, Simulations of phase space distributions of stormtime proton ring current, *J. Geophys. Res.*, *99*, 5745, 1994.

Cliver, E. W., and N. U. Crooker, A seasonal dependence for the geoeffectiveness of eruptive solar events, *Solar Phys.*, *145*, 347, 1993.

Cliver, E. W., V. Boriakoff, and K. H. Bounar, The 22-yr cycle of geomagnetic and solar wind activity, *J. Geophys. Res.*, *101*, 27091, 1996.

Cliver, E. W., Y. Kamide, and A. G. Ling, Mountains vs. valleys: The semiannual variation of geomagnetic activity, *J. Geophys. Res.*, *105*, 2413, 2000.

Cliver, E. W., Y. Kamide, A. G. Ling, and N. Yokoyama, Semiannual variation of the geomagnetic *Dst* index: Evidence for a dominant nonstorm component, *J. Geophys. Res.*, *106*, in press, 2001.

Coleman, P. J., E. J. Smith, L. Davis, Jr., D. E. Jones, The radial dependence of the interplanetary magnetic field: 1.0-1.5 AU, *J. Geophys. Res.*, *74*, 4803, 1969.

Crooker, N. U., and E. W. Cliver, Postmodern view of M-regions, *J. Geophys. Res.*, *99*, 23383, 1994.

Crooker, N. U., E. W. Cliver, and B. T. Tsurutani, The semiannual variation of great geomagnetic storms and the postshock Russell-McPherron effect preceding coronal mass ejecta, *Geophys. Res. Lett.*, *19*, 429, 1992.

Daglis, I. A., The role of magnetosphere-ionosphere coupling in magnetic storm dynamics, in *Magnetic Storms*, Geophys. Monogr. Ser. 98, edited by B. T. Tsurutani, W. D. Gonzalez, Y. Kamide, and J. K. Arballo, 107-116, Amer. Geophys. Union, Washington, D. C., 1997.

Daglis, I. A., Space Storms, Ring Current and Space-Atmosphere Coupling, this issue, 2001.

Daglis, I. A., S. Livi, E. T. Sarris, and B. Wilken, Energy density of ionospheric and solar wind origin ions in the near-Earth magnetotail during substorms, *J. Geophys. Res.*, *99*, 5691, 1994.

74

Daglis, I. A., R. M. Thorne, W. Baumjohann, and S. Orsini, The terrestrial ring current: origin, formation, and decay, *Rev. Geophys.*, *37*, 407, 1999.

Daglis, I. A., W. I. Axford, S. Livi, B. Wilken, M. Grande, and F. Søraas, Auroral ionospheric ion feeding of the inner plasma sheet during substorms, *J. Geomag. Geoelectr.*, *48*, 729-739, 1996.

Daglis, I. A., and W. I. Axford, Fast ionospheric response to enhanced activity in geospace: Ion feeding of the inner magnetotail, *J. Geophys. Res.*, *101*, 5047, 1996.

Dessler, A. J., and E. N. Parker, Hydromagnetic theory of magnetic storms, *J. Geophys. Res.*, *64*, 2239, 1959.

Dremukhina, L. A., Y. I. Feldstein, L. I. Alexeev, and V. V. Kalegaev, Structure of the magnetospheric magnetic field during magnetic storms, *J. Geophys. Res.*, *104*, 28,351, 1999.

Farrugia, C. J., M. P. Freeman, L. F. Burlaga, R. P. Lepping, and K. Takahashi, The Earth's magnetosphere under continued forcing: Substorm activity during the passage of an interplanetary magnetic cloud, *J. Geophys. Res.*, *98*, 7657, 1993.

Feldstein, Y. I., A. E. Levitin, S. A. Golyshev, L. A. Dremukhina, U. B. Vesrchezerova, T. E. Valchuck, and A. Grafe, Ring current and auroral electrojets in connection with interplanetary medium parameters during magnetic storms, *Ann. Geophys.*, *12*, 602, 1994.

Feldstein, Y. I., V. Y. U., Pisarsky, N. M. Rudneva, and A. Grafe, Ring current simulation in connection with interplanetary space condition, *Planet. Space Sci. Rev.*, *32*, 975, 1984.

Frank, L. A., Direct detection of asymmetric increase of extraterrestrial 'ring current' proton intensities in the outer radiation zone, *J. Geophys. Res.*, *72*, 3753, 1967.

Frank, L. A., Several observations of low-energy protons and electrons in the Earth's magnetosphere with OGO 3, *J. Geophys. Res.*, *72*, 1905, 1967.

Frank, L. A., On the extraterrestrial ring current during geomagnetic storms, *J. Geophys. Res.*, *75*, 1263, 1970.

Gonzalez, W. D., and B. T. Tsurutani, Criteria of interplanetary parameters causing intense magnetic storms (*Dst* < −100 nT), *Planet. Space Sci.*, *35*, 1101, 1987.

Gonzalez, W. D., J. A. Joselyn, Y. Kamide, H. W. Kroehl, G. Rostoker, B. T. Tsurutani, and V. M. Vasyliunas, What is a Geomagnetic Storm?, *J. Geophys. Res.*, *99*, 5771, 1994.

Gonzalez, W. D., B. T. Tsurutani, A. L. Clua de Gonzalez, F. Tang, E. J. Smith, and S.-I. Akasofu, Solar wind-magnetosphere coupling during intense geomagnetic storms (1978 - 1979), *J. Geophys. Res.*, *94*, 8835, 1989.

Grande, M., C. H. Perry, J. B. Blake, M. W. Chen, J. F. Fennell, and B. Wilken, Observations of iron, silicon, and other heavy ions in the geostationary altitude region during late March 1991, *J. Geophys. Res.*, *101*, 24707, 1996.

Hamilton, D. C., G. Gloeckler, F. M. Ipavich, W. Studemann, B. Wilken, and G. Kremser, Ring current development during the great geomagnetic storm of February 1986, *J. Geophys. Res.*, *93*, 14343, 1988.

Hultqvist, B., On the acceleration of positive ions by high-latitude, large amplitude electric field fluctuations, *J. Geophys. Res.*, *101*, 27111, 1996.

Hultqvist, B., R. Lundin, K. Stasiewicz, L. Block, P.-A. Lindquist, G. Gustafsson, H. Koskinen, A. Bahnsen, T. A. Potemra, and L. J. Zanetti, Simultaneous observations of upward moving field-aligned electrons and ions on auroral zone field lines, *J. Geophys. Res.*, *93*, 9765-9776, 1988.

Jorgensen, A. M., H. E. Spence, W. J. Hughes, and H. J. Singer, A statistical study of the global structure of the ring current, *J. Geophys. Res.*, in press, 2001.

Joselyn, J. A., and B. T. Tsurutani, geomagnetic sudden impulses and storm sudden commencements, *Eos Trans.*, Amer. Geophys. Union, *71*, 1808, 1990.

Kamide, Y., Is substorm occurrence a necessary condition for a magnetic storm?, *J. Geomag. Geoelectr.*, *44*, 109, 1992.

Kamide, Y., From discovery to prediction of magnetospheric processes, *J. Atmos. Solar-Terr. Phys.*, *62*, 1659, 2000.

Kamide, Y., Interplanetary and magnetospheric electric fields during geomagnetic storms: What is more important, steady state fields or fluctuating fields?, *J. Atmos. Solar-Terr. Phys.*, *63*, in press, 2001.

Kamide, Y., R. L. McPherron, W. D. Gonzalez, D. C. Hamilton, H. S. Hudson, J. A. Joselyn, S. W. Kahler, L. R. Lyons, H. Lundstedt, and E. Szuszczewicz, Magnetic storms: Current understanding and outstanding questions, in *Magnetic Storms*, edited by B. T. Tsurutani, W. D. Gonzalez, Y. Kamide, and J. K. Arballo, Geophys. Monograph 98, pp. 1-19, American Geophys. Union, Washington, D. C., 1997.

Kamide, Y., W. Baumjohann, I. A. Daglis, W. D. Gonzalez, M. Grande, J. A. Joselyn, R. L. McPherron, J. L. Phillips, E. G. D. Reeves, G. Rostoker, A. S. Sharma, H. J. Singer, B. T. Tsurutani, V. M. Vasyliunas, Current understanding of magnetic storms: Storm/substorm relationships. *J. Geophys. Res.*, *103*, 17705, 1998.

Klimas, A. J., D. N. Baker, D. A. Roberts, and D. H. Fairfield, A nonlinear dynamical analogue model of geomagnetic activity, *J. Geophys. Res.*, *97*, 12253, 1992.

Kosugi, T., and K. Shibata, Solar coronal dynamics and flares as a cause of interplanetary disturbances, in *Magnetic Storms,* Magnetic Storms, Geophys. Monogr. Ser. 98, edited by B. T. Tsurutani, W. D. Gonzalez, Y. Kamide, and J. K. Arballo, pp. 21-34, Amer. Geophys. Union, Washington, D. C., 1997.

Kozyra, J. U., V. K. Jordanova, R. B. Horne, and R. M. Thorne, Modeling of the contribution of electromagnetic ion cyclotron (EMIC) waves to storm time ring current erosion, in *Magnetic Storms*, Geophys. Monogr. Ser. 98, edited by B. T. Tsurutani, W. D. Gonzalez, Y. Kamide, and J. K. Arballo, 187-202, Amer. Geophys. Union, Washington, D. C., 1997.

Krimigis, S. M., and G. Haerendel and R. W. McEntire and G. Paschmann and D. A. Bryant, The active magnetospheric particle tracer explorers (AMPTE) program, *EOS Trans.*, *63*, 843, Amer. Geophys. Union, Washington, D. C., 1982.

Loewe, C. A., and G. W. Prölss, Classification and mean behavior of magnetic storms, *J. Geophys. Res.*, *102*, 14209, 1997.

Lundin, R., and B. Hultqvist, Ionospheric plasma-escape by high altitude electric fields: Magnetic moment pumping, *J. Geophys. Res.*, *94*, 6665, 1989.

Lundin, R., L. Eliasson, and K. Stasiewicz, Plasma energization on auroral field lines as observed by the Viking spacecraft, *Geophys. Res. Lett.*, *14*, 443, 1987.

Lundin, R., G. Gustafsson, A. I. Eriksson, and G. T. Marklund, On the importance of high-altitude low-frequency electric fluctuations for the escape of ionospheric ions, *J. Geophys. Res.*, *95*, 5905, 1990.

Lyons, L. R., A new theory for magnetosphere substorms, *J. Geophys. Res.*, *100*, 19069, 1995.

Lyons, L. R., and D. J. Williams, *Quantitative Aspects of Magnetospheric Physics*, D. Reidel Publishing Company, Dordrecht, 1984.

Lyons, L. R., G. T. Blanchard, J. C. Samson, R. P. Lepping, T. Yamamoto, and T. Moretto, Coordinated observations demonstrating external substorm triggering, *J. Geophys. Res.*, *102*, 27039, 1997.

Malin, S., Historical introduction to geomagnetism, in *Geomagnetism*, Vol. 1, edited by J. A. Jacobs, pp. 1-50, Academic Press, London, 1987.

McPherron, R. L., The role of substorms in the generation of magnetic storms, in *Magnetic Storms*, edited by B. T. Tsurutani, W. D. Gonzalez, Y. Kamide, and J. K. Arballo, Geophys. Monograph 98, American Geophysical Union, Washington, D. C., pp. 131-148, 1997.

McPherron, R. L., T. Terasawa, and A. Nishida, Solar wind triggering of substorm onset, *J. Geomag. Geoelec.*, *38*, 1089, 1986.

National Space Weather Program Strategic Plan, Office of the Federal Coordinator for Meteorological Services and Supporting Research, FCM-P30-1995, Washington, D. C., 1995.

Ness, N. F., The Earth's magnetotail, *J. Geophys. Res.*, *70*, 2989, 1965.

Nishida, A., *Geomagnetic Diagnosis of the Magnetosphere*, Springer Verlag, Heidelberg, 1978.

Olbert, S., G. L. Siscoe, and V. M. Vasyliunas, A simple derivation of the Dessler-Parker-Sckopke relation, *J. Geophys. Res.*, *73*, 1115, 1968.

Orlando, M., G. Moreno, M. Parisi, and M. Storini, The diurnal modulation of geomagnetic activity by the southward component of the geomagnetic field, *J. Geophys. Res.*, *100*, 19,565, 1995.

Parker, E. N., Newtonian development of the dynamical properties of ionized gases at low density, *Phys. Rev.*, *107*, 924, 1957.

Phillips, J. L., J. T. Gosling, and D. J. McCormas, Coronal mass ejections and geomagnetic storms: Seasonal variations, in *Solar-Terrestrial Predictions - IV*, edited by J. R. Hruska, M. A. Shea, D. F. Smart, and G. Heckman, pp. 242, Geological Survey of Canada, Ottawa, 1993.

Roederer, J. G., *Dynamics of Geomagnetically Trapped Radiation*, Springer-Verlag, Heidelberg, 1970.

Rostoker, G., Triggering of expansive phase intensifications of magnetospheric substorms by northward turnings of the interplanetary magnetic fields, *J. Geophys. Res.*, *88*, 6981, 1983.

Rostoker, G., S.-I. Akasofu, J. C. Foster, R. A. Greenwald, Y. Kamide, K. Kawasaki, A. T. Y. Lui, R. L. McPherron, and C. T. Russell, Magnetospheric substorms - definition and signatures, *J. Geophys. Res.*, *85*, 1663, 1980.

Russell, C. T., and R. L. McPherron, Semiannual variation of geomagnetic activity, *J. Geophys. Res.*, *78*, 92, 1973.

Russell, C. T., R. L. McPherron, and P. K. Burton, On the cause of geomagnetic storms, *J. Geophys. Res.*, *79*, 1105, 1974.

Sabine, E., On periodical laws discoverable in the mean effects of the large magnetic disturbances, *Phil. Trans. Roy. Soc. London*, *146*, 357, 1856.

Schatten, K. H., N. S. Ness, J. M. Wilcox, Influence of a solar active region on the interplanetary magnetic field, *Solar Phys.*, *5*, 240, 1968.

Schröder, W., Some aspects of the earlier history of solar-terrestrial physics, *Planet. Space Sci.*, *45*, 395, 1997.

Schulz, M., and L. J. Lanzerotti, *Particle Diffusion in Radiation Belts*, Springer-Verlag, Heidelberg, 1974.

Sckopke, N., A general relation between the energy of trapped particles and disturbance field near the Earth, *J. Geophys. Res.*, *71*, 3125, 1966.

Shelley, E.G., R. G. Johnson, and R. D. Sharp, Satellite observations of energetic heavy ions during a geomagnetic storm, *J. Geophys. Res.*, *77*, 6104, 1972.

Singer, S. F., A new model of magnetic storms and aurorae, *Eos Trans.*, *38*, Amer. Geophys. Union, 175, 1957.

Siscoe, G. L., The virial theorem applied to magnetospheric dynamics, *J. Geophys. Res.*, *75*, 5340, 1970.

Siscoe, G. L., Big storms make little storms, *Nature*, *390*, 448, 1997.

Siscoe, G. L., and N. U. Crooker, Diurnal oscillation of *Dst*: A manifestation of the Russell-McPherron effect, *J. Geophys. Res.*, *101*, 24,985, 1996.

Siscoe, G. L., and N.U. Crooker, A theoretical relation between *Dst* and the solar wind merging electric field, *Geophys. Res. Lett.*, *1*, 17, 1974.

Smith, E. J., L. Davis, Jr., P. J. Coleman, Jr., W. E. Jone, Magnetic field measurements near Mars, *Science*, *149*, 1241, 1965.

Smith, P. H., and R. A. Hoffman, Ring current particle distribution during the magnetic storms of December 16-18, 1971, *J. Geophys. Res.*, *78*, 4731, 1973.

Stern, D. P., A brief history of magnetospheric physics before the spaceflight era, *Rev. Geophys.*, *27*, 103, 1989.

Störmer, C., *The Polar Aurora*, Oxford University Press, London, 1955.

Sugiura, M., and S. Chapman, The average morphology of geomagnetic storms with sudden commencement, *Abhandl. Akad. Wiss. Göttingen. Math.-Phys., Kl. Sondeheft 4*, Göttingen, 1960.

Svalgaard, L., Interplanetary magnetic-sector structure 1926-1971, *J. Geophys. Res.*, *77*, 4027, 1972.

Taylor, J. R., M. Lester, and T. K. Yeoman, A superposed epoch analysis of geomagnetic storms, *Ann. Geophys.*, *12*, 612, 1994.

Tsurutani, B. T., W. D. Gonzalez, F. Tang, and Y. T. Lee, Great magnetic storms, *Geophys. Res. Lett.*, *19*, 73, 1992.

Tsurutani, B. T., W. D. Gonzalez, F. Tang, S.-I. Akasofu, and E. J. Smith, Solar wind southward B_z features responsible for major magnetic storms of 1978 - 1979, *J. Geophys. Res.*, *93*, 8519, 1988.

Yokoyama, N., and Y. Kamide, Statistical nature of geomagnetic storms, *J. Geophys. Res.*, *102*, 14215, 1997.

Van Allen, J. A., C. E. McIlwain, and G. H. Ludwig, Satellite observations of electron artificially injected into the geomagnetic field, *J. Geophys. Res.*, *64*, 874, 1959.

Vassiliadis, D., A. J. Klimas, D. N. Baker, and D. A. Roberts, A description of the solar wind-magnetosphere coupling based on nonlinear filters, *J. Geophys. Res.*, *100*, 3495, 1995.

Webb, D. F., E, W. Cliver, N. U. Crooker, O. S. St. Cyr, and B. J. Thompson, Relationship of halo coronal mass ejections, magnetic clouds, and magnetic storms, *J. Geophys. Res.*, *105*, 7491, 2000.

Wilken, B., I. A. Daglis, and S. Livi, Observations of geomagnetic storms by the CRRES satellite, EOS Trans., *American Geophys. Union*, *73*, 457, 1992.

Wrenn, G. L., Persistence of the ring current, 1958-1984, *Geophys. Res. Lett.*, *16*, 891, 1989.

Chapter 3

From the Discovery of Radiation Belts to Space Weather Perspectives

Commentary of historical events

Joseph F. Lemaire
*Institut d'Aéronomie Spatiale de Belgique, 3 Avenue Circulaire, 1180 Bruxelles
& Institut G. Lemaître, Université cath. Louvain, 1348 Louvain-La-Neuve, Belgium*

Abstract   The discovery of the Radiation Belts by the first American and Soviet spacecraft is recalled. It is shown how different features of this harsh radiation environment were discovered, and disclosed, step by step, as part of a tight race between the Americans and Soviets for space exploration. The discovery of the Radiation Belts was the first major milestone of the International Geophysical Year (IGY) scientific program. The series of statistical radiation belt models developed by NASA from numerous early satellite surveys of the magnetosphere are important assets. The models are used to evaluate the ionizing radiation fluence of future space missions and for instance the risk of Single Event Effects and of other more dramatic failures. These 'first generation' empirical models are now considered out of date and, anyway, are unable to describe day-to-day variability of the highly dynamical environment in outer space. It will be a major task for future international space weather programs to produce more comprehensive time dependent predictive models of the ever-changing space environment.

Keywords   Radiation belts, trapped particles, static and dynamical models

1.   INTRODUCTION

Perhaps because of the abstract way Physics and Science is taught at University there is a lack of interest among young people to read older original scientific papers. This is why we gradually lose sight of how major discoveries have been made in the past. The young scientist preparing a

I.A. Daglis (ed.), Space Storms and Space Weather Hazards, 79–102.
© 2001 *Kluwer Academic Publishers. Printed in the Netherlands.*

PhD or involved in a project with a rapidly approaching deadline does not have time to read 'old' papers. He wants his information as quickly as possible by reading the most recent review articles written by colleagues who, fortunately, sometimes remember and maintain an interest in the History of Science.

It is for these young scientists, as well as older colleagues who contributed to the discovery of the Radiation Belts, that the first part of this article has been prepared. Indeed, the discovery of the Radiation Belts has been a most challenging and politically sensitive endeavour at the dawn of magnetospheric research. The word 'Magnetosphere' was coined by Gold (1959) soon after the historical observation by the Americans and Soviets of 'geomagnetically trapped corpuscular radiation'.

My tutorial presented at the NATO-Advanced Study Institute (ASI) Space Storms and Space Weather Hazards meeting (Crete, June 2000) was entitled 'Radiation Belt Models'. The talk was introduced by a rather detailed historical account outlining the discovery of the Radiation Belts. Taking place during the 1957-58 International Geophysical Year (IGY), this is a most interesting part of the history of the early days of Space Age involving major scientific achievements.

The experimental discovery of the Radiation Belts took place during the Cold War in a challenging and politically sensitive race for space exploration. The Sputnik, Explorer, Lunik and Pioneer spacecraft were built and launched from the Soviet Union and United States under strict security. Experiments by both countries using military rockets and high altitude nuclear tests were also shrouded under strict secrecy. As a result of this, much information remained hidden for a long time, or, indeed, is still classified and unknown to the scientists who developed the scientific equipment.

During and after the NATO-ASI meeting a few new facts came to light. They have been incorporated into the following text so that they will not be lost. Possibly they can be used later by a professional historian interested in compiling a comprehensive and independent investigation of this exploratory epoch in the History of External Geophysics.

Due to lack of time and page limitation this historical essay will not address in detail another scientifically interesting and sensitive aspect related to the Radiation Belts discovery, the development of ideas and theories, proposed in the United States and the Soviet Union, explaining the satellite findings or foreshadowing them.

Partial overviews and accounts of this story can be found in the articles by Singer and Lenchek (1962), Williams (1971), D.P. Stern (1996), J.A. Van Allen (1997), C.E. McIlwain (1997), S.F. Singer and R.C. Wentworth (1997) and M. Walt (1997) in the Monograph *'Discovery of the Magnetosphere'* edited by Gillmor and Spreiter (1997), as well as in the paper by Panasyuk (1998).

2.   HISTORY OF THE RADIATION BELT DISCOVERY

The exotic idea to use artificial satellites to explore outer space and to observe charged particles artificially injected and trapped into the geomagnetic field was proposed before 1957 by S.F. Singer and others (Singer, 1952, 1956).   Use of satellites for geophysical studies had been officially adopted by an international committee preparing the International Geophysical Year 1957-58 (IGY) in the 50's. The IGY was a worldwide campaign of geophysical observations of the Earth. All geophysical disciplines were involved with this endeavour.   Both in America and the Soviet Union it was agreed that scientific satellites would be launched to observe the topside atmosphere and exosphere of the Earth.

In October and November 1957, the Soviet Union launched Sputniks 1 and 2 into orbit around the Earth. The latter spacecraft was equipped by S.N. Vernov of Moscow State University with Geiger-Müller counters to study the primary cosmic ray fluxes in the topside atmosphere at altitudes higher than those reached by stratospheric balloons or small rockets.

In January 1958, a four-stage Jupiter-C rocket developed by Werner von Braun for the U.S. Army launched the first American satellite Explorer 1 into orbit.   Explorer 1 contained similar scientific equipment to Sputnik, mainly, a Geiger-Müller counter provided by James Van Allen of the University of Iowa.   The count rate was recorded in real-time only when Explorer 1 passed one of the 16 receiving antennae distributed around the World.   The data tapes from these real-time passes showed the expected count rate of 30 count/sec corresponding to the Cosmic Ray background. However, some of the tapes showed the peculiar result of zero counts/sec for the entire duration (1 or 2 minutes) of a pass. The passes showing zero counts were all near apogee at an altitude of more than 1000 km, while the low-altitude passes at around 500 km showed the normal 30 counts/sec. This suggested the possibility of a problem with the counters.   Though this idea was soon discarded as the recorded counter temperature measured by Explorer I was normal (Hess, 1968).   This left the Iowa University group, formed by James Van Allen, Ernst Ray, George Ludwig and Carl McIlwain (then a post graduate student), with the problem of why there seemed to be no cosmic rays at high altitude.

However, Explorer 3, launched successfully on March 26, 1958, contained a Geiger counter with additional tape recorder (built by Ludwig) allowing complete orbits of data to be obtained. [The Iowa group also had a Geiger counter on the failed Explorer 2 mission (it ditched into the ocean).] Within a day of receiving the first complete orbits of data, Ray and Van Allen had analysed the data. At the beginning of an orbit the count rate was the normal cosmic-ray count rate. A few minutes later the count rate rapidly increased to 128 counts/sec, which was the maximum the tape recorder could read. Then, about 15 minutes later, the count rate dropped to zero. After a further 10 minutes it rose again to 128 counts/sec. At the end of the orbit, when Explorer 3 was near perigee, the count rate returned again to the cosmic ray background (this indicated that Explorer 3's instrument gave similar results to Explorer 1's and that it was working properly).

To understand these findings, McIlwain conducted tests with the prototype Geiger tube and circuit using a small X-ray machine. On March 28, 1958, he demonstrated that a true rate exceeding around 25,000 counts/sec would indeed result in an apparent rate of zero. The conclusion was then immediate: at higher altitudes the intensity was actually, at least, a thousand times greater than the intensity due to cosmic radiation. When the counters entered a region of very large particle fluxes the dead-time effect of the counters reduced the count rate essentially to zero. On learning this Ernest Ray made the famous remark 'My God, space is radioactive!'. The puzzle involving the apparent 'absence' of cosmic rays was solved. 'There are the usual number of cosmic rays present, plus a lot of some other radiation'. These are the words of Hess (1968) and Van Allen (1997) to report the most dramatic discovery of the IGY programme.

3.   IDENTIFICATION OF THE NATURE, THE PITCH ANGLE DISTRIBUTION AND ENERGY SPECTRA OF THIS 'OTHER RADIATION'

The next question asked was: 'What was the other radiation causing the counter to become saturated ?'. In his first public lecture on the satellite results, May 1, 1958, at a joint session of the American Physical Society and the National Academy of Sciences, Van Allen adopted the working hypothesis that radiation was trapped in the Earth's geomagnetic field and that it consisted of 'electrons and likely protons, with energies of the order of 100 keV and down, mean energies probably about 30 keV'. He estimated an omnidirectional flux of 10^8 to 10^9 cm^{-2} sec^{-1} of 40 keV (auroral type) electrons would be required to account for the observed count rate at altitudes of around 1500 km over the equator. However, in his talk and in response to a question at the end, Van Allen emphasized that there was no definitive identification of particle species and that the particles might be either penetrating protons or penetrating electrons (Van Allen, 1960a, 1997). A few months later the Iowa group showed that their hypothesis was correct.

The early measurements were made with Geiger counters, which cannot identify the particle that causes the discharge. Also, the pitch angle distribution of the particles was not measured. Without knowing the pitch angle distribution, it was not certain that the particles were trapped. The pitch angle distributions of the energetic particles was first measured onboard of Sputnik 3 launched in May 1958 (Galperin, 2001, personal communication). Krassovskiy and his colleagues measured fluxes of 10 keV (!) particles mostly in directions perpendicular to the magnetic field, and sometime parallel to this direction. These results were presented in Moscow at the International Conference on Cosmic Rays in 1959, but were in press only in 1960 (Krassovskiy et al., 1960). They are claimed to be the first experimental observation for a pitch angle distribution with an empty loss cone, and therefore as existence for trapped particles, according to the Ring Current theory of Singer (1956a & b).

In the US, the first measurement of the pitch angle distributions of energetic electrons (50-1000 keV) as a function of height (to 1045 km) were made with a magnetic spectrometer flown on a Javelin sounding rocket, launched from Wallops Island, Virginia, in July of 1959. Preliminary results were presented by M. Walt at the First International Space Science Symposium held in Nice, France (Walt et al., 1960), and by J. B. Cladis at the American Physical Society Meeting held in New York City (Cladis et al., 1960). The final results on pitch angles as well as the energy spectra of the electrons in the range 50-1000 keV were published one year later (Cladis et al., 1961). Integral fluxes of electrons > 1.5 MeV, and of protons > 1 MeV were also measured during this rocket flight.

The most detailed knowledge of the energy spectrum of trapped particles at the low altitude edge of the inner zone came from photographic emulsion measurements of Freden and White (1959a & b). This energy spectrum confirmed the one calculated by Singer (1958b & c). The identification of the bulk of the outer zone particles as low energy (10-50 keV) electrons came from the measurements of Krassovskiy et al. (1959, 1960, 1961) with the directional detectors onboard of Sputnik 3. It is not clear, however, whether auroral electrons were detected instead of or in addition to particles trapped in the outer zone. An energy spectrum of E^{-5} for < 50 keV, and E^{-3} for the more energetic particles was found from measurements with Lunik-1 (Vernov et al., 1959a). According to Stern (1996), some uncertainty about the composition of outer radiation belt persisted, however, until the publications of David and Williamson's (1963, 1966) proton observations, and the OGO-3 electron measurements by Frank (1967).

The idea of a charged particle population trapped in the geomagnetic field, however, was well understood already in 1958. The idea had been developed by S.F. Singer (a former co-worker of James Van Allen at the Johns Hopkins University, APL, until 1950). Singer visited Hannes Alfvén in the spring of 1953 at the Royal Institute of Technology. Alfvén was then vehemently critical of the popular magnetic storm theory of Chapman-Ferraro based on an electrically polarized particle beam coming from the Sun. Singer noticed that the original Terrella experiments by Birkeland, in which he shot electron beams at a magnetized sphere in a vacuum chamber, showed luminosity coming from Störmer's forbidden regions. This observation stimulated him to conclude that the imperfect vacuum allowed the electrons to be scattered, changing their (Störmer) invariant integral of motion and entering regions where they would spiral along magnetic field lines and become stably trapped like in the magnetic mirror machine.

He showed that these injected charged particles would also drift in longitude, eastwards for electrons and westward for positive ions. In 1955 he submitted a paper showing that the azimuthal drifts of protons and electrons created a completely stable ring current. This Ring Current encircling the Earth offered a new and much simpler explanation for the formation of magnetic storms than the theories of Chapman-Ferraro or Alfvén. Singer's theory for the Ring Current and magnetic storms was, at first, rejected by a *Journal of Geophysical Research* referee as being 'too fantastic'. Fortunately at this time, H. Landsberg asked him to prepare a review article on research from Earth satellites, which appeared in 1956 and contained a discussion of trapped-particle/ring current theory (Singer, 1956a & b). A complete discussion appeared in EOS, the Transactions of American Geophysical Union (Singer, 1957) and in Apel et al. (1962). This idea contained the ingredients necessary to describe a radiation belt. The trapped particles were supposed to have entered into the geomagnetic field by perturbing it and, at the same time, being scattered, occupying the region later referred to as the 'outer' Radiation Belt (Singer, 1957, 1958a). Only a few people remember this pioneering contribution by Singer. Nevertheless, he first understood the idea of the loss cone and he did show that the motion of many trapped ions and electrons produce a Ring Current with lifetimes of the order of one day.

In an e-mail correspondence dated 17 July 2000, S.F. Singer points out that after Van Allen's discovery, he immediately realised that these must be particles with appreciable lifetimes (in order to explain the high flux), namely high-energy (more than 50 MeV) protons. The particles the Explorer counters observed must have a large enough range to give a long enough atmospheric lifetime to account for the high count rates observed. These were just the sort of Cosmic Ray albedo particles he had been working with;

but if protons came out of the atmosphere, then they must also re-enter: i.e. short lifetime. That was when he first thought of NEUTRON albedo. The neutron albedo theory explicitly identified the nature of the particles in the inner belt as high energy (> 100 MeV) protons, while others, except S.N. Vernov, still speculated that they were about 100 keV particles.

In his two pioneering articles published in 1958 in Phys. Rev. Letters (Singer, 1958b & c) he published the calculated energy spectrum and spatial distribution of the trapped protons (Singer 1960, as well as Lenchek and Singer, 1962). This theory predicted an energy spectrum of the protons comparable with that which was measured one year later by Freden and White (1959a & b); it predicted also a theoretical pitch angle distribution not so far from what is needed to account for the spatial distribution of the observed fluxes. The lifetimes of trapped radiation belt particles determined by Coulomb scattering were evaluated by Wentworth et al. (1959). Although diffusion and transport processes change the spectrum of the injected protons and electrons at energies below 10-20 MeV, this source is still considered as the major continuous source of the high-energy protons forming the inner radiation belt.

As recalled in "My Adventures in the Magnetosphere" by Singer and Wentworth (1997), he predicted the breakdown of the second adiabatic invariant of high-energy protons (> 100 MeV) at an altitude of about 2 earth radii and therefore their limited survival time in the outermost region of the geomagnetic field; this led him to explain the existence of TWO belts.

4. THE RUSSIAN MEASUREMENTS

Using the Geiger counter instrument KS-5 (the first orbiting instrument for cosmic ray studies), Sputnik 2 (this was the satellite carrying the dog Laika) had detected trapped radiation near apogee over Australia. But since S.N. Vernov and A.E. Chudakov did not receive the data from the Australian receiving station, they did not see the subsequent rapid rise in intensity with altitude (the 'outer' Radiation Belts signature) until much later. In Sydney, Australia, scientists with Professor H. Messel, a noted cosmic ray researcher and head of the School of Physics at the University of Sydney, recorded the telemetry signals from Sputnik 2. But they did not have the telemetry code. Asked about this during the Cosmic Ray Congress in 1959, Messel said to Singer: 'They would not send us the code and we were not about to send them the data' (Hess, 1968). This is why in the November 23, 1984, issue of *Science*, Alex Dessler published an editorial titled 'The Vernov Radiation Belt (Almost)'.

An interesting and historical written testimony has been reproduced in Appendix 1, with permission of its author. It is a letter by Singer, mailed to Alex Dessler, 15 September 1986. It indicates that Vernov was not the only one who missed discovering the trapped radiation; indeed, according to Fred Singer, he missed it himself four times. As noticed by M. Walt, 28 November 2000 in an e-mail message to me: " ... great achievements in physics are often a matter of luck. Fred's failure to discover the radiation belt was certainly a case of bad luck".

Note, however, that low altitude data collected at the telemetry station in Russia showed count rates near 700 km altitude more than 40% in excess over that expected from the altitude dependence of the Cosmic Ray background (Vernov et al., 1959b). No significance was attached to this fact. Now it is clear that these were particles from the inner radiation belt.

Furthermore, on November 7, 1957, an anniversary of Revolution Day (!), Sputnik 2's Geiger-Muller counter measured a sharp increase in particle counts. Vernov et al. (1958) attributed this enhancement to the intrusion of solar particles caused by a weak magnetic disturbance. This was actually due to particle precipitation from the outer radiation belt (Panasyuk, 2000, personal communication). This story and a description of the life of Sergei Nikolayvich Vernov has recently been reported by Panasyuk (2000) in an article on the History of Science [published in *Science in Russia*].

Another Russian group led by Valerian I. Krassovskiy and Iosif S. Shklovsky, with whom Y.I. Galperin worked at the beginning of his scientific career, also contributed in 1958-1960 to the early observation of the Radiation Belts. There is a feeling that these results are insufficiently acknowledged (Krassovskiy et al., 1959). The third Soviet satellite (Sputnik 3) was launched on May 15, 1958, only two weeks after Van Allen's official announcement of his discovery at the joint meeting of the American Physical Society and National Academy of Science. This 'Cosmic Observatory' carried a large package of scientific instruments and had, as Sputnik 1 and 2, an orbital inclination of 65° (i.e. a significantly higher inclination than Explorer 1 or 3). This is why the Soviet spacecraft penetrated into the outer radiation belt more than two months before Explorer 4 was launched on July 26, 1958.

Singer (2000, personal communication) considers that S.N. Vernov should be acknowledged for the discovery of the radiation belt, and that he developed independently of him the Cosmic Ray Albedo Neutron Decay theory (CRAND). Nevertheless, even now that the Cold War is over, this discovery is not acknowledged worldwide. In Van Allen's (1997, p.245) detailed historical review, there is no reference to similar work and results of the two Russian groups. Indeed, here we read: 'On the basis of the first few weeks of data from Explorer 4 we [the Iowa group] had advised ARPA

[Advanced Research Project Agency] of our discovery of a minimum in the previously present radiation [i.e. before the firing of the three Argus nuclear bombs, 27 August, 30 August, 6 September, 1958] when intensity at constant altitude was plotted against latitude'. Of course, in August 1958 Van Allen was probably not yet aware of Russian findings, made two months earlier using measurements from the heavily instrumented Sputnik 3. Furthermore, the double structure of the trapped radiation that was also revealed by Explorer 4's data, was published by the Americans before the Soviet claim. Strangely, no data was presented in the paper by Vernov and Chudakov (1960) to document the Soviet claim (Hess, 1968, page 11).

This piece of History re-opens the issue of who, in scientific races, are remembered as the key initiators of important discoveries: the theoretician who had the idea first, the experimentalist who designed an instrument to check this idea and prove it to be correct, or, the author(s) whose paper passed the refereeing process and who, luckily, first published the results in open literature. In Geophysics it is the latter who wins this 'Guinness Book of Records' competition. All others who were required to set the stage in preparation for these discoveries, for example, the scientists and staffs behind building the Jupiter-C rocket that launched Explorer 1 into orbit, are not remembered, even though their experience and work were keys to success.

5. THE ARGUS EXPERIMENTS AND DRIFT SHELL CONCEPT

'In December 1957, Nicholas Christofilos, a Greek engineer-scientist at the Livermore Radiation Laboratory, proposed exploding one or more small nuclear fission bombs at high altitude (~ 200 km) to test [among other objectives] the injection of large numbers of energetic electrons ($E_e > 2$ MeV) into durably trapped orbits in the Earth's magnetic field' (Van Allen, 1997). This was the aim of the Argus project as outlined by Christofilos (1959).

Christofilos graduated from Athens Polytechnical University in 1938 and moved to New York in 1949. In his spare time he computed charged particle trajectories in complex static electric and magnetic fields. This work led in 1955 to a patent related to the concept of strong focusing which became the central concept of modern accelerators. In the early 1950s he notified the Army that many charged particles, due to the dipole magnetic field, could be trapped around the Earth (i.e., in Störmer's forbidden zones). This is also when he proposed that an artificial radiation belt, due to beta decay, could be created by exploding a nuclear bomb at high altitude (Papadopoulos, 2000, personal communication). The note he sent was filed, but re-examined in 1956 after he was taken seriously following his strong focusing patent (Papadopoulos, 2000, personal communication). A complementary line of thought was that the Argus tests would provide the United States with the necessary competence to detect high-altitude nuclear bomb tests by the Soviet Union or by other countries (Van Allen, 1997, p.243).

Explorer 4 was built in less than three months to measure the results of the Argus tests. It was launched on July 26, 1958. A comprehensive description of the Argus experiments has been given by Walt (1997) who was then working on this project at Lockheed Missiles Systems Divisions.

Yet another particle population was injected on July 9, 1962 by the explosion of a Hydrogen bomb in the inner belt region. Starfish was the code name of this project of the Defence Atomic Support Agency (DASA) and the Atomic Energy Commission (AEC). The Starfish electrons persisted for five years, underscoring the long lifetime of the natural particles in the inner belt. Before the international test ban took place, the former Soviet Union also exploded three larger bombs, but these tests occurred on more distended field lines, and the artificial radiation belts lasted only several weeks (White, 1966).

6. MODELLING THE RADIATION BELTS

The results of Explorer 1 already demonstrated conclusively that the distribution of the magnetic field intensity, B, controls the distribution of trapped particles along magnetic field lines as a function of altitude. Trapped particles bounce back and forth along field lines, mirroring at the same value of B at all longitudes. Therefore, the particle flux measured at a given value of B and at a similar geomagnetic latitude should be independent of longitude and local time, as it is (Yoshida et al., 1960).

Furthermore, the lower edge of the trapped radiation belt measured by Explorers 1 and 3 seemed to be controlled by the atmosphere, i.e. the distance in which the radiation increases by a factor of e is similar to the distance in which the atmospheric density decreases by a factor of e. If the lifetimes of trapped particles are controlled by the atmosphere the trapping time should be approximately inversely proportional to the atmospheric density (as is seen by Explorer 1 and 3 observations).

Explorer 4 was equipped with four different detectors with four different energy thresholds, designed to handle large fluxes (smaller cross-sections) and also to investigate the nature and spectra of trapped particles. This mission provided over an extended region of space a map of the distribution of trapped radiation. It also gave an integral range spectrum (Van Allen et al., 1959a). Around 1500 km altitude, energy fluxes larger than 50 erg/cm^2/s/ster were detected through the thin foil (1 mg/cm^2). It was still uncertain whether the particles detected were protons or electrons.

Explorer 4 also monitored the three Argus nuclear experiments (Van Allen, 1959b). The explosions, carried out during the summer of 1958, produced artificial belts of trapped electrons resulting from the beta-decay of fission fragments. The decay of these belts was followed for several weeks until they were no longer distinguishable from the natural radiation belt. A small fraction of these electrons were stably trapped in well-defined magnetic drift shells corresponding to Störmer's forbidden regions. The apparent mean lifetime was of the order of three weeks depending on the altitude of the explosions. Worldwide survey of these artificial radiation belts provided results of basic importance: a full geometrical description of the locus of trapping of the injected particles. Additionally, the Iowa group found that the physical nature of the Argus radiation, as characterized by their four Explorer 4 detectors, was quite different from that of the pre-Argus radiation (i.e. the natural radiation belt).

A workshop on the interpretation of Argus observations was convened at Livermore, February 1959. The physical principles of geomagnetic trapping were greatly clarified at this workshop. Prior to this it was not clear how, with the irregular nature of the real geomagnetic field and the existence of both radial and longitudinal drift resulting from gradients in the magnetic field intensity, a thin radial drift shell of trapped electrons could maintain a durable integrity for several weeks. This was clarified by T. Northrop in a tutorial lecture (Northrop and Teller, 1960).

In addition to the first adiabatic invariant of Alfvén governing trapping along a given magnetic field line, Northrop invoked the second and third adiabatic invariants of cyclic motion to account for the observations of the Argus drift shells. These invariants had been proven previously by Rosenbluth and Longmire (1957) and applied to plasma confined in laboratory magnetic fields. This led McIlwain (1961) to the concept of a drift-shell uniquely determined by the first and second adiabatic invariants, or, equivalently, by (1) the magnetic field intensity B at the mirror point of a trapped particle and (2) the virtual equatorial radial distance L of this magnetic shell. The concept of (B,L) drift shells greatly aided in the reduction of particle distributions from three to two spatial dimensions. This assisted the mapping of the Explorer flux measurements (as well as all subsequent satellite charged particle measurements).

McIlwain's L-parameter has permeated the entire subsequent literature of magnetospheric physics. It is still routinely in use despite unsuccessful attends to introduce alternative L-parameters, for example, an L*-parameter determined by the third adiabatic invariant, i.e. the magnetic flux encompassed by a drift shell (Roederer, 1972, 1996; Schulz and Lanzerotti, 1974). Other attempts have been made to promote the use of Euler potentials [Stern, 1968] and other canonical co-ordinates, as explained in a review by Schulz (1996). But this did not meet success.

7. THE CONTRIBUTIONS OF THE PIONEER AND LUNIK MISSIONS

Any historical account of the discovery of the Radiation Belts has to recall the important results contributed by the Pioneer 3 mission. Pioneer 3 was launched on December 6, 1958. This spacecraft (like Lunik 1 which was launched on the January 2, 1959) was intended as a lunar probe. But, although Pioneer 3 did not make it to the moon, it was the first to traverse the centre of the Radiation Belts measuring its entire spatial extent. Pioneer 3 carried two Geiger tubes [Van Allen and Frank, 1959]. Its trajectory crossed the inner and outer radiation belts at L values of 1.5 and 3.5 respectively. The trajectory and a sketch of both radiation belts is shown in fig. 1 [also used to illustrate the jacket of the book 'Discovery of the Magnetosphere' by Gillmor and Spreiter (1997)].

Combining data from the Explorer missions and Pioneer 3 led to the first complete map (as sketched in fig. 1) of the trapped radiation (for electrons and protons penetrating 1 gr/cm^2 of shielding). This picture confirmed the concept of an inner and outer radiation zone. The inner high intensity zone was that seen by both Explorer 1 and 3. The lower ends of the outer zone were first seen by Sputnik 3 and Explorer 4. The concept of inner and outer zones, still used nowadays, is, however, of limited merit. The flux of protons has only one peak in the inner region while electrons have, for most of the time, two peaks whose locations depend on energy as well as many other factors still requiring investigation through co-ordinated campaigns of observations within international Space Weather programs.

In January 1959 and September 1959 respectively, Lunik 1 and 2, sent out into deep space, also penetrated through the outer Radiation Belt. They carried scintillators and Geiger counters capable of measuring particles of E > 45 keV, E > 450 keV and E > 4.5 MeV. Interpreting the results as electrons, Vernov and Chudakov (1960) found a flux of electrons of 10^5 electrons/cm^2/s/ster of energy E > 0.5 – 1 MeV and a very few high-energy electrons of E > 4.5 MeV (indicating that there are very few particles of very high energy in the outer belt). The peak flux of electrons with energy exceeding 20 keV was estimated as 10^9 electrons/cm^2/s/ster.

At the COSPAR meeting, Nice, January 1960, Van Allen (1960b) presented a contribution suggesting omnidirectional fluxes of approximately

10^{11} electrons/cm^2/s for electrons of energy > 40 keV in the heart of the outer zone.   Such intense energetic electron flux should have produced a considerable negative current on board the ion traps of Lunik 1 and 2 (Gringauz et al, 1960). Since no such current was observed, Gringauz concluded that the maximum flux of energetic electrons (> 20 keV) in the outer radiation belt had to be less than 10^7-10^8 electrons/cm^2/s. These more conservative values have been proven to be the more correct (see p. 8 in Lemaire and Gringauz, 1998).

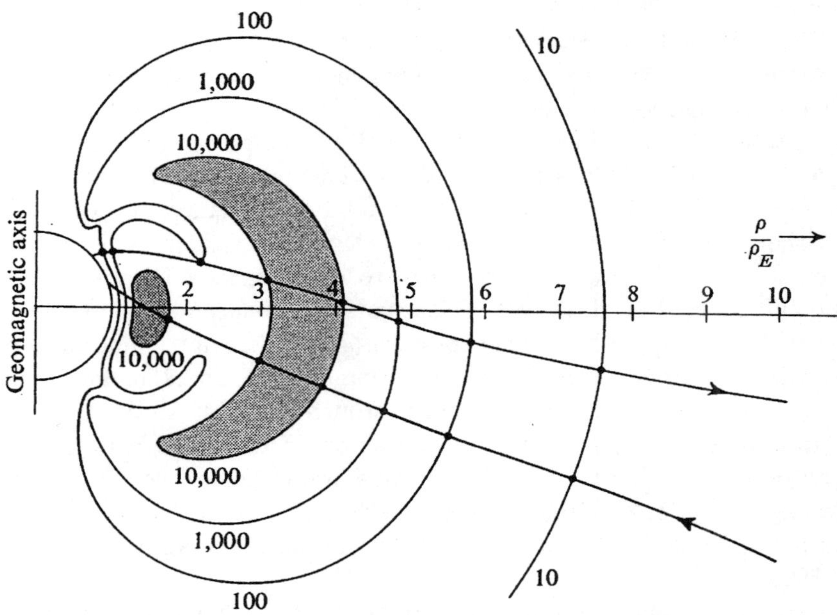

Figure 1. Van Allen's first map of the radiation belts showing the high count rate inner and outer zones. The contours are labelled by the count rate of a Geiger counter of 1 cm^2 covered by 1 gm/cm^2 of lead. The outbound and inbound trajectory of Pioneer 3, launched December 6, 1958, is also shown.

Comparing Lunik 1 and 2 results showed time variations in the electron flux and energy spectrum, as well as in the radial distance of the peak electron flux from one traversal of the outer zone to the next (Vernov and Chudakov, 1960).

A survey of Radiation Measurements made aboard Russian spacecraft in Low-Earth Orbit can be found in the report by Benton and Benton (1999).

8. MORE COMPREHENSIVE OBSERVATIONS OF THE RADIATION BELTS

That the outer zone electron flux is time-variable was also shown by Pioneer 4 observations (launched March 3, 1959). Fig. 2 shows the typical changes in the equatorial fluxes of relativistic electrons on two different drift shells measured during 1965 by the Explorer 26 satellite. The rapid increases at times of magnetic disturbances and the tendency to decay with about a two-week time constant at L= 5 & 4 are clearly seen. In 1966 McIlwain proposed five distinct physical processes, which cause these time variations in the outer zone energetic electrons distribution (see McIlwain, 1996).

The variability of electron fluxes over a wider range of L-values is illustrated even better by fig. 3, showing more modern observations of the temporal changes of electrons flux above different energy thresholds. These fluxes were measured between August 1990 and October 1991 by detectors on board of CRRES (Blake et al., 1992). This picture shows that electrons are impulsively injected or accelerated at all radial distances: i.e. on all (B,L) drift shells. They are energized by time dependent induction electric fields or betatron acceleration, as conclusively demonstrated by the numerical simulations of Li et al. (1993). Massive flux enhancements, however, are observed less frequently at the deepest drift shells than at the outermost ones. Nevertheless, between the CRRES orbit numbers 500 and 750, during one of the hugest injection/acceleration events (on March 24, 1991), the flux of electrons at all energies up to 15 MeV was enhanced by orders of magnitude on all drift shells down to L < 3. On such rare instances the slot region has filled up and for a few days afterward there is no way to distinguish between an inner and outer radiation belt in the equatorial flux distribution (Blake et al., 1992).

This comprehensive figure, compiled by J.B. Blake (1998, personal communication), also illustrates that the trapping time at L = 2 - 3 is smaller than at lower or higher L-values, presumably by the combined action of Coulomb collisions, plasmaspheric-hiss, lightning-generated-whistlers, manmade powerful VLF-transmitters and wave-particle scattering mechanisms (Walt and MacDonald, 1964; Abel and Thorne, 1998). This L-range is precisely the location where the slot region is forming time and time again after these deep injection/acceleration events. Therefore, in all statistical/empirical models of the Radiation Belts obtained by binning and averaging observations collected over long periods of time there is a minimum flux of electrons between L = 2 − 3; the location of this minimum depends slightly on the energy of the electrons.

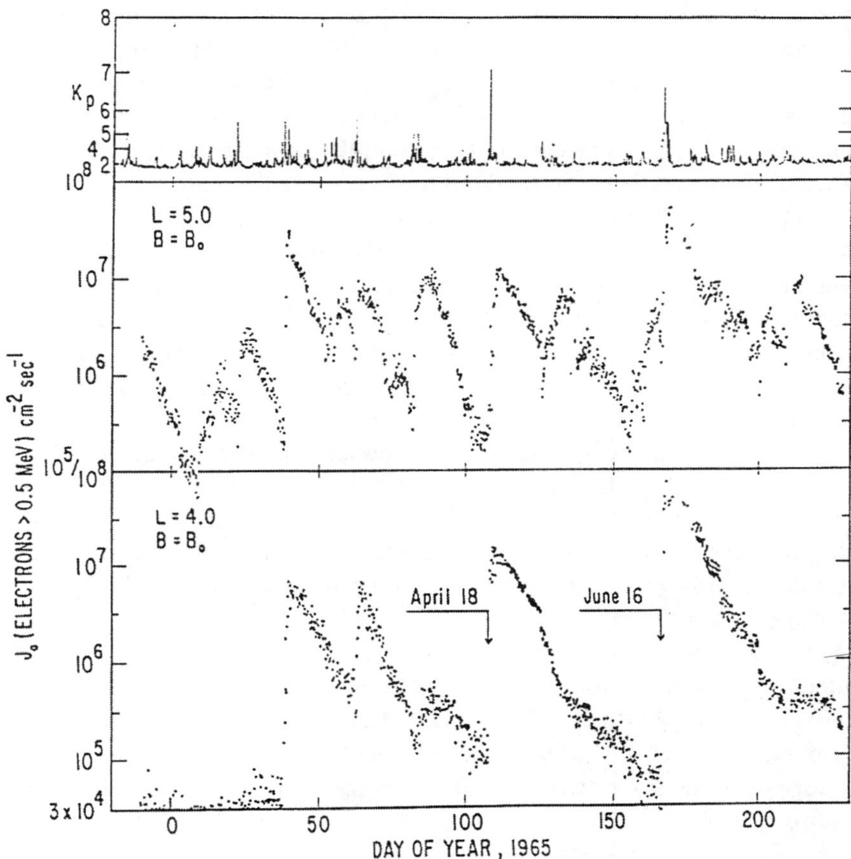

Figure 2. Time variations in the omnidirectional flux of outer zone electrons with energy greater than 0.5 MeV at L=4.0 and 5.0, as measured by the Explorer 26 satellite (McIlwain, 1996)

The four panels in fig. 3 dramatically illustrate that the energy spectrum and spatial distribution of electrons are highly dynamical in the whole magnetosphere. Paulikas and Blake (1976), Blake et al. (1997) and Li et al. (2001) showed that the flux relativistic electrons observed in the outer zone are correlated with the upstream solar wind velocity and interplanetary magnetic field. A comprehensive and up-to-date review of our understanding of the Radiation Belts has been prepared by Panasyuk (2001) for this issue.

The statistical Radiation Belt models AE-8 and AP-8, respectively for the energetic electrons and protons, have been constructed by Jim Vette and his team at GSFC/NASA, between mid-60's and 1976 (Vette, 1991a & b). Two versions of these static models are available, respectively for minimum and maximum solar activity conditions. These NASA models remain standards and are still currently employed to calculate the fluxes of energetic electrons and ions in the magnetosphere, the ionising dose or to calculate the shielding thickness for a future spacecraft. The strength and limitations of these early NASA models have been pointed out in several articles and reports including the TREND Technical Notes (Lemaire et al., 1990). More elaborated dynamical Radiation Belt and Ring Current models need to be developed along the lines of the CRESSRAD, CRRESELE and CRRESPRO models contributed more recently at Phillips Laboratory Space Physics Division, for six separate ranges of the Ap15 geomagnetic activity index, and for two different states of the magnetosphere : quiet and active (see Gussenhoven et al., 1996).

Time dependent theoretical/physical models like the SALAMMBÔ code (Beutier et al., 1995; Bourdarie et al., 1996) need also to be developed in parallel with statistical ones; they should be validated by well co-ordinated multi-spacecraft Space Weather programs, e.g. the ***ESA Space Weather Programme Studies*** and the '***Living with a Star***' program which is currently under discussion in the US.

Space Radiation Environment Models are available from different Data Centers and Services, including the ESA *SPace ENvironment Information Service (SPENVIS)* which is hosted at BIRA-IASB, Brussels, and described by Heynderickx, et al., (1999); the Internet address of this public Website and Service to the scientific and industrial communities is given below in the list of references.

The first multi-satellite project has already been conceived almost 40 years ago in Russia and launched in 1964. It was a mission of 4 spacecraft: two pairs of satellites ELECTRON-1 & 3 on polar orbits with apogees at 7000 km, and ELECTRON-2 & 4 launched half a year later, also into polar orbits, but with apogees at 60 000 km altitude. However, in 1964 the magnetosphere was significantly polluted by hard radiation injection from the Starfish nuclear explosion.

Multi-satellite missions like CLUSTER or of up to a hundred small platforms had been suggested in 1967 by Carl McIlwain who commented in a letter to S.N. Vernov: 'Most of the important phenomena involve simultaneous variations in space and time. In some cases simultaneous measurements made at two well-chosen locations will provide unambiguous results' (McIlwain, personal communication, 1990). This appears the way adopted to explore near-earth space in future and to investigate Space-Weather issues.

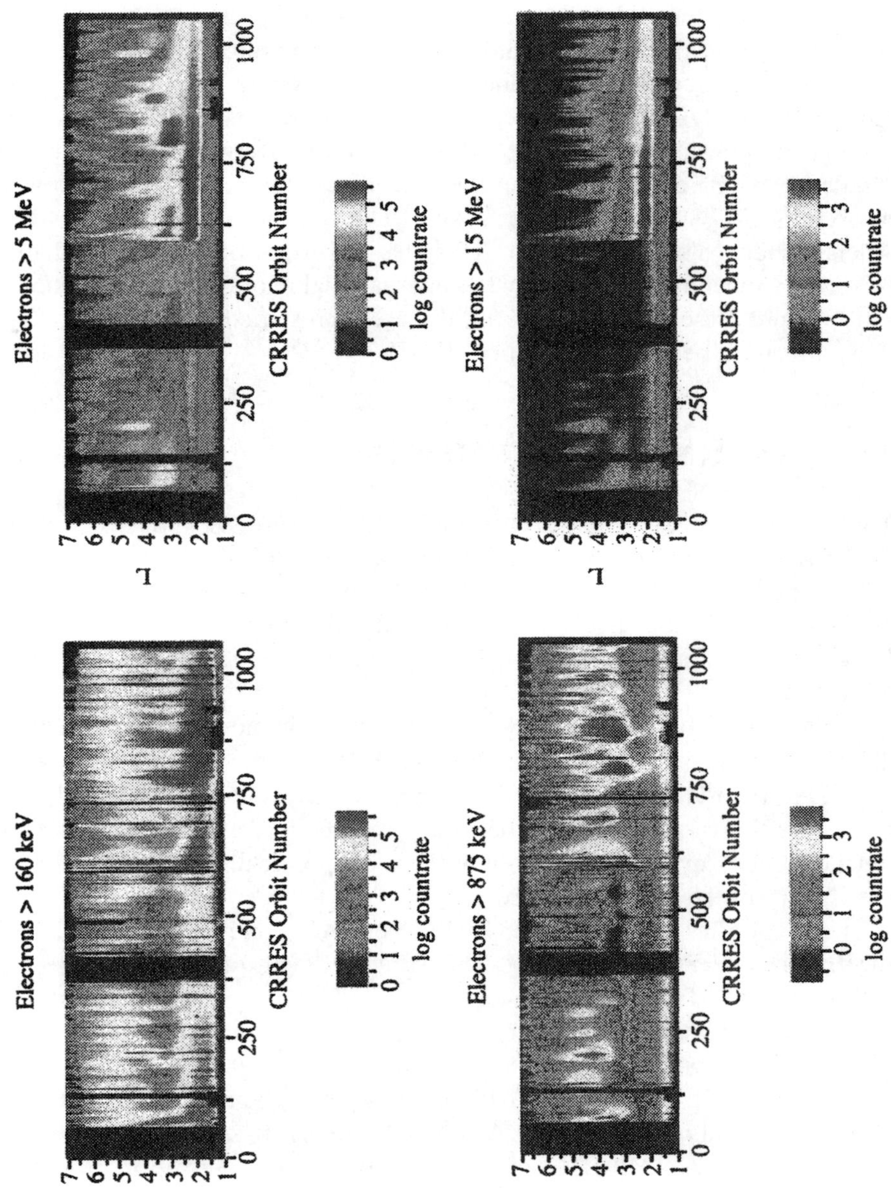

Figure 3. The L-dependence of the integral count rate of energetic electrons above four different energy thresholds as observed during the CRRES mission. Shown for E_e > 160 keV (upper left), > 875 keV (lower left), > 5 MeV (upper right) and (lower right) > 15 MeV (J.B. Blake, personal communication, 1998). (Color plate, see p. 479)

9.   POSTFACE

Before closing this 'historical review' on the discovery of the Radiation Belts, let me recall the story that prompted the community to use the word 'belt' rather than 'region', or 'zone'. In his early reports, Van Allen used the term 'radiation' or 'geomagnetically trapped corpuscular radiation'. At the press conference following the May 1, 1958, meeting at the National Academy of Sciences, he described the distribution of the radiation as encircling the Earth. A reporter asked: 'Do you mean like a belt?'; Van Allen replied: 'Yes, like a belt'! This was the origin of the term 'radiation belt'. At a meeting sponsored by the International Atomic Energy Agency in Europe during the Summer of 1958, Robert Jastrow first used the term 'Van Allen radiation belt' (see Van Allen, 1997).

10.   ACKNOWLEDGEMENTS

This chapter has been significantly enhanced as a result of inputs sent to me after my presentation of a tutorial lecture at the NATO-ASI meeting Space Storms and Space Weather Hazards, held in Crete, June 18-29, 2000. I acknowledge the useful contributions from J.B. Cladis, Y.I. Galperin, M.I. Panasyuk, D. Papadopoulos, C.E. McIlwain, S.F. Singer, D.P. Stern, M. Walt and D.J. Williams.

I thank them for the time they have spent to remember and clarify certain historical facts, and offering additional elements not reported in earlier historical reports. We hope that these elements could be useful to a professional Historian who, in the future, may be interested in compiling an independent essay reporting the important and politically sensitive events of the History of the Magnetosphere.

This work has been supported by the Research Contract BL/10/B07 from the Belgian "Services Fédéraux des Affaires Scientifiques, Techniques et Culturelles" (SSTC).

The author wishes also to acknowledge ESA for its contractual support [ESTEC-TRP contracts 9011/88/NL/MAC, 9828/92/NL/FM, 10725/94/NL/JG, 11711/95/NL/JG]. At BIRA-IASB we wish to express our appreciation to Eamonn Daly, head of the *Space Environments and Effects Analysis Section* at ESTEC-TOS-EMA, for his initiative and leadership in promoting Radiation Belt and Space Weather studies within the European scientific community.

Personally, I also wish to thank the Staff of the Belgian Institute for Space Aeronomy, specially Daniel Heynderickx, and Andrew Orr for their suggestions to enhance and improve this manuscript.

11. REFERENCES

Apel, J.R., Singer, S.F. and Wenthworth, R.C., Effects of trapped particles on the geomagnetic field, Advances in Geophysics, Vol. 9, pp. 131-189, (Ed H.E. Landsberg) Academic Press, New York, 1962.

Benton, E.R., and Benton, E.V., A Survey of Radiation Measurements Made Aboard Russian Spacecraft in Low-Earth Orbit, NASA/CR-1999-209256, Marshall Space Flight Center, Alabama 35812, 1999.

Beutier, T., and Boscher, D., SALAMMBÔ: A three-dimensional simulation of the proton radiation belts, J. Geophys. Res., 100, 17181-17188, 1995.

Blake, J.B., Baker, D.N., Turner, N., Ogilvie, K.W., and Lepping, R.P., Correlation of changes in the outer-zone relativistic–electron population with upstream solar wind and magnetic field measurements, Geophys. Res. Lett., 24, 927-929, 1997.

Blake, J.B., Kolasinski, W.A., Fillius, R.W., and Mullen, E.G., Injection of electrons and protons with energies of tens of MeV into L < 3 on 24 March, 1991, Geophys. Res. Letters, 19, 821-824, 1992.

Bourdarie, S., Boscher, D., Beutier, T., Sauvaud, J.-A., and Blanc, M., Magnetic storm modeling in the Earth's electron belt by the Salammbô code, J. Geophys. Res., 100, 27171-27176, 1996.

Christofilos, N.C., The Argus Experiment, J. Geophys. Res., 64, 869-875, 1959.

Cladis, J. B., Chase, L. F., Imhof, W. L., and Knecht, D. J., Energy spectra and spatial distributions of electrons trapped in the earth's magnetic field, Bull. Am. Phys. Soc., 5 [2], Paper N4, January 27, 1960.

Cladis, J. B., Chase, L. F., Imhof, W. L., and Knecht, D. J., Energy Spectrum and Angular Distributions of Electrons Trapped in the Geomagnetic Field, J. Geophys. Res., 66, 2297-2312, 1961.

Davis, L.R., and Williamson, J.M., Low energy trapped protons, Space Research, 3, 365-375, 1963.

Davis, L.R., and Williamson, J.M., Outer zone protons, in *Radiation Trapped in the Earth's Magnetic Field*, Edited by B.M. McCormac, pp. 110-125, D. Reidel, Norwell, Mass., 1966.

Dessler, A.J., The Vernov Radiation Belt (Almost), Science, Vol. 226, p. 915, 1984.

Frank, L.A., On the extraterrestrial ring current during geomagnetic storms, J. Geophys. Res., 72, 3753-3767, 1967.

Gillmor, C.S. and Spreiter, J.R., Preface in "Discovery of the Magnetosphere, Eds. C. Steward Gillmor, J.R. Spreiter, History of Geophysics 7, AGU, Washington D.C., V-VI, 1977.

Gringauz, K.I., Bezrukikh, V.V., Ozerov, V.D., and Rybchinsky, R.E., A study of the Interplanetary ionized gas, high-energy electrons and corpuscular radiation of the Sun, employing three-electrode charged particle set up traps on the second Soviet space rocket, Doklady Adademiya Nauk. SSSR, 131, 1302-1304, 1960; translated in 1960, Soviet Physics Doklady, 5, 361-363, 1960; published again in 1962, Planetary and Space Science, 9, 103-107, 1962.

Gussenhoven, M.S., Mullen, E.G., and Brautigam, D.H., Phillips Laboratory Space Physics Division Radiation Models, in "Radiation Belts: Models and Standards, Eds. Lemaire, J.F., Heynderickx, D., Baker, D.N., AGU Geophysical Monograph 97, Washington D.C., p. 93-101, 1996.

Freden, S.C. and White, R.S., Protons in the Earth's magnetic field, Phys. Rev. Letters, 3, 9-10, 1959a.

Freden, S.C. and White, R.S., Particle fluxes in the inner radiation belt, J. Geophys. Research, 65, 1377-1383, 1979b.

Hess, W.N., The radiation Belt and Magnetosphere, Blaisdell Publishing Company, London, pp. 548, 1968.

Heynderickx D., Quaghebeur B., Speelman E., and Daly E.J., ESA's SPace ENVironment Information System (SPENVIS): A WWW Interface to Models of the Space Enviroment and its Effects, http://www.spenvis.oma.be/spenvis/ , AIAA 2000-0371, 38th Aerospace Sciences Meeting & Exhibit, Reno, NV, 10-13 January 2000.

Krassovskiy,V.I., Shklovsky, I.S., Galperin, Y.I. and Svetlitskiy, E.M., Discovery in the upper atmosphere from the Sputnik-3 of electrons with energy about 10 keV, Dokl. AN SSSR (Soviet Physics Doklady to the USSR Academy of Sciences), AN SSSR, Moscow, 78-81, 1959.

Krassovskiy,V.I., Shklovsky, I.S.,. Galperin Y.I. and Svetlitskiy, E.M., On the high-energy corpuscles in the upper atmosphere, Proceedings of the International Conference on Cosmic Rays, Moscow, 1960. Vol.3, The Earth's Radiation Belt, AN SSSR, Moscow, 64-79, 1960.

Krassovskiy,V.I., Shklovsky, I.S., Galperin, Y.I., Svetlitskiy, E.M., Kushnir, Y.M. and Bordovskiy, G.A., Discovery in the upper atmosphere of the electrons with energy about 10 keV, Iskusstvennye Sputniki Zemli (Artificial Earth's satellites), N6, 113-126, 1961.

Lemaire, J.F., Heynderickx, D., Baker, D.N., (eds.), AGU Geophysical Monograph 97, Washington D.C., 1996.

Lemaire, J.F., Gringauz, K.I.,, with contributions from Carpenter, D.L. and Bassolo, V., The Earth's Plasmasphere, Cambridge University Press, Cambridge, UK, pp. 350, 1998.

Lemaire, J., Roth, M., Wisemberg, J., Domange, P., Fonteyn, D., Lesceux, J.M., Loh, G., Ferrante, G., Garres, C., Bordes, J., McKenna-Lawlor, S., Vette, J.I.& Daly, E.J., Development of improved models of the Earth's Radiation Environment, Technical Notes 1 to 6 and Final Report, Contract ESTEC No.8011/88/NL/MAC, BIRA-IASB, Brussels, July 1990.

Lenchek, A.M. and Singer, S.F., Geomagnetically trapped protons from Cosmic-Ray Albedo Neutrons, J. Geophys. Res., 67, 1263-1286, 1962.

Li, X., Roth, I., Temerin, M. Wygant, J.R., Hudson, M.K. and Blake, J.B., Simulation of the prompt energization and transport of radiation belt particles during the March 24, 1991 SSC, Geophys. Res. Letters, 20, 2423-2426, 1993.

Li, X., Temerin, M., Baker, D.N., Reeves, G.D., Larson, D., Quantitative Prediction of Radiation Belt Electrons at Geostationary Orbit Based on Solar Wind Measurements, Geophys. Res. Lett., 2001.

Paulikas, G.A., and Blake, J.B., Modulation of trapped energetic electrons at 6.6 Re by the direction of the interplanetary magnetic field, Geophys. Res. Lett., 3, 227, 1976.

McIlwain, C.E., Coordinates for mapping the distribution of magnetically trapped particles, J. Geophys. Res., 66, 3681-3691, 1961.

McIlwain, C.E., Processes acting upon outer zone electrons, in "Radiation Belts: Models and Standards, Eds. Lemaire, J.F., Heynderickx, D., Baker, D.N., AGU Geophysical Monograph 97, Washington D.C., p. 15-26, 1996.

McIlwain, C.E., Music and the Magnetosphere, in "Discovery of the Magnetosphere, Eds. C. Steward Gillmor, J.R. Spreiter, History of Geophysics 7, AGU, Washington D.C., p. 129-142, 1997.

Northrop, T.G. and Teller, E., Stability of the adiabatic motion of charged particles in the Earth's field, Phys. Rev., 117, 215-225, 1960.

Panasyuk, M.I., "S.N. Vernov at the foundation of national space physics", Acta Astronomica, 43, 51-56, 1998.

Panasyuk, M.I., Breakthrough into outer space, Science in Russia, 4, 61-66, 2000.

Panasyuk, M.I., Cosmic Ray and Radiation Belt Hazards for Space Missions, (this volume), 2001.

Roederer, J.G., Geomagnetic field distortions and their effects on radiation belt particles, Rev. Geophys. and Space Phys., 10, 599-630, 1972.

Roederer, J.G., Introduction to trapped particle flux mapping in "Radiation Belts: Models and Standards", eds. J.F. Lemaire, D. Heynderickx and D.N. Baker, Geophysical Monograph 97, AGU, Washington D.C., p. 149-151, 1996.

Rosenbluth, M.N. and Longmuire, C.L., Stability of plasma confined by magnetic fields, Ann. Phys., 1, 120-140, 1957.

Schulz, M., Canonical coordinates for Radiation-Belt Models, in "Radiation Belts: Models and Standards", eds. J.F. Lemaire, D. Heynderickx and D.N. Baker, Geophysical Monograph 97, AGU, Washington D.C., p. 153-160, 1996.

Schulz, M. and Lanzerotti, L.J., Particle diffusion in the radiation belt, 215 pp., Springer-Verlag, New York, 1974.

Singer, S.F., A minimum orbiting unmanned satellite of the Earth (MOUSE), J. Brit. Interplan. Soc., 11, 61, 1952.

Singer, S.F., Geophysical Research with Artificial Earth Satellites, pp. 302-367 in Advances in Geophysics, vol. 3, Ed. H.E. Landsberg, Academic Press, New York, 1956a.

Singer, S.F., Trapped orbits in the Earth's dipole field, Bull. Am. Phys. Soc. Series II, 1, 229(A), 1956b.

Singer, S.F., A new Model of Magnetic Storms and Aurorae, Trans. Am. Geophys. Union, 38, 175, 1957.

Singer, S.F., Role of Ring Current in Magnetic Storms, Trans. Am. Geophys. Union, 39, 532, 1958a.

Singer, S.F., Radiation Belt and trapped cosmic ray albedo, Phys. Rev. Letters, 1, 171-173, 1958b.

Singer, S.F., Trapped Albedo theory of the Radiation Belt, Phys. Rev. Letters, 1, 181-183, 1958c.

Singer, S.F., On the nature and origin of the Earth's Radiation Belts, in Space Research I (Ed. H.K. Kallman-Bijl), Proceedings of the First International Space Science Symposium, Nice, 1960, North-Holland Publish. Co, Amsterdam, pp. 797-820, 1960.

Singer, S.F., and Lenchek, A.M., "Geomagnetically Trapped Radiation", Chapter III (pp. 245-335) in *Progress in Elementary Particle and Cosmic Ray Physics*, Vol VI (ed. J.G. Wilson and S.A. Wouthuysen). North Holland Publishing Company, Amsterdam, 1962.

Singer, S.F. and Wentworth, R.C., My adventures in the Magnetosphere (with Addendum: A Student's Story, in "Discovery of the Magnetosphere", Eds. C. Steward Gillmor, J.R. Spreiter, History of Geophysics 7, AGU, Washington D.C., p. 165-184, 1997.

Shklovsky, I.S., Krassovskiy V.I. and Galperin Y.I., On the nature of the corpuscular radiation in the upper atmosphere, Izv. AN SSSR, ser. Geophys., (Proceedings of USSR Academy of Science), N 12, 1799-1806, 1959.

SPENVIS, The Space Environment Information Service, BIRA-IASB, Brussels, http://www.spenvis.oma.be/spenvis/ , 2000.

Stern, D.P., Euler potentials and geomagnetic drift shells, J. Geophys. Res., 73, 4373-4378, 1968.

Stern, D.P., A brief History of magnetospheric physics during the Space Age, Reviews of Geophysics, 34, 1-31, 1996.

Störmer, C., Sur les trajectoires des corpuscules électriques dans l'espace sous l'action du magnétisme terrestre, Chapitre IV, Arch. Sci. phys. et naturelles, 24, 317-364, 1907.

Störmer, C., The polar aurora, Oxford at Clarendon Press, 1955.

Van Allen, J.A., The geomagnetically trapped corpuscular radiation, J. Geophys. Res., 64, 1683-1689, 1959.

Van Allen, J.A. The first public lecture on the discovery of the geomagnetically-trapped radiation, State University of Iowa Report, 60-13, 1960a.

Van Allen, J.A., Origin and nature of the geomagnetically-trapped radiation, Space Research I, ed. H.K. Kallman-Bijl, p. 749-750, 1960b.

Van Allen, J.A., Observations of high intensity radiation by satellites 1958 alpha and 1958 gamma [Explorer I and III], IGY Satellite Report, 13, 1-22, National Academy of Sciences, Washington D.C., 1961.

Van Allen, J.A., Energetic particles in the Earth's external magnetic field, in "Discovery of the Magnetosphere", Eds. C.S. Gillmor and J.R. Spreiter, History of Geophysics 7, AGU, Washington D.C., pp. 235-264, 1997.

Van Allen, J.A., and Frank, L.A., Radiation around the Earth to a radial distance of 107, 400 km, Nature, 183, 430-434, 1959.

Van Allen, J.A., Ludwig, G.H., Ray, E.C. and McIlwain, C.E., Observation of high intensity radiation by Satellites 1958 Alpha and Gamma, Jet Propulsion, Sept. 588-592, 1958.

Van Allen, J.A., McIlwain, C.E. and Ludwig, G.H., Radiation observations with satellite 1958ε.X, J. Geophys. Res., 64, 271-286, 1959a.

Van Allen, J.A., McIlwain, C.E. and Ludwig, G.H., Satellite observations of radiation artificially injected into the geomagnetic field, J. Geophys. Res., 64, 877-892, 1959b.

Vernov, S.N., Artificial satellite measurements, of cosmic radiation, Doklady Akad. Nauk. SSSR, 120, p. 1231-1233, 1958 (in Russian).

Vernov, S.N., Chudakov, A.E., Vakulov, P.V. and Logachev, Yu.L., The study of the terrestrial corpuscular radiation and cosmic rays during the flight of a cosmic rocket, Doklady Akad. Nauk SSSR, 125, p. 304-307, 1959 (in Russian).

Vernov, S.N., Grigorov, N.L., Logachev, Yu.I., and Chudakov, A.E., Dok. Akad. Nauk SSSR, 120, 1231-1233, 1958, Study of terrestrial corpuscular radiation and cosmic rays during the flight of a cosmic rocket, Soviet Phys. Doklady, 4, 338, 1959.

Vernov, S.N., Chudakov, A.E., Terrestrial corpuscular radiation and cosmic rays, in *Space Research I (*Ed. H.K. Kallman-Bijl), Proceedings of the First International Space Science Symposium, Nice, 1960, North-Holland Publish. Co, Amsterdam, pp. 751-796, 1960.

Vette, J.I., The NASA/National Space Science Data Center Trapped Radiation Environment Model Program (TREMP) (1964-1991), NSSDC/WDC-A-R&S 91-29, NASA Goddard Space Flight Center, Greenbelt, Maryland, November, 1991a.

Vette, J.I., The AE-8 Trapped Electron Model Environment, NSSDC/WDC-A-R&S 91-24, NASA Goddard Space Flight Center, Greenbelt, Maryland, November, 1991b.

Walt, M., Chase, L. F., Cladis, J. B., Imhof, W. L., and Knecht, D. J., Energy spectra and altitude dependence of electrons trapped in the earth's magnetic field, in *Space Research I* (Ed. H.K. Kallman-Bijl), Proceedings of the First International Space Science Symposium, Nice, 1960, North-Holland Publish. Co, Amsterdam,, pp. 910-920, 1960.

Walt, M., From nuclear physics in space physics by way of high altitude nucler tests, in "Discovery of the Magnetosphere, Eds. C. Steward Gillmor, J.R. Spreiter, History of Geophysics 7, AGU, Washington D.C., pp. 253-264, 1997.

Walt, M. Sources and Loss Processes for Radiation Belt Particles, in "Radiation Belts: Models and Standards, Eds. Lemaire, J.F., Heynderickx, D., Baker, D.N., AGU Geophysical Monograph 97, Washington D.C., pp. 1-14, 1996.

White, R.S., The Earth's Radiation Belts, Phys. Today, 19, 25-38, Oct. 1966.

Williams, D.J., Charged particles trapped in the Earth's magnetic field, Landsberg, H.E. and Van Mieghem, J. (eds.), Advances in Geophysics, 15, 137-218, 1971.

Wentworth, R.C. MacDonald, W.M., and Singer, S.F., Lifetimes of trapped radiation belt particles determined by Coulomb scattering, Phys. Fluids, 2, 499-599, 1959.

Yoshida, S., Ludwig, G.H. and Van Allen, J.A., Distribution of trapped radiation in the geomagnetic field, J. Geophys. Res., 65, 807-813, 1960.

1.　　APPENDIX

1.1　　Copy of a letter written by S.F. Singer to A.J. Dessler (personal communication, 2000)

George Mason University, 15 September 1986

Dr. A. J. Dessler
Department of Space Physics
Rice University
Houston, TX 77005

Dear Alex:

For nearly two years now I have been meaning to write you and comment you on the editorial "The Vernov Radiation Belt (Almost)," which appeared in the Nov 23, 1984 issue of Science.

Your point is absolutely correct. Vernov lost his priority to the discovery of the radiation belt because of Russian secrecy. Vernov's instrument on Sputnik-2 recorded radiation belt particles six months before Van Allen's in Explorer-l. But the Sputniks elliptic orbit penetrated the belt significantly only in the Southern hemisphere, and the Russians did not release the telemetry code to anyone.

Prof. Harry Messel, a noted cosmic-ray researcher and head of the School of Physics at the University of Sidney, told me the whole story in a Moscow hotel room (the Hotel Moskva, I believe) during the Cosmic Ray Congress in 1959. He recorded the Sputnik signal every time it passed over Australia, but they wouldn't send him the code. When they finally asked for a copy of the recorded data, he told them to go to hell (as only Harry Messel could). Harry, you must remember, is a Ukrainian from Canada; he told the story with great glee.

But the full story is a little more complicated. Vernov <u>did</u> record the radiation belt but never interpreted his results properly. I have analyzed the matter in a review article on "Geomagnetically Trapped Radiation," published in <u>Progress in Elementary Particle and Cosmic Ray Physics</u> Vol. VI. (North Holland Publ., Amsterdam, 1962). I enclose pp. 249—258, "Historical Introduction," and draw your attention to p. 254. Vernov et al reported in 1958 a 40% increase in count rate between 500 and 700 km. But only 12% can be due to cosmic rays; the rest must be radiation belt particles. Of course, had they gotten data up to Sputniks's apogee altitude of 1680 km, then there would have been no doubt.

But Vernov is not the only one who missed discovering the trapped radiation. I don't know about others, but I am certainly one of them ---.four times to be exact!

1) In a 1950 Aerobee firing off Peru, I measured the east-west asymmetry of cosmic ray primaries, mostly relativistic protons. But I also measured the ionizing efficiency of the particles and found a component of high ion densities (presumably low-energy protons) with a reversed E-W asymmetry. (These were trapped protons; I later developed a theory for their E—W asymmetry (see p. 274), eventually confirmed by Heckman's observations.)

My 1950 notebook indicates that I considered albedo protons emanating from and curving back into the atmosphere as an explanation. But statistics of the data were not good enough to draw firm conclusions. Some details are given in another review article on "The Primary Cosmic Radiation and its Time Variation" in <u>Progress...</u> Vol IV (1958), pp. 263-276. (See esp. p. 264).

2) In the summer of 1950 I flew thin-walled Geiger counters in balloons launched from an icebreaker between Boston and Thule. In the auroral zone, off Labrador, the count rate went

crazy. I concluded that I was seeing noise from high—voltage discharge in the instrument, as the air pressure reached a certain low value. I never published the results; but evidently I was seeing trapped electrons of the outer belt. I should have either had a student like Carl McIlwain, or flown thick—walled counters along with the thin—walled variety.

3) By 1956 I was quite sure about the existence of trapped radiation (although I had not yet thought of the neutron albedo mechanism). I designed a 4-stage balloon-launched rocket for the Air Force OSR, to go to 4000 miles altitude. I then got the contract to supply a scientific payload, a simple Geiger counter. The Air Force called the project Far Side and diddled a lot. But right after Sputnik they tried to launch it in a great hurry from Eniwetok. I never learned officially why the project failed; all I know is that I never received any data from my instrument. Too bad; because I had published an article in <u>Missiles</u> and <u>Rockets</u> magazine, around 1957, that we would measure trapped radiation in the Far Side project.

I was one of the contenders for a spot on Explorer-l, with an experiment to measure meteoric erosion, using a Geiger counter. It would have seen trapped radiation, but the experiment got bumped. End of story.

I think this is the first time I have written all this down, or even thought about it in a coherent way. Your editorial stimulated all this; I know how Vernov must have felt.

I suppose I owe most of my radiation belt insights to Hannes Alfven, from whom I learned a great deal about charged particle motion. Even earlier, John Wheeler at Princeton taught me some useful things about ergodic motion of particles in a trapping region. Someday I'll document the evolution of the ideas and theory a little better. For the time being, the enclosed will have to do.

My best wishes to you,

Cordially

S. Fred Singer, Visiting Eminent Scholar

SFS/clk

1.2   A note written by S.F. Singer after a discussion with M. Walt during the Fall AGU meeting in San Francisco. It concerns certain statements in Singer's letter to A.J. Dessler, dated 15 September 1986 (personal communication,)

San Francisco, Dec 18, 2000

Postscript: Martin Walt kindly pointed out to me that the Aerobee rocket experiment may not have seen trapped protons since they would have been removed with the "geomagnetic anomaly" to the east of Peru.

Also, the Geiger counters in the auroral zone balloon flights were likely seeing X-rays produced by electrons; but one cannot be sure that these auroral electrons were indeed trapped.

S. Fred Singer

Chapter 4

The Interplanetary Causes of Magnetic Storms, Substorms and Geomagnetic Quiet

Bruce T. Tsurutani
Jet Propulsion Laboratory, California Institute of Technology
Pasadena, CA 91109, USA

Abstract   The current knowledge of the interplanetary and solar causes of superstorms ($D_{ST} \leq$ -350 nT), major magnetic storms ($D_{ST} \leq$ -100 nT), recurring substorms and HILDCAAs will be summarised. The causes of geomagnetic quiet during both solar maximum and solar minimum will also be reviewed. The discussion will start with geomagnetic activity during the solar maximum portion of the solar cycle and then that of the declining phase. A newly identified type of aurora, caused by interplanetary shocks, will be discussed. Such auroras may occur at other planets as well.

Keywords   Space storms, substorms, CMEs, magnetic clouds, magnetic reconnection, adiabatic motion, interplanetary shocks, aurora, interplanetary magnetic field, plasma instabilities, plasma waves.

1.   INTRODUCTION

I will start this lecture with a few illustrative questions to help orient the beginner to the field of space physics. The questions are meant to be provocative ones, in that they address some common misconceptions held by the uninitiated. The answers will be given without explanations for the time being. Explanations to the answers will all be contained within the text of this chapter.

1. Do Coronal Mass Ejections (CMEs) from the sun (which hit the Earth's magnetosphere) cause geomagnetic storms at Earth?   <u>Answer:</u> Sometimes.

103

I.A. Daglis (ed.), Space Storms and Space Weather Hazards, 103–130.
© 2001 *Kluwer Academic Publishers. Printed in the Netherlands.*

2. If the Interplanetary Coronal Mass Ejections (ICMEs) have southwardly directed magnetic fields, *then* will this be sufficient to cause magnetic storms? <u>Answer</u>: No, not necessarily.

3. (Corollary to 2): Since the ICME fields will be equally northwardly and southwardly directed, then shouldn't at least half of all ICMEs (which impinge upon the Earth) cause magnetic storms? <u>Answer</u>: No.

4. Since the answers to the previous three questions were "No" or "Sometimes", then why are CMEs/ICMEs important at all? Answer: For *fast* ICMEs, the solar ejecta material and their upstream sheaths (behind the shocks) contain *intense* magnetic fields giving them a *statistically higher probability* of the right conditions to generate magnetic storms.

5. What are the necessary interplanetary conditions for the generation of large magnetic storms? <u>Answer</u>: Intense hours-long duration, southwardly directed magnetic fields ($B_Z < -10$ nT, $\tau > 3$ hrs).

6. What are the solar and interplanetary causes of the very biggest magnetic storms (superstorms)? <u>Answer</u>: At this time we are not sure. But I will make some speculations based on what we presently know.

References: 1. Tsurutani et al., 1988a and references therein. 4. Tsurutani et al. (1988b). 5. See Gonzalez and Tsurutani (1987); Gonzalez et al. (1994), Kamide et al. (1998a,b). 6. Tsurutani et al. (1992, 1999).

2.   MAGNETIC STORM HISTORY

Mankind has certainly detected the effects of magnetic storms before written history took place. Red auroras are easily observable at midlatitudes during intense storms. Using lodestones, strong terrestrial magnetic deflections would have been obvious when the observer was under the auroral electrojet. One of the first published papers on magnetic storms was written by Baron Alexander von Humboldt in *Annalen der Physik* (1808). In the paper, von Humboldt described the results of an experiment performed from his home in Berlin, Germany, on 21 December 1806. Every half hour he and a colleague used a microscope to observe magnetic declinations of small magnetic needles. This was done from midnight to early morning. Von Humboldt noted that there were auroras overhead. He also noticed that when the northern lights disappeared at dawn, the magnetic needle deflections died out. He called this geomagnetic activity interval a "Magnetisches Ungewitter", or a "magnetic storm". This is the origin of the name of the phenomenon. (Von Humboldt was the first to study the geography, geology, and climatology of South America, leading to modern geophysics. He also founded the "Magnetische Verein" whose members included Gauss and Weber.)

These magnetic variations that von Humboldt saw and reported on are associated with the auroral electrojet which typically flows at a ~100 km altitude and at ~65° magnetic latitude in the local midnight sector. The currents have nominal intensities of ~10^6 Amperes during substorms. During intense geomagnetic activity (magnetic storms), the electrojet moves to lower latitudes and can be even more intense. The auroral electrojet during extreme events may cause fields at the Earth's surface to be as large as several thousand nanotesla (nT), or Earth magnetic field deviations of up to ~10%.

What is standardly used as a signature of a magnetic storm is an index formed from the output of four or more ground-based magnetometers located at or near the magnetic equator. This average deflection (taking out diurnal variations due to Sun-lit ionospheric current systems) is called the D_{ST} index, first proposed and constructed by Sydney Chapman. This index was adapted by a working group of the 1975 Grenoble IUGG.

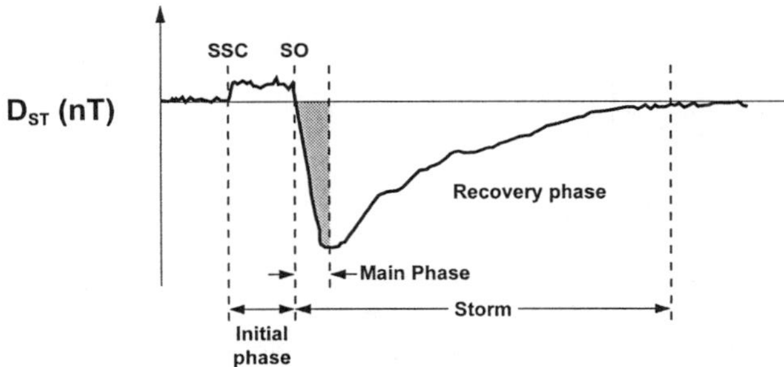

Figure 1. A profile of an idealised magnetic storm during solar maximum, as measured by ground based near-equatorial magnetometers (D_{ST} index).

The magnetic field magnitude profile of a magnetic storm occurring near solar maximum is shown in Figure 1. The storm has three phases: an "initial phase" where the magnetic field increases anywhere from +10 to +50 nT, a "main phase" where the field magnitude decreases by 100 (or more) nT, and a recovery phase where the field gradually recovers to the ambient value. The initial phase typically starts suddenly (<5min duration) and lasts an indeterminate amount of time. It may or may not be followed by a storm main phase (these first two phases, the initial and main phases, will be

106

shown to be caused by different physical phenomena). The main phase can be as short as an hour or as long as a day. The recovery phase typically lasts 7 to 10 hours (Chapman and Bartels, 1940).

The sudden sharp jump in the Earth's field at the onset of the initial phase (Storm Sudden Commencement or SSC) is caused by the abrupt increase in the solar wind ram pressure at interplanetary shock (Araki et al., 1988). The plasma density (and magnetic field) across the shock increases by a value which is approximately the shock Mach number (Kennel et al., 1985). For typical interplanetary shocks, the Mach number ranges from 1 to about 3 (Tsurutani and Lin, 1985). Although the shock thickness is only ~ seconds in width, after the shock hits the magnetosphere, the compressional wave travels at the magnetosonic wave speed from multiple points of the outer magnetosphere. Thus, the SSC temporal width measured at the surface of the Earth is much broader, typically ~mins wide (Araki et al., 1977).

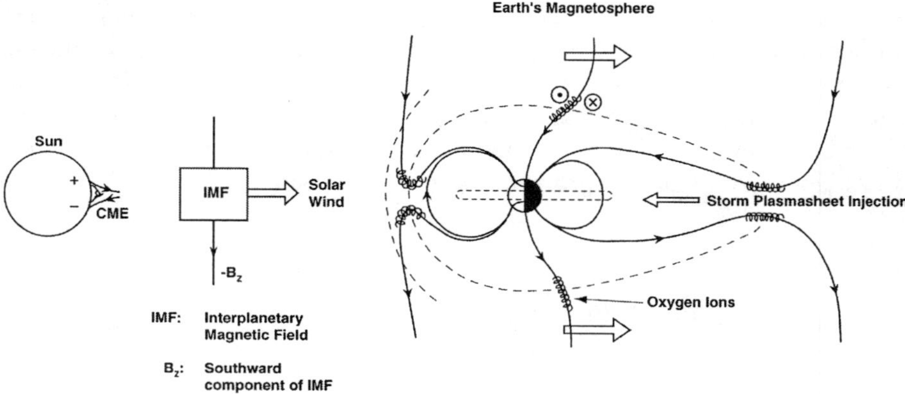

Figure 2. A schematic of magnetic reconnection between interplanetary magnetic fields and the Earth's field. Note that, in this scenario, plasma injection into the magnetosphere occurs near midnight, explaining the local time dependence of auroras.

The storm main phase is caused by magnetic interconnection between interplanetary magnetic fields and the Earth's field (Dungey, 1961; Gonzalez and Mozer, 1974). This process is most efficient when the interplanetary fields are directly opposite to that of the Earth's field at the magnetopause. or a southward direction. This is shown in Figure 2, an adaptation of a schematic from Dungey (1961). The interconnected field lines are dragged back by the solar wind plasma and reconnect in the nightside magnetotail. When the fields are reconnected once more, the release of magnetic tension causes the entrained plasma to be sling-shot from the tail towards the Earth to the near midnight sector of the magnetosphere. The energetic plasma on the closed magnetic field lines exhibits three adiabatic motions (Alfvén and

Fälthammer, 1963; Northrop, 1961): 1) a particle gyromotion about the magnetic field, 2) a bounce motion up and down the field within the "magnetic bottle", and 3) azimuthal motions around the dipole field. These three particle motions are illustrated in Figure 3. For singly charged particles, electrons drift from midnight towards dawn due to the presence of magnetic curvature and field gradients, and the ions drift from midnight toward dusk due to the same causes. Because the oppositely charged particles drift in opposite directions, these drifts form a current. This ring of current decreases the Earth's magnetic field (a diamagnetic current), and is called the "ring-current".

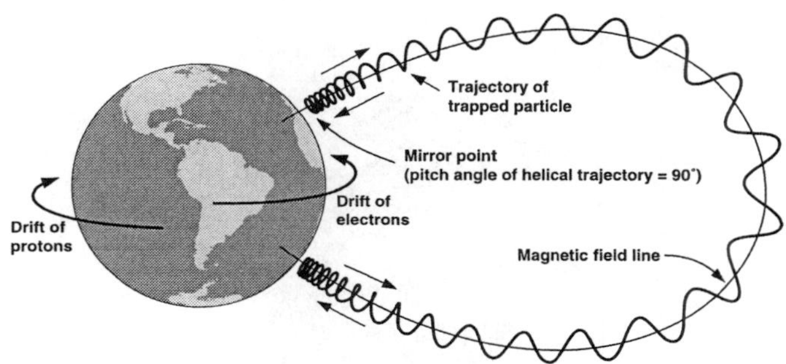

Figure 3. Charged particle motions in the magnetosphere. The three adiabatic motions are illustrated.

The near-equatorial field decrease that occurs during the storm main phase is caused by the formation and injection of this ring-current deep into the magnetosphere. Dessler and Parker (1959) and Sckopke (1966) have shown that the magnitude of the field decrease is linearly related to the total particle kinetic energy of the ring-current. Other current systems can certainly contribute significantly to D_{ST}, but it is currently being debated as to how much an effect this is (Campbell 1999; Kamide et al., 1999; Singer et al. 2000).

There are two regions associated with fast ICMEs where the magnetic fields might be sufficiently intense to cause a magnetic storm main phase: a magnetic cloud region within the ICME (Klein and Burlaga, 1982) and the interplanetary sheath (Tsurutani et al., 1988b; Tsurutani and Gonzalez, 1997), located upstream (antisunward) of the ICME and behind the shock.

An example of a magnetic cloud event causing a magnetic storm is shown in Figure 4. It is the well-studied January 10, 1997 event (Fox et al., 1998). D_{ST} is noted to decrease coincident with the intense, smoothly varying southward magnetic field of the cloud, supporting the hypothesis that the solar wind energy transfer mechanism is indeed magnetic reconnection.

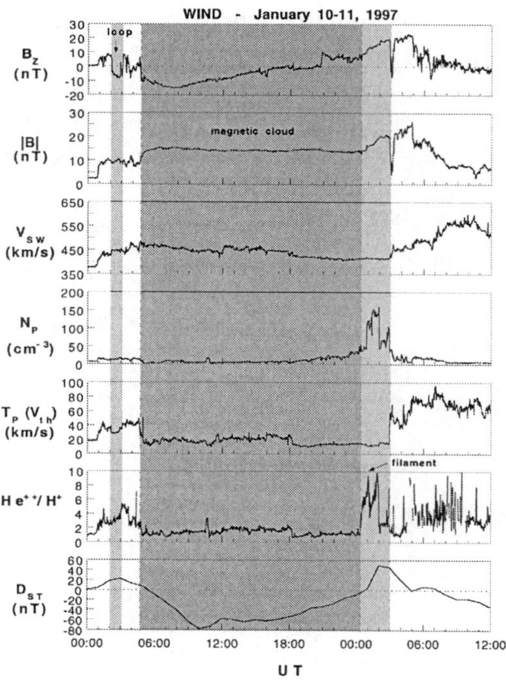

Figure 4. The January 10, 1997 example of a magnetic cloud event causing the main phase of a magnetic storm. Magnetic clouds are parts of ICMEs/driver gases.

The storm recovery phase is associated with the loss of the ring-current particles from the magnetosphere. Physical processes for the loss are: convection of plasma out the dayside magnetopause, charge-exchange with atmospheric neutral particles, Coulomb collisions, and wave-particle resonant interactions. See Kozyra et al. (1997) for a general discussion of wave-particle loss processes. Further, it has been noted that the physics of particle losses is extremely complex. The "decay time" depends on the particle energy, species, pitch angle and location, thus there is an infinite number of τ values, not simply a single "7 to 10 hour" value as stated earlier. Daglis (this book) has pointed out that during the peak phase of the storm, oxygen ions dominate the ring-current energy densities. These particles are lost most rapidly with time scales of ~1-2 hours.

An example of an interplanetary sheath B_S event leading to a major magnetic storm is shown in Figure 5. The shock is indicated by the dashed vertical line. There is intense interplanetary B_S just behind the interplanetary shock. The former causes the storm main phase. The IMF B_S increase behind the shock is most probably due to shock compression of the upstream slow stream IMF B_S.

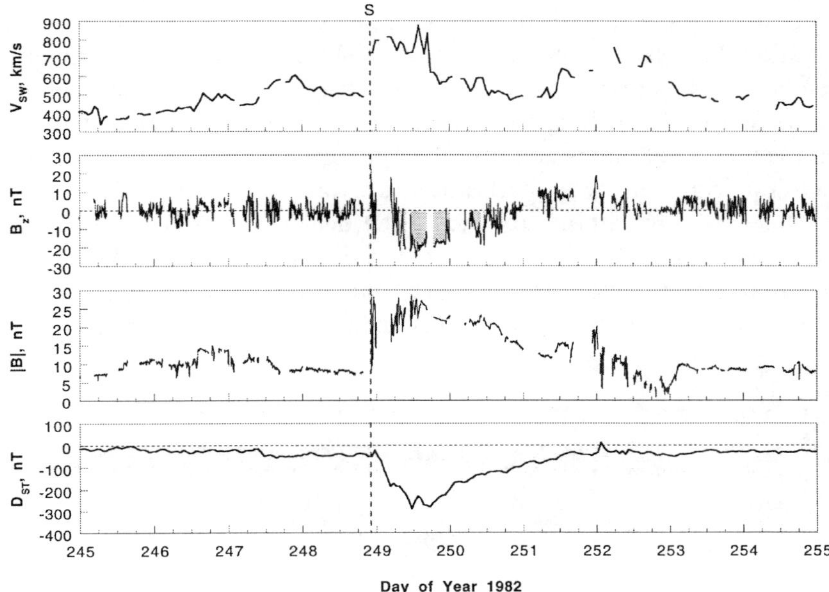

Figure 5. The September 1982 interplanetary shock event. Solar wind velocity, and magnetic field z-component and magnitude measured by ISEE-3 are shown, along with D_{ST} data. The southward magnetic fields creating the storms are sheath fields.

3.   AURORAS

What causes auroras during magnetic storms? When the plasma is sling-shot into the magnetosphere, both the electrons and ions are "compressed" such that their perpendicular temperatures become higher than their parallel temperatures. Such anisotropies lead to instabilities like the loss-cone instability (Kennel and Petschek, 1966). One consequence of such instabilities is the growth of electromagnetic plasma waves called chorus, shown in Figure 6 (Tsurutani and Smith, 1974). The waves through cyclotron resonant interaction pitch-angle scatter the particles (Tsurutani and Lakhina, 1997), leading to their loss to the upper atmosphere/ionosphere. The precipitating particles lose their kinetic energy through collisional excitation processes. Resultant excited ionospheric atoms and molecules decay to their ground state giving off characteristic auroral light.

Strong cross-magnetospheric convection electric fields with concomitant field-aligned potentials are also a consequence of magnetic reconnection and strong magnetic field distortions (Haerendel, 1994). Parallel electric fields above the ionosphere lead to the downward acceleration of electrons to energies of 1-10 keV and from their loss, to the formation of auroral arcs. A schematic taken from Elphic et al. (1998) showing upward and downward current systems, is given in Figure 7. An image of a long auroral arc taken from the Space Shuttle is given in Figure 8. The accelerated electrons come down magnetic field lines and lose their energy by collisional excitation. A red auroral fringe at the highest altitudes is due to a 6300 Å line from the metastable decay of atomic oxygen. The decay is present above 200 km altitude where the collisional de-excitation time is longer than the ~200s for the natural (metastable) decay. The blue-green oxygen light at lower altitudes is a mixture of oxygen 5577 Å and nitrogen 3914 Å lines.

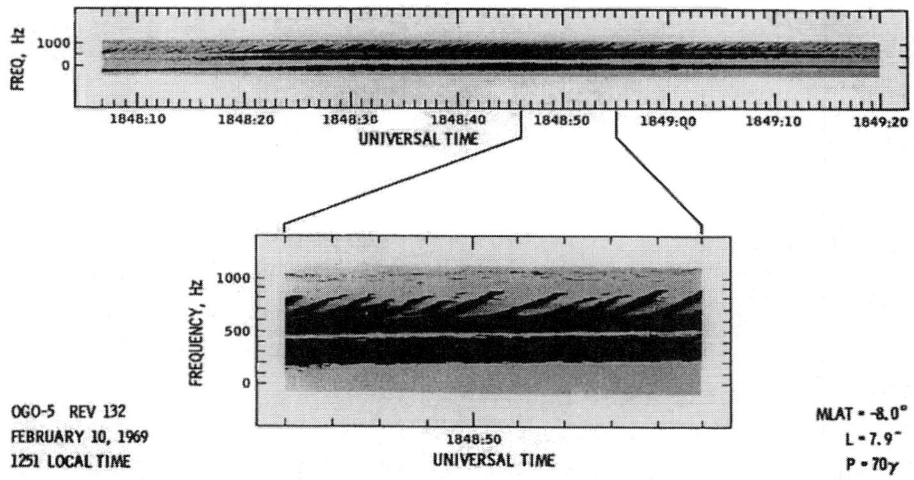

Figure 6. Magnetospheric electromagnetic whistler mode chorus emissions. These emissions are produced by the electron loss-cone instability.

During magnetic storms, pure red (6300 Å) auroras are also produced at lower than normal latitudes. The exact physical mechanism is unknown at this time (see other articles of this book). These red auroras occur during the storm recovery phase. Figure 9 is an example of an event that was seen at the Jet Propulsion Laboratory's Table Mountain Observatory (near Los Angeles, California) during an intense magnetic storm on April 12, 1981 (courtesy of J. Young).

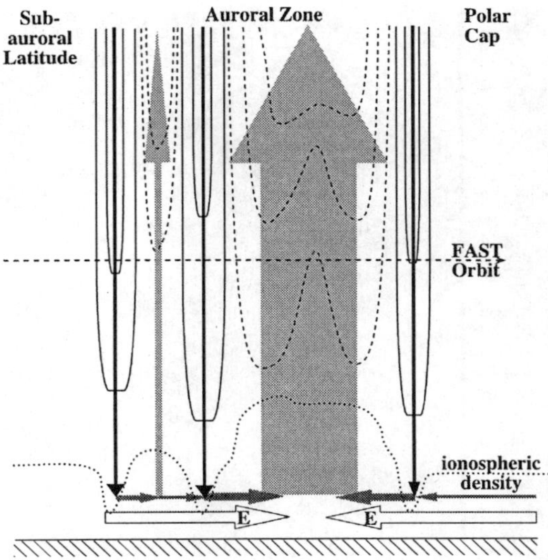

Figure 7. A schematic of field-aligned currents and the acceleration of electrons which form auroral arcs.

4. SOLAR MAXIMUM

4.1 Super-intense Storms

What causes super-intense storm events that can produce ground power outages, major satellite damage and satellite losses? There are a number of possibilities, but unfortunately we do not know for certain. We have data on too few events to really understand all of the causes at this time. However, we can make some reasonable speculations.

4.1.1 High Velocity CMEs

Single, violent CME events could lead to superintense storms. Gonzalez et al. (1998) have shown that there is a statistical relationship between the peak magnetic field magnitude within an ICME at 1 AU and its velocity (Fig. 10). This empirical relationship is most likely due to the CME release and acceleration mechanism occurring near the Sun. However, computer simulations need to be performed to verify this speculation.

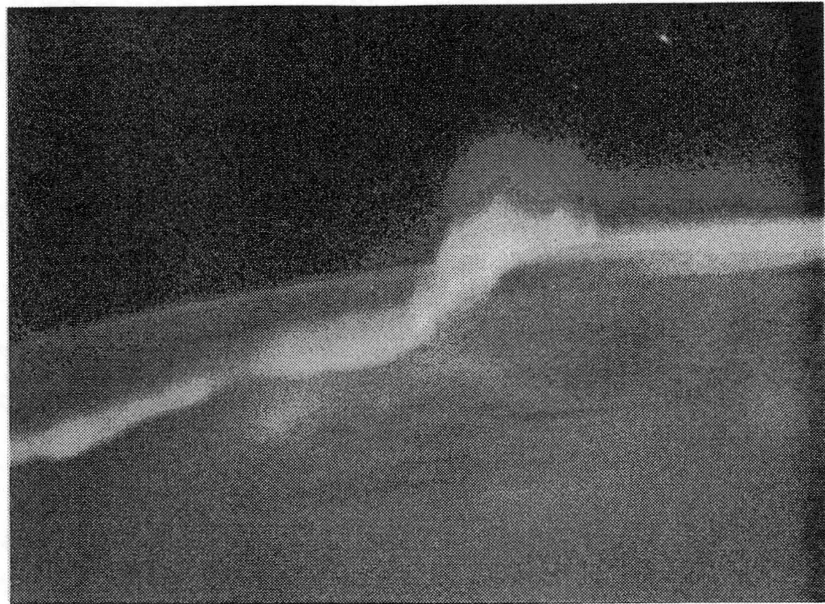

Figure 8. An auroral arc photographed from the Space Shuttle. (Color plate, see p. 480)

Figure 9. A photograph of a red aurora taken during the April 1981 magnetic storm. The photograph was taken near Los Angeles, California. (Color plate, see p. 480)

Not shown in Figure 10 are the particularly high fields and velocities of the August 1972 ICME event (see discussion in Tsurutani et al., 1992). This general V_{SW} - |B| relationship holds for this extreme event as well.

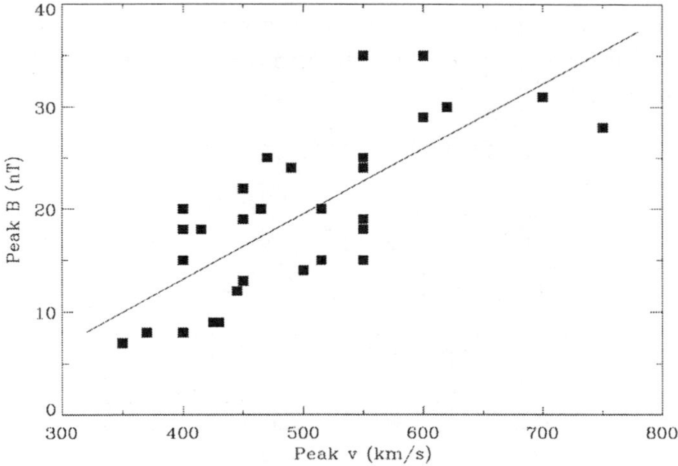

Figure 10. An empirical relationship between the interplanetary magnetic field strength and the speed of magnetic clouds.

4.1.2   Multiple ICMEs (Multiple Flaring at the Sun)

Shock compression of high intensity magnetic fields is a process which, under the right conditions, can lead to even higher magnetic field strengths. Figure 11, taken from Lepping et al. (1997) shows one such "double" event. The compression at point C is most likely shock compression within a magnetic cloud. However, we note that the plasma beta (β equals plasma thermal pressure divided by magnetic pressure) within clouds is generally lower than the present case (Tsurutani and Gonzalez 1997; Farrugia et al., 1997), so events similar to this one should be rare (note that the compression is present only where the β is somewhat high). For low beta plasmas, the magnetosonic wave speeds can be comparable or even higher than the solar wind speeds, so shock waves in magnetic clouds will become evanescent. There should be little or no magnetic compression for these cases.

A more probable mechanism will be shock compression of sheath plasmas. The August 1972 event was an event of this type. This is shown as Figure 12, adapted from Smith et al. (1976). Note that at Pioneer 10 distances (2.2 AU), there are 2 forward shocks and one reverse shock. The two fast forward shocks are presumably due to two fast CME injection events occurring at the Sun. The first forward shock compresses the ambient

114

magnetic field from ~2 nT to ~8 nT and the second increases the field further from ~8 nT to ~16 nT. These are the highest magnetic field strengths of this compound interplanetary event, higher than the cloud field (the magnetic cloud is present from 12 UT day 220 to 16 UT day 221). The field within the magnetic cloud is primarily northwardly directed. The result from the interaction of the cloud with the Earth's magnetosphere was geomagnetic quiet (not shown).

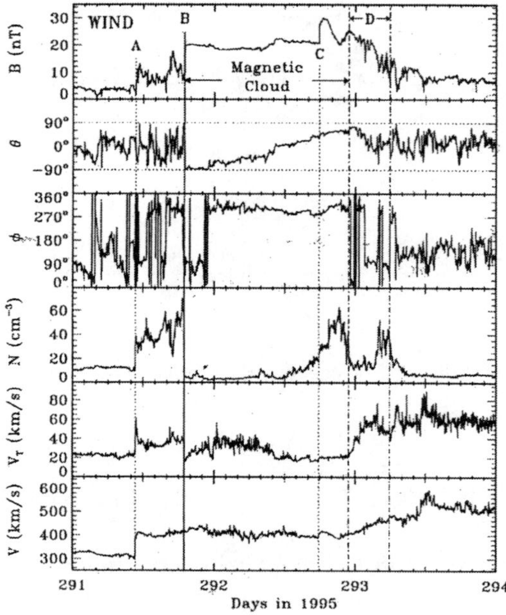

Figure 11. An unusual case of a compressive wave (c) within an interplanetary magnetic cloud.

4.1.3   Multiple Magnetic Storms

Storms that occur in quick succession can increase the total ring current energy to give the appearance of a particularly large storm. Figure 13 shows the results of a superposed epoch analyses by Yokoyama and Kamide (1997) and Kamide et al. (1998b), for an examination of single storms and double storms. The top panel gives the AL index for single and double storm events. On the bottom are the IMF B_S events corresponding to the top panel events. For the double storm events where the second storms are more intense (of the two), the IMF B_S is approximately equal for the two events, indicating that there is some form of nonlinearity within the system. One possible explanation is that the plasma sheet becomes "primed" by hot oxygen ions (Kozyra et al., 2000) during the first storm, leading to a much more intense second event even though the interplanetary driver is essentially the same as for the first event. The interplanetary drivers of the two storms of double storms are: a) southward B_Z associated with the sheath and b) the magnetic cloud of the fast ICME (for the second storm). Thus if the sheath and the magnetic cloud fields associated with a fast ICME are both directed southward, the composite, "double storm" will be more intense than one might expect from the IMF B_S values.

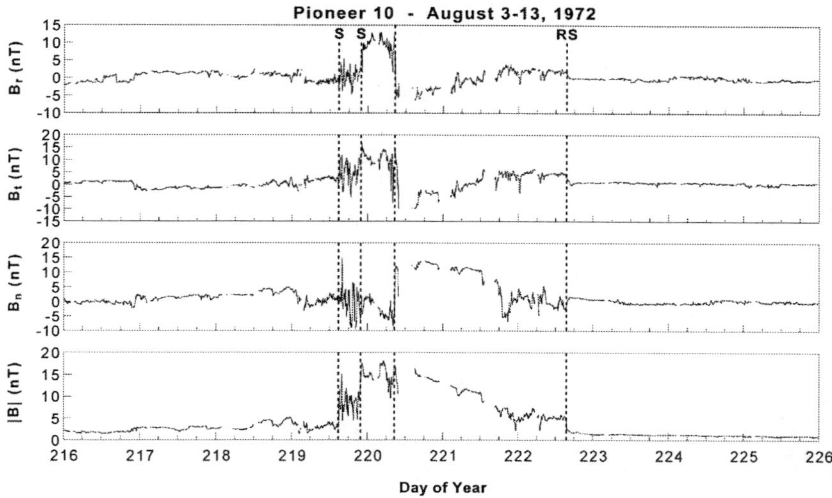

Figure 12. A shock within an interplanetary sheath event (double forward shocks). (taken from Tsurutani et al., 1992).

4.2   Geomagnetic Quiet

Intense, northward interplanetary magnetic fields like the August 1972 event lead to extreme geomagnetic quiet. For magnetic cloud cases where there are equal north and south IMF B_Z portions, the southward B_Z parts cause storm main phases and the northward B_Z parts cause geomagnetic quiet.

Table 1 gives the "efficiency" of solar wind coupling for 11 events where $B_N > +10$ nT and $T > 3$ hrs (during a solar maximum time period). Most of these events were portions of magnetic clouds. It was found that the average coupling efficiency for these 11 events was ~3 x 10^{-3}, i.e., ~0.3% of the incident solar wind ram energy gets into the magnetosphere. This efficiency is approximately 30 times less than during magnetic reconnection (IMF B_S) events.

116

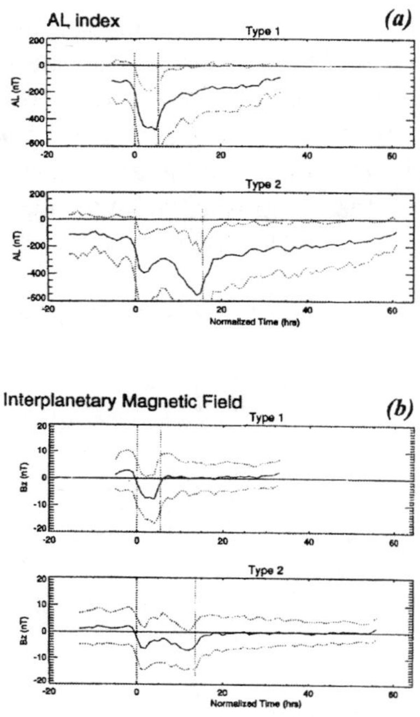

Figure 13. Superposed epoch analyses of single storms and "double" storm events.   The interplanetary B_z components are given in Panel (b).

TABLE 1. U_T and energy transfer efficiency

Date	UT	dE_{mag}/dt (erg s^{-1})	η efficiency
18 Dec 1978	0100	5.3×10^{17}	4.0×10^{-3}
21 Feb 1979	1200	2.2×10^{17}	1.0×10^{-3}
3 Apr 1979	1230	2.2×10^{17}	2.2×10^{-3}
5 Apr 1979	0300	6.0×10^{17}	3.0×10^{-3}
5 Apr 1979	1230	9.0×10^{17}	2.0×10^{-3}
29-30 May 1979	2130	4.5×10^{17}	7.1×10^{-3}
20 Aug 1979	0830	4.5×10^{17}	3.4×10^{-3}
18-19 Sep 1979	2400	1.1×10^{17}	1.3×10^{-3}
6 Oct 1979	1800	3.3×10^{17}	1.7×10^{-3}
7 Oct 1979	0800	1.1×10^{18}	1.1×10^{-3}
11 Nov 1979	1800	3.0×10^{17}	2.1×10^{-3}

5.   DECLINING PHASE OF THE SOLAR CYCLE

5.1   Corotating Streams - CIRs

During the declining phase of the solar cycle, corotating high-speed streams emanating from coronal holes dominate geomagnetic activity.   During this phase of the solar cycle, polar coronal holes expand in spatial extent and have portions that migrate toward and sometimes cross the ecliptic plane. These latter cases lead to solar wind streams which engulf the Earth's magnetosphere once per ~27 days.   The streams thus cause ~27 day recurrence of small geomagnetic storms and recurrences of High Intensity Long Duration Continuous AE Activity (HILDCAA) events (Tsurutani and Gonzalez, 1987).

When the high-speed solar wind catches up with the slower speed solar wind, the interaction leads to a compression in plasma and magnetic fields. These compression regions are called Corotating Interaction Regions or CIRs.

Figure 14 shows a high-speed solar wind/slow speed solar wind interaction on January 24-27, 1974. The high-speed solar wind proper is to the right of the vertical dashed line, the undisturbed slow solar wind is on the far left of the figure. The interaction region is in the middle. The intense magnetic field region (shaded in the next to the bottom panel) is the CIR. The resultant small magnetic storm is shown in the bottom panel (D_{ST}).

At distances of 1 AU from the Sun, CIRs typically do not have fast forward shocks (Tsurutani et al., 1995).   Therefore there is no SSC associated with the storm initial phase for these events. The storm main phases are small and irregular in profile (in comparison to solar maximum/magnetic cloud related events shown earlier). The causes of the irregularly shaped D_{ST} indices are noted in the B_Z data. The B_Z component is highly fluctuating. The lack of a long, continuous southward B_Z leads to the small intensity of the magnetic storm, even though the B_S magnitudes are sometimes quite high.

5.2   HILDCAAs

The recovery phase of the storm in Figure 14 is quite long. The peak D_{ST} value of ~ -65 nT occurs at ~21:30 UT on day 25. The D_{ST} value is still depressed by the end of day 26.

118

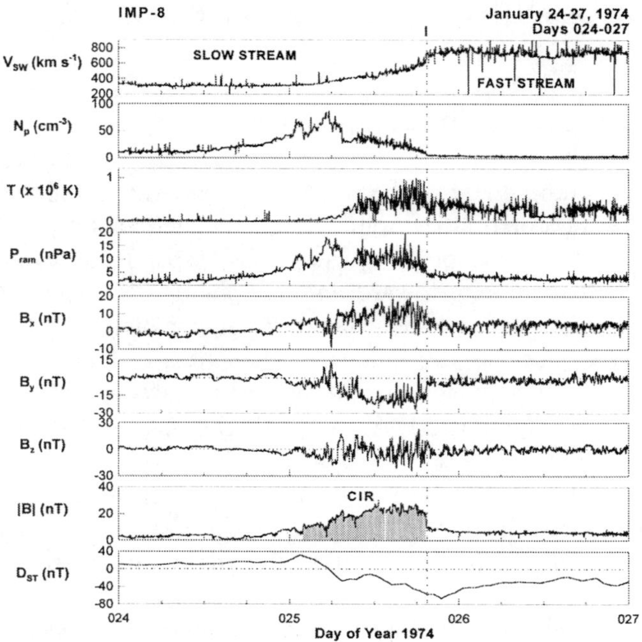

Figure 14. An example of a CIR. The high-speed stream (on the right) is colliding with a low-velocity high density heliospheric current sheet (HCS) plasma sheet, forming a compressed magnetic field region.

The D_{ST} indices for all of 1974 are shown in Figure 15. There are only 3 large storm events with $D_{ST} < -100$ nT. Each of these have been shown to be caused by ICMEs and/or their upstream sheaths (Tsurutani et al., 1995). The many smaller (recurrent) storms during 1974 are associated with CIRs interactions with the magnetosphere. However, what is particularly noteworthy in the figure are the intense AE events in each of the long storm "recovery phases". The recovery phases can last weeks or longer. The D_{ST} recoveries are associated with the high AE values. The average AE value for 1974 was 283 nT, whereas it was only 225 nT for 1979 (solar maximum)! *Thus averaging over a year, corotating streams can be more geoeffective in transferring solar wind energy into the magnetosphere than ICMEs during solar maximum.* This is because the substorms associated with the high-speed streams are occurring continuously (during the solar cycle declining phase), whereas magnetic storms are sporadic during solar maximum.

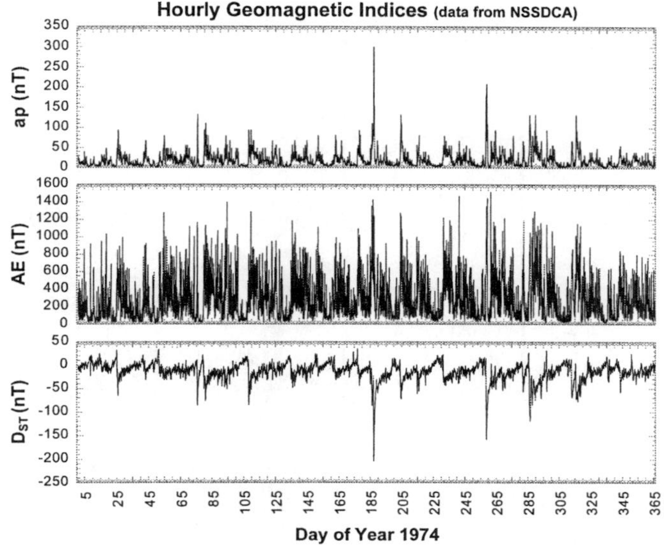

Figure 15. Ap, AE and D_{ST} for 1974. HILDCAAs are the high-intensity AE intervals following the small 27 day recurring magnetic storms. HILDCAAs are near-continuous substorms created by the Alfvén wave B_S fluctuations.

What is the interplanetary cause of these long duration storm recovery phases and high AE values? The answer is given in Figure 16. The interplanetary B_Z is highly fluctuating in this high velocity stream event. With every southward field turning, there is an increase in AE and decrease in D_{ST}. The southward field turnings cause magnetic reconnection and plasma injections into the nightside magnetosphere. There are slight D_{ST} decreases at each of these injections. These periods of continuous substorm activity are called HILDCAAs, and the sporadic injection of plasma into the magnetosphere is the reason why the ring current does not appear to "decay". What is actually happening is the HILDCAAs are related to sporadic, low-intensity particle injections into the outer portions of the ring-current, thus the lack of an overall "decay" (Tsurutani et al., 1995).

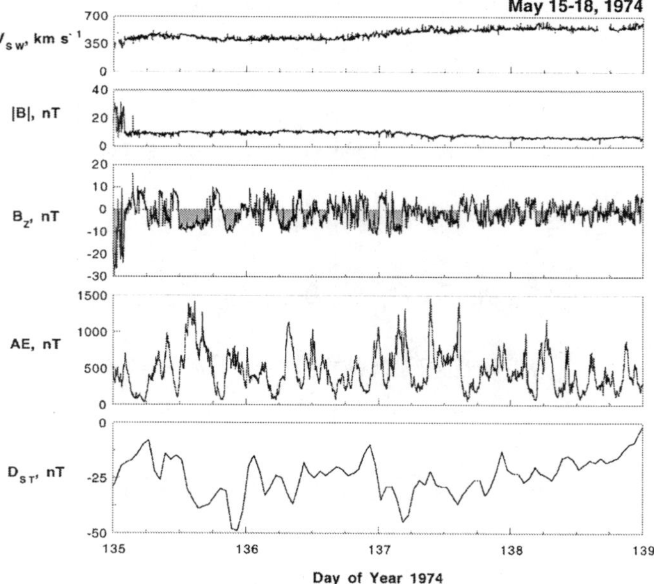

Figure 16. Interplanetary Alfvén waves, AE increases and D_{ST} decreases. The causes of long recovery phases in recurrent storms is due to sporadic magnetic reconnection (the IMF B_z component of the Alfvén waves), and consequential (substorm) injections of plasma into the magnetosphere.

The continuous presence of magnetospheric chorus plasma waves (and other wave modes) associated with HILDCAAs has been invoked (Horne and Thorne, 1998; Summers et al., 1998; 2000) to explain the magnetospheric relativistic electron events (Baker et al., 1989; Li et al., 1997) present in these intervals of high-speed solar wind streams. These relativistic electrons have been related to the possible failure of a Canadian telecommunication satellite.

What causes the interplanetary B_Z fluctuations? The NASA/ESA Ulysses mission has provided us with answers. Figure 17 shows the magnetic field and plasma components taken over the solar north pole within a high-speed solar wind stream (coming from a polar coronal hole). Continuous fluctuations are noted in all of the magnetic field and velocity components. When these fluctuations are analysed (by performing cross-correlations between the B and V components), it is found that the components are highly correlated at zero lag. Belcher and Davis (1971) have demonstrated that this indicates that these are Alfvén waves. The waves are determined to be propagating away from the Sun (determined by the sign of the correlation coefficient). Thus the Alfvén waves present in the high-speed streams lead to the B_Z fluctuations within the CIRs (the waves are compressed) leading to the irregularly shaped storm main phase, and also are the fluctuations that cause HILDCAAs in the storm "recovery phases".

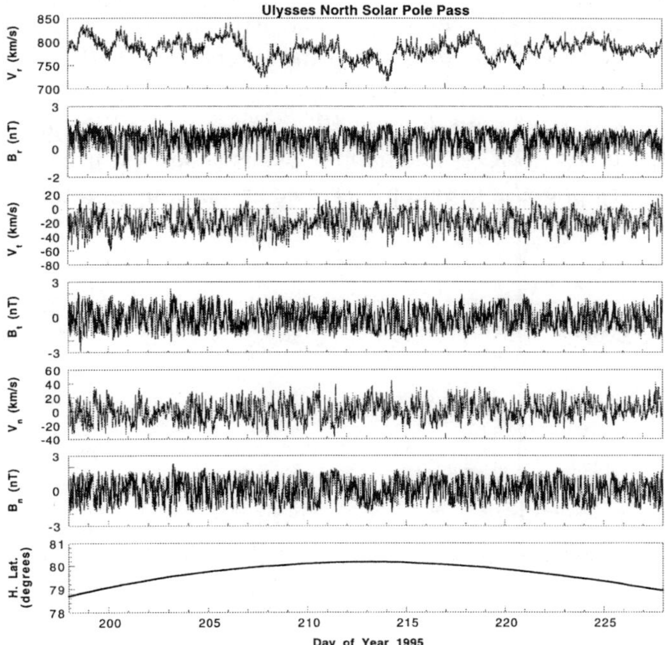

Figure 17. Alfvén waves measured in a high-speed stream (of coronal hole origin).

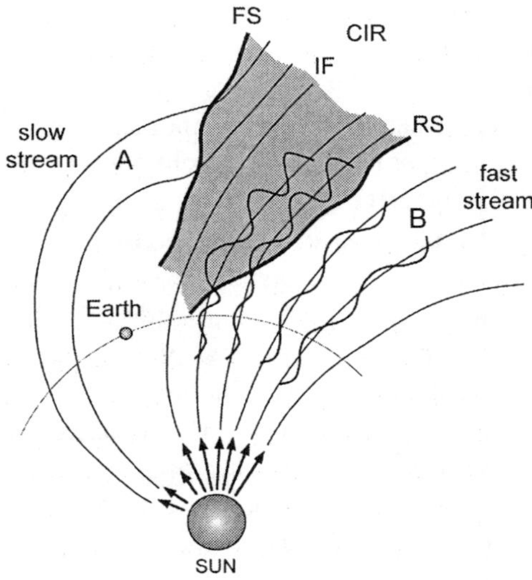

Figure 18. A schematic of a high-speed/slow-speed stream interaction and CIR formation.

For visualisation purposes, a two-dimensional schematic of the high-speed/slow-speed solar wind stream interactions is shown in Figure 18. Note the interface (IF) between the slow-speed stream and high-speed stream is a tangential discontinuity. The Alfvén waves of the high-speed stream are amplified by compression at the reverse shock (RS).

In summary, a profile of magnetic storms during the declining phase of the solar cycle is given in Figure 19. The initial phase does not start suddenly (there is no SSC). The main phase is small and irregularly shaped, and the recovery phase is irregularly shaped and of long duration.

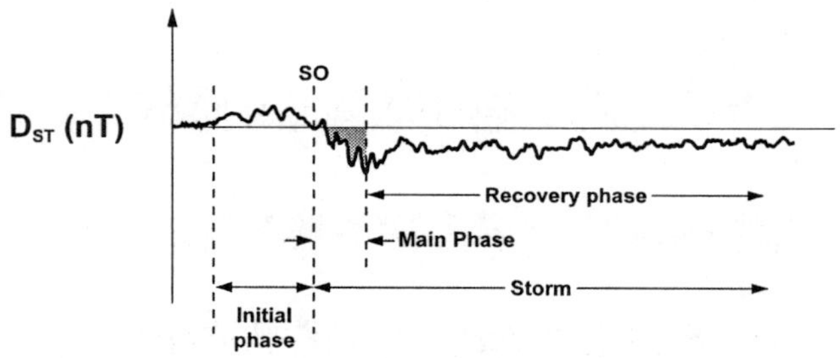

Figure 19. An idealised profile of a magnetic storm in the declining phase of the solar cycle.

5.3   Geomagnetic Quiet

What causes geomagnetic quiet during the declining phase of the solar cycle? Figure 15 can be used to identify such regions. AE and Ap have particularly low values at the trailing ends of the streams and are located at times occurring after the HILDCAA events. D_{ST} is generally positive, indicative of high plasma density regions (increased ram pressure). Figure 20 shows the solar wind plasma, magnetic field and D_{ST} indices for the entire year 1974. Using the positive D_{ST} events as markers, we find that the geomagnetically quiet intervals occur at the ends of the high-speed streams and at the beginnings of the heliospheric current sheet plasma sheet regions (see sector boundary markers at top). The ends of high speed streams are characterised by low plasma velocities, low plasma densities, low magnetic field magnitudes and the absence of B_Z fluctuations. The second region, the heliospheric current sheet plasma sheet, is characterised by high plasma densities. Both of these regions contribute to geomagnetic quiet during the declining phase of the solar cycle. The cause of the geomagnetically quiet intervals is the lack of magnetic reconnection between the interplanetary medium and the magnetosphere.

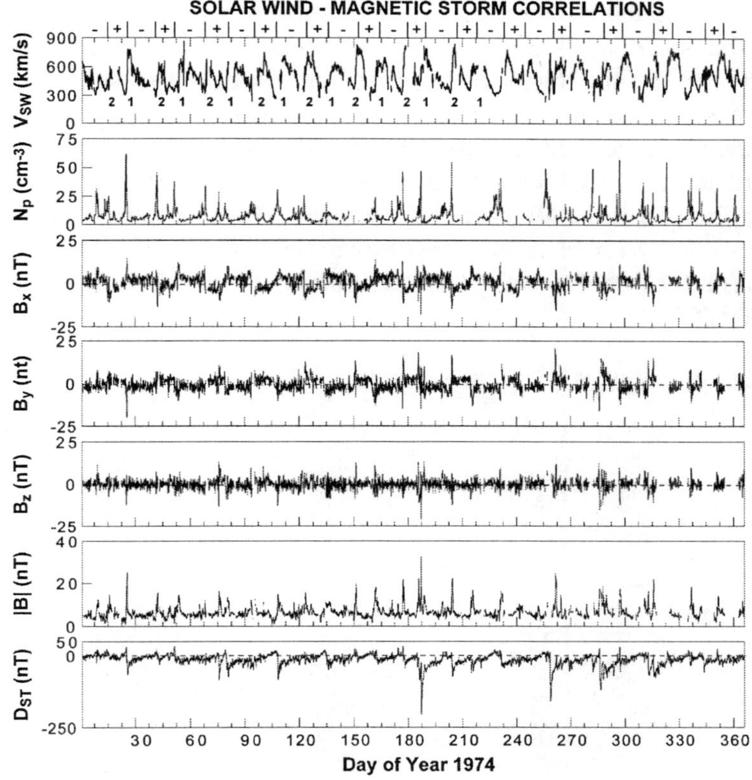

Figure 20. The solar wind plasma, magnetic field, and D_{ST} indices for year 1974.

6.   SHOCK-AURORAS AT EARTH, JUPITER AND SATURN

It has recently been shown that the shocks found ahead of fast ICMEs cause energy transfer directly into the (dayside) magnetosphere, rather than by transport first to the magnetotail and then to the nightside magnetosphere in the case of magnetic reconnection discussed earlier. When the ram pressure pulses (associated with interplanetary shocks) compress the Earth's magnetosphere, dayside auroras result almost instantaneously. Figure 21 is an example of a shock-aurora event taken by Polar UVI imaging instrument. The images are the LBH long wavelength images displayed in magnetic local time (MLT) co-ordinates. The north pole is at the centre, and 60° latitude local noon is at the top in each panel. Dawn is to the right and dusk to the left. The time sequence goes from the top left to the right. Each image is separated by ~3 min 4 s. The January 10, 1997 event is shown. Using the solar wind speed measured by WIND and the shock speed calculated from the Rankine-Hugoniot conservation equations, the shock arrival time at the magnetopause was calculated. The arrival time was determined to occur between the second and third images of the figure. At the third image, 01:03:48 UT, there is a brightening of the aurora on the dayside from 10 to 12 MLT at ~75° latitude. This brightening is within the auroral oval. With time, the brightening spreads towards both dawn and dusk until the whole oval (less the midnight sector) is intensified (by 0113:00 UT).

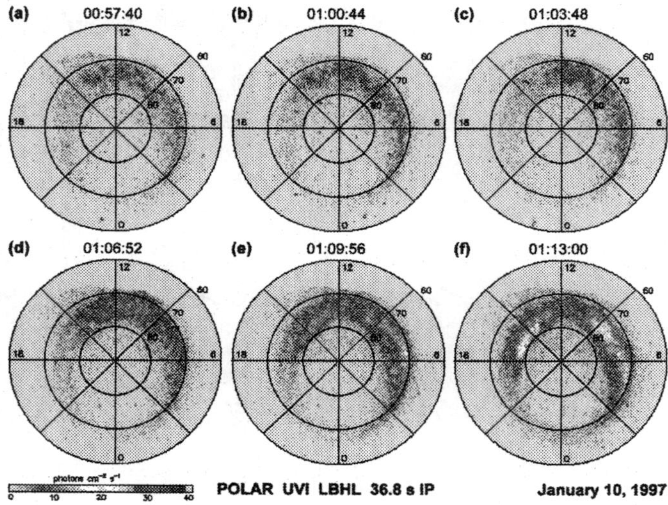

Figure 21. A dayside aurora caused by an interplanetary shock (a shock-aurora) (taken from Zhou and Tsurutani, 1999). (Color plate, see p. 482)

Nine interplanetary shock events were detected in 1997 when Polar had clear viewing of the dayside auroral zone. For these cases, "dayside" auroras occurred each time. The velocities of the aurora in the ionosphere are shown in Table 2, Column 2. Three events are listed. The velocities range from 6 to 11 km s^{-1}, much higher than the standard <1 km s^{-1} detected for substorm or storm nightside auroras. If the ionospheric velocities are extrapolated to the equatorial plane of the same magnetic field lines, the velocities (at the magnetopause) will be 280 to 370 km s^{-1}. These speeds are quite similar to that of the measured solar wind speeds for these events (see Column 4). Thus the speed of the aurora as it propagates from noon to dawn and dusk is associated with the antisunward propagation of the shock pressure pulse along the magnetopause boundary.

magnetopause) will be 280 to 370 km s^{-1}. These speeds are quite similar to that of the measured solar wind speeds for these events (see Column 4). Thus the speed of the aurora as it propagates from noon to dawn and dusk is associated with the antisunward propagation of the shock pressure pulse along the magnetopause boundary.

TABLE 2. Ionospheric auroral "speed", mapped into the magnetosphere, and solar wind velocity

Event	Ionospheric V (km/s)	Mapped V* (km/s)	Observed $V_{sh/sw}$ (km/s)	Spacecraft Position (R$_e$)
10 Jan 1997	6 (dusk)	280	300	I-T (Sheath) (-19, 19, 10)
1 Oct 1997	10 (dusk)	370	460	IMP-8 (SW) (10, 32, -3)
10 Dec 1997	11 (dawn)	365	360	GT (SW) (-4, -25, -0.5)

* Assuming a dipole field of L=10

Shock created auroras are fainter than those of substorm auroras, but because of the much greater latitudinal extent of the former, the energy deposition rate is ~5 times greater than that of a moderate substorm (Tsurutani et al., 2001a).

The specific mechanisms for solar wind energy transfer into the magnetosphere are uncertain at this time. Two possibilities have been suggested in the literature and are schematically illustrated in Figure 22a and b. In Figure 22a, the interplanetary shock compresses outer zone magnetospheric magnetic fields and pre-existing plasma. The heating of the plasma in the direction perpendicular to the field leads to temperature anisotropies and the loss cone instability. The loss of these energetic charged particles to the ionosphere would result in a diffuse aurora. A second mechanism (Figure 22b) is that shock compression of the outer zone dayside magnetosphere creates field-aligned potentials that accelerate electrons into the ionosphere. The precipitating electrons will create auroral arcs.

To determine whether one, both, or none of these mechanisms are correct, it would be extremely useful to have ground based observations to determine what types of auroral forms are created by interplanetary shock compression. Unfortunately, ground-based observations have not been reported for these types of events to date.

Perhaps our first test of these models might occur with auroral observations at Jupiter. Jovian UV auroras have been detected from observations made using the Hubble Space Telescope (HST) (Prangé et al., 1993; Clarke et al., 1998). An example of a Jovian polar aurora is shown in Figure 23. Note that there are two auroral rings. The brightest one occurs at L ≈ 20-30 (Prangé et al., 1997) and a fainter one poleward of this. This latter feature could correspond to the magnetopause boundary layer, similar to the situation at Earth.

Following Haerendel (1994), the potential drops along the Jovian magnetic fields have been calculated (Tsurutani et al., 2001b). Input values used for the calculation were a magnetopause field strength of 5 nT, plasma density of 0.1 cm^{-3}, a measured magnetopause/boundary layer width of ~7000 km (Sonnerup et al., 1981), and a shock "perturbation" field of 5 nT. Using the above numbers, a parallel potential drop of ~50 kV was determined. It happens that ~50 keV electrons are needed to explain the Jovian aurora spectroscopic measurements (H. Waite, private communication, 2000), so this mechanism may indeed explain the higher latitude auroral ring.

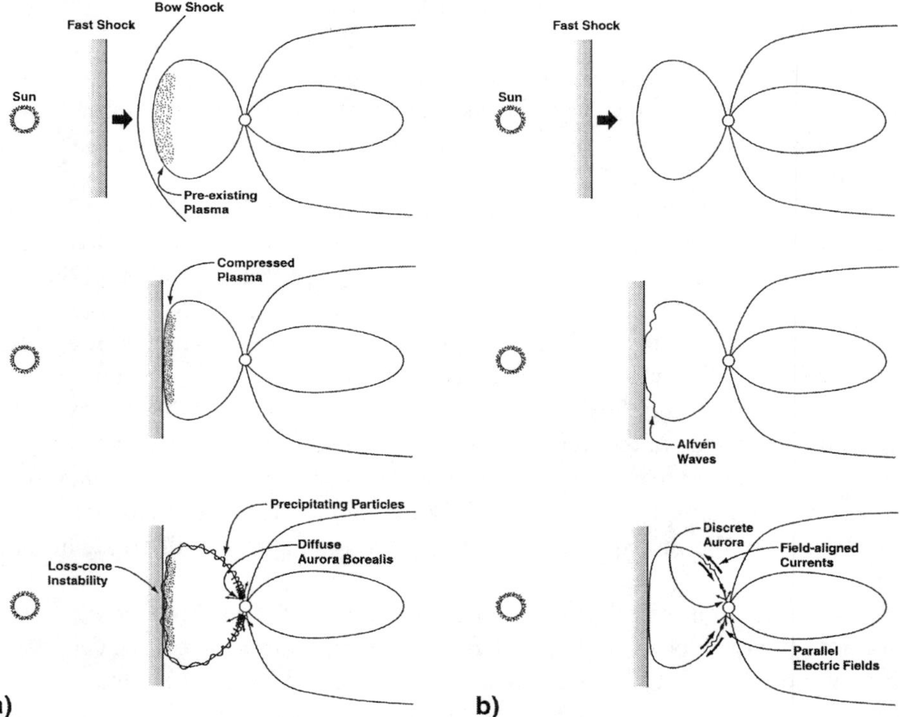

Figure 22. Two possible mechanisms for shock-aurora formation.

In the near future, Cassini will fly past Jupiter, measuring the interplanetary medium and imaging the aurora. Galileo will be inside the magnetosphere determining the state of the radiation belts and imaging instruments will also be observing the aurora. HST will be viewing Jupiter's UV aurora as the Cassini flyby takes place. Jovian polar auroras sometimes episodically reach intensities of more than a megarayleigh (10^{12} photons cm^{-2} s^{-1}), 10 to 100 times more intense than that for the Earth's auroras (J. Clarke, personal communication, 2000). It will be interesting to see if these particularly intense auroras are caused by interplanetary shocks and if so, if they are discrete or diffuse auroras.

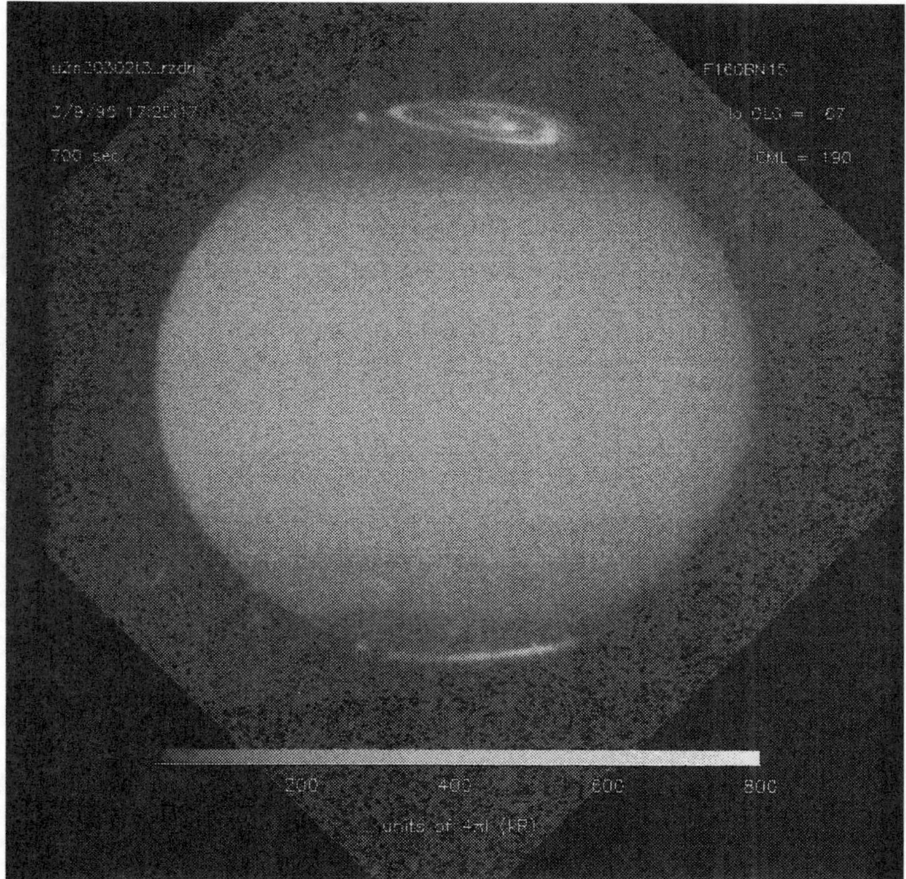

Figure 23. Jovian UV aurora. There are two auroral rings, in this instance.
(Color plate, see p. 481)

7.   ACKNOWLEDGEMENTS

Portions of the research reported here were performed at the Jet Propulsion Laboratory, California Institute of Technology, Pasadena, California under contract with the National Aeronautics and Space Administration, Washington DC. The author wishes to thank W.D. Gonzalez and K. Papadopoulos for critical readings of the manuscript.

8. REFERENCES

Alfvén, H., and C.-G. Fälthammer, *Cosmic Electrodynamics*, Oxford, 1967.

Araki, T., Global structure of geomagnetic sudden commencements, *Planet. Space Sci.*, **25**, 373, 1977.

Araki, T., A physical model of geomagnetic sudden commencement, in *Solar Wind Sources of Magnetospheric Ultra Low Frequency Waves*, edited by M. Engebretson, K. Takahashi, and M. Scholer, 81, 183, Amer. Geophys. Un. Press, Washington D.C., 1994.

Baker, D.N., et al., Relativistic electrons near geostationary orbit: Evidence for internal magnetospheric acceleration, *Geophys. Res. Lett.*, **16**, 559, 1989.

Belcher, J.W., and L. Davis Jr., Large amplitude Alfvén waves in the interplanetary medium, 2, *J. Geophys. Res.*, **76**, 3534, 1971.

Campbell, W.H., Comment on "Current understanding of magnetic storms: Storm-substorm relationships" by Y. Kamide et al., *J. Geophys. Res.*, **104**, 7049, 1999.

Chapman, S., and J. Bartels, *Geomagnetism, 1*, Clarendon, Oxford, 1940.

Clarke, J.T., G. Ballester, J. Trauger, J. Ajello, W. Prior, K. Tobiska, J.E.P. Connerney, G.R. Gladstone, J.H. Waite, L.B. Jaffel, and J.-C. Gerard, Hubble Space Telescope imaging of Jupiter's UV aurora during the Galileo orbiter mission, *J. Geophys. Res.*, **103**, 20217, 1998.

Dessler, A.J., and E.N. Parker, Hydromagnetic theory of magnetic storms, *J. Geophys. Res.*, **64**, 2239, 1959.

Dungey, J.W., Interplanetary magnetic field and the auroral zones, *Phys. Res. Lett.*, **6**, 47, 1961.

Farrugia, C.J., L.F. Burlaga, and R.P. Lepping, Magnetic clouds and the quiet storm effect at Earth, in *Magnetic Storms*, edited by B.T. Tsurutani, W.D. Gonzalez, Y. Kamide and J.K. Arballo, AGU Press, Wash. D.C., **98**, 91, 1997.

Fox, N.J., M. Paredo, and B.J. Thompson, Cradle to grave tracking of the January 6-11, 1997 Sun-Earth Connection Event, *Geophys. Res. Lett.*, **25**, 2461, 1998.

Gonzalez, W.D., and F.S. Mozer, A quantitative model for the potential resulting from reconnection with an arbitrary interplanetary magnetic field, *J. Geophys. Res.*, **79**, 4186, 1974.

Gonzalez, W.D., and B.T. Tsurutani, Criteria of interplanetary parameters causing intense magnetic storms ($D_{ST} < -100$ nT), *Planet Space Sci.*, **35**, 1101, 1987.

Gonzalez, W.D., J.A. Joselyn, Y. Kamide, H.W. Kroehl, G. Rostoker, B.T. Tsurutani, and V.M. Vasyliunas, What is a geomagnetic storm?, *J. Geophys. Res.*, **99**, 5771, 1994.

Haerendel, G., Acceleration from field-aligned potential drops, *Astrophys. J. Suppl.*, **90**, 765, 1994.

Horne, R.B., and R.M. Thorne, Potential waves for relativistic electron scattering and stochastic acceleration during magnetic storms, *Geophys. Res. Lett.*, **25**, 3011, 1998.

Kamide, Y., W. Baumjohann, I.A. Daglis, W.D. Gonzalez, M. Grande, J.A. Joselyn, R.L. McPherron, J.L. Phillips, E.G.D. Reeves, A.S. Sharma, H.J. Singer, B.T. Tsurutani, and V.M. Vasyliunas, Current understanding of magnetic storms: Storm-substorm relationships, *J. Geophys. Res.*, **103**, 17705, 1998a.

Kamide, Y., N. Yokoyama, W. Gonzalez, B.T. Tsurutani, I.A. Daglis, A. Brekke, and S. Masuda, Two-step development of geomagnetic storms, *J. Geophys. Res.*, **103**, 6917, 1998b.

Kamide, Y., W. Baumjohann, I.A. Daglis, W.D. Gonzalez, M. Grande, J.A. Joselyn, R.L. McPherron, J.L. Phillips, E.G.D. Reeves, G. Rostoker, A.S. Sharma, H.J. Singer, B.T. Tsurutani, and V.M. Vasyliunas, Reply, *J. Geophys. Res.*, **104**, 7051, 1999.

Kennel, C.F., J.P. Edmiston, and T. Hada, A quarter century of collisionless shock research, in *Collisionless Shocks in the Heliosphere: A Tutorial Review*, AGU Press, Wash. D.C., **34**, 1, 1985.

Kennel, C.F., and H.E. Petschek, Limit on stably trapped particle fluxes, *J. Geophys. Res.*, **71**, 1, 1996.

Klein, L.W., and L.F. Burlaga, Interplanetary magnetic clouds at 1 AU, *J. Geophys. Res.*, **87**, 613, 1982.

Kozyra, J.U., V.K. Jordanova, R.B. Horne, and R.M. Thorne, Modeling of the contribution of electromagnetic ion cyclotron (EMIC) waves to stormtime ring-current erosion, in *Magnetic Storms*, edited by B.T. Tsurutani, W.D. Gonzalez, Y. Kamide, and J.K. Arballo, AGU Press, Wash. D.C., **98**, 187, 1997.

Lepping, R.P., L.F. Burlaga, A. Szabo, K.W. Ogilvie, W.H. Mish, D. Vassiliadis, A.J. Lazarus, J.T. Steinberg, C.J. Farrugia, L. Janoo, and F. Mariani, The wind magnetic cloud and events of Oct. 18-20, 1998: Interplanetary properties and triggering for geomagnetic activity, *J. Geophys. Res.*, **102**, 14049, 1997.

Li, X., et al., Multisatellite observations of the outer zone electron variation during the November 3-4, 1993 magnetic storm, *J. Geophys. Res.*, **102**, 14123, 1997.

Northrup, T.G., The guiding center approximate to charged particle motion, *Ann. Phys.*, **15**, 79, 1961.

Prangé, R., M. Dougherty, and V. Dols, Identification d'une aurorae, "current-driven" sur Jupiter, á l'aide d'observations corréleés avec HST et Ulysses, in *Compte-redus du Séminaire scientifique du GdR Plasmae*, ed. D. Hubert, DESPA - Observatoire du Paris, 38, 1993.

Prangé, R., S. Maurice, W.M. Harris, D. Rego, and T. Livengood, Comparison of IUE and HST diagnostics of the Jovian Aurorae, *J. Geophys. Res.*, **102**, 9289, 1997.

Sckopke, N., A general relation between the energy of trapped particles and the disturbance field near the Earth, *J. Geophys. Res.*, **71**, 3125, 1966.

Smith, E.J., The August 1972 solar terrestrial events: Interplanetary magnetic field observations, *Space Sci. Rev.*, **19**, 661, 1976.

Summers, D., R.M. Thorne, and F. Xiao, Relativistic theory of wave-particle resonant diffusion with application to electron acceleration in the magnetosphere, *J. Geophys Res.*, **103**, 20487, 1998.

Summers, D., and C.-Y. Ma, A model for generating relativistic electrons in the Earth's inner magnetosphere based on gyroresonant wave-particle interactions, *J. Geophys. Res.*, **105**, 2625, 2000.

Tsurutani, B.T., and R.P. Lin, Acceleration of >47 keV ions and >2 keV electrons by interplanetary shocks at 1 AU, *J. Geophys. Res.*, **90**, 1, 1985.

Tsurutani, B.T., and W.D. Gonzalez, The cause of high intensity long-duration continuous AE activity (HILDCAAs): Interplanetary Alfvén wave trains, *Planet. Space Sci.*, **35**, 405, 1987.

Tsurutani, B.T., B.E. Goldstein, W.D. Gonzalez, and F. Tang, Comment on "A new method of forecasting geomagnetic activity and proton showers", by A. Hewish and P.J. Duffet-Smith, *Planet. Space Sci.*, **36**, 205, 1988a.

Tsurutani, B.T., W.D. Gonzalez, F. Tang, S.-I. Akasofu, and E.J. Smith, Origin of interplanetary southward magnetic storms near solar maximum (1978-1979), *J. Geophys. Res.*, **93**, 8519, 1988b.

Tsurutani, B.T., W.D. Gonzalez, F. Tang, and Y.T. Lee, Great magnetic storms, *Geophys. Res. Lett.*, **19**, 73, 1992.

Tsurutani, B.T., W.D. Gonzalez, F. Tang, Y.T. Lee, M. Okada, and D. Park, Reply to L.J. Lanzerotti, *Geophys. Res. Lett.*, **19**, 1993, 1992.

Tsurutani, B.T., W.D. Gonzalez, A.L.C. Gonzalez, F. Tang, J.K. Arballo, and M. Okada, Interplanetary origin of geomagnetic activity in the declining phase of the solar cycle, *J. Geophys. Res.*, **100**, 21717, 1995.

Tsurutani, B.T., and W. D. Gonzalez, The efficiency of "viscous interaction" between the solar wind and the magnetosphere during intense northward IMF events, *Geophys. Res. Lett.*, **22**, 663, 1995.

Tsurutani, B.T., and W.D. Gonzalez, The interplanetary causes of magnetic storms: A review, in *Magnetic Storms*, edited by B.T. Tsurutani, W.D. Gonzalez, Y. Kamide and J.K. Arballo, AGU Press, Wash. D.C., **98**, 77, 1997.

Tsurutani, B.T., and G.S. Lakhina, Some basic concepts of wave-particle interactions in collisionless plasmas, *Rev. Geophys.*, **35**, 491, 1997.

Tsurutani, B.T., Y. Kamide, J.K. Arballo, W.D. Gonzalez, and R.P. Lepping, Interplanetary causes of great and superintense magnetic storms, *Phys. Chemistry Earth*, **24**, 101, 1999.

Tsurutani, B.T., X.-Y. Zhou, J.K. Arballo, W.D. Gonzalez, G.S. Lakhina, V. Vasyliunas, J.S. Pickett, T. Araki, H. Yang, G. Rostoker, T.J. Hughes, R.P. Lepping, and D. Berdichevsky, Auroral zone dayside precipitation during magnetic storm initial phases, *J. Atmosph. Solar Terr. Phys.*, in press, 2001a.

Tsurutani, B.T., X.-Y. Zhou, V.M. Vasyliunas, G. Haerendel, and J.K. Arballo, Interplanetary shocks, magnetopause boundary layers and dayside auroras, to appear in *Surveys in Geophys.*, 2001b.

Von Humboldt, A., Magnetische Ungewitter, *Annales der Physik*, **29**, 25, 1808.

Yokoyama, N., and Y. Kamide, Statistical nature of geomagnetic storms, *J. Geophys. Res.*, **102**, 14215, 1997.

Zhou, X.-Y., and B.T. Tsurutani, Rapid intensification and propagation of the dayside aurora: Large scale interplanetary pressure pulses (fast shocks), *Geophys. Res. Lett.*, **26**, 1097, 1999.

Chapter 5

Interplanetary Magnetic Field Dynamics

The key to space weather monitoring

Norman F. Ness
Bartol Research Institute, University of Delaware
Newark, DE 19716, USA

Abstract   The role of the IMF in the dynamics of solar wind interaction with the geomagnetosphere was proposed by Dungey in 1961. Early in-situ IMF measurements in the 60's confirmed the association of the magnitude and direction of the IMF with the Kp magnetic activity index. The IMF is a primary force in controlling our local space weather because of the physical process known as reconnection.

Keywords   Interplanetary magnetic field (IMF), Interplanetary Monitoring Probe (IMP), archimedean, sector, source surface, heliospheric current sheet (HCS), Kp, interaction regions, coronal mass ejections, co-rotating interaction regions (CIRs), magnetic clouds, eruption, anisotropy, shocks, discontinuities, geomagnetic storms, sudden storm commencement (SSC).

1.   INTRODUCTION

Geomagnetic disturbances or storms and polar aurora and their semi-regular recurrence with a periodicity of 27 days had long ago suggested the connection of Earth's space environment to events on the sun, such as solar flares. The exact nature of that connection eluded specificity for many decades and only recently could be carefully studied. This was made possible as a result of the space age and spacecraft borne instrumentation.

Both in-situ measurements of the interplanetary medium as well as certain remote sensing observations of the Sun, impossible to make from ground-based observatories because of Earth's atmosphere, have contributed to the dramatic and revolutionary new views of this Sun-Earth connection. See Crooker and Cliver (1994) for a historical review of recurrent storms

I.A. Daglis (ed.), Space Storms and Space Weather Hazards, 131–155.
© 2001 *Kluwer Academic Publishers. Printed in the Netherlands.*

and the mysterious M-Regions, and the autobiographical report by Akasofu (1996).

It was long suspected that the Sun was a source of ionized and magnetized ejecta, a plasma, which was propelled into interplanetary space by an unspecified process. But whether or not the Sun was only an intermittent source of such material, which interacted with Earth, was unclear and left to pioneering speculation using comet tails as detectors/indicators until the Space Age began in 1957.

The important fundamental role played by the Interplanetary Magnetic Field (IMF) in the dynamics of the Earth's magnetosphere was first considered explicitly by Dungey (1961). He had been encouraged by his thesis advisor, Fred Hoyle, to study the implications of merging of the IMF with the geomagnetic field. This study was done prior to the actual first clear detection of the magnetized solar wind in the early 1960's by Mariner II (Neugebauer and Snyder, 1962) as it transited from Earth to Venus. For a recent theoretical treatment of the solar wind as the "driver" of space weather, see Tsinganos (2001).

Dungey's model depended upon the solar magnetic field being carried into interplanetary space by the electrically conducting, ionized solar coronal atmosphere, the solar wind. The merging process had been suggested earlier by the physicist, R. Giovanelli, to explain the observed energetic particles detected after the occurrence of solar flares. The development of physical and quantitative aspects of this solar flare mechanism for the transport of particles and their energization and dissipation were then studied theoretically by Parker (1957) and Sweet (1958).

Petschek (1964) and others addressed the problem of merging of the IMF with the geomagnetic field invoking magnetohydrodynamic aspects somewhat differently than the Parker-Sweet mechanism. Petschek (1996) has reflected upon these early days of theorizing about the solar wind interaction with the geomagnetic field. Gonzalez *et al.* (1994) have considered how a geomagnetic storm is defined and Tsurutani and Gonzalez (1993) and Bothmer and Schwenn (1995) have discussed the causes of such storms.

The seminal idea of Dungey's work was to define the concept of bi-modal magnetosphere structures, closed and open, as illustrated in Figure 1. Since then a number of detailed cartoon approximations of the merging process and magnetosphere structure have been put forward with a renaming of the merging mechanism to a re-connection process. Numerical simulations are also in vogue at this time. Early studies of the aerodynamic and MHD interactions of the solar wind with the geomagnetic field were presented by Levy *et al.* (1964), Axford *et al.* (1965) and Siscoe (1966). An interesting utilization of those studies was the Avco-Everett Research

Laboratories (AERL) 1963 Christmas card portrayal (see Figure 2) since the work had been supported by that private sector firm. The original task in the 1960's of AERL had been the development of survivable nose cones for missiles re-entering Earth's atmosphere as part of the US National Security program during the Cold War.

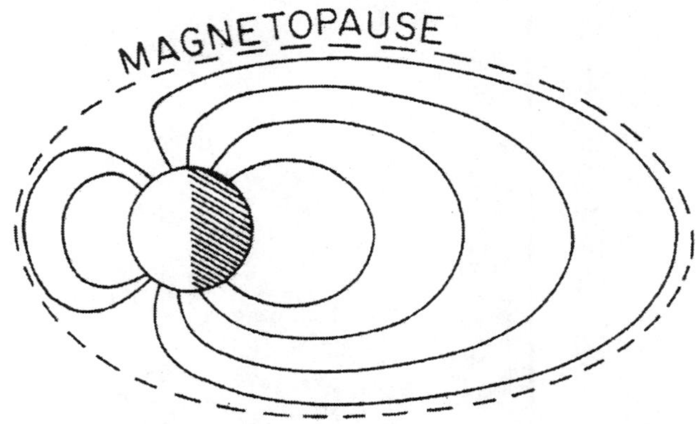

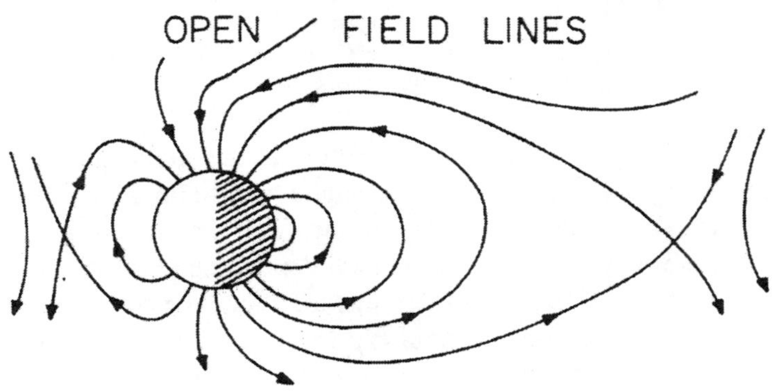

Figure 1. Pre-space era concepts of geomagnetosphere. Upper panel illustrates "closed" magnetosphere. Lower panel illustrates "open" magnetosphere with connection of IMF occurring at sub-solar region and in polar regions.

134

Figure 2. Avco Everett Research Laboratory Christmas Card for 1963. Based upon Levy, Petschek and Siscoe (1964) (courtesy of Harry Petschek).

The closed geometry of Figure 1, in which the geomagnetic field is isolated from interplanetary space by the magnetopause boundary, is actually an over-simplification. Re-connection occurs near the sub-solar region of solar wind interaction or in the polar regions depending upon whether or not the IMF is directed Southward or Northward with respect to the Earth's rotation axis. An early and comprehensive theoretical overview of the merging or reconnection process was given by Vasyliunas (1975). An overview text on dynamics of the IMF has been written by Burlaga (1995).

2.　　EARLY IMF MEASUREMENTS

One of the major challenges in the early exploration of the physical properties of space, 1957-1962, was the accurate measurement of the IMF because it is so weak at 1 AU. Extrapolating solar magnetic fields, according to the model developed by Parker (1958), led to estimates of a few nT (note that the magnetic field ranges from 30,000 to 72,000 nT on the surface of Earth.) Special techniques and instruments were developed to address this challenge. An early review of the first decade of magnetic field studies by spacecraft (S/C) was published by Ness (1970), surveying all S/C from 1959-1969 carrying magnetometers.

1960-62	Pioneer V, Mariner II	Earliest Efforts
1963-73+	The 10 IMPs (Interplanetary Monitoring Platforms)	Explorers 18, 21, 28, 33, 34, 35, 41, 47, 50*
	Pioneers 6, 7 and 8	Near 1 AU (0.9-1.1 AU)
1964+	Mariners 4, 5, 10, PVO, HEOS	Mars, Venus and Mercury
1972+	Pioneers 10, 11	Jupiter and Saturn 5 AU and beyond
1974+	Helios A, B	Inner Heliosphere 0.3-1.0 AU
1977+	Voyagers 1*, 2*	1 to 70 AU and beyond and higher heliographic latitudes
1980s	ISEE 1, 2 and 3	1 AU
1990s+	WIND*, ACE*, ULYSSES*	1-5 AU and high heliographic latitudes over both poles

Table 1. History of in-situ IMF Studies by Spacecraft (* still operating).

Table 1 presents a time sequence summary of all S/C, which have made major contributions to the study of the IMF. See Ness (2000) and Parker (2000) for comprehensive reviews of the experimental problems with and successes of S/C studies of the IMF in the heliosphere.

The first accurate data was obtained by the first Interplanetary Monitoring Probe (IMP) or Explorer 18, in 1963-64 (Ness *et al.,* 1964). This S/C was the first of 10 highly successful IMPs launched between 1963-1973 to investigate the particle and field environment of Earth and Moon (NASA TM-80758, 1980). IMP-1 was placed into a highly elliptical orbit with apogee half way to the Moon in order to avoid, if possible, any effects associated with the solar wind interaction with the geomagnetic field.

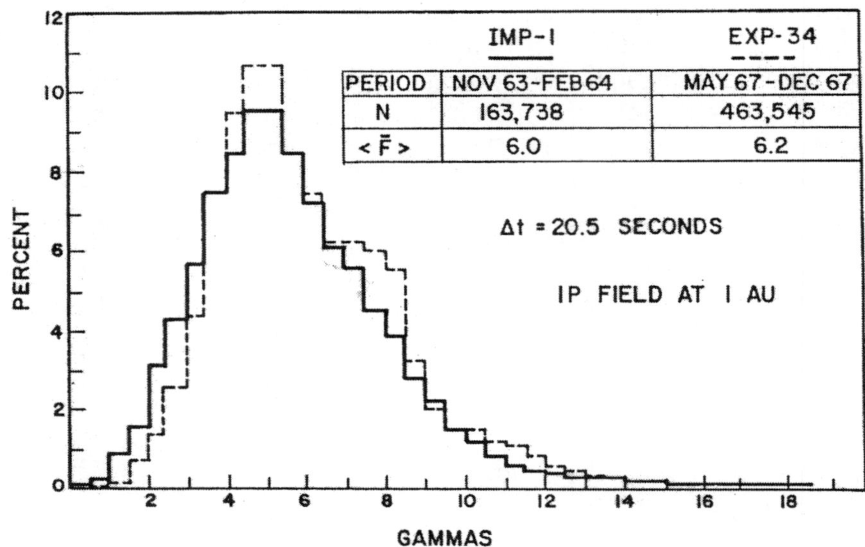

Figure 3. Histogram of early studies of IMF magnitude by IMP-1 and IMP-4 S/C in 1963 and 1967 (Ness, 1969).

The distribution function of the field intensity measured by IMP-1 and the follow-on IMP-4 (Explorer 4) are shown in Figure 3 (Ness, 1969). These 2 histograms provide evidence that the most probable field strength at those times, 1963-64 to 1967, was 6.1 ± 0.1 nT. This was in good agreement with the estimates using Parker's solar wind model.

More importantly, the directional properties of the field were also consistent with Parker's model of an Archimedean spiral geometry with the field near the ecliptic. Data taken by the Pioneer 6 S/C in 1965-1966 are shown in Figure 4 (Burlaga and Ness, 1968) with projections in the ecliptic and orthogonal to it. The right panel shows that for the elevation angle, the field tends to be close to the ecliptic. In the left panel, the azimuthal direction, the field tends to lie at either 135° or 315° with respect to the sunward direction. The convention was established that fields within the range 70° to 210° are considered positive polarity because when extrapolated back to the Sun, the field would be directed radially outwards. Negative polarity refers to the range 250° to 30°.

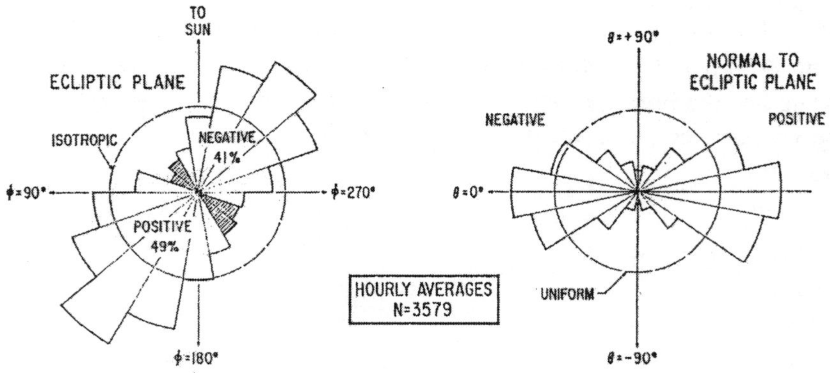

Figure 4. Directional distribution function for interplanetary magnetic field as observed by Pioneer 6, 16 December 1965-29 September 1966 (Burlaga and Ness, 1968).

3. IMF SECTOR STRUCTURE

Additional study of the IMP-1 data indicated there was a simple ordering or regularity of the directional properties of the IMF which became readily obvious when the directions, + or -, were plotted in a circular diagram as shown in Figure 5 (Wilcox and Ness, 1965; Ness and Wilcox, 1965). Here the coordinate system rotates with the Sun with the synodic period of 27 days. This diagram led to the concept of unipolar sectors of the IMF in which the polarity sense remained constant over extended periods of time. During this period of IMP-1 data, there were 4 sectors of almost equal duration or extent.

Subsequent studies of this sector structure have shown that it evolves with shifting boundaries and sometimes there are only 2 sectors rather than 4. Correlations of photospheric magnetic field directions with this IMP-1 sector pattern revealed a time lag of 4.5 days (Ness and Wilcox, 1964; 1966). This was consistent with the delay time for convection of the solar magnetic field by the solar wind at the average speed of 385 km/sec., consistent with that actually observed.

Continued studies of the sector structure in the ecliptic and close to 1 AU were conducted by a number of S/C (Ness and Wilcox, 1967). Of special interest to the large scale structure of these sectors was their extent in radial distance from the Sun. Figure 6 (Behannon, 1978) presents the results from

138

the heliocentric orbit of the 1973 Mariner 10 Venus Mercury mission. During this period, December 1973-April 1974, there were only 2 sectors identified. That was also reflected in observations at 1 AU. Thus, the concept emerged of the polarity sectors originating at the Sun being preserved as the field was convected outwards into the more distant heliosphere.

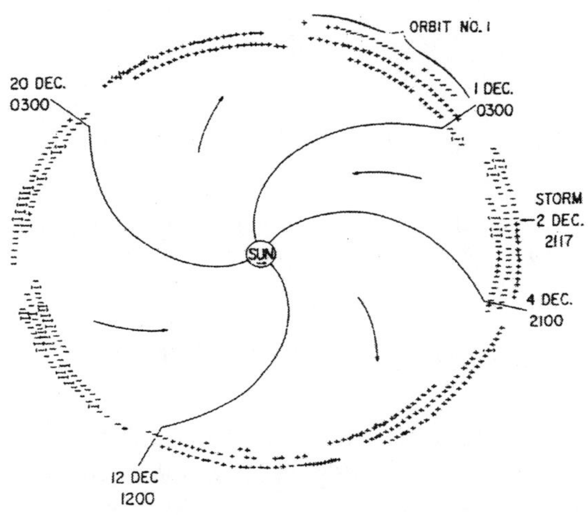

Figure 5. Sector structure of the interplanetary magnetic field derived from the IMP-1 observations (Ness and Wilcox, 1965; Wilcox and Ness, 1965). The + and - signs indicate the direction of IMF along spiral angle during successive 3-hr intervals.

4.   HELIOSPHERIC CURRENT SHEET (HCS)

Continued studies by Schulz (1973) of this problem of extending the photospheric field introduced the concept of a current sheet separating the oppositely directed dipolar solar magnetic field. This HCS was proposed as the explanation for the observed sectoring of unipolar source regions. Subsequently, smaller unipolar source regions were identified and referred to as coronal holes (Hundhausen 1972; 1977).

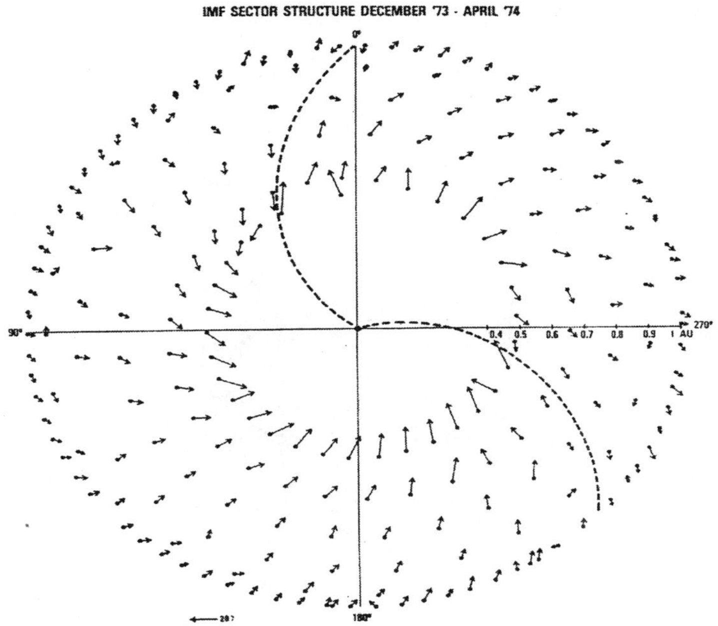

Figure 5. Sector structure of IMF in the inner heliosphere observed by Mariner 10 along its heliocentric trajectory in 1973-1974 (Behannon, 1978) while enroute to Mercury encounter in March 1974.

It was not until the Pioneer 11 S/C was re-directed into an encounter with Saturn following its 1975 encounter with Jupiter that this HCS model was placed on a firm observational basis. Figure 8, taken from Smith *et al.* (1978) presents a simplified version of the 3-D geometry of the extension of the solar magnetic field. (A poetic analogy of the shape of the HCS with a ballerina skirt was made by Hannes Alfvén.)

Since the initial theoretical and experimental work was done, the causal relationship of the structure of the HCS to the IMF has been studied closely. Successful projections of solar observations into interplanetary space predicting the sector structure have been routinely done (Burlaga *et al.*, 1981; Hoeksema, 1989; Suess *et al.*, 1993).

The heliolatitude extent of the HCS and its solar cycle variation has been studied by ULYSSES (Schulz, 1995; Smith *et al.*, 1993; Balogh *et al.*, 1995; Forsyth *et al.*, 1996) validating the overview of the structure of the heliosphere in 3-D geometry and the dynamical variations of the HCS. An important parameter now commonly used to describe the HCS is the tilt angle. This is essentially a measure, at the Sun, of the latitude extent and, thereby, the amount of curvature of the HCS in the heliosphere.

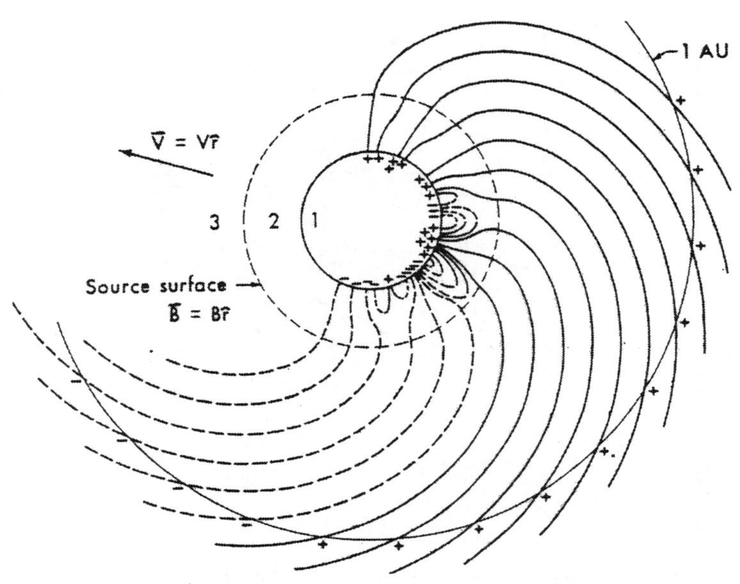

Figure 6. Schematic of the source surface model. Photospheric magnetic field observed in region 1 by Mt. Wilson. Closed field lines (loops) exist in region 2. Currents on source surface eliminate the transverse components of IMF there. (Schatten et al., 1969)

5. GEOMAGNETIC DISTURBANCES AND THE IMF

An early study of the relationship of geomagnetic activity and the IMF to the newly discovered sectors was conducted using data from the IMP-1 S/C. Figure 9 shows the results of this study by Wilcox and Ness (1965), which indicated that both Kp and IMF were higher near the start of a new IMF sector and slowly decreased as the sectors passed Earth. It should be noted that Snyder *et al.* (1963) had found a correlation of Kp with solar wind speed in the Mariner II study of data. But the sector structure of the IMF was not identified in the Mariner II data due, in part, to S/C field contamination in the measurements.

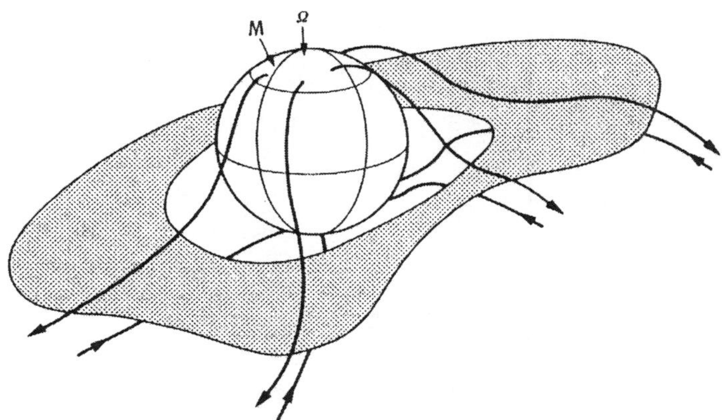

Figure 7. Sketch of the current sheet responsible for sector structure. The axis of the current sheet, M, is tilted relative to the solar rotation axis, Ω (Smith et al., 1978)

In a further attempt to understand these relationships more clearly and identify the primary magnetic field parameters responsible for geomagnetic activity, Schatten and Wilcox (1967) undertook a more comprehensive study of data from the IMP S/C. They used IMF magnitude and orientation with respect to ecliptic as independent parameters. Figure 10 presents the results of their study, which in the left panel indicates a positive correlation of Kp with IMF magnitude, on average.

A more interesting relationship was found when the latitude of the IMF was taken as the independent parameter. A negative correlation was obtained showing that as the latitude decreased from +90 (i.e., northward) to -90 (i.e., southward), Kp increased. This result was the first direct experimental validation of the concept of the important role of the IMF in the dynamics of the geomagnetosphere. Fairfield (1967) showed that substorms began with the southward turning of the IMF. Additional studies by Feynman (1976) confirmed this relationship. The increased levels of Kp for more southward IMF strongly suggested that the concepts of merging or re-connection of the IMF were significant in the physical processes causing variations of the geomagnetosphere.

142

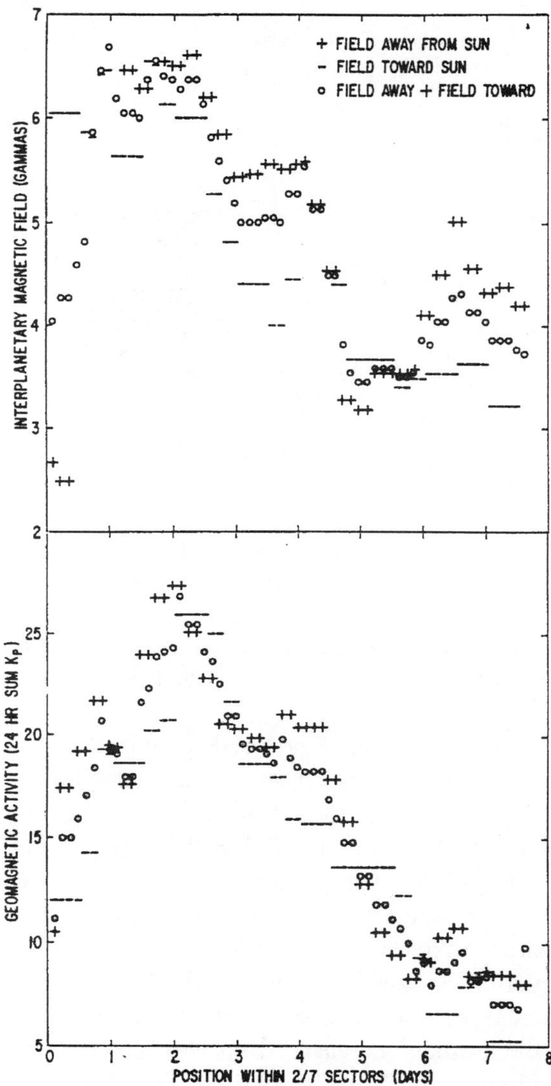

Figure 8. Superposed epoch graphs of IMF magnitude and Kp index within interplanetary sectors. Clearly evidenced is a coherent variation with increased fields and activity near the leading position of the sector (Wilcox and Ness, 1965).

The well-known semi-annual variation of geomagnetic activity was studied and shown to be a result of the relatively large (23.7°) obliquity of Earth's rotation axis (Russell and McPherron, 1973) and the concomitant seasonal variation in the relative orientation of the nearly axial geomagnetic dipole axis (inclination 11.4°).

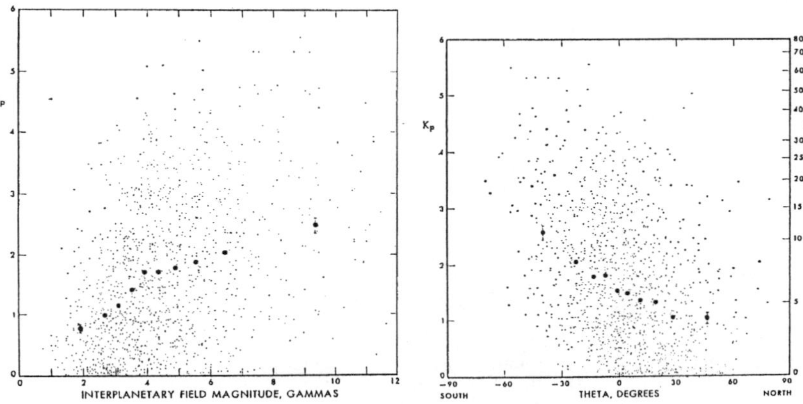

Figure 9. Scatterplot of Kp with IMF magnitude (left panel) and θ, the angle between the interplanetary magnetic field and the ecliptic (right panel) (Schatten and Wilcox, 1967). Data from IMP-1 in 1963-64.

6. INTERACTION REGIONS, CORONAL MASS EJECTIONS

As ongoing in-situ studies of the interplanetary medium from the 1960's continued into the 1970's, solar activity was increasing. The number of active regions on the Sun increased as did the number and frequency of high-speed solar wind streams, i.e., speeds > 400 km/sec. One of the conclusions of the IMP-1 studies identifying sectors was that higher speed solar wind streams from the Sun had lifetimes that extended well beyond a solar rotation period of 27 days. These were referred to as co-rotating streams. The interaction of long lived high speed streams led to the concept of Co-rotating Interaction Regions or CIR's (Burlaga and Barouch, 1976).

These CIR's and other co-rotating interplanetary disturbances, such as shocks, were studied by multiple spacecraft (Burlaga and Klein, 1986). This work and others (Burlaga and Scudder, 1975a) revealed the complex

144

geometry of the IMF caused by solar wind dynamics. Figure 11 (Burlaga and Scudder, 1975b) illustrates the effects of a flare associated stream interacting with a pre-existing CIR. Neupert and Pizzo (1974) identified solar coronal holes as the source of recurrent geomagnetic disturbances. Substantial and very significant progress was made in studying interplanetary dynamics with the SPACELAB mission, which observed and identified the phenomenon descriptively referred to as Coronal Mass Ejections or CMEs (Hundhausen, 1988).

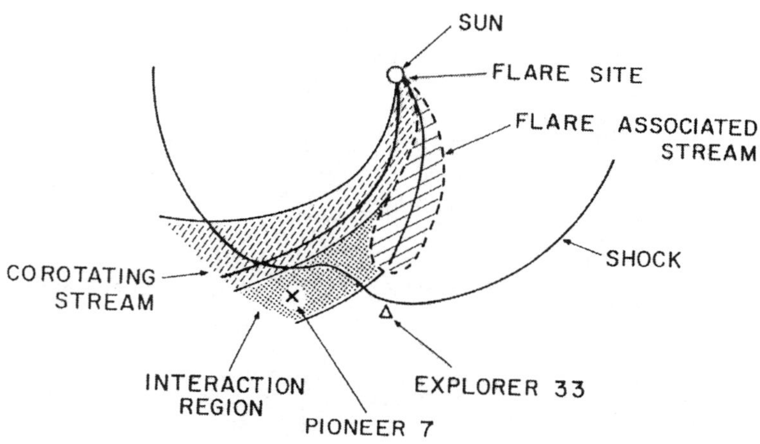

Figure 10. Interaction of a shock with a corotating stream and interaction region (Burlaga and Scudder, 1975b).

7.   MAGNETIC CLOUDS

Since the identification and subsequent studies of CME's, many authors (i.e., Webb, 1995; Bothmer and Schwenn, 1995) have emphasized CME's as the primary solar events whose parameters govern the dynamics of not only the interplanetary medium but also the geomagnetosphere. This view omits any reference to the important role played by the IMF, which as shown earlier, has been known to be closely correlated with geomagnetic activity.

Clarification and resolution of this dilemma was the identification of the IMF configuration first referred to as a magnetic cloud by Burlaga *et al.* (1981) and elaborated more fully upon by Burlaga *et al.* (1982) and Burlaga (1988 and 1991). Figure 12 illustrates the ideal geometry of a magnetic cloud with a helical field configuration within the cloud. Goldstein (1983) suggested that this configuration might well be a force-free flux rope. The typical average physical properties of a magnetic cloud at 1 AU are summarized in Table 2.

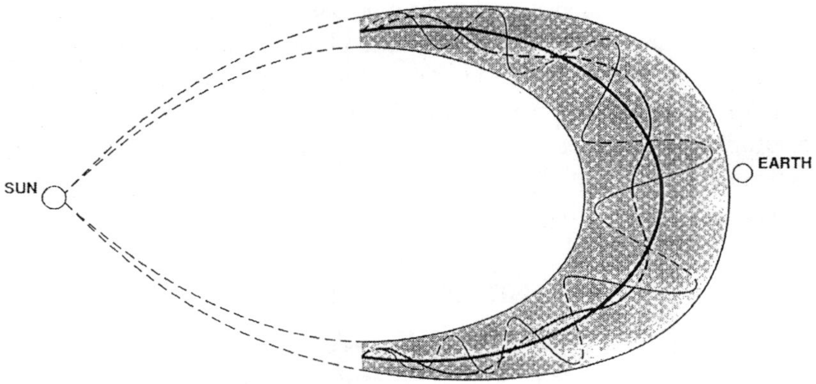

Figure 11. A sketch of the geometry of a magnetic cloud and the field lines in the magnetic clouds, which are helices viewed here in projection. The figure, drawn by Amy Burlaga, is reproduced from Burlaga et al., 1990.

Table 2. Salient Properties of Magnetic Clouds ($\leq$ 1 AU)

1. Structures with scale size . 0.25 AU
2. $|\mathbf{B}|$ greater than $<|\mathbf{B}|>$
3. $\mathbf{B}$ rotates smoothly/monotonically through a large angle $.\pi$
4. Proton temperature is low
5. Proton $\beta \ll 1$
6. May be described as force-free with variable alpha
7. If alpha = constant, is lowest energy configuration
8. Stability not proven if alpha = constant

When such a magnetic cloud structure passes Earth, the slowly changing field configuration leads to southward field orientations for extended periods. A study by Lepping *et al.* (1997) of such a cloud is shown in Figure 13. The cloud is first identified at about 1900 UT of day 291 and lasts at least through day 292. The top three panels are the magnetic field in magnitude (B), latitude (θ) and longitude (Φ), N is number density, V_T is proton thermal velocity, given by $\sqrt{(2kT/m)}$, and V (km/sec) is flow speed. The magnetic field is given in GSM coordinates. The vertical lines marked

146

"interface" refer to the probable begin-time (considered boundary D_1) and end-time (D_2) of a stream interface bordering on the rear of the magnetic cloud (Lepping *et al.*, 1997).

Theoretical and observational studies of magnetic clouds are now a common project for any student of geomagnetic disturbances. Chen and Garren (1993) proposed a model solar structure for the development, "eruption" and propagation of a magnetic cloud from the Sun's corona into interplanetary space. The event is initiated by the formation of a toroidal flux loop in the solar corona from a bipolar field in the solar photosphere. A recent study by Bravo *et al.* (1999) has shown that magnetic clouds are almost always associated with CMEs and their work summarizes the characteristics of such clouds.

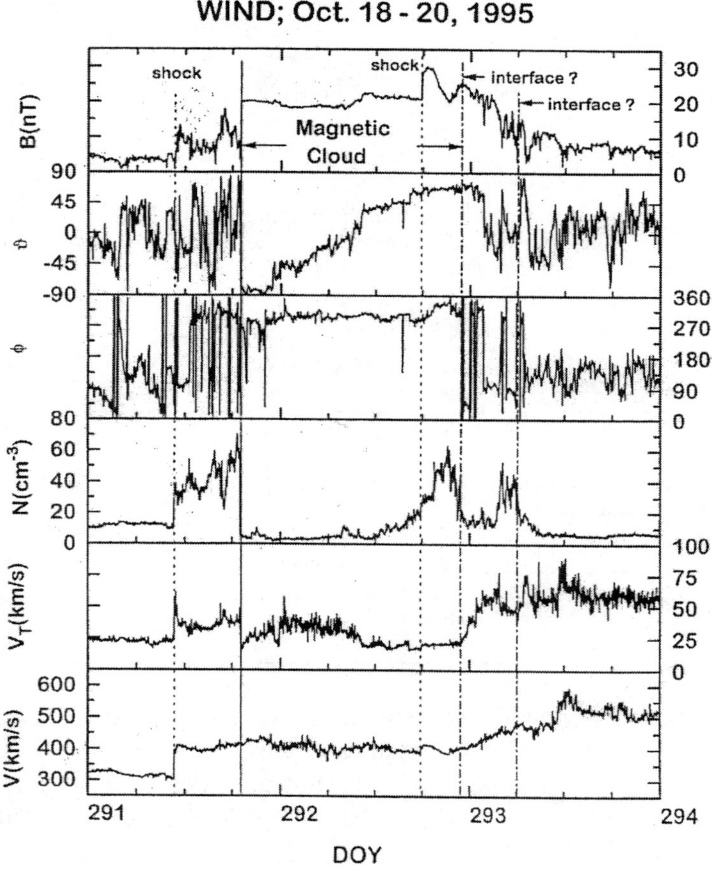

Figure 12. Magnetic field and plasma observations of the magnetic cloud and surrounding flows measured by the WIND S/C October 18-21. (Lepping et al., 1997)

Many studies of magnetic clouds and related terrestrial disturbances as specific events have well established that this IMF magnetic cloud structure is one of the causes, if not the primary cause, of geomagnetic storms and substorms. Studies of CMEs (Burlaga *et al.*, 1982; Gosling, 1990; Bothmer, 1999) have indicated that they are closely associated phenomena. Bothmer and Schwenn (1994, 1996 and 1998) have studied the evolution of clouds and their close relationship with CMEs. A CME often carries with it a magnetic cloud. But, the most basic parameter causing large disturbances of the geomagnetosphere is the southward geometry of the field (Bothmer and Schwenn, 1995). Further study of the composition of the plasma within a cloud has given evidence of solar prominence material (Burlaga *et al.*, 1998).

Bothmer and Rust (1997) studied the solar cycle variations of magnetic clouds. Farrugia *et al.* (1998) studied the geoeffectiveness of different cloud structures. Rust (1997, 1999) has examined the helicity of both solar and interplanetary fields and this additional property of the IMF may prove important in the dynamics of the geomagnetosphere. Kahler *et al.* (1999a; 1999b) have carried the study of field topology even further by investigating not only the helicity but also the chirality of a number of magnetic clouds. The Ulysses spacecraft has allowed the study of CMEs at high heliographic latitudes and elaborated upon the magnetic field characteristics (Bothmer *et al.*, 1996 and Malandraki *et al.*, 2000).

Detailed examinations of the fluctuations of the IMF have been underway now for many years related to studies of the modulation of galactic cosmic rays entering the heliosphere and also the propagation of solar cosmic rays. Studies by Leamon *et al.* (1999) and Smith *et al.* (1999) of IMF fluctuations have shown that within a magnetic cloud there is considerable anisotropy of the magnetic fluctuations and pressures (See Figure 14 showing ACE/SWEPAM observations of the radial component of the wind speed V_R (km/sec), proton ß, the ratio of magnetic power in the perpendicular and parallel components ($P_\perp/P_\parallel$) of the fluctuations relative to the mean field. These are computed over 3-hour intervals in the inertial (circles) and dissipation (squares) subranges [Smith *et al.*, 1999]). These may also be important parameters driving certain characteristics of magnetic activity in the geomagnetosphere.

8.   SHOCKS, DISCONTINUITIES AND GEOMAGNETIC STORMS

In addition to the observed 27 day periodicity of geomagnetic storms, it was also noticed that many were distinguished by a very rapid increase of the

148

magnetic field on Earth's surface prior to the longer term, large decrease associated with the formation of the terrestrial ring current. These are called Sudden Commencement (SSC) storms and are the result of a sudden change in the solar wind momentum flux interacting with and rapidly compressing the magnetosphere. The SSC event is often driven by an interplanetary shock wave generated by a disturbance at the sun or which evolves from the interactions associated with high-speed streams in interplanetary space. Feynman (1976) presented a then contemporary view of the IMF and a subset of disturbances known as sub-storms.

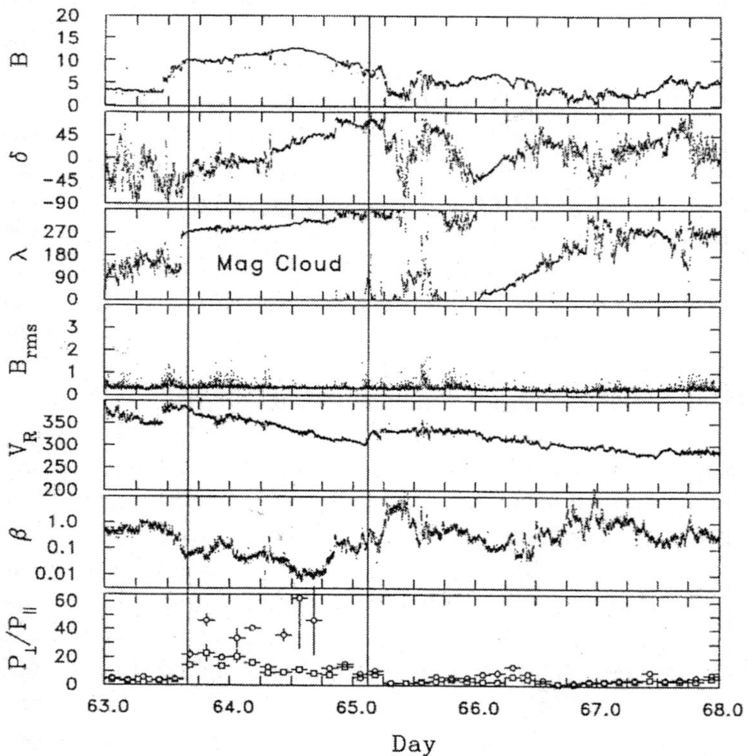

Figure 13. ACE data for March 4-8, 1998: MAG = magnitude B (nT), latitude δ and longitude λ, RMS based on 16-s means.

There are other structures in the magnetized solar wind than shocks, which may affect the state of the geomagnetosphere. These MHD structures include both propagating and non-propagating forms. Table 3 summarizes the physical characteristics of these MHD discontinuities, whose origins are still under study.

A. Contact	$B_n \neq 0, V_n = 0$	$<B> = <P> = 0; <n> \neq 0$
B. Tangential	$B_n = V_n = 0$	$<P_{total}> = 0$
C. Rotational	$B_n, V_n \neq 0$	$<B> = <P> = <n> = 0$
D. Fast Shock	$B_n, V_n \neq 0$	$<B>, <n> \gg 0 ; <V> < 0$
E. Slow Shock	$B_n, V_n \neq 0$	$<n> \gg 0 ; <B>, <V> < 0$

Table 3. Classification of IMF-MHD Discontinuities. Non-Propagating A, B. Forward and Reverse Propagating C, D, E

Early studies of IMF discontinuities were conducted by Burlaga (1969) and Burlaga and Ness (1968; 1969). The radial dependence from 0.3 to 1.0 AU was investigated by Mariani *et al.* (1973) and Behannon (1978). A recent study by Sperveslage *et al.* (2000) surveyed the range from 0.3 to 17 AU. It appears that these structures originate close to the Sun and evolve in different ways as the solar wind convects the plasma away from the Sun. Some of the discontinuities evolve, in part, by propagating either towards or away from the sun. The role of these structures, except for the interplanetary shock waves, with respect to geomagnetic disturbances is presently under study (Sitar and Clauer, 1999).

9. EPILOGUE

After 40 years of space exploration, the direct connection of transient events on the Sun with our terrestrial magnetosphere and its dynamics has been well established. This relationship is commonly described as Space Weather. This indicates the maturing of the initial exploratory discoveries to a level of sufficient understanding that we can begin predicting the changing state of the geomagnetosphere in response to upstream interplanetary conditions (Feynman and Gabriel, 2000) and their long term variations (Stamper *et al.*, 1999). The individual features of the IMF and solar wind that have been important milestones along the way are listed chronologically in Table 4.

Continuing studies of the re-connection process by spacecraft have tried to observe the long sought *in-situ* evidence for the reconnection process in action (Paschman *et al.*, 1979; Sonnerup *et al.*, 1981; Kessel *et al.*, 1996; Phan *et al.*, 2000; Scudder *et al.*, 2000). These observations have occurred at the magnetopause and, when possible, used a complex suite of instrumentation to reveal the microstructural and rapidly time varying features of this process. Theoretical numerical studies of the response of the geomagnetosphere to northward fields is also an active area of research (Song *et al.* 1999) as quantitative models have been ever more comprehensive. In this high technology era, more "assets" of our society are space based, so the effects of space weather will be ever more important to our daily activities.

1964+	Sectors Within the Archimedean Spiral Structure
1966+	Discontinuities, Alfvén Waves and Turbulence
1968+	Solar Source Surface Models of IMF
1973+	Heliospheric Current Sheet, Coronal Mass Ejections and Holes
1981+	Magnetic Clouds

Table 4. Significant Discoveries and Developments in Studies of IMF Dynamics

Continuing studies of the re-connection process by spacecraft have tried to observe the long sought *in-situ* evidence for the reconnection process in action (Paschman *et al.*, 1979; Sonnerup *et al.*, 1981; Kessel *et al.*, 1996; Phan *et al.*, 2000; Scudder *et al.*, 2000). These observations have occurred at the magnetopause and, when possible, used a complex suite of instrumentation to reveal the microstructural and rapidly time varying features of this process. Theoretical numerical studies of the response of the geomagnetosphere to northward fields is also an active area of research (Song *et al.* 1999) as quantitative models have been ever more comprehensive. In this high technology era, more "assets" of our society are space based, so the effects of space weather will be ever more important to our daily activities.

10.   ACKNOWLEDGEMENTS

I appreciate the participation and contributions of my colleagues in the studies reported upon here, especially Dr. L. F. Burlaga at NASA-GSFC. This work is supported in part by JPL contract 959167 from the Voyager project and Caltech sub-contract PC251439 for support of the ACE Magnetic Field Experiment. I apologize to those authors whose research publications have dealt with the many physical processes associated with the Sun-Earth connections but are not referenced explicitly herein. My goal was to highlight the historical developments with representative publications. Space limitations precluded otherwise.

11.   REFERENCES

Akasofu, S.-I., 1996. Search for the unknown quantity in the solar wind: A personal account. *J. Geophys. Res.* 101, 10531-10540.

Axford, W.I., Petschek, H.E. and Siscoe, G.L., 1965. Tail of the magnetosphere. *J. Geophys. Res.* 65, 1231.

Balogh, A., Smith, E.J., Tsurutani, B.T., Southwood, D.J., Forsyth, R.J. and Horbury, T.S., 1995. The heliospheric magnetic field over the south polar region of the Sun. *Science* 268, 1007-1010.

Behannon, K.W., 1978. Heliocentric distance dependence of the interplanetary magnetic field. *Rev. Geophys. Space Res.* 16, 125-145.

Bothmer, V., 1999. Magnetic field structure and topology within CMEs in the solar wind. In: Habbal, S.R., Esser, R., Hollweg, J.V. and Isenberg, PA. (Eds.), *CP471, Solar Wind Nine*, The American Institute of Physics, 119-126.

Bothmer, V., Desai, M.I., Marsden, R.G., Sanderson, T.R., Trattner, K.J., Wenzel, K.-P., Gosling, J.T., Balogh, A., Forsyth, R.J. and Goldstein, B.E., 1996. Ulysses observations of open and closed magnetic field lines within a coronal mass ejection. *Astron. Astrophys.* 316, 493-498.

Bothmer, V. and Rust, D.M., 1997. The field configuration of magnetic clouds and the solar cycle. In: *Coronal Mass Ejections*. Geophysical Monograph 99, American Geophysical Union, Washington, 139-146.

Bothmer, V. and Schwenn, R., 1994. Eruptive prominences as sources of magnetic clouds in the solar wind. *Space Science Reviews* 70, 215-220.

Bothmer, V. and Schwenn, R., 1995. The interplanetary and solar causes of major geomagnetic storms. *J. Geomag. Geoelectr.* 47, 1127-1132.

Bothmer, V. and Schwenn, R., 1996. Signatures of fast CMEs in interplanetary space. *Adv. Space Res.* 17(4/5), 319-322.

Bothmer, V. and Schwenn, R., 1998. The structure and origin of magnetic clouds in the solar wind. *Ann. Geophysicae* 16, 1-24.

Bravo, S., Blanco-Cano, X. and López, C., 1999. Characteristics of interplanetary magnetic clouds in relation to their solar association. *J. Geophys. Res.* 104, 581-591.

Burlaga, L.F., 1969. Directional discontinuities in the interplanetary magnetic field. *Solar Physics* 7, 54-71.

Burlaga, L.F., 1988. Magnetic clouds and force-free fields with constant-alpha. *J. Geophys. Res.* 93, 7217-7224.

Burlaga, L.F., 1991. Magnetic clouds. In: Schwenn, R. and Marsch, E. (Eds.), *Physics of the Inner Heliosphere II*. Springer-Verlag, Berlin, Heidelberg.

Burlaga, L.F., 1995. *Interplanetary magnetohydrodynamics*. Oxford Univ. Press.

Burlaga, L.F. and Barouch, E., 1976. Interplanetary stream magnetism: Kinematic effects. *Astrophys. J.* 203, 257.

Burlaga, L.F., Fitzenreiter, R., Lepping, R., Ogilvie, K., Szabo, A., Lazarus, A., Steinberg, J., Gloeckler, G., Howard, R., Michels, D., Farrugia, C., Lin, R.P. and Larson, D.E., 1998. A magnetic cloud containing prominence material: January 1997. *J. Geophys. Res.* 103, 277-285.

Burlaga, L.F., Hundhausen, A.J. and Zhao, X.-P., 1981. The coronal and interplanetary current sheet in early 1976. *J. Geophys. Res.* 86, 8893-8898.

Burlaga, L.F. and Klein, L.W., 1986. Configurations of corotating shocks in the outer heliosphere. *J. Geophys. Res.* 91, 8975.

Burlaga, L.F., Klein, L.W., Sheeley, Jr., N.R., Michels, D.J., Howard, R.A., Koomen, M.J., Schwenn, R. and Rosenbauer, H., 1982. A magnetic cloud and a coronal mass ejection. *Geophys. Res. Lett.* 9, 1317-1320.

Burlaga, L.F., Lepping, R.P. and Jones, J., 1990. Global configuration of a magnetic cloud. In: Russell, C.T., Priest, E.R., and Lee, L.C. (Eds.), *Physics of Magnetic Flux Ropes*, Geophysical Monograph 58, American Geophysical Union, Washington, 373.

Burlaga, L.F. and Ness, N.F., 1968. Macro- and microstructure of the interplanetary magnetic field. *Can. J. Physics* 46, S962-965.

Burlaga, L.F. and Ness, N.F., 1969. Tangential discontinuities in the solar wind at 1 AU. *Solar Physics* 9, 467.

Burlaga, L.F. and Scudder, J., 1975a. Corotating streams and interaction regions. *J. Geophys. Res.* 80, 4044-4059.

Burlaga, L.F. and Scudder, J., 1975b. Motion of shocks through interplanetary streams. *J. Geophys. Res.* 80, 4004-4010.

Burlaga, L.F., Sittler, E., Mariani, F. and Schwenn, R., 1981. Magnetic loop behind an interplanetary shock: Voyager, Helios and IMP 8 observations. *J. Geophys. Res.* 86, 6673-6684.

Chen, J. and Garren, D.A., 1993. Interplanetary magnetic clouds: Topology and driving mechanism. *Geophys. Res. Lett.* 20, 2319-2322.

Crooker, N.U. and Cliver, E.W., 1994. Postmodern view of M-regions. *J. Geophys. Res.* 99, 23383-23390.

Dungey, J.W., 1961. Interplanetary magnetic field and the auroral zones. *Phys. Rev. Lett.* 6, 47.

Fairfield, D.H., 1967. Polar magnetic disturbances and the interplanetary magnetic field. *Space Research VIII*, p. 107.

Farrugia, C.J., Scudder, J.D., Freeman, M.P., Janoo, L., Lu, G., Quinn, J.M., Arnoldy, R.L., Torbert, R.B., Burlaga, L.F., Ogilvie, K.W., Lepping, R.P., Lazarus, A.J., Steinberg, J.T., Gratton, F.T. and Rostoker, G., 1998. Geoeffectiveness of three wind magnetic clouds: A comparative study. *J. Geophys. Res.* 103, 17261-17278.

Feynman, J., 1976. Substorms and the interplanetary magnetic field. *J. Geophys. Res.* 81, 5551-5555.

Feynman, J. and Gabriel, S.B., 2000. On space weather consequences and predictions. *J. Geophys. Res.* 105, 10543-10564.

Forsyth, R.J., Balogh, A., Horbury, T.S., Erdös, G., Smith, E.J. and Burton, M.E., 1996. The heliospheric magnetic field at solar minimum: Ulysses observations from pole to pole. *Astron. Astrophys.* 316, 287-295.

Goldstein, H., 1983. On the field configuration in magnetic clouds. In: Neugebauer, M. (Ed.), *CP-2280, Solar Wind Nine*, NASA Conf. Publ. NASA, Washington, 731-733.

Gonzalez, W.D., Joselyn, J.A., Kamide, Y., Kroehl, H.W., Rostoker, G., Tsurutani, B.T., and Vasyliunas, V.M., 1994. What is a geomagnetic storm? *J. Geophys. Res.* 99, 5771-5792.

Gosling, J.T., 1990. Coronal mass ejections and magnetic flux ropes in interplanetary space. In: Russell, C.T., Priest, E.R. and Lee, L.C. (Eds.), *Physics of Magnetic Flux Ropes*, Geophysical Monograph 58, American Geophysical Union, Washington, 344.

Hoeksema, J.T., 1989. Extending the sun's magnetic field through the three-dimensional heliosphere. *Adv. Space Res.* 9:4, 141-152.

Hundhausen, A.J., 1972. *Coronal expansion and solar wind.* Springer-Verlag, New York.

Hundhausen, A.J., 1977. An interplanetary view of coronal holes. In: Zirker, J.B. (Ed.) *Coronal Holes and High Speed Streams.* Colorado Associated University Press, Boulder, 225.

Hundhausen, A.J., 1988. The origin and propagation of coronal mass ejections. In: Pizzo, V.J., Holzer, T.E. and Sime, D.G. (Eds.) *NCAR Technical Note 306 and Proc., Solar Wind Six,* Boulder, 181.

Kahler, S.W., Crooker, N.U. and Gosling, J.T., 1999a. A magnetic polarity and chirality analysis of ISEE 3 interplanetary magnetic clouds. *J. Geophys. Res.* 104, 9911-9918.

Kahler, S.W., Crooker, N.U. and Gosling, J.T., 1999b. The polarities and locations of interplanetary coronal mass ejections in large interplanetary magnetic sectors. *J. Geophys. Res.* 104, 9919-9924.

Kessel, R.L., Chen, S.-H., Green, J.L., Fung, S.F., Boardsen, S.A., Tan, L.C., Eastman, T.E., Craven, J.D. and Frank, L.A., 1996.   Evidence of high-latitude reconnecting during northward IMF: Hawkeye observations. *Geophys. Res. Lett.* 23, 583-586.

Leamon, R.J., Smith, C.W., Ness, N.F. and Wong, H.K., 1999. Dissipation range dynamics: Kinetic alfven waves and the importance of β_e. *J. Geophys. Res.* 104, 22331-22344.

Lepping, R.P., Burlaga, L.F., Szabo, A., Ogilvie, K.W., Mish, W.H., Vassiliadis, D., Lazarus, A.J., Steinberg, J.T., Farrugia, C.J., Janoo, L. and Mariani, F., 1997. The Wind magnetic cloud and events of October 18-20, 1995: Interplanetary properties and as triggers for geomagnetic activity. *J. Geophys. Res.* 102, 14049-14063.

Levy, R.H., Petschek, H.E. and Siscoe, G.L., 1964. Aerodynamic aspects of magnetospheric flow. *AIAA J.* 2, 2065-2076.

Malandraki, O., Sarris, E.T., and Trochoutsos, P., 2000. Probing the magnetic topology of coronal mass ejections by means of Ulysses/HI-SCALE energetic particle observations. *Ann Geophysicae* 18, 129-140.

Mariani, M., Bavassano, B., Villante, U. and Ness, N.F., 1973. Variations in the occurrence rate of discontinuities in the interplanetary magnetic field. *J. Geophys. Res.* 78, 8011.

NASA Technical Memo TM-80758, 1980. Interplanetary Monitoring Platform Engineering History and Achievements. NASA-GSFC, Greenbelt, MD.

Ness, N.F., 1969. The magnetic structure of interplanetary space. In: Bozoko, G., Gombosi, E., Sebestyen, A. and Somogyi, A. (Eds.), *Proceedings of Budapest XI International Cosmic Ray Conference*, 41-83.

Ness, N.F., 1970. Magnetometers for space research. *Space Science Reviews XI*, 111-222.

Ness, N.F., 2000. Spacecraft studies of the interplanetary magnetic field. *J. Geophys. Res.* 105 (accepted August, 2000).

Ness, N.F., Scearce, C.S. and Seek, J.B., 1964. Initial results of the IMP-1 magnetic field experiment. *J. Geophys. Res.* 69, 3531-3569.

Ness, N.F. and Wilcox, J.M., 1964. Solar origin of the interplanetary magnetic field. *Phys. Rev. Letters* 13, 461-464.

Ness, N.F. and Wilcox, J.M., 1965. Sector structure of the quiet interplanetary magnetic field. *Science* 148, 1592-1594.

Ness, N.F. and Wilcox, J.M., 1966. Extension of the photosphere field into interplanetary space. *Astrophys. J.* 143, 23-31.

Ness, N.F. and Wilcox, J.M., 1967. Interplanetary sector structure, 1962-1966. *Solar Physics* 2, 351-359.

Neugebauer, M. and Snyder, C.W., 1962. The mission of Mariner II: Preliminary results, solar plasma experiment. *Science* 138, 1095-1097.

Neupert, W.N. and Pizzo, V., 1974. Solar coronal holes as sources of   recurrent geomagnetic disturbances. *J. Geophys. Res.* 79, 3701.

Parker, E.N., 1957. Sweet's mechanism for merging magnetic fields in conducting fluids. *J. Geophys. Res.* 62, 509-520.

Parker, E.N., 1958. Dynamics of the interplanetary gas and magnetic fields. *Astrophys. J.* 128, 664-675.

Parker, E.N., 2000. The history of the early work on the heliospheric magnetic field. *J. Geophys. Res.* 105, in press.

Paschmann, G. *et al.,* 1979. Plasma acceleration at the Earth's magnetopause: Evidence for reconnection. *Nature* 282, 243-246.

Petschek, H.E., 1964. Magnetic field annihilation. In: Hess, W.N. (Ed.) *AAS-NASA Symposium of the Physics of Solar Flares,* NASA Spec. Publ. 15, 425-437.

Petschek, Harry E., 1996. Glimpses of space physics in the 1960's and 1990's. *J. Geophys. Res.* 101, 10511-10519.

154

Phan, T.D., Kistler, L.M., Klecker, B., Haerendel, G., Paschmann, G., Sonnerup, B.U.Ö., Baumjohann, W., Bavassano-Cattaneo, M.B., Carlson, C.W., DiLellis, A.M., Fornacon, K.-H., Frank, L.A., Fujimoto, M., Georgescu, E., Kokubun, S., Moebius, E., Mukai, T., Øieroset, M., Paterson, W.R., Reme, H., 2000. Extended magnetic reconnection at the Earth's magnetopause from detection of bi-directional jets. *Nature* 404, 848-850.

Russell, C.T. and McPherron, R.L., 1973. Semiannual variation of geomagnetic activity. *J. Geophys. Res.* 78, 92-108.

Rust, D.M., 1997. Helicity conservation. In: *Coronal Mass Ejections*, Geophysical Monograph 99, American Geophysical Union, Washington, 119-125.

Rust, D.M., 1999. Magnetic helicity in solar filaments and coronal mass ejections. In: *Magnetic Helicity in Space and Laboratory Plasmas*, Geophysical Monograph 111, American Geophysical Union, Washington, 221-227.

Schatten, K.H., Ness, N.F. and Wilcox, J.M., 1968. Influence of a solar active region on the interplanetary magnetic field. *Solar Physics* 5, 240-256.

Schatten, K.H. and Wilcox, J.M., 1967. Response of the geomagnetic activity index Kp to the interplanetary magnetic field. *J. Geophys. Res.* 72, 5185-5191.

Schatten, K.H., Wilcox, J.M. and Ness, N.F., 1969. A model for the determination of interplanetary and coronal magnetic fields. *Solar Physics* 6, 442-455.

Schulz, M., 1973. Interplanetary sector structure and the heliomagnetic equator. *Astrophys. Space Sci.* 34, 371-383.

Schulz, M., 1995. Fourier parameters of heliospheric current sheet and their significance. *Space Sci. Rev.* 72, 149-152.

Scudder, J.D., Mozer, F.S., Maynard, N.C., Puhl-Quinn, P.A., Ma, Z.W. and Russell, C.T., 2000. Fingerprints of collisionless reconnection I: evidence for hall MHD scales. *J. Geophys. Res.* (Submitted May, 2000).

Siscoe, G.L., 1966. A unified treatment of magnetospheric dynamics. *Planet. Space Sci.* 14, 947-967.

Sitar, R.J. and Clauer, C.R., 1999. Ground magnetic response to sudden changes in the interplanetary magnetic field orientation. *J. Geophys. Res.* 104, 28343-28350.

Smith, C.W., Leamon, R.J., Ness, N.F., Burlaga, L.F., Tokar, R.L. and Skoug, R.M., 1999. Magnetic Fluctuation Properties of Interplanetary Magnetic Clouds, In: Kieda, D., Salamon, M., and Dingus, B. (Eds.) *Proceedings of the 26th International Cosmic Ray Conference*, Salt Lake City, 6, 452-455.

Smith, E.J., Neugebauer, M., Balogh, A., Bame, S.J., Erdös, G., Forsyth, R.J., Goldstein, B.E., Phillips, J.L. and Tsurutani, B.T., 1993. Disappearance of the heliospheric sector structure at Ulysses. *Geophys. Res. Lett.* 20, 2327-2330.

Smith, E.J., Tsurutani, B.T. and Rosenberg, R.L., 1978. Observations of the interplanetary sector structure up to heliographic latitudes of   16 degrees: Pioneer 11. *J. Geophys. Res.* 83, 717.

Snyder, C.W., Neugebauer, M. and Rao, U.R., 1963. The solar wind velocity and its correlation with cosmic-ray variations and with solar and geomagnetic activity. *J. Geophys. Res.* 68, 6361-6372.

Song, P., DeZeeuw, D.L., Gombosi, T.I., Groth, C.P.T. and Powell, K.G., 1999. A numerical study of solar wind - magnetosphere interaction for northward interplanetary magnetic field. *J. Geophys. Res.* 104, 28361-28378.

Sonnerup, B.U.Ö, Paschmann, G., Papamastorakis, I., Sckopke, N., Haerendel, G., Bame, S.J., Asbridge, J.R., Gosling, J.T. and Russell, C.T., 1981. Evidence for magnetic field reconnection at the Earth's magnetopause. *J. Geophys. Res.* 86, 10049-10067.

Sperveslage, K., Neubauer, F.M., Baumgärtel, K. and Ness, N.F., 2000. Magnetic holes in the solar wind between 0.3 AU and 17 AU. *Nonlinear Processes in Geophysics* 7, 191-200.

Stamper, R., Lockwood, M., Wild, M.N. and Clark, T.D.G., 1999. Solar causes of the long-term increase in geomagnetic activity. *J. Geophys. Res.,* 104, 28325-28342.

Suess, S.T., McComas, D.J. and Hoeksema, J.T., 1993. Prediction of the heliospheric current sheet tilt: 1992-1996. *Geophys. Res. Lett.* 20, 161-164.

Sweet, P.A., 1958. The neutral point theory of solar flares. In: Lehnert, B. (Ed.), *Electromagnetic Phenomena in Cosmical Physics, IAU Symposium 6*, Cambridge University Press, New York, 123-134.

Tsinganos, K. MHD modeling of space weather drivers (this volume), 2001.

Tsurutani, B.T. and Gonzalez, W.D., 1993. On the solar and interplanetary causes of geomagnetic storms. *Phys. Fluids B* 5(7), 2623.

Vasyliunas, V.M., 1975. Theoretical models of magnetic field line merging. *Revs. Geophys. Space Sci.* 13, 303-336.

Webb, D.F., 1995. Coronal mass ejections: the key to major interplanetary and geomagnetic disturbances. *Reviews of Geophysics*, Supplement, U.S. National Report to International Union of Geodesy and Geophysics 1991-1994, 577-583.

Wilcox, J.M. and Ness, N.F., 1965. A quasi-stationary co-rotating structure in the interplanetary medium. *J. Geophys. Res.* 70, 793-5806.

Chapter 6

Bayesian Classification of Geoeffective Solar Wind Structures

Real-time prediction of large geomagnetic storms

James Chen
Plasma Physics Division, Naval Research Laboratory
Washington, DC 20375, U.S.A.

Abstract   "Space weather" refers to the condition of geospace, a plasma-filled region primarily consisting of the magnetosphere and the ionosphere. Space weather is controlled by the solar wind impinging on the outer boundary of the magnetosphere. Sporadic eruptions at the Sun such as coronal mass ejections and solar flares produce solar wind structures that can cause disturbances in the Earth's plasma environment, leading to adverse consequences on technological systems. Perhaps the most damaging of such "space weather effects" are severe geomagnetic storms characterized by intense disturbances that are long-lasting and global in geographic scale, encompassing both high-latitude and low-latitude regions around the Earth. Large geomagnetic storms are relatively infrequent occurrences but can seriously disrupt space-borne as well as ground-based susceptible systems including communications networks and electric power grids. Storms are often accompanied by increased populations of high-energy charged particles in geospace that can jeopardize satellites and astronauts. Thus, accurate and timely prediction of large storms is one of the most important end products of space weather research. The present prediction methods in operation, which are based on solar observations, are inaccurate because the trajectories and the magnetic fields of the ejecta from solar eruptions cannot be accurately predicted. This article describes the basics of a new approach to making real-time predictions of large geomagnetic storms. The approach is based on recognizing, in real-time solar wind data, quantifiable physical features that allow one to estimate the duration and geoeffectiveness of the solar wind that has yet to arrive at the detector. The results of an extensive test of the method using the archival WIND data from $1995-1999$ indicate that high prediction accuracy ($\geq 70-80\,\%$) and moderately long warning times (several hours to more than ten hours) are achievable.

157

I.A. Daglis (ed.), Space Storms and Space Weather Hazards, 157–176.
© 2001 *Kluwer Academic Publishers. Printed in the Netherlands.*

Keywords   Space weather, forecasting, Bayesian classification, magnetic storms, solar physics, solar wind, magnetosphere.

1.   INTRODUCTION

The condition of geoplasma space, referred to as "space weather," is governed by the solar wind energy entering the magnetosphere in the form of plasma particles and interplanetary magnetic fields. Figure 1 shows a schematic of geospace in relation to the Sun and the solar wind. The solar wind (SW) is a continuous outflow of tenuous plasmas emanating from the Sun. The typical SW at 1 astronomical unit (AU) has average density of 5 cm^{-3} and speed of ≈ 400 km/s (the slow wind) to ≈ 600 km/s (the fast wind). The interplanetary magnetic field (IMF) fluctuates with amplitude of ≈ 5 nT (5×10^{-5} gauss). Under these "typical" or "quiet" SW conditions, the Earth's magnetosphere is quiescent and may be said to be in the "ground state," with the energy input and loss/dissipation roughly in dynamic equilibrium.

The magnetosphere, however, can undergo periods of highly disturbed conditions, resulting in myriad effects in near-Earth space and on the ground. Among the most severe conditions are geomagnetic storms. They are characterized by the presence of an intense ring current in the magnetosphere, resulting in significant modification to the Earth's magnetic field around the globe. Such changes in the geomagnetic fields, frequently accompanied by aurorae at high latitudes, occur over large geographic areas encompassing hundreds to thousands of kilometers and extending from high- to low-latitude regions. The intensification of the ring current results from increased energy input into the magnetosphere from the solar wind. The observable effects (such as aurorae and deflection of the horizontal ground magnetic field in low latitudes) and the underlying phenomenon of geomagnetic storms have been of keen scientific interest since before the era of satellite observations [e.g., *Birkeland*, 1908; *Chapman*, 1935]. (The connection between geomagnetic field changes observed at Kew and solar eruptions was speculated upon when solar flares were first discovered [*Carrington*, 1860; *Hodgson*, 1860].) A review of geomagnetic storms can be found in *Gonzalez et al.* [1994], and a comprehensive range of related issues are discussed in the recent monograph, *Magnetic Storms* [*Tsurutani et al.*, 1997]. Developing physical understanding of the phenomenon and capabilities to accurately predict impending storms is a major objective of space physics.

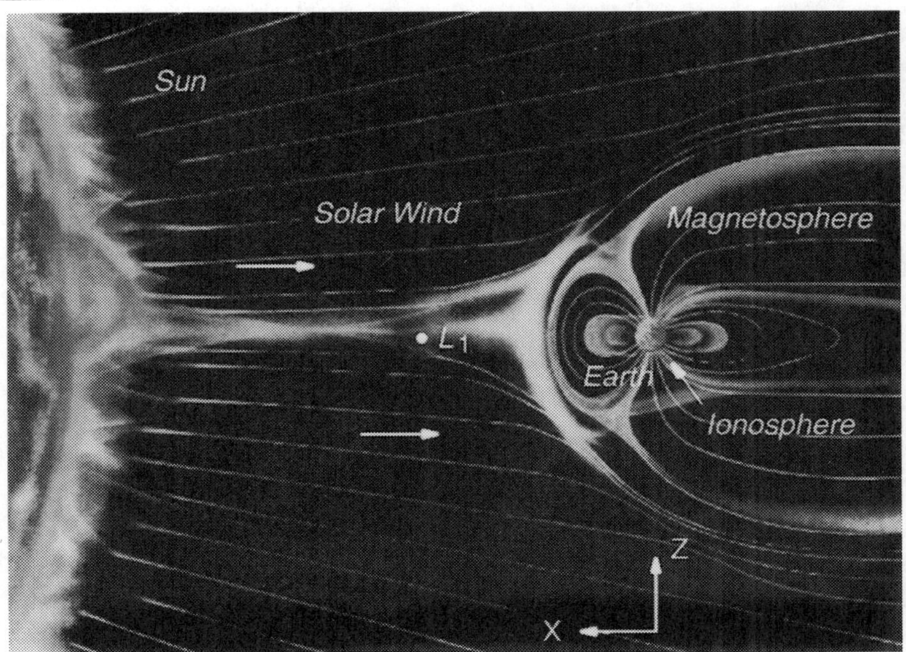

Figure 1. Schematic of the Sun and geospace. Plasma coupling between the Sun and Earth, mediated by the solar wind, plays a key role in space weather. The ionosphere (≥ 100 km from the ground) is immediately above the neutral atmosphere. The Earth's dipole field, distended by the magnetospheric currents on the nightside, is depicted. The Earth is 1 AU $\cong 1.5 \times 10^{13}$ km from the Sun. The solar radius (Rs) is 7×10^{5} km, and one Earth radius (Re) is 6371 km. The $L1$ point is ≈ 215 Re upstream of the Earth. The coordinate axes show the respective directions, with $+y$ pointing out of the figure. The coordinate system is centered at the Earth. (Original unannotated artwork: source unknown) (Color plate, see p. 482)

In recent years, it has become clear that large disturbances in geomagnetic fields can cause deleterious effects on space-borne and ground-based technological infrastructure over wide geographic areas, including communications networks, electric power grids, and satellite systems [e.g., *Lanzerotti*, 1983; *Allen et al.*, 1989; *Kappenman and Albertson*, 1990]. Large storms can also result in significant Joule heating and expansion of the neutral atmosphere, disrupting the ability to manage satellite orbits. The integrity of satellites and astronaut safety can be jeopardized by significantly enhanced fluxes of energetic particles often associated with high levels of geomagnetic disturbances. These adverse effects are becoming more pronounced now because of increased miniaturization of micro-electronic circuits, reliance on satellite systems, space exploration, and more extended power grids and long-line communication systems [*Lanzerotti*, 2001]. An extensive discussion on the space weather effects on power grids is given in *Kappenman* [2001].

The SW cause of large geomagnetic storms is well known: long durations of strong southward IMF imposed on the magnetosphere [e.g., *Rostoker and Falthammar*, 1967; *Hirshberg and Colburn*, 1969; *Russell et al.*, 1974; *Perreault and Akasofu*, 1978; *Gonzalez and Tsurutani*, 1987]. Such drivers of large storms result from sporadic solar eruptions, e.g., coronal mass ejections (CMEs), and take $2-4$ days to reach the Earth. However, the properties and trajectories of ejecta are difficult to predict with accuracy. As a result, the current predictions of geomagnetic storms, which are based on observations of solar eruptions (e.g., as indicated by X-ray emissions), generate significant numbers of false negatives and false positives [*Joselyn*, 1995]. In addition, the duration and the severity of storms cannot be predicted. Thus, the development of an accurate method of predicting large storms is a prominent and high-impact goal of space weather research.

To overcome the low accuracy rates arising from using only solar observations as input, a number of prediction methods have been proposed that use as input the SW data obtained immediately upstream of the Earth, for example, at the $L1$ Lagrange point approximately 215 Re toward the Sun on the Sun-Earth line (Figure 1), where a satellite can be positioned in gravitationally stable equilibrium. One earth radius (Re) is 6371 km.

One class of methods seek to determine the nonlinear relationship between the SW impinging on the Earth and the geomagnetic response as represented by the AE and Dst indices, which are measures of geomagnetic activity. These quantities are obtained from ground measurements of magnetic field in high- and low-latitude regions, respectively (see below). One approach is to treat the magnetosphere as a low-dimensional dynamical system and use the historical SW and AE index time-series to reconstruct the dynamics [e.g., *Sharma et al.*, 1993; *Price and Prichard*, 1993; *Vassiliadis et al.*, 1995]. The low-dimensional dynamical model is then applied to new SW time-series to predict the geomagnetic response. In another approach, neural networks are trained on historical SW data and geomagnetic indices and are then applied to new input SW time-series [e.g., *Lundstedt*, 1992; *Hernandez et al.*, 1993; *Wu and Lundstedt*, 1996; *Wu et al.*, 1998].

In these methods, the basic idea is to predict the response of the magnetosphere in terms of AE or Dst index to the SW input. This concept relies on the fact that the driver SW can be detected before the SW impinges on the magnetosphere. For such prediction methods the forecasting time is roughly the SW transit time from the monitoring position to the magnetosphere. This time is approximately 1 hour if the SW speed is 400 km/s and is significantly shorter (30 minutes or less) for fast CME ejecta causing large storms, which often have speeds in the range of $600 - 800$ km/s or sometimes faster.

Valdivia et al. [1996] proposed a nonlinear method that uses the real-time *Dst* index as input and advances in time the predicted *Dst* index using the calculated *Dst* value from the preceding time step. The nonlinear filter is determined from historical SW and *Dst* data, but for prediction purposes, this method uses only near-real time *Dst* data. The actual *Dst* index is the hourly average of ground measurements, so that only the evolution of storms already in progress on the ground can be predicted.

A different approach to using real-time SW data as input for prediction is to infer the physical features of geoeffective SW structure upstream of a SW monitoring satellite [*Chen et al.*, 1996, 1997]. Here, the term *geoeffective* is used to characterize SW structures that cause large storms, to be defined more precisely later in Section 2.3. This approach takes advantage of the fact that SW drivers of large storms are clearly distinguishable from the nongeoeffective SW and have long spatial correlation (up to $\approx 1/2$ AU) corresponding to $10 - 20$ hours. Using a method based on Bayesian classification, it was found that $\approx 70\%$ of large storms exceeding a specified threshold in *Dst* can be correctly predicted while producing few false positives [*Chen et al.*, 1997]. Recent improvement indicates that accuracy in excess of 80% may be obtained. The advance forecasting time is moderately long, ranging from a few hours to more than ten hours. This time is determined by the size of the geoeffective SW structure that has yet to arrive, rather than the transit time from the observing platform. This method will be discussed in the next section.

The *Dst* index is the traditional measure of the storm intensity. It is the deflection of the horizontal magnetic field measured by several ground stations at mid- to low-latitudes around the globe. The deflection of the low-latitude magnetic field is caused by the reduction in the north-south field due to the enhanced ring current. The *Dst* index shows the hourly average of this magnetic field relative to that under quiet conditions, defined to be $Dst = 0$ nT. (Note that $Dst = 0$ level is determined monthly.) The more negative *Dst* is, the more intense a storm is. The *Dst* index represents a time-integrated response of the magnetosphere to variations in the SW [*Burton et al.*, 1975] and responds to changes in the SW that are sustained for a few hours or longer. The *AE* index is a measure of auroral electrojets affecting the high-latitude magnetic field on the ground. The *AE* index shows geomagnetic disturbances on time scales on minutes to tens of minutes responding to fluctuations in the SW.

The relationship between the SW and the storm intensity has been an important issue. *Burton et al.* [1975] found an empirical formula expressing $dDst/dt$ in terms of Dst, B_z, V_x, and the SW ram pressure along with some

162

parameters. This formula has been modified and improved by *Fenrich and Luhmann* [1998] and *O'Brien and McPherron* [2000].

Note that the storms of interest are the nonrecurrent type. There are storms that recur with the solar rotation period of ≈ 27 days. They are associated with the Earth crossing of magnetic sectors corresponding to the "open-field" regions (coronal holes) on the Sun [*Neupert and Pizzo*, 1974; *Sheeley et al.*, 1976]. Large storms tend to be caused by sporadic CMEs and are difficult to predict.

2. CLASSIFICATION OF SOLAR WIND STRUCTURES

The prediction method consists of two basic steps: (i) estimate the magnetic field structure of the SW upstream of the observer and (ii) predict the occurrence, duration, and strength of geoeffective SW structures. In order to do this we must first quantitatively characterize the geoeffective SW features.

2.1 Geoeffective Solar Wind Features

The SW drivers of large geomagnetic storms are clearly identifiable in SW data. Figure 2 shows the SW data observed by IMP-8 during $13-15$ January 1988, displaying five-minute averages of B (panel a) defined by

$$B \equiv \left(B_y^2 + B_z^2\right)^{1/2},$$

B_z (panel b), the Earthward speed $|V_x|$ (panel d), and the proton density n (panel e). Here, is in the east-west direction, and $+\mathbf{z}$ points to the north, as illustrated in Figure 1. Prior to about 23:00 UT, 13 January 1988, B_z fluctuates about 0 nT with $B \approx 5$ nT on the average, varying on a time scale of minutes to tens of minutes (with superposed faster small-amplitude fluctuations). The Earthward speed, V_x, is about 450 km/s, and the density is $n \cong 5$ cm^{-3}. These values are typical of the slow SW.

Starting around 23:00 UT, 13 January, B increases to slightly more than 30 nT in $16-18$ hours and then decreases to the background values in ≈ 18 hours. In Figure 2b, we further see that the B_z component varies slowly (in comparison with minutes) having two uninterrupted periods of unipolar B_z. Figure 2d shows that the SW speed rapidly increases from about 450 km/s to about 750 km/s, and the density jumps to more than 10 cm^{-3}, indicating that there is a shock. It is significant that B_z remains southward without any interruption for about 18 hours. During this period, the magnitude exceeds 10 nT, reaching ≈ 20 nT.

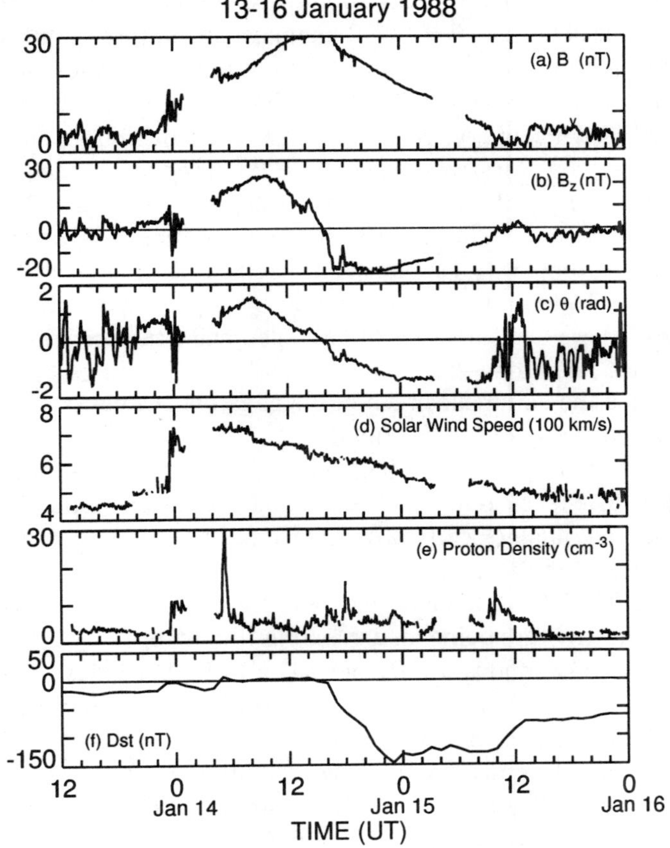

Figure 2. Solar wind and the *Dst* index. The SW was observed on 13−15 January 1988 by IMP 8 located at $L1$. Panels (a)-(e) are 5-min averaged data, provided by the UCLA Space Physics Group, and the hourly *Dst* index is courtesy of the World Data Center C2, Kyoto University. A shock driven by a magnetic cloud arrived at about 00:00 UT, 14 January 1988. Each tick mark on the horizontal axis is 2 hours. There are two large gaps in the data.

The magnetic field and speed profiles between $\approx 00{:}00$ UT, 14 January and $\approx 10{:}00$ UT, 15 January show a prototypical "magnetic cloud" [*Burlaga et al.*, 1981]. A magnetic cloud is well described as a magnetic flux rope [*Burlaga*, 1988], exhibiting slow rotation of the magnetic field vector through large angles. This is shown by the quantity θ (Figure 2c), which is the angle subtended by the $y-z$ projection of the magnetic field and the ecliptic plane and is defined by

164

$$\theta(t) \equiv \sin^{-1}(B_z / B), \tag{1}$$

where all the quantities are understood to be functions of t. Here, the sign convention is such that $\theta = +\pi / 2$ denotes an entirely northward $y - z$ projection of the field vector. Figure 2c shows that θ rotates smoothly from $+\pi / 2$ (northward) at about 8:00 UT, 14 January, to $-\pi / 2$ (southward) at about 0:00 UT, 15 January, corresponding to the beginning and end of the cloud as defined by *Burlaga* [1988]. The B_z component reaches about $+28$ nT during the north-polarity period and decreases to about -20 nT during south-polarity period. The latter period, the trailing half of the magnetic cloud (from $\approx 16{:}00$ UT, 14 January till 8:00 UT, 15 January), imposed a long period (≈ 16 hours) of strong southward B_z component ($\approx$ -20 nT) on the magnetosphere, causing a fairly large geomagnetic storm as indicated by the *Dst* values reaching -150 nT. By comparing Figures 2b and 2f, it is easy to see that the commencement of the negative *Dst* period corresponds to the southward turning of the B_z field.

The above definition of the beginning and the end of a cloud identifies the cloud with the current channel embedded inside a flux rope, which includes the poloidal (locally azimuthal) field preceding (following) the beginning (end) of the cloud [*Chen*, 1996]. The linear decrease in V_x between $\approx 4{:}00$ UT, 14 January and $\approx 12{:}00$ UT, 15 January, is consistent with the entire structure being an expanding flux rope.

After a strongly southward B_z period ends the *Dst* index recovers (tends toward zero) typically over several hours to a few days, indicating the time scale on which the ring current returns to the quiescent state. The specific recovery profile depends on the SW condition after the storm. In the example shown in Figure 2, B_z remains strongly negative for about half a day, but V_x slows down to about 500 km/s. The *Dst* index remains at about -150 nT during this time, corresponding to a moderately disturbed state. After the magnetic cloud passes, the density decreases to $1-2$ cm^{-3}, and *Dst* stays at about -50 nT, because of the fluctuating but negative B_z field.

Figures 2a and 2b show two features of the magnetic cloud that are prominently and quantifiably different from those of the background SW: (i) in the undisturbed solar wind θ varies between $\pm \pi /2$ on time scales of minutes to tens of minutes (in the 5-min averaged data). In contrast, inside a magnetic cloud the time scale of variation is many hours (16 hours in this example), one to two orders of magnitude slower; (ii) the magnitude of the southward IMF is strong (≈ 20 nT versus ≈ 5 nT in the background SW). With an average speed of $V_x \cong 600$ km/s and duration of ≈ 36 hours, the size of the flux rope is approximately 1/2 AU at the Earth. Thus, geoeffective SW features correspond to large and magnetically dominant structures that are clearly distinguishable from nongeoeffective SW.

Figure 2f shows that the *Dst* index turns strongly negative in response to southward IMF (i.e., $B_z < 0$) sustained over a few hours or longer. It has been known that *Dst* can be approximated by a time integral of $B_z < 0$ [*Burton et al.*, 1975] so that it is relatively insensitive to rapid variations in B_z. Thus, the intensity of the storm depends not only on the magnitude but also on the duration of $B_z < 0$. (We will not discuss the influences of V_x and n in this chapter.) The precise time of storm commencement is affected by the state of the magnetosphere prior to the southward turning of the IMF and the internal magnetospheric dynamics as well as the B_y field.

2.2 Prediction of Upstream IMF

The fact that geoeffective SW features are quantifiable and distinct from the background SW gives rise to the possibility that they can be recognised as a pattern. The fact that they are magnetically organised suggests that the data from the leading-edge of a geoeffective SW structure contain structural information pertaining to the entire structure, representing a correlation distance of perhaps up to 1/2 AU upstream of the SW detector (e.g., Figure 2). In particular, it may be possible to estimate the magnetic field of the geoeffective SW that has yet to arrive. The stronger and the longer lasting the southward B_z component is, the more severe the resulting storm is likely to be. Thus, the drivers of large storms are perhaps the most prominent and therefore most distinguishable structures in the SW. This is indeed the case. In this section, we will discuss a method of predicting large geomagnetic storms that utilizes these recognizable physical features of geoeffective SW structures. The method was first proposed by *Chen et al.* [1996, 1997], and the reader is referred to these papers for details. The basics of the method are reviewed below, and the results of some recent tests are discussed.

The first step is to define SW events. Because geoeffectiveness is mainly determined by the north-south polarity and the magnitude of B_z, we define a SW event as beginning when B_z turns zero at time t_1. The event ends when B_z next turns zero at time t_2, which is the beginning of the next event, and so on. Thus, the SW is taken to consists of a series of B_z events, each of which is characterized by two features: the duration $\tau \equiv t_2 - t_1$ and B_{zm}, defined as the maximum B_z excursion from $B_z = 0$ during the event.

At any time t, the SW detector is in a SW event. The average rate of rotation of the magnetic field in the $y - z$ plane computed from the beginning of the event is given by

$$\dot{\theta}(t) \equiv \theta(t) / \Delta t_z, \tag{2}$$

where $\theta(t)$ is defined by equation (1), $\Delta t_z \equiv t - t_z$ is the time that has elapsed since the beginning of the B_z event at $t_z = t_1$. The subscript z is used to denote quantities defined for B_z events, to be distinguished from those obtained from B_y data. The duration of the B_z event is estimated by

$$\tau(t) = \left| \pi / \theta(t) \right| /. \tag{3}$$

That is, τ' is the estimated time it takes the $y - z$ projection of the magnetic field vector to rotate through π. The estimate is made at time t.

Throughout this chapter, we will use primed symbols to denote quantities that are obtained using only the data available at the time of estimation. Unprimed quantities are obtained from historical data and constitute the prior knowledge of the physical system.

Having observed the value B_z at time t, we estimate the maximum B_z value, B_{zm}, that will occur sometime during the event by

$$B'_{zm}(t) = B_z(t) / f_z(t), \tag{4a}$$

where $f_z(t)$ is a specified form function for B_z events. We will use

$$f_z(t) = \sin(\pi \Delta t_z / \tau'). \tag{4b}$$

The sine function is used for convenience but was found to give reasonable estimates for B_z profiles. This prescription can fail if the time scale of magnetic field rotation varies significantly within an event.

The above algorithm is quite general and is designed to estimate the profile of the B_z component of any unipolar period. If a SW structure is a magnetic flux rope (e.g., a magnetic cloud), the orientation of the flux-rope axis with respect to the ecliptic can be determined by considering the B_y component. First, applying equations (1)-(4) to the B_y data, we can derive analogous expressions for the estimated unipolar B_y duration, τ'_y, and the maximum B_y value, B'_{ym}. We then define

$$\Lambda(t) \equiv B'_{ym} / B'_{zm}, \tag{5}$$

where B'_{ym} and B'_{zm} are understood to be values estimated at time t using equation (4a) and its B_y analogue. It is straightforward to show that $|\Lambda| > 1$ implies that the flux-rope axis is approximately in the ecliptic plane and that if $|\Lambda| < 1$, the axis is roughly normal to the ecliptic.

The pair, $\mathbf{X} = (B_{zm}, \tau)$, is a vector, called a feature vector, and defines a feature space. The ability to predict storms is predicated on being able to recognise geoeffective SW structures in this feature space.

2.3   Bayes Theorem

The basic idea is to use only the real-time SW data (no ground data) and the prior knowledge of the SW to calculate the probability that the event being observed, including the estimated part that is still upstream of the detector, is geoeffective. For this purpose, we need to determine the probability distribution functions (PDFs) to characterize the geoeffective and nongeoeffective SW events. This is accomplished by examining the historical SW data and measuring the duration τ and B_{zm} of each event. First, we must define a criterion for deciding whether a SW event is geoeffective: if a SW event leads to Dst values less than -80 nT for two consecutive hours or longer, we classify it as geoeffective; otherwise the SW event is defined to be nongeoeffective. Figure 3 shows the geoeffective (solid line) and nongeoeffective (dashed line) probability distribution functions according to this two-class classification scheme. Here, each PDF is normalized to unity. This figure has been constructed by adding geoeffective events identified in the WIND/MFI data set during 1995–1999 to the set previously obtained by *Chen et al.* [1997]. In this figure, there are 80 geoeffective and over 8600 nongeoeffective events. Figure 3 provides the *a priori* probabilistic knowledge of the geoeffectiveness of SW events.

To determine from the measured quantities the probability that a given SW event being observed is geoeffective, we must know a number of conditional probabilities. For example, we need to know the probability that certain estimated values of τ' and B'_{zm} will be obtained *given* that a geoeffective (nongeoeffective) event is encountered. Such a PDF can be written as $P(B'_{zm},\tau'\,|\,c_i,\xi')$, where c_i refers to the geoeffective class ($c = e$) or nongeoeffective class ($c = n$). Here, we have defined $\xi' \equiv \Delta t_z / \tau'$ as the fraction of the estimated duration τ' to which the current time t corresponds.

The next step is to determine the probability that the actual values τ, B_{zm}, and class $c = e$ will be obtained having found the estimated values $(\tau'(t), B'_{zm}(t))$ at the current time t. This probability can be obtained by Bayes' theorem [e.g., *Grimmett and Stirzaker*, 1992]:

$$P\big((B_{zm},\tau)\cap e\,|\,B'_{zm},\tau';\xi'\big)$$

$$= \frac{P(B'_{zm},\tau'\,|\,(B_{zm},\tau)\cap e;\xi')P(B_{zm},\tau\,|\,e)P(e)}{\sum_j P(B'_{zm},\tau'\,|\,c_j;\xi')P(c_j)} \tag{6}$$

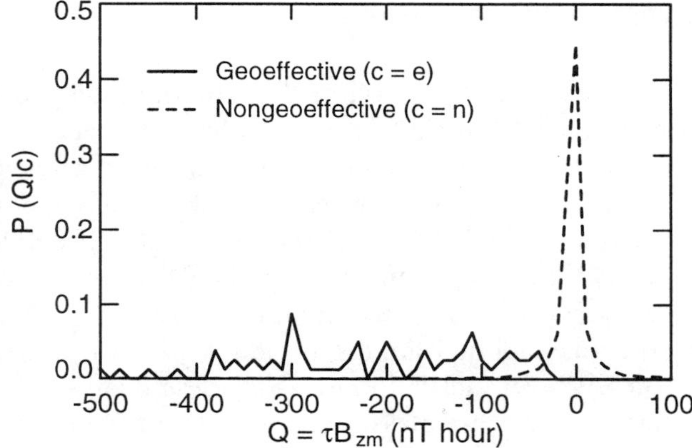

Figure 3. Probability distribution functions of geoeffective and nongeoeffective solar wind events. Geoeffective SW events are those that produce $Dst \leq -80$ nT. There are 80 geoeffective and over 8600 nongeoeffective events.

where the summation in the denominator is over different classes (c_j), and $\cap$ denotes the intersection of two sets. (The rigorous derivation of this equality is given in Appendix A of *Chen et al.* [1997].) For the nongeoeffective class, simply replace e with n. It is important to remember that the PDFs on the right in equation (6) are determined from existing data and have the following meaning: $P(c_i)$ is the unconditional probability of the class c_i, and $P(B_{zm}, \tau \mid e)$ is the conditional probability of finding events with $\mathbf{X} = (B_{zm}, \tau)$ given that the classification is geoeffective.

The PDF $P(B'_{zm}, \tau' \mid (B_{zm}, \tau) \cap e; \xi')$ is the probability of obtaining the estimates $B'_{zm}(t)$ and $\tau'(t)$ at $\xi' = t/\tau'$ given that the event is geoeffective with B_{zm} and τ. The final result, $P((B_{zm}, \tau) \cap e \mid B'_{zm}, \tau'; \xi')$, gives at time t the probability density that an event being examined is geoeffective with the actual values B_{zm} and τ, having obtained the estimates $B'_{zm}(t)$ and $\tau'(t)$.

In practice, one would deal with probability rather probability density. To convert probability density to probability, we integrate equation (6) over some ranges of τ and B_{zm}. For concreteness, we will use

$$P_1 \equiv P(e \mid B'_{zm}, \tau'; \xi') \equiv \int P((B_{zm}, \tau) \cap e \mid B'_{zm}, \tau; \xi') dB_{zm} d\tau \qquad (7)$$

where the integration is carried out over $-\infty < B_{zm} < +\infty$ and $-\infty < \tau < +\infty$. Thus, P_1 is the probability that a SW event is geoeffective with some values of B_{zm} and τ, having obtained the estimates of B'_{zm} and τ'. The integration ranges can be chosen according to particular applications.

3.   PERFORMANCE STATISTICS OF THE PREDICTION METHOD

The most important issue is how accurate the predictions are when real-time SW data are used as input. At the present time, no real-time SW data with which to test the method are available. It has been tested retrospectively, however, using the IMF data obtained by ISEE 3 during 1978 and the WIND/MFI instrument from 1995 to 1999 and is being tested using the near-real time Advanced Composition Explorer (ACE) satellite data at the Space Environment Center (SEC), NOAA, Boulder, CO. Here, we will show some examples of hits, misses, and marginal cases.

First, we must define positive and negative predictions: If P_1 exceeds 0.5 for two hours or longer, it is taken to be a positive prediction; otherwise, it is negative. The geoeffectiveness threshold is $Dst = -80$ nT as defined above.

Figure 4 shows the WIND MFI data (B_y and B_z) and SWE data (V_x) for a four-day period in October 1998. Figures 4a and 4b shows a long-duration strong southward B_z event on 19 October (DOY 292), consistent with a magnetic cloud with its axis nearly normal to the ecliptic. The SW ahead of the magnetic cloud was undisturbed. Figures 4d and 4e show the estimated τ' and B'_{zm} as functions of time. At any time t, the spacecraft is in a SW event, and $\tau'(t)$ and $B'_{zm}(t)$ pertain to the event being sampled at t. Where the magnetic field is fluctuating with small amplitude, the estimated τ' is short and B'_{zm} is small. The τ' plot shows sharp spikes that arise from locally slow variations in B_z. If the average value of B_z remains nearly unchanged for an extended period of time, τ' shows gradual rise, as can be prominently seen on 18 and 19 October. The algorithm to compute B'_{zm} turns each B_z event into a step function. Figures 4d and 4e show that τ' and B'_{zm} correctly describe the actual data. Figure 4f gives the probability P_1 for the computed values of τ' and B'_{zm} given in Figures 4d and 4e. It shows that even though τ' has spikes, the computed probability of exceeding the Dst threshold, $Dst = -80$ nT, is zero until $\approx 04{:}00$ UT, 19 October (Figure 4g). The Dst index reaches and exceeds the specified threshold at about 06:00 UT, attaining -112 nT around 16:00 UT. The computed probability is nearly unity, indicating that the event was clearly geoeffective, outside the overlap of the two PDFs shown in Figure 3.

The SW speed (Figure 4c) remained at about 400 km/s so that it was clearly the southward B_z that was the main contributor to the storm.

170

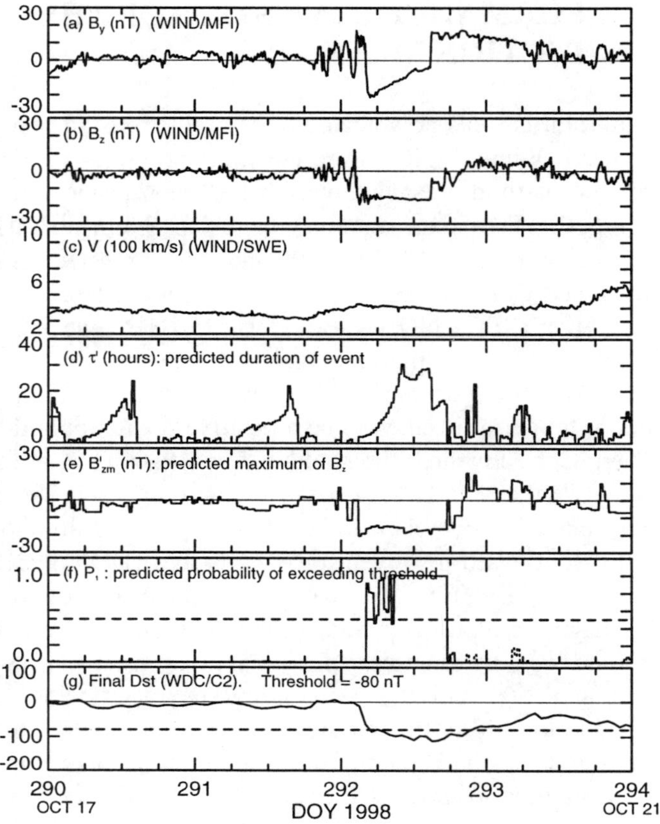

Figure 4. Correct prediction. B_y and B_z indicate a magnetic cloud on 19 October, with its axis nearly normal to the ecliptic. The magnetosphere was quiescent ($Dst \cong 0$) except on 19 October when Dst fell to about -110 nT.

Figure 5 shows the SW data for a four-day period in January 1997, when a "marginal" storm of minimum $Dst \cong -80$ nT occurred in response to a magnetic cloud. A partial halo CME was observed on 6 January 1997 by the Large Angle and Spectrometric Coronagraph (LASCO) on the Solar and Heliospheric Observatory (SOHO) satellite. The magnetic cloud is the presumed SW counterpart of the CME. The field is consistent with a magnetic flux rope with its axis nearly in the ecliptic plane and perpendicular to the Sun-Earth line.

Prior to the magnetic cloud, the solar wind was undisturbed, having left the Sun before the eruption. The estimated τ' and B'_{zm} are shown in Figures 5d and 5e, in the same format as that of Figure 4. The computed probability P_1 exceeds 0.5 for about three hours, so it is a positive prediction according to our definition. The value, however, is only about 0.7, indicating that this event is inside the overlap of the two PDFs in Figure 3. The *Dst* index reaches only -78 nT, so the prediction result is counted as a technical false positive in our binary testing procedure. Nevertheless, the algorithm correctly identified the SW event as a marginal case. We have found that for marginal storms the method predictably yields P_1 that hovers around 0.5.

As an example of false negative, we show a three-day period in April 1995 in Figure 6. The predicted probability P_1 (Figure 6f) during the first half of 7 April shows the behaviour typical of marginal storms discussed above. This response was correct given the moderately strong ($B_{zm} \cong -10$ nT) and long period (≈ 10 hours) of southward B_z; the *Dst* index reached the minimum value, -83 nT, for only one hour on 7 April. The southward B_z event ended shortly thereafter. Subsequently, however, the SW speed increased, and the B_z field fluctuated with large amplitude, remaining negative on the average. This caused the already disturbed magnetosphere to develop a moderately large storm with *Dst* falling to about -150 nT during 19:00 hour UT, 7 April 1995. The predicted P_1 was essentially zero during this period. The reason why this storm was missed is that it was caused by strong fluctuating B_z field coupled with a fast SW impinging on the already disturbed magnetosphere. These physical features are not accounted for in the present implementation.

We have used examples of hits, misses, and false alarms from the recent test using the WIND MFI data from the period 1995–1999 and illustrated the behavior of the prediction algorithm. The results from an earlier, more limited test using ISEE 3 data (days 225–365, 1978) have been previously published in detail [*Chen et al.*, 1997]. In this earlier test, five out of six storms exceeding the *Dst* = -80 nT threshold were correctly identified, with one miss, and one false alarm. In the recent and more extensive test, during the 1995–1999 period for which there were WIND/MFI data, there were 31 storms exceeding the specified threshold, of which 25 storms were correctly predicted and six were missed. There were three false alarms. Thus, there were a total of 34 significant predictions, of which 25 (73%) were correct, 6 (17%) were misses, and 3 (9%) were false alarms. It is significant to note, however, that of the six misses, three were technical misses in which marginal storms were in fact correctly recognised. One of the false alarms, shown in Figure 5, is a marginal storm that was also correctly recognised. These were all counted as incorrect predictions in our binary test, but the algorithm actually produced the correct responses. Thus, it is possible that the accuracy rate can be raised even further if the marginal cases can be correctly and predictably identified.

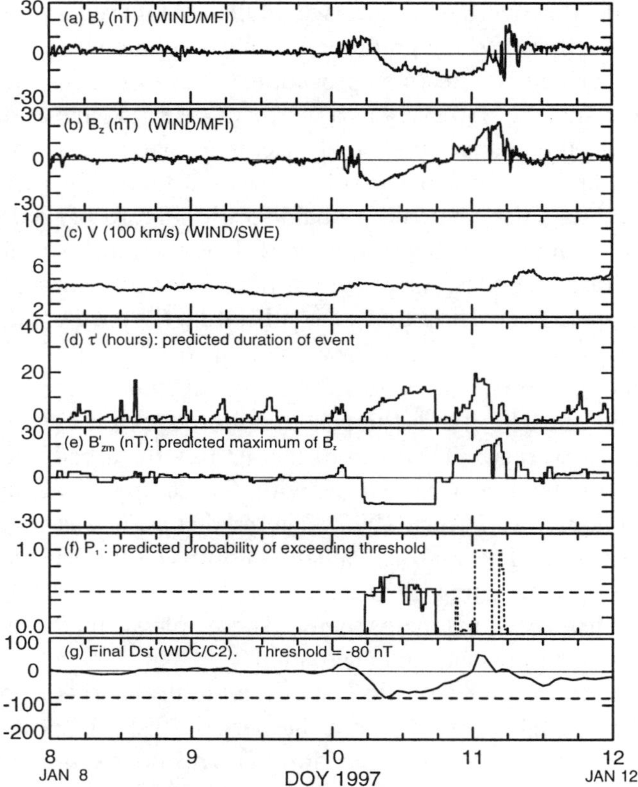

Figure 5. A "marginal" storm and SW on 8 – 11 January 1997(WIND/MFI data). B_y and B_z indicate the occurrence of a magnetic cloud on 10 January. The *Dst* index shows that the magnetosphere was quiescent except on 10 January when *Dst* fell to about −78 nT. This storm and the associated CME were studied earlier [*Chen et al.*, 1999].

4.   SUMMARY AND DISCUSSION

This chapter has described a method of accurately predicting the occurrence, severity, and duration of large geomagnetic storms. The algorithm is based on quantifying and recognising geoeffective physical features in the SW upstream of the magnetosphere, i.e., in the SW that has yet to arrive. This distinguishes the method from those that use the input SW or geomagnetic index time series to predict the response of the magnetosphere. Because the SW ahead of a geoeffective period (e.g., a CME ejecta) contains little information concerning what is to follow, the warning time obtainable from these methods is intrinsically limited to the SW transit time from the monitoring position to the magnetosphere, which is significantly less than 1 hour for large storms caused by fast CMEs. For the Bayesian method described here based on estimating and recognising the geoeffective SW upstream of the monitoring position, the forecasting time is potentially much longer, several hours to possible more than ten hours. Such long warning times are possible because the SW structure causing large storms have strong IMF with long (up to 1/2 AU) spatial and temporal correlation. Such storms are relatively rare but are the most damaging of space weather effects. The present method is particularly advantageous for predicting large storms.

The results of the recent tests show that significantly higher accurate rates than those of the present methods based only on remote-sensing solar observations may be consistently achieved. The method is currently being tested at SEC in a more operational environment. The near-real time SW data from the ACE satellite positioned at $L1$ are used for this purpose. The initial results appeared to be consistent with those obtained using the WIND data described in this chapter.

The discussion of Figure 5 points to a number of areas where the algorithm may be improved. For example, it is well known that the initial state of the magnetosphere critically determines its response to a given SW input. Thus, the time history of magnetospheric activity needs to be incorporated. Increased speeds coupled to a southward B_z field can enhance the geoeffectiveness. The interaction of the geoeffective solar ejecta (e.g., magnetic clouds) with the ambient SW can modify the IMF strength and profile, so that the SW preceding or following the ejecta can be important. For example, *Fenrich and Luhmann* [1998] found that a significant fraction of magnetic clouds with north-south polarity tend to have stronger southward IMF because they are compressed by faster SW behind them. This is in spite of the fact that the flux rope expansion would tend to reduce the field in the trailing half of a magnetic cloud. This is favourable for the present method because the leading northward B_z segment can provide more than 10 hours of forecasting time before B_z turns southward.

Figure 2 shows an example of such a magnetic cloud: Application of the method would have provided a warning time of ≈ 15 hours [*Chen et al.*, 1997]. In addition, a better understanding of the ejecta-SW interactions and their consequences on the IMF profile imposed on the magnetosphere can help improve the prediction results where the incident SW structures show complexity arising from interactions with the SW. Some simulation studies of CME-SW interactions [e.g., *Cargill et al.*, 1995, 1996; *Odstrcil and Pizzo*, 1999] are beginning to uncover the physical processes involved in the propagation of CMEs through the interplanetary medium.

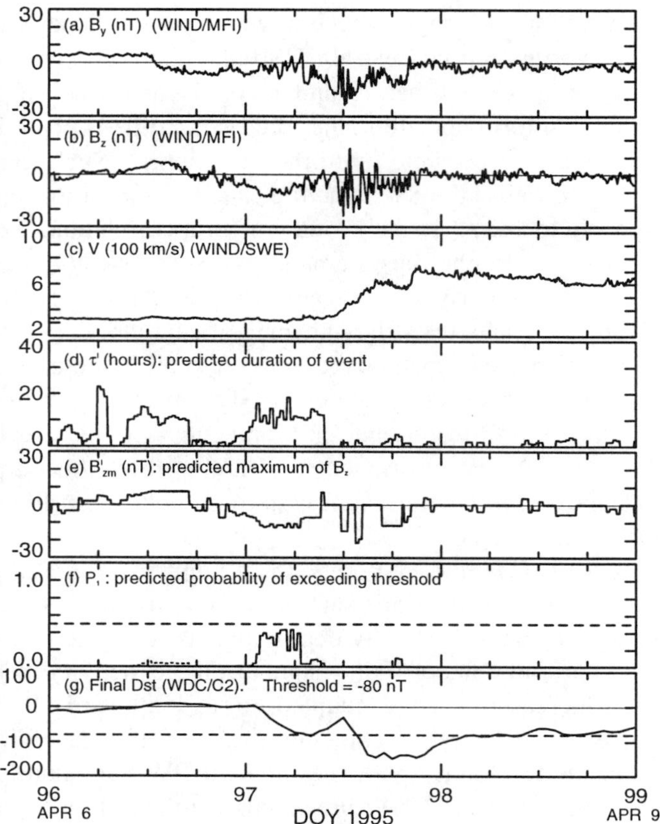

Figure 6. A false negative. 7 April 1995. The *Dst* index shows that the magnetosphere was quiescent prior to 7 April when *Dst* fell to about −83 nT and then to −150 nT. The second large storm was completely missed.

In this chapter, we have discussed the basic tenets of the new method. We have used only B_y and B_z. The geoeffectiveness of a SW structure, however, depends additionally on the SW speed and density. Preliminary studies have shown that inclusion of the SW speed as an additional feature can improve the prediction accuracy especially in marginal cases. There are also separate efforts to predict the SW speed (i.e., the fast SW) based on observations of large-scale solar surface magnetic fields [*Wang and Sheeley*, 1990; *Wintoft and Lundstedt*, 1999]. The fast SW tends to be more closely associated with small to moderate storms but does not by itself cause large storms. Large storms can be caused by strong southward IMF in the slow SW, but prolonged and strongly southward IMF structures are typically associated with fast CMEs. Thus, the SW speed may be a useful feature to incorporate into the prediction algorithm.

5.   ACKNOWLEDGEMENTS

This work was supported by ONR. I gratefully acknowledge the use of the IMP 8 data provided by NASA/GSFC IMP 8 team and the UCLA Space Physics Data Center, the IMF and SW data provided by the WIND/MFI and WIND/SWE teams, respectively, of GSFC, and the *Dst* index provided by the World Data Center-C2, Kyoto University.

6.   REFERENCES

Allen, J., H. Sauer, L. Frank, and P. Reiff, Effects of the March 1989 Solar Activity, *EOS, 70*, 1479, 1989.

Birkeland, K., *The Norwegian Aurora Polaris Expedition 1902-1903*, A. Aschehoug, Christiania, 1908.

Burton, R. K., R. L. McPherron, and C. T. Russell, An empirical relationship between interplanetary conditions and *Dst, J. Geophys. Res., 80*, 4204, 1975.

Cargill, P. J., J. Chen, D. S. Spicer, and S. T. Zalesak, Magnetohydrodynamic simulations of the motion of magnetic flux tube, *J. Geophys. Res., 101*, 855, 1996.

Cargill, P. J., J. Chen, D. S. Spicer, and S. T. Zalesak, Geometry of Interplanetary Magnetic Clouds, *Geophys. Res. Lett., 22*, 647, 1995.

Carrington, R. C., Description of a singular appearance seen on the Sun on September 1, 1859, *Mon. Not. R. Astron. Soc., 20*, 13, 1860.

Chapman, S., The electric current-systems of magnetic storms, *Terrest. Magn., 40*, 349, 1935.

Chen, J., P. J. Cargill, and P. J. Palmadesso, Real-time identification and prediction of geoeffective solar wind structures, *Geophys. Res. Lett., 23*, 625, 1996.

Chen, J., P. J. Cargill, and P. J. Palmadesso, Predicting solar wind structures and their geoeffectiveness, *J. Geophys. Res., 102*, 14701, 1997.

Chen, J., S. P. Slinker, and P. J. Cargill, Prediction of geomagnetic storms associated with the halo CMEs of January and April 1997, in *Proc. Workshop on Space Weather*, ESA WPP-155, p. 403, 1999.

Fenrich, F. R., and J. G. Luhmann, Geomagnetic response to magnetic clouds of different polarity, *Geophys. Res. Lett., 25*, 2999, 1998.

Gonzalez, W. D., and B. T. Tsurutani, Criteria of interplanetary parameters causing intense magnetic storms (*Dst* < −100 nT), *Planet. Space Sci., 35*, 1101, 1987.

Gonzalez, W. D., J. A. Joselyn, Y. Kamide, H. W. Kroehl, G. Rostoker, B. T. Tsurutani, and V. M. Vasyliunas, What is a geomagnetic storm?, *J. Geophys. Res., 99*, 5771, 1994.

Grimmett, G. R., and D. R. Stirzaker, *Probability and Random Processes*, p. 22, Clarendon Press, Oxford, UK, 1992.

Hernandez, J. V., T. Tajima, and W. Horton, Neural net forecasting for geomagnetic activity, *Geophys. Res. Lett., 20*, 2707, 1993.

Hirshberg, J., and D. S. Colburn, Interplanetary field and geomagnetic variations-A unified view, *Planet. Space Sci., 17*, 1183, 1969.

Hodgson, R., On a curious appearance seen in the Sun, *Mon. Not. R. Astron. Soc., 20*, 16, 1860.

Joselyn, J. A., Geomagnetic activity forecasting: The state of the art, *Rev. Geophys., 33*, 383, 1995.

Kappenman, J. G., and V. D. Albertson, Bracing for the geomagnetic storms, *IEEE Spectrum*, March, 1990.

Kappenman, J. G., An introduction to power grid impacts and vulnerabilities from space weather, this volume, 2001.

Lanzerotti, L. J., Geomagnetic induction effects in ground-based systems, *Space Sci. Rev.*, *34*, 347, 1983.

Lanzerotti, L. J., Space weather effects on communications, this volume, 2001.

Lundstedt, H., Neural networks and predictions of solar-terrestrial effects, *Planet. Space Sci.*, *40*, 457, 1992.

Neupert, W. M., and V. Pizzo, Solar coronal holes as sources of recurrent geomagnetic disturbances, *J. Geophys. Res.*, 79, 3701, 1974.

O'Brien, T. P., and R. L. McPherron, An empirical phase space analysis of ring current dynamics: Solar wind control of injection and decay, *J. Geophys. Res.*, *105*, 7707, 2000.

Odstrcil, D., and V. J. Pizzo, Distortion of the interplanetary magnetic field by three-dimensional propagation of coronal mass ejections in a structured solar wind, *J. Geophys. Res.*, *104*, 28225, 1999.

Perreault, P. and S.-I. Akasofu, A study of geomagnetic storms, *Geophys. J. R. Astron. Soc.*, *54*, 547, 1978.

Price, C. P., and D. Prichard, The non-linear response of the magnetosphere: 30 October 1978, *Geophys. Res. Lett.*, *20*, 771, 1993.

Rostoker, G., and C.-G. Falthammar, Relationship between changes in the interplanetary magnetic field and variations in the magnetic field at the Earth's surface, *J. Geophys. Res.*, 72, 5853, 1967.

Russell, C. T., R. L. McPherron, and R. K. Burton, On the cause of geomagnetic storms, *J. Geophys. Res.*, *79*, 1105, 1974.

Sharma, A. A., D. V. Vassiliadis, and K. Papadopoulos, Reconstruction of low-dimensional magnetospheric dynamics by singular spectrum analysis, *Geophys. Res. Lett.*, *20*, 335, 1993.

Sheeley, N. R., Jr., J. W. Harvey, and W. C. Feldman, Coronal holes, solar wind streams, and recurrent geomagnetic disturbances: 1973–1976, *Solar Phys.*, *49*, 271, 1976.

Tsurutani, B. T., W. D. Gonzalez, Y. Kamide, and J. K. Arballo, editors, *Magnetic Storms*, Geophys. Monogr. 98, American Geophysical Union, Washington, DC, 1997.

Valdivia, J. A., S. A. Sharma, and K. Papadopoulos, Prediction of magnetic storms by nonlinear models, *Geophys. Res. Lett.*, *23*, 2899, 1996.

Vassiliadis, D., A. J. Klimas, D. N. Baker, and D. A. Roberts, A description of solar wind-magnetosphere coupling based on nonlinear filters, *J. Geophys. Res.*, *100*, 3495, 1995.

Wang, Y. M., and N. R. Sheeley, Jr., Solar wind speed and coronal flux-tube expansion, *Astrophys. J.*, *355*, 726, 1990.

Wintoft, P., and H. Lundstedt, A neural network study of the mapping from solar magnetic fields to the daily average solar wind velocity, *J. Geophys. Res.*, *104*, 6729, 1999.

Wu, J.-G., and H. Lundstedt, Prediction of geomagnetic storms from solar wind data using Elman recurrent neural networks, *Geophys. Res. Lett.*, *23*, 319, 1996.

Wu, J.-G., H. Lundstedt, P. Wintoft, and T. R. Detman, Neural network models predicting the magnetospheric response to the 1997 January halo-CME event, *Geophys. Res. Lett.*, *25*, 3031, 1998.

Chapter 7

Coronal Mass Ejections at the Sun and in Interplanetary Space

A review of CMEs in the solar corona and in the solar wind

Peter J. Cargill
Space and Atmospheric Physics, The Blackett Laboratory, Imperial College
London SW7 2BZ, United Kingdom

Abstract This chapter reviews the properties of coronal mass ejections (CMEs) in the solar corona, solar wind and interplanetary space. CMEs are now widely believed to be responsible for the most severe geomagnetic storms and are consequently the major solar driver of space weather. As seen in coronagraphs, CMEs involve an expulsion of solar plasma (and magnetic field) into interplanetary space at speeds that lie anywhere between 100 and 1500 km/s. The largest CMEs are as energetic as a major solar flare, but are not caused by large flares. Instead, it is now clear that CMEs and flares are both phenomenon arising from a large-scale destabilisation of the coronal magnetic field. A common (but not unique) CME scenario involves the eruption into space of a three-part structure comprising of an inflated helmet streamer (which leads), and a prominence cavity and associate prominence (which trail). The prominence cavity is believed to be a large magnetic flux rope. In the interplanetary medium, CMEs are detected at 1 AU 2 – 3 days after they leave the Sun. They are often preceded by an interplanetary shock, but their speeds are generally within 100 km/s of the ambient solar wind speed, indicating that a significant interaction between CME and solar wind has occurred. A significant fraction (30-40%) of interplanetary CMEs (ICMEs) have a geometry consistent with a magnetic flux rope, and are referred to as magnetic clouds. The interplanetary flux rope is likely to be the same structure as formed the prominence cavity at the Sun. Magnetic clouds are distinguished by the appearance in many cases of prolonged (many hours) periods of Southward interplanetary magnetic field, and are hence responsible for major geomagnetic storms.

I.A. Daglis (ed.), Space Storms and Space Weather Hazards, 177–207.

178

Keywords   Coronal mass ejections, geomagnetic disturbances, magnetic clouds, magnetic flux ropes, coronagraphs, SOHO.

1.   INTRODUCTION

Coronal mass ejections (CMEs) are almost certainly the most important solar phenomena from the point of view of space weather, but in the overall context of solar physics research, they have had rather a brief history. Although it was clear from eclipse photographs, and especially the first ground-base coronagraphs, that the solar atmosphere undergoes frequent eruptions that eject large volumes of plasma and magnetic field into the corona, it was only with the advent of space-based coronagraphs that the ubiquity of these eruptions became clear. Indeed the year 2001 marks the 30th anniversary of the first space-based observation of a CME (Tousey, 1973).

While it has been known for many years that some solar flares could be associated with geomagnetic disturbances, it was only recently that the true association between CMEs and geomagnetic activity was established (e.g. Gosling, 1993). Gosling et al. (1991) showed that major geomagnetic storms (defined as having a Kp index of 7- or greater) had an excellent association with the interplanetary manifestations of CMEs, and with interplanetary shocks. CMEs are geo-effective because they can contain long periods (many hours) of southward interplanetary magnetic field (IMF), which enhances the transfer of energy from the solar wind to the magnetosphere. The shocks associated with CMEs are also able to accelerate promptly large numbers of energetic protons (e.g. Reames, 1999).

These considerations have lead to a downplaying of the importance of solar flares as a cause of major geomagnetic activity. When CMEs were first detected at the Sun in considerable numbers by the coronagraph on the Skylab observatory, it was argued by some workers that they were simply the large-scale coronal response to a large solar flare (e.g. Dryer et al., 1979). Work by Harrison (1986) and Harrison et al. (1990) using data from the Solar Maximum Mission (SMM) showed that this was incorrect: in the cases that Harrison looked at, the initial phase of a large flare (as seen in hard X-rays with energies in the range 10-30 keV) occurred after the outward motion of the CME had commenced. In fact, data from the current generation of space missions (SOHO, Yohkoh) suggests that flares and CMEs are related intimately, with both being the consequence of a large-scale destabilisation of the magnetic field in the Sun's corona. Indeed, the fact that the solar corona is highly dynamic has gradually become clear over the past three decades of observations by space-based coronagraphs, with ejections of mass taking place on a wide range of spatial and temporal scales. Instruments observing the solar disk in the EUV and in X-rays have shown that the coronal activity seen by the coronagraphs often has distinct manifestations at these wavelengths.

CMEs have been detected in interplanetary space for many years using in-situ measurements of magnetic fields and particle populations. In many cases, a clear connection has been established between a CME seen at the Sun, and one detected in the interplanetary medium. This has become especially true since the detection of Earthward-directed CMEs by the Large Angle Spectrometric Coronagraph (LASCO) on the SOHO satellite (see especially Webb et al., 2000). An important aspect of the present generation of CME observations has been the presence of a dedicated monitoring satellite (Advanced Composition Explorer: ACE) located at the L1 point, 215 R_e upstream of the Earth. This has, for the first time, given scientists the ability to observe an Earthward-directed CME leave the Sun, and 2 – 4 days later measure its plasma and magnetic field properties at 1 AU, and makes at least the benchmarking of space weather forecasting tools a possibility.

This chapter aims to describe the properties of CMEs at the Sun and in interplanetary space, drawing from spacecraft data (both remote sensing and in-situ observations). The emphasis is not on showing attractive pictures (these are available elsewhere), but on describing measurements of fundamental physical quantities (e.g. velocities, densities and magnetic fields), and how these may be interpreted in terms of robust physics-based theories. Section 2 addresses CMEs at the Sun. Observations from the three generations of space-based coronagraphs are reviewed, as are the theories that try to account for these eruptions. Section 3 addresses the interplanetary aspect of CMEs, and shows how one can relate what leaves the Sun to what is seen in space. An important aspect is the use of data from the entire heliosphere to constrain models. Section 4 addresses the future prospects for CME research, and how CMEs fit into space weather.

2. CORONAL MASS EJECTIONS AT THE SUN

A working definition of a CME as observed at the Sun has been proposed by Hundhausen (1993): "an observable change in coronal structure that (1) occurs on a timescale between a few minutes and several hours and (2) involves the appearance of a new discrete, bright white-light feature in the coronagraph field of view". It is sensible to add to this definition the fact that a CME involves the expulsion of plasma and magnetic field from the solar corona into interplanetary space. This is best illustrated by an example. Figure 1 shows a CME observed on August 18 1980 by the coronagraph on the Solar Maximum Mission (SMM). The pre-eruption coronal configuration consists of a helmet streamer, which gradually inflates (panel 1). At 11:54 UT a bright rim appears as the streamer vanishes, and the rim moves out into interplanetary space. Behind it (at 12:15 UT), the prominence material follows as the dense bright blob. Between the bright rim and the prominence is a darker region, believed to be an organised magnetic structure responsible for the prominence support. Finally, in panel 4 the white rim has moved out of the field of view and the prominence is all that remains visible. For further discussion of this event, see Illing and Hundhausen (1986). For the author, this set of observations summarises much of what is important about CMEs. There is a definite ejection of material (and hence magnetic field) into space, there is the suggestion that organised magnetic structures, prominences are involved, and the whole process is clearly "eruptive" or "explosive" in the sense that it occurs on the transit timescale of Alfvén and sound waves. It is clearly not a series of equilibrium configurations.

CMEs at the Sun are best observed by coronagraphs. While there are still a number of ground-based coronagraphs operating, such as the High Altitude Observatory (HAO) facility on Mauna Loa and the new Mirror Coronagraph in Argentina (MICA), the major scientific advances in the past 30 years have been made from space due to the larger potential field of view. There have been four major space-based coronagraphs in the past 30 years. They differ mainly in their field of view, and sensitivity, with the newest generation obtaining images both closer to the Sun and farther into space, all with greater sensitivity. The Skylab manned observatory operated between 1973 and 1974 (near the minimum of the solar cycle) and the coronagraph on board had a field of view from $1.5 - 5\ R_s$. The Solwind coronagraph on the P78-1 spacecraft operated between 1978 and 1985 (the maximum and declining phase of the cycle) and had a field of view from 2.5 and $10\ R_s$. SMM operated in 1980 (the maximum) and 1984-1989 (the minimum and ascending phase of the cycle) and its coronagraph had a field of view from $1.6 - 5\ R_s$, and the three coronagraphs that comprise LASCO (C1, C2, C3) on the SOHO spacecraft have operated since 1996 (the ascending and maximum of the present cycle) and have fields of view from $1.1 - 3$, $1.5 - 6$ and $4 - 30\ R_s$ respectively. It is to be hoped that LASCO can continue to operate until the minimum of the cycle in 2005/2006 in order to provide a data set from a complete solar cycle with the same instrument.

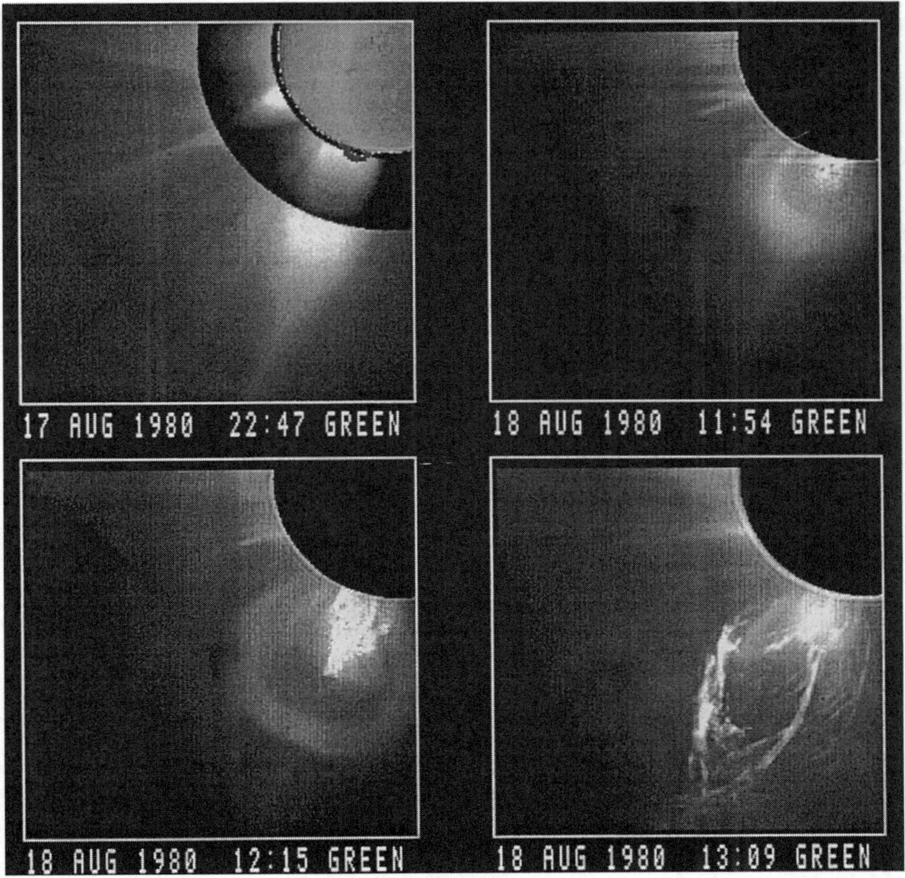

Figure 1. The four panels show the CME observed by the SMM coronagraph on August 18, 1980. The first panel shows a large coronal streamer on the East limb of the Sun, which has gradually become inflated. In the second panel, the CME has commenced, with the streamer structure disappearing, and a bright rim is moving out from the Sun. In the third panel, the bright rim has moved further out, behind it is a dark structure (the cavity) and behind that is cool, dense prominence material. In the final panel, the prominence is now moving through the corona. [Figure courtesy of High Altitude Observatory.]

The remainder of this section addresses the solar aspects of CMEs. Their properties can be split up into a number of categories: morphology (Section 2.1), rate, location and size (Section 2.2), physical properties (mass, velocity etc.: Section 2.3). In addition, we address two important results from the LASCO observations: the paradigm that magnetic flux ropes are a major part of CMEs (Section 2.4) and the observations of Earthward-directed (halo) CMEs (Section 2.5). We then comment on signatures of CMEs seen on the solar disk in X-rays and the extreme ultraviolet (Section 2.6) and finally discuss briefly our theoretical understanding of what causes CMEs (Section

2.7). We also refer the reader to the excellent review of CMEs in the SMM epoch (Hundhausen, 1999), and the first major statistical survey of the SOHO epoch (St Cyr et al., 2000).

2.1   Morphology

CMEs come in all shapes and sizes, and seem to resist a clear categorisation, with results from different instruments showing major apparent discrepancies. Some of the terms that arise in the literature include "loops, clouds, bubbles, halos and spikes". Yet for each epoch, a paradigm for CMEs has emerged, even though that particular type of CME is sometimes a small fraction of the total sample. For example, a survey of Skylab results (Munro and Sime, 1985) examined 77 major events seen during 1973/74, and split them into 6 categories (plus a seventh, called "unclassifiable"). The largest class were referred to as loops (20 events), and comprised a bright loop-like structure of material moving outward, with a rarefaction region often immediately behind the loop. This led to a dominant paradigm of the "loop transient". It is interesting to note that the analogous results for the first 3 years of the Solwind instrument (Howard et al., 1985) found that 2% of CMEs qualified as loops. However, the vast majority of their major events were categorised as "curved fronts" (40%), which are effectively loops without a trailing edge, so the pictures were perhaps not too inconsistent.

The most important morphological paradigm to emerge in recent years has been the "three part CME structure" (e.g. Illing and Hundhausen, 1985, 1986; Low, 1996; Hundhausen, 1999). The earliest manifestation of this appears to have been the CMEs of August 5 and 18 1980, which were both observed by SMM. It was noted that the pre-eruption configuration consisted of a prominence, above which was a dark region, known as the prominence cavity, with both being located at the base of a large helmet streamer (see upper left panel of Figure 1). As the eruption proceeded (see remainder of Figure 1), a three-part structure corresponding to these regions was seen moving into space. Further evidence for this configuration has been found in numerous other SMM observations, in more recent eruptions seen by LASCO (e.g. Chen et al., 2000), and in EUV and X-ray observations (Dere et al., 1997; Nitta and Akiyama, 1999).

It was suggested by Low (1996: see also Hundhausen, 1999) that the prominence cavity was a large magnetic flux rope, with the prominence being suspended at the bottom. [A flux rope is an organised region of magnetic field, topologically distinct from the surrounding field and plasma, that has field-aligned electric currents, and hence a twisted magnetic field.] Other than the prominence, the flux rope contains only low-density plasma, and so appears as a dark region (cavity) in coronagraph images. [This picture becomes especially attractive when one takes into account that a significant fraction of interplanetary CMEs also appear to be magnetic flux ropes.] The cause of eruption in this scenario is unknown, but Hundhausen (1993) has noted that the pre-eruption streamers undergo inflation when looked at with synoptic maps, with the inflation being terminates by the eruption. These were named "bugles" (Hundhausen, 1993), and led to the suggestion that streamer inflation and subsequent blow-out was an important ingredient of CMEs: this view has been challenged recently by Subramanian et al (1999). Of course it should also be pointed out that many CMEs do not have this 3-part structure.

2.2   Rate, location and size

The rate of CMEs depends on the phase of the solar cycle. Events occur at a rate of roughly 0.5 per day at solar minimum and between 2 and 3 a day at solar maximum (Hundhausen, 1993; Webb and Howard, 1994; St Cyr et al., 2000). This does not seem to have changed between solar cycles 22 and 23.

CMEs typically come from near the equator at solar minimum, and occur over a wider range of latitudes at solar maximum (Hundhausen, 1993; St Cyr et al., 2000), on occasions occurring north or south of 60°. According to Hundhausen (1993), the distribution of latitudes does not correspond to those of sunspots of flares, but does correspond to larger structures such as prominences and a bright background corona and streamers.

CMEs are massive structures. Hundhausen (1993) noted that the distribution of angular widths peaks at around 40°, with a tail extending up to 100°. St Cyr et al (2000) found a similar result, but with a more extensive tail, with events having widths between 100 and 360°. It is almost certain that these wide CMEs are Earthward-directed events filling a large part of the sky. Such "halo CMEs" will be discussed later in Section 2.5.

2.3   Mass, velocity, acceleration and energy

Estimates of the average mass and energy of a CME has varied little between the various data sets. Hundhausen (1997) quotes 4×10^{15} g for Solwind and 2.5×10^{15} g for SMM events. A survey of a limited class of LASCO events by Vourlidas et al. (2000) suggests a number in the same region. If we recall that the total solar wind mass loss is about 10^{17} g per day, we see that CMEs are a small contributor to the total solar mass (and angular momentum) loss.

CME velocities span a wide range of values, and are usually defined as being in the plane of the sky. SMM results based on 673 CMEs showed a peak at about 350 km/s, but with a distribution extending below 100 km/s and up to 2000 km/s (Hundhausen et al., 1994; Hundhausen, 1997). The peak is somewhat lower than that measured by Solwind (Howard et al., 1985). It was noted that there were significant annual variations, but these were not related to the solar cycle in any meaningful way. It is also notable that the peak of the distribution lies below typical interplanetary solar wind speeds: we will return to this point in Section 3. Two extensive studies of CME speeds have been carried out using the LASCO database. Using similar methods to that of Hundhausen et al., (1994), St Cyr et al., (2000) examined the speed of 640 CMEs occurring between 1996 and 1998. Their distribution of CME speeds was very similar to that found by Hundhausen et al., (1994), with a peak near 300 km/s, and a long tail extending above 1500 km/s. However, Sheeley et al., (1999) used a differencing method to study the motion of CMEs, and, using a more limited sample, claimed that CMEs fell into two different types (see also MacQueen and Fisher, (1983), for a similar conclusion based on Mauna Loa observations). These were: gradual CMEs with a speed in the range 400 – 600 km/s and apparently associated with prominence eruptions and secondly, fast CMEs with speeds in excess of 750 km/s, and associated with flares. The reason for the difference between these two analysis is unclear at this time.

A fundamental paradigm concerning CMEs is that they are accelerated from the solar corona and into the solar wind by magnetic forces. While the absence of magnetic field measurements in the corona obviously limits an assessment of the reality of this claim, high quality data concerning the motion and especially the acceleration of CMEs is useful. An early study used results from the Mauna Loa observatory (MacQueen and Fisher, 1983) and considered motion as far as 2.4 R_s. They noted that flare associated events showed the highest speeds (see above), but little acceleration within the field of view, whereas those associated with prominence eruption exhibited large acceleration.

The high sensitivity and large field of view of the SOHO coronagraphs has made possible the study of CME acceleration to large radial distances. Simnett et al., (1997) showed two CMEs where outward acceleration was clearly operative over a large distance in the C2 and C3 fields of view. However, St Cyr et al., (2000) were able to ascertain acceleration reliably in only 17% of CMEs. When averaged over the whole C2 and C3 field of view they found accelerations in the range $1.4 – 49$ m/s^2. They found no cases of deceleration. Sheeley et al. (1999) found accelerations of similar magnitudes to St Cyr et al, but also found cases where deceleration occurred. These latter events were mostly associated with fast CMEs moving perpendicular to the plane of the sky, whereas fast CMEs in the plane of the sky travelled at roughly constant velocity. Sheeley et al suggested that the CMEs moving perpendicular to the plane of the sky were in fact being decelerated at much larger radial distances (> 50 R_s), which would not be seen in the field of view when the CME is in the plane of the sky.

The results concerning acceleration and deceleration are somewhat puzzling. A CME moving through the solar corona is subjected to three forces: an outward Lorentz force, an inward gravitational force, and a net drag force that parameterises the interaction of the body with the ambient solar wind: this can either be outward or inward depending on the relative motion of the CME with respect to the wind (e.g. Cargill et al., 1995; Chen, 1996). Therefore one would expect to see fast CMEs undergo a swift and drastic deceleration unless they are either very dense (in which the drag force is ineffective), or they are being strongly driven by magnetic forces within the coronagraph field of view.

Finally, in order to assess the processes responsible for the CME motion, estimates of the energies are needed. The kinetic and potential energies are straightforward to obtain from coronagraph data. The three data sets mentioned above give an average kinetic energy of close to 3×10^{30} ergs: SMM gave a average potential energy of 5.4×10^{30} ergs, and LASCO somewhat less (Vourladis et al., 2000). However, in view of the fact that CMEs are almost certainly driven by forces associated with the magnetic field, it is the magnetic energy that is of major interest. Unfortunately, direct measurements of coronal magnetic fields are impossible due to the swamping of the Zeeman effect by thermal line broadening. Attempts to estimate the magnetic energy in CMEs by the extrapolation of interplanetary magnetic field data back to the Sun (Vourlidas et al., 2000) raise more questions than they answer.

2.4   Magnetic flux ropes in CMEs

The three-part CME paradigm proposed in the SMM era has naturally prompted a search for flux rope-like structure in coronagraph data. Perhaps the earliest report of such a structure was by Illing and Hundhausen (1983), and although it was identified at the time as evidence for magnetic reconnection, Dere et al., (1999) suggest that a flux rope is a better interpretation. There have been a number of detailed studies that have used LASCO data to identify magnetic flux ropes in CMEs which exhibit evidence of magnetic connection to the Sun (Chen et al., 1997b, 2000; Wood et al., 1999; Dere et al., 1999). The feature that is sought is a circular intensity pattern that moves out through the corona. A variety of image enhancement and differencing methods have identified these structures (e.g. Dere et al., 1999; Wood et al., 1999), and it now appears clear that flux ropes do indeed exist as part of CMEs in the inner corona. Indeed Dere et al.,

186

(1999) estimated that in the first two years of operation, between 25 and 50% of CMEs had helical structure.

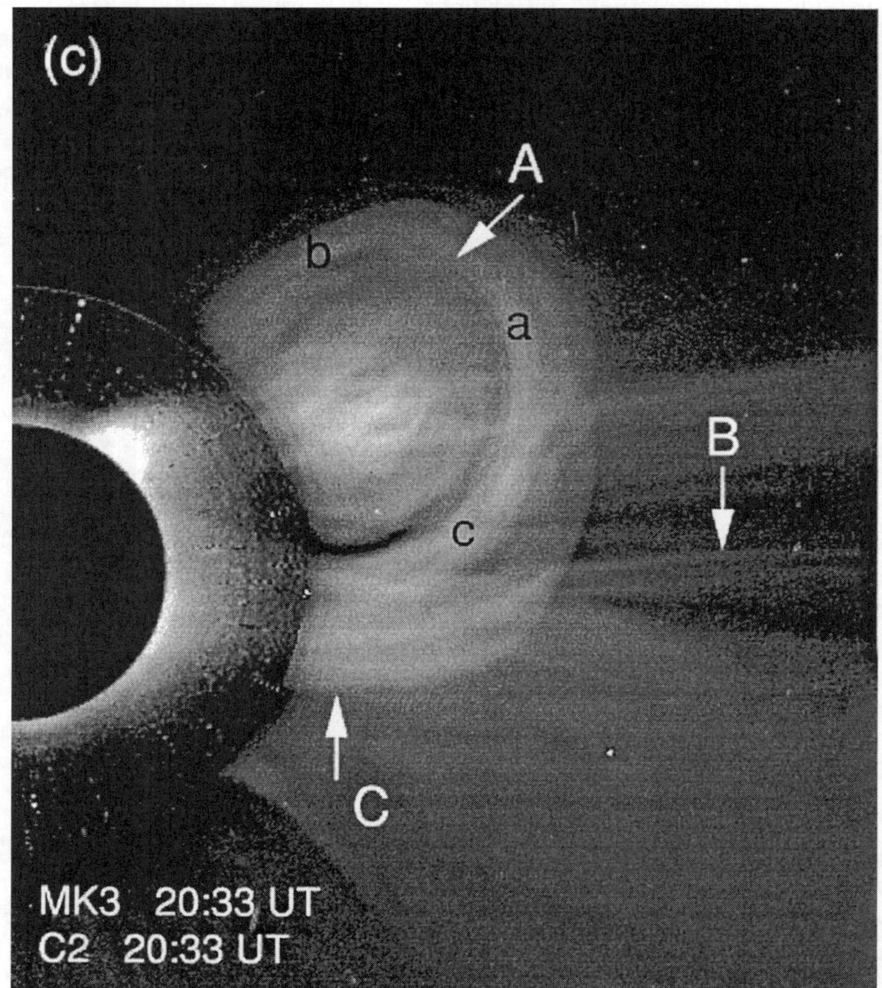

Figure 2: The CME observed on September 9, 1997 (Chen et al., 2000). The CME first became visible in the LASCO field of view at approximately 20:00, and this image is roughly 1 hour later. The inner part of the image is from the HAO coronagraph on Mauna Loa, and the outer part from the LASCO C2 coronagraph. Note the almost circular structure upper right of the centre of the picture. This is evidence for the presence of a magnetic flux rope. (Figure courtesy of J. Chen and the LASCO consortium.]

2.5   Halo CMEs

From the space weather viewpoint, one would like to be able to observe Earthward-directed CMEs. For several reasons, it is much easier to observe CMEs directed perpendicular to the Sun-Earth line than along it (e.g. Hundhausen et al., 1994). However, one might expect a CME directed Earthward to be visible around a large fraction of the field of view of the coronagraph. The first detection of such an Earthward-directed CME was reported by Howard et al., (1982), using data from the Solwind coronagraph. They noted that such CMEs should appear as a "halo" around the Sun, and such halo CMEs have been detected extensively by the LASCO coronagraphs on SOHO (e.g. Webb et al., 2000). Figure 3 shows a typical example with a faint structure being evident around much of the solar limb. It should be stressed that from a physical viewpoint, halo CMEs cannot be different from any others (although see comments in Section 2.3).

From the space weather viewpoint, the major interest in halo CMEs is the possibility that warnings of hazardous space weather can be issued when a halo CME meeting a certain criterion is observed. Using criterion proposed by Thompson et al., (1999b), St Cyr et al (2000) identified 92 halo CMEs in their data set, of which roughly half were not Earthward directed, since their point of origin was probably on the backside of the Sun. During this time interval, there were 21 major geomagnetic storms (defined as having a Kp index ≥ 6), and 15 of these could be associated reasonably with a halo CME. Of the 6 "missed predictions", 3 occurred in SOHO data gaps, so the success rate was 15/18 (St Cyr et al., 2000). However, the worrying feature of St Cyr et al.'s analysis was the fact that 25 of the halo CMEs did not lead to major geomagnetic activity: in other words they were false alarms. This suggests that (a) new criteria need to be developed when issuing alerts based on halo CMEs and/or (b) halo CMEs should trigger a lower level warning as opposed to an alert.

2.6   Non-coronagraph detection of CMEs

Eruptive solar phenomena have been observed without coronagraphs for many years. Eruptive prominences, surges, sprays, and a variety of jets are some of the more common features. An excellent review of the topic can be found in Hudson and Cliver (2000), who cover a fuller range of phenomena than there is space for here. In the past decade, results from the Yohkoh and SOHO missions have revealed a number of possible proxies on the solar disc indicative of CME onset.

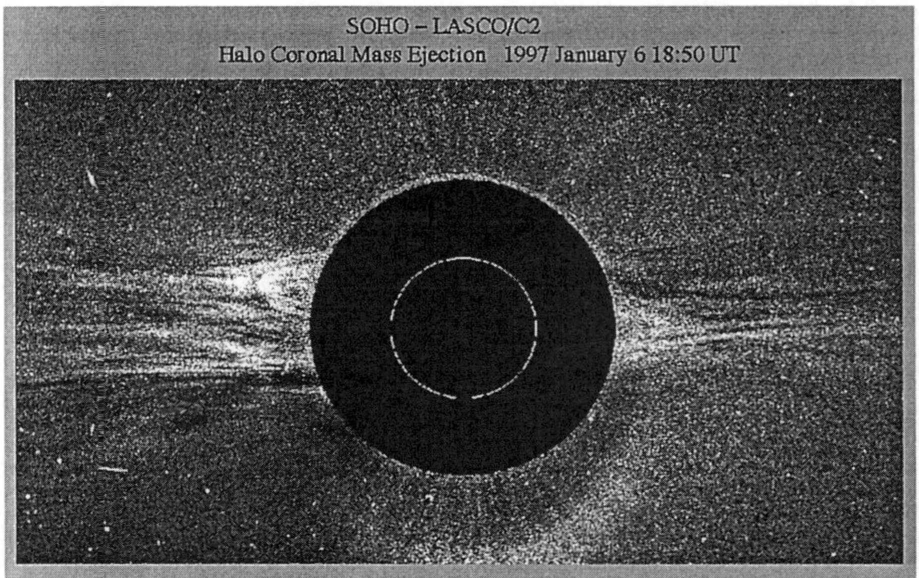

Figure 3: A halo CME. (Figure courtesy of the LASCO consortium).

Since a CME involves the ejection of mass into interplanetary space, it is reasonable to look for a decrease in the coronal emission at and/or near the site of the eruption. For example, Sterling and Hudson (1997) noted a sudden depletion (or dimming) of the corona as seen by the broad-band soft X-ray telescope on Yohkoh after the CME on April 7, 1997. They argued that roughly 10^{14} g had been ejected, that the CME involved the ejection of a magnetic flux rope and that the dimming could be viewed as being a "transient coronal hole" (Hudson and Webb, 1997; see also Rust and Hildner, 1976). Analysis of the same event using data from the narrow-band EIT instrument on SOHO drew a similar conclusion (Zarro et al., 1999). The generality of coronal dimming was confirmed in the more extensive study of Thompson et al., (2000), who looked at a number of fast (> 639 km/s) CMEs during the active period in April/May 1998.

A discovery of SOHO has been the existence of a class of extremely fast-moving "coronal waves" that appear to be associated with CME onset. These were discussed initially in the context of CMEs on April 7 and May 12, 1997 (Thompson et al., 1998, 1999a). Waves that propagate rapidly across the solar disc (Moreton waves) have been observed using ground-based instrumentation for many years (e.g. Moreton, 1961), and are assumed to be non-linear magnetosonic waves generated by an impulse of some description. Thompson et al were able to track the (almost isotropic) wave front for over an hour, and establish an approximate average propagation speed of 245 ± 40 km/s. This seems rather low for a coronal magnetosonic speed, implying either a high density, or low field strength. It is interesting to note that Sterling and Hudson (1997) were unable to see a wave in their broad-band X-ray images from Yohkoh. One possible reason is that the SXT instrument measures much higher temperatures than EIT: a second was poor data coverage. Clearly further studies of the ubiquity of "coronal Moreton waves" are required.

The final recent discovery concerns patterns in X-ray emission (presumably somehow reflecting magnetic field topologies) that are observed prior to CME onset. Canfield et al. (1999: see also Rust and Kumar, 1996 and Hudson et al., 1998) reported that active regions characterised by X-ray emission having a "S" or "inverse-S" topology (often referred to as a sigmoidal topology) have a higher probability of eruption than ones without this topology. [A sigmoid is defined as a single, twisted flux rope observed in projection against the solar disk as a bright S (Glover et al., 2000).] However, recent work of Glover et al (2000), which made use of additional data from SOHO, as well as a more complete Yohkoh data set, suggests that the picture is somewhat more complicated. Glover et al showed that it was possible to "mis-define" active regions as being sigmoidal when either all that being seen was a projection effect, or when the sigmoidal nature vanished at high spatial resolution. On recategorisation, the "true sigmoids" still showed a tendency to erupt, but with a lower probability than in the original analysis. A study of a large data set is badly needed to clarify these discrepancies.

2.7 Theories of CME onset

The advances in the interpretation of coronagraph observations have not been matched by an increased theoretical understanding of what causes the eruption. Two classes of models are currently under consideration. In the first (see Klimchuk, 2000, for a review), a pre-existing closed coronal structure undergoes shearing of its footpoints, leading to an increase in the magnetic energy in the corona. The essential concept is that once the shear is increased to a certain value, the coronal structure loses equilibrium and/or sustains an internal instability, and erupts. For simple coronal geometries (e.g. a simple force-free bipolar arcade), it turns out that the energy of magnetic field with all field lines open to infinity exceeds that of any closed structure (Aly, 1984), hence making this scenario untenable. Aly's result ignores coronal pressure gradients (e.g. Wolfson and Dlamini, 1999), magnetic flux rope geometries embedded in the arcade (e.g. Amari et al., 2000), and more complex non-bipolar field configurations (Antiochos et al., 1999). Of these, only the last appears to provide a viable way forward.

The second class of model (e.g. Chen, 1996, 2000) removes the above restrictions by a direct driving of the CME by currents generated below the photosphere. The scenario involves the existence of a pre-eruption coronal flux rope (the prominence cavity) with an overlying magnetic field (the streamer). Magnetic flux is injected into the flux rope from below the photosphere on a timescale of a few hours. The flux rope responds by moving away from the Sun. A reasonable agreement with coronagraph observations can be obtained (Chen et al., 2000). The model has been criticised (e.g. Klimchuk, 2000) on the grounds that the flux injection would lead to unreasonably large photospheric velocities, although no specific numbers have ever been derived from the model equations by the critics. Despite this question, the model is attractive for its simple bypassing of the constraints quoted above for the other class of models.

3. CMEs IN INTERPLANETARY SPACE

To observe CMEs in the interplanetary medium, one generally relies on in-situ sampling of particles and magnetic fields by spacecraft. [There are other possible techniques using radio signatures from CME-associated shock waves and other sites of energetic electron production, as well as newer methods of all-sky imaging to be flown on the SMEI and STEREO missions. We do not discuss these in this chapter, nor do we consider possible ground-based methods of detection and tracking that rely on interplanetary scintillations.] Monitoring of CMEs in interplanetary space near the Earth is essential for understanding and forecasting space weather. Ideally the monitor should be as far upstream as possible, but in reality this usually involves an orbit near the L1 point. For the viewpoint of space weather, the most important physical parameters to be measured at 1 AU are the topology of the CME magnetic field (especially the duration and strength of any Southward component), and the magnitude of the solar wind velocity and density (and any associated shock waves). The latter is important because (1) the electric field across the polar cap is $\propto V_x B_z$ (in GSE co-ordinates) and (2) the position of the magnetopause is determined by the magnitude of the solar wind ram pressure.

In this Section, we first discuss some generic properties of Interplanetary CMEs (ICMEs: Section 3.1), then examine a specific subset (magnetic clouds) that are of particular importance for space weather (Section 3.2). Section 3.3 outlines important processes that determine the properties of a CME at 1 AU, Section 3.4 describes initial efforts at assessing the 3-D structure of ICMEs and Section 3.5 we discuss the detection of CMEs in high-speed solar wind by the Ulysses spacecraft.

3.1 Basic Concepts

A typical ICME is shown in Figure 4. The panels show (from top to bottom) the plasma density, temperature, the three magnetic field components in GSE co-ordinates, the total magnetic field, and the Dst (magnetic storm) index. This ICME was detected by the ISEE-3 satellite in December 1980, and as can be seen in the Dst index, produced a major geomagnetic storm. The main features to note are the presence of a leading shock wave, and especially the enhanced magnetic field and smooth rotation of the GSE y and z components. The z component is especially important. The GSE z direction is defined as pointing from South to North. A strong negative-z (Southward) IMF leads to major magnetic reconnection at the sub-solar point, and hence the transfer of energy from the solar wind into the Earth's magnetosphere and from the viewpoint of space weather is the most important property of an ICME.

Examination of solar wind time series reveals many such examples. Indeed ICMEs have been detected in the ecliptic plane at a range of distances: (0.3 AU by the Helios mission and 5 AU by the Voyager and Ulysses missions). Their properties in fact are similar at all of these distances. Most numerous are the observations close to the Earth. The IMP-series of satellites has provided in-situ measurements of ICMEs for over 30 years. ISEE-3 provided an excellent data set during its time spent at the L1 point (1978-1982), and more recently, WIND (since 1995) and ACE (since 1997) have provided coverage upstream of the Earth. Nevertheless there are large gaps in coverage since the late 1960s, and there are many major geomagnetic storms for which no interplanetary data is available, the major geomagnetic storm of March 1989 being the most obvious example. Given the importance of space weather, it is unlikely such a situation will ever arise again.

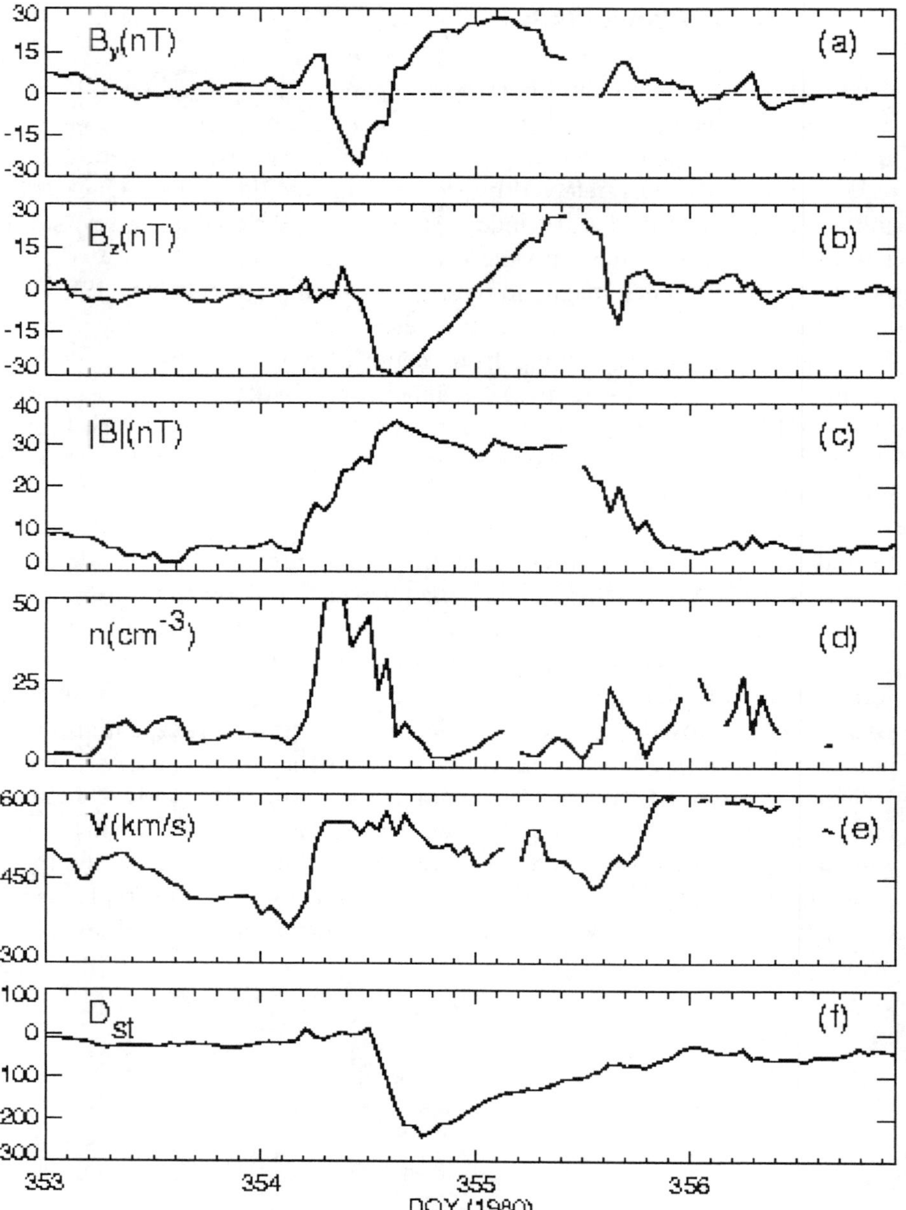

Figure 4: The y and z magnetic field components (in GSE coordinates), the total magnetic field strength, plasma density, velocity and the Dst index, for an ICME measured in December 1980. Note the leading shock wave, the long period of smooth magnetic field enhancement, the extended Southward IMF, and the associated depression of the Dst index. This ICME is an example of a magnetic cloud. [Data courtesy of OMNIweb.]

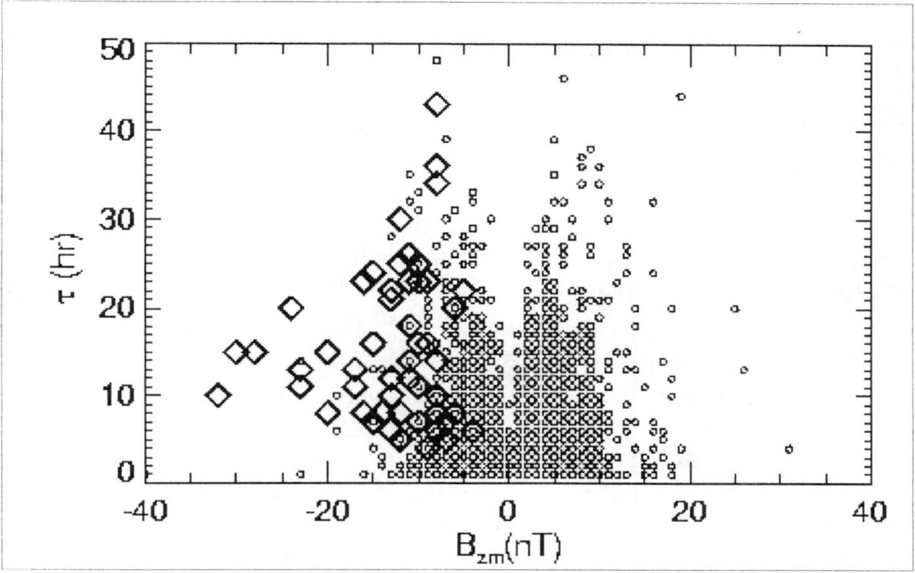

Figure 5: The relation between the duration of periods between changes of sign of B_z, and the maximum (and minimum) values of B_z within each period. Diamonds correspond to events associated with a Dst index below -80nT for 2 hours or more, circles denote all other events. Note that long periods with a strong field are required for major geomagnetic activity. [From Chen et al., 1997.]

It is well known that the most important property that the solar wind magnetic field needs in order to give rise to geomagnetic storms is long periods of Southward IMF. Figure 5 shows the distribution of the duration of periods of southward IMF as a function of the minimum magnetic field in each interval (Chen et al., 1997). Approximately 8 years of data from the OMNI database were used. The intervals were divided up depending on whether they were associated with a sustained period of negative Dst (-80 nT or less for 2 hours or more: diamonds) or not (circles). Clearly the solar wind magnetic field distribution breaks down into a fairly clear division between events which are geo-effective, and events which are not. Many (but not all) of the geo-effective events are CMEs, identified as such either by the smooth field rotation characteristic of a magnetic cloud (see below), or on the basis of counter-streaming heat flux electrons (Gosling et al., 1987).

In another study using 10 years of Pioneer Venus Orbiter (PVO) data from 0.7 AU, Lindsay et al., (1995) examined the change in the IMF associated with both CMEs and solar wind stream interactions. They found that while both classes of event produced enhanced magnetic fields, it was CMEs that were responsible for the largest values and most sustained periods of southward IMF.

3.2   Magnetic clouds: an important interplanetary manifestation of CMEs

Burlaga et al., (1981) and Klein and Burlaga (1982) noted that a significant fraction of ICMEs had smooth magnetic field profiles that changed on timescales of hours and lower than usual plasma temperatures (as is shown in the example in Figure 4). They named ICMEs with such a property "magnetic clouds". Roughly 30-40% of ICMEs are magnetic clouds (Gosling, 1990), and their importance from the point of view of space weather is considerable, since the smoothly-changing magnetic field often leads to an IMF that is Southward for many hours. Magnetic clouds are vast structures, often being 0.25 AU in diameter, and taking a day to pass by the Earth.

Magnetic clouds have been interpreted as being large magnetic flux ropes (Burlaga, 1988) attached at both ends to the Sun (Figure 6). Since it is unlikely that such an organised structure could form spontaneously from a turbulent solar wind, its origin must be solar. The most attractive picture is that the magnetic cloud is the residue of a large solar loop-like structure, and is in fact the prominence cavity discussed in the three-part CME model in Section 2. The first statistical association that backed up this conjecture was presented by Wilson and Hildner (1984, 1986). [It should be stressed that flux ropes can be created in the low corona by other means (e.g. Gosling et al., 1995), but the connection of an erupting prominence cavity to a magnetic cloud remains attractive to many workers.] The evidence for the attachment of both ends to the Sun partly comes from the evidence of bi-directional heat flux electrons (Gosling et al., 1987), although there are exceptions that make this method not entirely foolproof (McComas et al., 1994). Alternative topologies for magnetic clouds include closed plasmoids (e.g. Vandas et al., 1993), though this has been questioned on grounds that the predicted magnetic field structure of a plasmoid is not seen in spacecraft data, and on the potential source region of the bi-directional heat flux electrons (e.g. Burlaga et al., 1990).

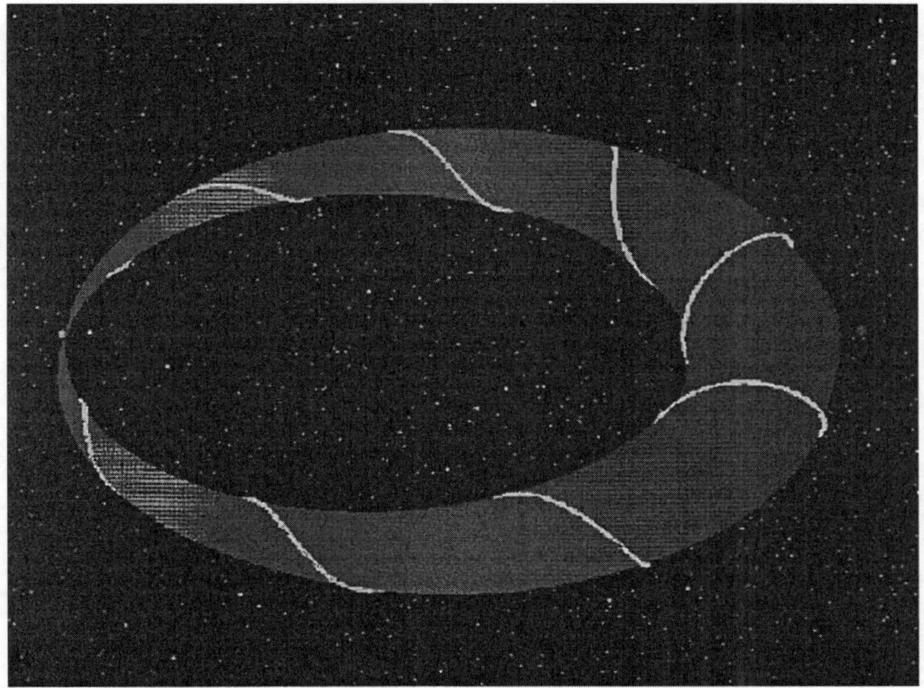

Figure 6: A sketch of a CME in the interplanetary medium. It has a structure corresponding to a magnetic flux rope (with twisted magnetic field lines as shown), with both footpoints rooted at the Sun. (Figure courtesy of J. Chen.)

3.3   Processes that determine ICME structure

The structure of a magnetic cloud (or in fact any CME) at 1 AU is determined by both its initial state, and by its interaction with the solar wind en route from the Sun. The solar wind consists of both a quasi-steady component, as well as shocks, discontinuities and multi-speed streams, all of which can interact with the flux rope. Indeed the fact that an organised structure like a magnetic cloud can survive to 1 AU suggests that it contains significant inherent robustness.

The clearest evidence that significant interaction does take place can be seen by a comparison of the distribution of observed CME speeds at the Sun, and those detected by spacecraft in the heliosphere. Gopalswamy et al., (2000) have carried out such a comparison for 28 CMEs seen in the SOHO epoch, using data from the LASCO instrument and the ACE spacecraft, and show that while CMEs at the Sun cover a wide range of speeds (100 – 1000 km/s), at 1 AU the speeds are bunched between 350 and 550 km/s. [Note

196

that most of the CMEs in this study were magnetic clouds.] Hence, the interaction between the CME and the solar wind tends to bring their velocities closer together, with slow CMEs being accelerated and fast ones being decelerated. Lindsay et al. (1999) have carried out a similar study using data from the SMM and Solwind coronagraphs, and interplanetary data from Pioneer Venus Orbiter. They reach similar conclusions to Gopalswamy et al, but also note a positive correlation between the CME speed at the Sun and the maximum total field strength in the interplanetary CME. Intriguingly, they can establish no real correlation between the Southward IMF and CME speed.

Clearly the motion of a ICME through the solar wind involves an interaction between two distinct magnetised plasmas. For a fast CME, the CME tries to push the solar wind (and its associated magnetic field) out of the way, and gives up energy in so doing. The full details of the interaction are complex, and involve many phenomena such as field line draping (e.g. McComas et al., 1988), magnetic reconnection (McComas et al., 1994), and interaction with interplanetary current sheets (e.g. Crooker et al., 1998b). The process is reversed for slow CMEs. The fast solar wind then tries to push the CME out of its way, and loses kinetic energy to the CME, hence reducing the velocity difference.

There have also been theoretical efforts to quantify the important processes, mostly through the use of MHD codes (see for example Cargill et al., 2000; Odstricil and Pizzo, 1999, Vandas et al., 1995; Wu et al., 1999 and references therein for each group's earlier work). Cargill et al. (1995, 1996) showed how, in a Cartesian geometry, the interaction between a flux rope and an ambient magnetised medium could be parameterised in terms of a simple aerodynamic drag force ($\propto C_D V_{SW}(V_{CME} - V_{SW})|V_{CME} - V_{SW}|$ in an obvious notation), where C_D is a standard aerodynamic coefficient with a value of order unity. This result appears to also hold in a spherical geometry (Schmidt and Cargill, 2000, unpublished results). Cargill et al., (1996) also showed that magnetic reconnection between the CME and ambient magnetic fields was inevitable. This is also evident in Vandas et al. (1995), but the authors did not comment on it.

These interactions also lead to distorted CME shapes. It is commonly assumed that magnetic clouds can be modelled to lowest order as cylindrically symmetric flux ropes at all points in their life. For example, Lepping et al. (1990) fit a large number of magnetic clouds seen by ISEE-3 to a Bessel function field model with a limited number of free parameters. However, simulations (e.g. Cargill et al., 1995, 2000; Vandas et al., 1995, Wu et al., 1999) show that magnetic clouds undergo significant distortion as a part of the aerodynamic drag they experience in the interplanetary medium, so that cylindrical symmetry is unlikely to be a good assumption.

Another complication is the complex interplanetary stream and magnetic field structure. Crooker et al., (1998a) used a sample of 14 magnetic clouds from the ISEE_3 data set to argue that there is a close link between the ICME structure and the presence of interplanetary sector structure. This is currently an active research topic, and has not as yet been addressed in numerical models using magnetic cloud geometry. [It needs to be stated that numerical modelling of magnetic clouds with a realistic IMF is computationally challenging. It is highly desirable to initialise the magnetic field in a MHD simulation with an exact expression for the vector potential. Then if the numerical scheme satisfied the $\nabla.\mathbf{B} = 0$ condition to machine accuracy, the initial conditions cannot be allowed to introduce any spurious monopoles. Until recently, there were no analytical solutions of a flux rope in a radial magnetic field, thus simulations could not be initialised properly. In particular, Vandas et al., (1995) and Wu et al., (1999) include an artificial divergence cleaning technique into initial conditions that approximate the required state. A solution of the required type has recently been obtained (Schmidt, 2000), and will be implemented in MHD models in the near future.]

A final issue of interest for magnetic clouds is the sense of rotation of the magnetic field when it reaches 1 AU. From the viewpoint of forecasting, it is useful to know whether to expect a northward or southward IMF to come first: indeed this is an essential part of one magnetic storm forecasting scheme (Chen et al., 1997). Bothmer and Rust (1997) and Bothmer and Schwenn (1998) noted that in the 1974-1982 timeframe, magnetic clouds tended to arrive at the Earth with a southward IMF leading, but towards the end of this period (i.e. around the time when the global field of the Sun reversed), the opposite was seen. Mulligan et al (1998) used 10 years of PVO data to carry out a more extensive analysis of this hypothesis. They found that between 1979 and 1988, the preferred field orientation did indeed change from southward leading to northward leading. It will be fascinating to see if this continues to hold with data from the present generation of solar wind monitors.

3.4   Information on 3-D structure of ICMEs

Information concerning the multi-dimensional structure of ICMEs requires data from more than one spacecraft. There are many cases where a CME was observed by a spacecraft near the L1 point, and later on nearer to the Earth (ISEE-3 and IMP-8 is the best example of such a conjunction). However, their very proximity, as well as close alignment along the Sun-earth line rules such observations out as a major source of useful information. One

198

requires separated spacecraft, and such conjunctions are generally unplanned. The author is aware of at least ten examples: Two involve the Ulysses mission, and eight the NEAR spacecraft.

Hammond et al. (1995) reported observations of a CME by Ulysses (at 5 AU) and Geotail (at 1 AU), at a time when Ulysses was 20° S and 50° W of Geotail. The portion of the CME seen by Ulysses was travelling much faster (200 km/s), but, despite this, there are recognisable similarities in the magnetic field profiles (the ICME is a magnetic cloud). A picture is presented of a CME with parts in both high and low speed wind, presumably with the part in the low speed wind being accelerated due to magnetic tension forces associated with the field line connection to the high-speed wind. A second event was reported by Gosling et al. (1995) using data from Ulysses (at 3.53 AU, and 54° S) and IMP-8. At Ulysses, this was an over-expanding CME (see Section 3.5), with a forward-reverse shock pair, while at IMP-8 there was only a leading shock. The difference in the ICME appearance at each latitude must be due to differences in the interplanetary medium each part of the ICME experiences. However, in this case, there was less similarity between the magnetic field profiles.

Another opportunity arose for such a comparison during 1997 when the Wind and NEAR spacecraft were separated by 0.18 – 0.63 AU, and by 1° – 33° in azimuth (Mulligan et al., 1999). Four magnetic clouds were seen at a range of spacecraft separations. A limitation of the analysis is the absence of solar wind plasma measurements from NEAR (hence the restriction to well-organised magnetic structures). When the spacecraft were close together (0.18 AU and 1°), the leading shock positions differed somewhat, but the major magnetic field structures were readily recognisable in both data sets, though there were very noticeable differences. As the spacecraft separation increased, the differences became unmistakable. For example, at a separation of 0.28 AU and 5°, the polarity of the x and z magnetic field components in the ICME were different, though the sense of field rotation appeared to be the same. The dissimilarities increased in the other two examples. In one case, although the sense of rotation was the same, the field components appear to be reversed when NEAR and Wind are compared. In the final case, Wind saw two ICMEs whereas NEAR saw only one. This case had the largest separation (0.63 AU and 33°), so it is unclear whether the two spacecraft saw the same event. Further spacecraft conjunctions, as well as a modelling effort are badly needed to resolve the issues raised here.

3.5   CMEs in high-speed solar wind

Observations at different heliocentric distances permit one to study how ICMEs evolve as they move away from the Sun, and hence constrain theories pertaining to this evolution. The same is true for ICMEs seen out of the ecliptic plane, especially in regions of pure high-speed solar wind. In this regard, the Ulysses mission has (and is) providing a unique and probably unrepeatable data set. During 1993-98, Ulysses spent considerable time in regions of purely high-speed solar wind, only passing through regions of low-speed wind as it carried out its fast latitude scan in 1995. (The transition from high to low-speed wind involved passage through an ICME: Forsyth et al., 1996). Ulysses detected a significant number of CMEs in regions of high speed wind, with detections being made at $54°$ S. If one accepts the current paradigm of an origin of CMEs near the equatorial streamer belt, presumably in low speed solar wind, one needs to ask the question of how CMEs can move from slow wind to fast wind. Schmidt and Cargill (2000b) have performed MHD simulations of the penetration of a magnetic cloud from low- to high-speed solar wind. They showed that, so long as the poleward velocity was large enough, the cloud moved readily through the velocity shear. [This should be contrasted with the result in a Cartesian geometry where a flux rope bounces off a velocity shear without penetration (Schmidt and Cargill, 2000a).]

Ulysses also detected a class of ICME that appear to be unique to the high-speed solar wind. In these events, the CME is characterised by a very low plasma density, and a relatively strong forward and reverse shock pair. Figure 7 shows an example of such a CME, seen at 4.6 AU and a latitude of $54°$ S, which also happens to be a magnetic cloud. [This observation testifies yet again to the robustness of the magnetic flux rope in the solar wind.] An interpretation originally proposed by Gosling et al. (1994) was that these CMEs were "over-expanding", having begun life with a large excess of plasma pressure, which had expended its energy into creating the shock pairs. This has been confirmed by hydrodynamic (Gosling and Riley, 1996) and multi-dimensional MHD simulations (Cargill et al., 2000) who also demonstrated that the over-expansion was essential for the maintenance of the flux rope geometry at large distances.

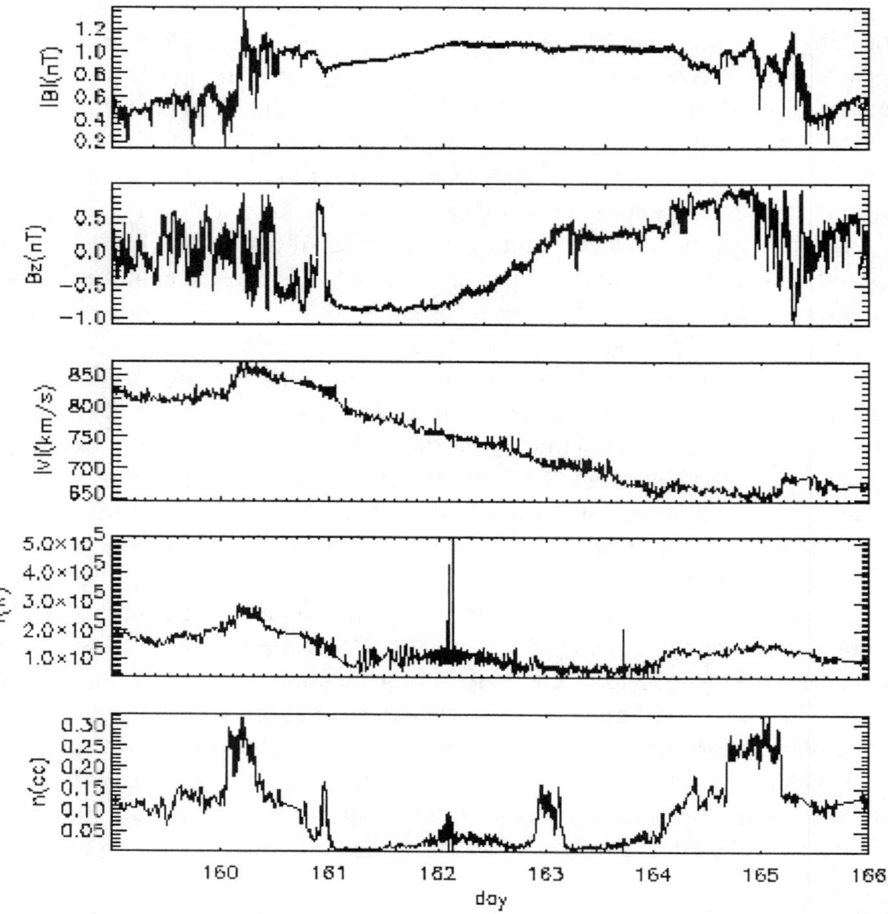

Figure 7: An over-expanding ICME observed by the Ulysses spacecraft on days 158-166 1993 at 4.64 AU and S32.5° latitude. The plot shows the total and North/South magnetic field, the plasma velocity, ion temperature and density. Note the forward and reverse shock, the very low plasma density in the CME, and the magnetic cloud structure. [Data courtesy of A. Balogh (magnetic field) and D. McComas (plasma).]

4.   CONCLUSIONS AND FUTURE PROSPECTS

4.1   Where CME research stands

We have learned a lot about the nature of CMEs in the past 30 years, and will continue to expand our knowledge while SOHO operates throughout the declining phase of the present solar cycle, and while satellites such as Ulysses and ACE continue to make measurements in interplanetary space. While some physical parameters are well documented (mass, velocity, rate, size), some are still controversial (acceleration) and some will never be determined with any reasonable accuracy (magnetic energy near the Sun). In addition, we have no real feel for the three dimensional structure of a CME, either in the inner solar corona, or in interplanetary space. We also have no understanding of the process(es) that initiate(s) a CME.

It is the last of these questions that is the most basic. Is there a single mechanism that is responsible for all solar ejecta above a certain size, with the broad range of masses, velocities etc. being accounted for by different initial conditions and the local plasma and magnetic field environment through which the CME must escape? It seems unlikely. Then, one must ask questions about flux ropes. It is seemingly integral to our present ideas, but we do not know where it comes from (through the photosphere, or formed in the corona, or formed during the eruption). One must also ask questions about the theories. The viewpoint that CME energy is stored in the corona due to slow motion of photospheric footpoints of coronal field lines has been around for 40 years. The impression gained by the author in writing this review is that progress on this topic has been at the rate of microns per year, with the Aly (1984) result proving to be almost insuperable. Perhaps it is time to consider other ideas, as discussed in Section 2.7.

Then there is the question of how the CME/ICME and solar wind interact with each other. This may seem to be an easier problem, but really it is not. One would expect fast CMEs to undergo prompt deceleration, yet that does not happen, at least inside 30 R_s (Sheeley et al., 1999). So does this mean that CMEs are strongly driven to beyond 30 R_s in order to overcome drag forces? Again most (but not all: Chen, 1996) theories do not address this. Interplanetary data reveals tremendous complexity and continual surprises: the recent work of Mulligan et al. (1999) and Kahler et al., (1999) being excellent examples of new, exciting results.

The need to look at large data sets (as opposed to single events) cannot be overestimated. The immense efforts documented in Hundhausen (1999) and St Cyr et al. (2000) provide us with the best overviews of CMEs we have. The analysis of data from SOHO, Yohkoh, ACE, Wind and Ulysses (as well as NEAR and other non-CME-focussed missions) will take the best part of the next decade, and will almost certainly resolve some outstanding questions by 2010 (but not, in my view, the all-important question of the cause of the eruption). We also need to be aware of information in older data sets that perhaps are not thought of as being useful for CME research. The important achievements from the ten year PVO data set by the UCLA group is an excellent example of what can be done. Particularly important are searches for multiple spacecraft measurements of ICMEs.

4.2   Where next with space missions

The most important future space mission from the viewpoint of CMEs is the NASA twin spacecraft Solar Terrestrial Relations Observatory (STEREO) mission, scheduled for launch in 2004. This mission comprises two spacecraft that will lead and trail the Earth, and will gradually separate to provide a stereo view of the Sun. Each spacecraft will have instruments that will image the solar disk, coronagraphs, as well as in-situ fields and particles packages. By use of stereo imaging techniques, STEREO will provide a three-dimensional picture of the structure of CMEs. In addition, STEREO can act as a space weather monitor by determining the velocity and direction of Earthward-directed (halo) CMEs.

Understanding the origin of CMEs on the solar surface and in the inner corona requires a different range of instrumentation. While STEREO will provide a picture of the large-scale structure of CMEs, what is required here are instruments to measure the (often subtle) plasma flows and field changes in the photosphere and low corona that precede CME onset. The need here is thus for high quality measurements of photospheric fields and flows, coupled with good spectroscopy in the EUV. In addition, one requires X-ray images in order to place the other measurements in their global coronal context. The ISAS Solar-B mission, scheduled for launch in 2005, is equipped to make such measurements. While CME onset is not one of its major scientific goals, and, like STEREO, it is flying at a time of solar minimum, the CME rate of 0.5 per day at that time of the cycle should provide an adequate data base provided focussed observing campaigns are carried out.

Thus the future of CME research is quite healthy from the viewpoint of space missions designed to observe their solar origin. The same cannot be said for interplanetary studies. The need here is for widely-spaced (> 0.05 AU) multi-point (several spacecraft) measurements, and, as noted above, the few that have been published have been largely serendipitous alignments of spacecraft (e.g. Ulysses and Geotail; Wind and NEAR). The results that are available are both fascinating and puzzling. However, as a result of the paucity of data, our three dimensional knowledge of interplanetary CMEs is obtained (1) from applying inappropriate (and unrealistic) simple models to time series obtained at a single point, (2) through numerical simulations that may have unrealistic initial and boundary conditions, (3) from theories that put the need for an analytic solution above the realities of the interplanetary medium and (4) speculation. Unfortunately the prospects of the required space mission in the short and medium term is very unlikely. Difficulties include (but are not limited to) the lack of possible easily accessible stable orbits, and the perception that interplanetary CMEs are quite well understood. It is the opinion of the author that such a multi-satellite mission would produce major surprises.

4.3   CMEs and space weather forecasting

Finally, the issue of forecasting space weather using observations of CMEs is of considerable interest. A zeroth order goal would appear to be the forecasting at 1 AU of the arrival time, CME speed, shock (if any) strength, duration and intensity of any Southward IMF, and the CME orientation, given information from the Sun about the eruption. There have been some proposals based on empirical formula recently concerning the arrival time (Gopalswamy et al., 2000), as well as rules concerning the sense of magnetic field rotation (Bothmer and Rust, 1997). It is conceivable that the former of these could be extended to include information on shock strength. However, when forecasting using solar input, one rapidly runs into the fact that a lack of measurements of the coronal magnetic field makes it all but impossible to forecast IMF strength at 1 AU. Since the strength of geomagnetic activity (e.g. the Dst index) depends on the IMF strength (in the case of Dst the dependence is linear at the onset of a magnetic storm: Burton et al., 1975), lack of a prediction of the IMF is a serious problem. The problem at hand is encapsulated in the results of St Cyr et al (2000) discussed above. More realistic perhaps is to use the solar observations as a warning of hazardous space weather, and to rely on quantities obtained by an upstream monitor to issue alerts. An example of such a method can be found in Chen et al. (1996, 1997a).

5.   REFERENCES

Aly, J.J., (1984), On some properties of force-free magnetic fields in infinite regions of space, *Astrophys.*, **283**, 349.

Amari, T., Luciani, J.F., Mikic, Z. and Linker, J., (2000), A twisted flux rope model for coronal mass ejections and two-ribbon flares, *Astrophys. J. Lett.*, **529**, L49.

Antiochos, S.K., DeVore, C.R. and Klimchuk, J.A., (1999), A model for coronal mass ejections, *Astrophys. J.*, **510**, 485.

Bothmer, V. and Rust, D.M., (1997), The field configuration of magnetic clouds and the solar cycle, in N.U. Crooker et al (Eds), *Coronal Mass Ejections*, AGU Monograph 99, AGU (Washington DC), p.139.

Bothmer, V. and Schwenn, R., (1998), The structure and origin of magnetic clouds in the solar wind, *Annales Geophys.*, **16**, 1.

Burlaga, L.F., (1988), Magnetic clouds: constant alpha force-free configurations, *J. Geophys. Res.*, **93**, 7217.

Burlaga, L.F., Sittler, E., Mariani, F. and Schwenn, R., (1981), Magnetic loop behind and interplanetary shock: Voyager, Helios and IMP8 observations, *J. Geophys. Res.*, **86**, 6673.

Burlaga, L.F., Lepping, R.P.and Jones, J.A., (1990), Global configuration of a magnetic cloud, in C.T. Russell et al. (Eds), *Physics of magnetic flux ropes*, AGU Monograph 58, AGU (Washington DC), p 373.

Burton, R.K., McPherron, R.L.and Russell, C.T., (1975), An empirical relationship between interplanetary conditions and Dst, *J. Geopyhys. Res*, **80**, 4204.

Canfield, R.C., Hudson, H.S. and McKenzie, D.E., (1999), Sigmoidal morphology and eruptive solar activity, *Geophys. Res.Lett.*, **26**, 627.

Cargill, P.J., Chen, J., Spicer, D.S. and Zalesak, S.T., (1995), Geometry of interplanetary magnetic clouds, *Geophys. Res. Lett.*, **22**, 647.

Cargill, P.J., Chen, J., Spicer, D.S. and Zalesak, S.T., (1996), MHD simulations of the motion of magnetic flux tubes through a magnetized plasma, *J. Geophys. Res,* **101**, 4855.

Cargill, P.J., Schmidt, J., Spicer, D.S. and Zalesak, S.T., (2000), The magnetic structure of over-expanding CMEs, *J. Geophys. Res,* **105**, 7509.

Chen, J., (1996), Theory of prominence eruption and propagation: interplanetary consequences, *J. Geophys. Res.,* **101**, 27,499.

Chen, J., (2000), Physics of coronal mass ejections: a new paradigm for solar eruptions, *Space Sci Revs.,* in press.

Chen, J., Cargill, P.J. and Palmadesso, P.J., (1996), Real-time identification and prediction of geoeffective solar wind structures, *Geophys. Res.Lett.*, **23**, 625.

Chen, J., Cargill, P.J. and Palmadesso, P.J., (1997a), Predicting geoeffective solar wind structures, *J. Geophys. Res.,* **102**, 14,701.

Chen, J. et al., (1997b), Evidence of an erupting magnetic flux rope: LASCO coronal mass ejection of 1997 April 13, *Astrophys. J. Letters,* **490**, L191.

Chen, J. et al., (2000), Magnetic geometry and dynamics of the fast coronal mass ejection of 1997 September 9, *Astrophys. J.,* **533**, 481.

Crooker, N.U., Gosling, J.T. and Kahler, S.W., (1998), Magnetic clouds at sector boundaries, *J. Geophys. Res.,* **103**, 301.

Crooker, N.U. et al., (1998b), Sector boundary transformation by an open magnetic cloud, *J. Geophys. Res.,* **103**, 26,859.

Dere, K.P. et al., (1997), EIT and LASCO observations of the initiation of a CME, *Solar Phys.,* **175**, 601.

Dere, K.P., Brueckner, G.E., Howard, R.A., Michels, D.J. and Delaboudiniere, J.P., (1999), LASCO and EIT observations of helical structure in CMEs, *Astrophys. J.,* **516**, 465.

Dryer, M., Wu, S.T., Steinolfson, R.S. and Wilson, R.M., (1979), Magnetohydrodynamic models of coronal transients in the meridional plane: II Simulation of the coronal transient of 1973, August 21, *Astrophys. J.,* **227**, 1059.

Forsyth, R.J., Balogh, A., Horbury, T.S., Erdos, G., Smith, E.J. and Burton, M.E., (1996), The heliospheric magnetic field at solar minimum: Ulysses observations from pole to pole, *Astron. Astrophys.,* **316**, 287.

Gopalswamy, N. et al., (2000), Interplanetary acceleration of coronal mass ejections, *Geophys. Res.Lett.,* **27**, 145.

Gosling, J.T., (1990) Coronal mass ejections and magnetic flux ropes in interplanetary space, in C.T. Russell et al. (Eds), *Physics of magnetic flux ropes*, AGU Monograph 58, AGU (Washington DC), p 343.

Gosling, J.T., (1993), The solar flare myth, *J. Geophys. Res.*, **98**, 18,937.

Gosling, J.T., McComas, D.J., Phillips, J.L. and Bame, S.J., (1991), Geomagnetic activity associated with Earth passage of interplanetary shock disturbances and coronal mass ejections, *J. Geophys. Res.*, **96**, 7831.

Gosling, J.T. et al., (1994), A forward-reverse shock pair in the solar wind driven by over-expansion of a CME: Ulysses observations, *Geophys. Res.Lett.*, **21**, 237.

Gosling, J.T. et al., (1995), A CME-driven solar wind disturbance observed at both low and high heliographic latitudes, *Geophys. Res. Lett.*, **22**, 1753.

Gosling, J.T., Birn, J. and Hesse, M., (1995), Three-dimensional magnetic reconnection and the magnetic topology of coronal mass ejection events, *Geophys. Res.Lett.*, **22**, 869.

Gosling,J.T. and Riley, P., (1996), The acceleration of slow coronal mass ejections in the high-speed solar wind, *Geiphys. Res. Lett.*, **23**, 2867.

Hammond, C.M. et al., (1995), Latitudinal structure of a CME inferred from Ulysses and Geotail observations, *Geophys. Res. Lett.*, **22**, 1169.

Harrison, R.A., (1986), Solar coronal mass ejections and flares, *Astron. Astrophys.*, **162,** 283.

Harrison, R.A., Hildner, E., Hundhausen, A.J., Sime, D.G. and Simnett, G.M., (1990), The launch of solar coronal mass ejections: results from the coronal mass ejection onset program, *J. Geophys. Res.*, **95**, 917.

Howard, R.A., Michels, D.J., Sheeley, N.R. and Koomen, M.J., (1982), The observation of a coronal transient directed at Earth, *Astrophys. J. Letters.*, **263**, L101.

Howard, R.A., Sheeley, N.R., Koomen, M.J. and Michels, D.J., (1985), Coronal mass ejections: 1979-1981, *J. Geophys. Res.*, **90**, 8173.

Hudson, H.S. and Cliver, E.W., (2000), Observing CMEs without coronagraphs, preprint.

Hudson, H.S. and Webb, D.F., (1997) Soft X-ray signatures of coronal mass ejections, in N.U. Crooker et al (Eds), *Coronal Mass Ejections*, AGU Monograph 99, AGU (Washington DC), p.27.

Hudson, H.S., Lemen, J.R., St Cyr, O.C., Sterling, A.C. and Webb, D.F., (1998), X-ray coronal changes during halo CMEs, *Geophys. Res.Lett*, **24**, 2481.

Hundhausen, A.J., (1997), Coronal mass ejections. in J.R. Jokipii et al (Eds), *Cosmic winds in the heliosphere*, Univ. Arizona Press, p.259.

Hundhausen, A.J., (1999), Coronal mass ejections, in K.T. Strong et al. (Eds), *The many faces of the Sun*, Springer, p143.

Hundhausen, A.J., (1993) Sizes and locations of coronal mass ejections: SMM observations from 1980 and 1984-1989, *J. Geophys. Res.*, **98,** 13,177.

Hundhausen, A.J., Burkepile, J.T. and St Cyr, O.C., (1994), Speeds of coronal mass ejections: SMM observations from 1980 and 1984-1989, *J. Geophys. Res.*, **99**, 6543.

Illing, R.M.E. and Hundhausen, A.J., (1983), Possible observation of a disconnected magnetic structure in a coronal transient, *J. Geophys. Res.*, **88**, 10,951.

Illing, R.M.E. and Hundhausen, A.J., (1985), Observations of a coronal transient from 1.2 to 6 solar radii, *J. Geophys. Res.*, **90**, 245.

Illing, R.M.E. and Hundhausen, A.J., (1986), Disruption of a coronal streamer by an eruptive prominence and coronal mass ejection, *J. Geophys. Res.*, **91**, 10,210.

Klein, L.W. and Burlaga, L.F., (1982), Interplanetary magnetic clouds at 1 AU, *J. Geophys. Res.*, **67**, 613.

Klimchuk, J.A., (2000) Theory of coronal mass ejections, in *AGU Chapman conference on space weather*, in press.

Lepping, R.P., Jones, J.A. and Burlaga, L.F., (1990), Magnetic field structure of interplanetary magnetic clouds at 1 AU, *J. Geophys. Res.*, **95,** 11,957.

Lindsay, G.M., Russell, C.T. and Luhmann, J.G., (1995), Coronal mass ejection and stream interaction region characteristics and their potential geomagnetic effectiveness, *J. Geophys. Res.*, **100**, 16,999.

Lindsay, G.M., Luhmann, J.G., Russell, C.T. and Gosling, J.T., (1999), Relationships between CME speeds from coronagraph images and interplanetary characteristics of associated ICMEs, *J. Geophys. Res.,* **104**, 12,515.

Low, B.C., (1996), Solar activity and the corona, *Solar Phys.,* **167**, 217.

MacQueen, R.M. and Fisher, R.R., (1983), The kinematics of solar inner coronal transients, *Solar Phys.,* **89**, 89.

McComas, D.J., Gosling,J.T., Winterhalter, D. and Smith, E.J., (1988), Interplanetary magnetic field draping around fast coronal mass ejecta in the outer heliosphere, *J. Geophys. Res.,* **93**, 2519.

McComas, D.J., Gosling, J.T., Hammond, C.M., Moldwin, M.B., Phillips, J.L. and Forsyth, R.J., (1994), Magnetic reconnection ahead of a coronal mass ejection, *Geophys, Res. Lett.,* **21**, 1751.

Moreton, G.F., (1961), Fast-moving disturbances on the surface of the Sun, *Sky and Telescope,* **21**, 145.

Mulligan, T., Russell, C.T. and Luhmann, J.G., (1998) Solar cycle evolution of the structure of magnetic clouds in the inner heliosphere, *Geophys. Res.Lett.,* **25**, 2959.

Mulligan, T. et al., (1999) Intercomparison of NEAR and Wind interplanetary coronal mass ejection observations, *J. Geophys. Res.,* **104**, 28,217.

Munro, R.H. and Sime, D.G., (1985), White-light coronal transients observed from Skylab May 1973 to February 1974: a classification by apparent morphology, *Solar Phys.,* **97**, 191.

Nitta, N. and Akiyama, S., (1999) Relation between flare-associated X-ray ejections and coronal mass ejections, *Astrophys. J. Lett,* **525**, L57.

Odstrcil, D. and Pizzo, V.J., (1999), Distortion of the interplanetary magnetic field by three-dimensional propagation of coronal mass ejections in a structured solar wind, *J. Geophys. Res.,* **104**, 28,225.

Reames, D.V., (1999), Solar energetic particles: is there time to hide?, *Radiation measurement,* **30**, 297.

Rust, D.M. and Hildner, E., (1976), Expansion of an X-ray coronal arch into the outer corona, *Solar Phys.,* **48**, 381.

Rust, D.M. and Kumar, A., (1996), Evidence for helically kinked magnetic flux ropes in solar eruptions, *Astrophys.J.Lett.,* **464**, L199.

St Cyr, O.C. et al., (2000) Properties of coronal mass ejections: SOHO LASCO observations from January 1996 to June 1998, *J. Geophys. Res.* **105**, 18,169.

Schmidt, J.M., (2000), Flux ropes embedded in a radial magnetic field: analytic solutions for the external magnetic field, *Solar Phys.,* in press.

Schmidt, J.M. and Cargill, P.J., (2000a), The evolution of magnetic flux ropes in sheared plasma flows, *J.Plasma Phys.,* **64**, 41.

Schmidt, J.M. and Cargill, P.J., (2000b), Magnetic cloud evolution in a multi-speed solar wind, *J.Geophys. Res.,* submitted.

Sheeley, N.R., Walters, J.H., Wang, Y.M. and Howard, R.A., (1999), Continuous tracking of coronal outflows: two kinds of coronal mass ejections, *J. Geophys. Res.,* **104**, 24,739.

Simnett, G.M. et al., (1997), LASCO observations of disconnected magnetic structures out to beyond 28 solar radii during coronal mass ejections, *Solar Phys.,* **175**, 685.

Sterling, A.C. and Hudson, H.S., (1997), Yohkoh SXT observations of X-ray dimming associated with a halo coronal mass ejection, *Astrophys. J.Lett,* **491**, L55.

Subramanian, P., Dere, K.P., Rich, N.B. and Howard, R.A., (1999) The relationship of coronal mass ejections to streamers, *J. Geophys. Res.,* **104**, 22,321.

Thompson, B.J., Plunkett, S.P., Gurman, J.B., Newmark, J.S., St Cyr, O.C. and Michels, D.J., (1998), SOHO/EIT observations of an Earth-directed coronal mass ejection on May 12, 1997, *Geophys. Res.Lett.,* **25**, 2465.

Thompson, B.J., Plunkett, S.P., Gurman, J.B., Newmark, J.S., St Cyr, O.C. and Michels, D.J., (1999a), SOHO/EIT observations of the 1997 April 7 coronal transient: Possible evidence of coronal Moreton waves,*Astrophys.J .Lett.,* **517**, L151.

Thompson, B.J. et al., (1999b), The correspondence of EUV and white light observations of CMEs, in *Sun-Earth plasma connections,* **Geophys. Mon., 109**, ed. J.L. Burch et al., AGU, Washington DC, p31.

Thompson, B.J., Cliver, E.W., Nitta, N., Delannee, C. and Delaboudiniere, J.-P., (2000), Coronal dimming and energetic CMEs in April-May 1998, *Geophys. Res.Lett.,* **27**, 1431.

Tousey, R., (1973), The solar corona, *Adv. Space Res.,* **13**, 713.

Vandas, M., Fischer, S., Pelant, P. and Geranios, A., (1993), Spheroidal models of magnetic clouds and comparison with spacecraft measurements, *J. Geophys. Res.,* **98**, 11467.

Vandas, M., Fischer, S., Dryer, M., Smith, Z. and Detman, T., (1995), Simulation of magnetic cloud propagation in the inner heliosphere in two-dimensions: 1. loop perpendicular to the ecliptic plane, *J. Geophys. Res.,* **100**, 12,285.

Vourladis, A., Subramanian, P., Dere, K.P. and Howard, R.A., (2000) LASCO measurements of the energetics of coronal mass ejections, *Astrophys. J.*, **534**, 456.

Webb, D.F. and Howard, R.A., (1994), Solar cycle variation of coronal mass ejections and the solar wind mass flux, *J. Geophys. Res.*, **99**, 4201.

Webb, D.F., Cliver, E.W., Crooker, N.U., St Cyr, O.C. and Thompson, B.J., (2000), Relationship of halo coronal mass ejections, magnetic clouds and magnetic storms, *J. Geophys. Res.,* **105**, 7491.

Wilson, R.M. and Hildner, E., (1984), Are interplanetary magnetic clouds manifestations of coronal transients at 1 AU?, *Solar Phys.,* **91**, 169.

Wilson, R.M. and Hildner, E., (1986), On the association of magnetic clouds with disappearing filaments, *J. Geophys. Res.,* **91**, 5867.

Wolfson, R. and Dlamini, B., Magnetic shear and cross-field currents: roles in the evolution of the pre-cprpnal mass ejection corona, *Astrophys. J.* **526**, 1046.

Wood, B.E., Karovska, M., Chen, J., Brueckner, G.E., Cook, J.W. and Howard, R.A., (1999), Comparison of two coronal mass ejections observed by EIT and LASCO with a model of an erupting flux rope, *Atrophys. J.,* **512**, 484.

Wu, S.T., Guo, W.P., Michels.,D.J, and Burlaga, L.F., (1999), MHD description of the dynamical relationships between a flux rope, streamer, coronal mass ejection, and magnetic cloud: analysis of the January 1997 Sun-Earth connection event, *J. Geophys. Res.,* **104**, 14,789.

Zarro, D.M., Sterling, A.C., Thompson, B.J., Hudson, H.S. and Nitta. N., (1999), SOHO/EIT observations of EUV dimming associated with a halo coronal mass ejection, *Astrophys. J.Lett.,* **520**, L139.

Chapter 8

Measurements of Energetic Particles in the Radiation Belts

Radiation detectors

Berndt Klecker

Max-Planck-Institut für extraterrestrische Physik
85740 Garching bei München, Germany

Abstract

The techniques for the measurement of plasma and energetic particles in the near-Earth space environment have evolved rapidly since the advent of the space age about 50 years ago. Since then hundreds of different techniques have been developed, optimised for different energy ranges and types of particles as, for example, electrons and ions. This instrumentation is based on well known physical principles, first used in instruments for e.g. laboratory plasma physics or high-energy particle accelerators. However, because of the stringent constraints on mass and power that could be accommodated onboard space missions, it took a long time to develop the technology that made the implementation of these techniques onboard satellites possible. This review provides an overview of some of the basic techniques for the determination of plasma parameters and the measurement of the energetic particle population in space. Most of the techniques used so far determine the particle distributions by in situ measurement of the particle parameters velocity, mass, and ionic charge. However, to obtain a more global perspective of the near-Earth environment either remote sensing techniques or multi-spacecraft missions with simultaneous in situ measurements onboard a large number of spacecraft are needed.

Keywords

Plasma detectors, energetic particle detectors, in situ measurements, remote sensing measurements.

I.A. Daglis (ed.), Space Storms and Space Weather Hazards, 209–230.

1. INTRODUCTION

In this tutorial lecture presented at the NATO Advanced Study Institute (ASI) Space Storms and Space Weather Hazards meeting (Crete, June 2000) basic techniques for the measurement of plasma and energetic particles in space will be reviewed. It is the purpose of this review to provide an overview of some techniques and their limitations, in particular for young scientists. The emphasis of the review is therefore on basic principles and their application, with references to the literature where more information on the technical details may be found.

Early measurements resulting, for example, in the discovery of the trapped particle populations in the radiation belts, were made with energetic (> 50 keV) particle detectors. These were primarily Geiger – Mueller (GM) counters, scintillation counters, and emulsions, i.e. well known laboratory devices that also could be used onboard the early Earth-orbiting spacecraft with little modification. For an historical review of the early discoveries see, for example, Lemaire (2001) in this volume, and references therein. Since these early days the techniques for the measurement of plasma and energetic particles in the near-Earth space environment have evolved rapidly and hundreds of different techniques have been developed, optimised for different energy ranges and types of particles as, for example, electrons and ions. This instrumentation is generally based on well-known physical principles. However, because of the stringent constraints on mass and power that can be accommodated onboard space missions, it took a long time to develop the technology that made the implementation of these techniques onboard satellites possible. This tutorial cannot possibly cover all these techniques. It will rather concentrate on some representative measurement principles that are also used in modern instrumentation.

It should be noted here that modern instruments for the determination of plasma and energetic particle parameters typically consist of many subunits that may be grouped into the following sections: (1) the 'sensor' part, that includes collimators or other optics and the 'detector' that provides electronic signals for every particle; (2) the 'electronics', consisting of the low- and high voltage power supplies, an 'analog' part for the amplification of the signals, and a 'digital' part where logical combinations of these signals are used for an onboard classification of, for example, energy and mass; (3) the data processing unit that usually contains a processor, a command unit and interface to the spacecraft. This review will be limited to the 'sensor' part, although the efficient implementation of the measurement techniques depends heavily on the technological development of electronics and data processing units.

In section 2 methods for the in situ measurement of plasma and energetic particles will be discussed. Section 3 provides a short overview of remote sensing techniques that have been developed recently to provide a more global view of the near Earth environment. Section 4 provides an outlook to the next step in the investigation of the magnetosphere of the Earth with multi-spacecraft missions.

2.   IN-SITU MEASUREMENTS

In this section we concentrate on some representative techniques for the in situ measurement of space plasma and energetic particles. The energy limit between plasma and energetic particles is somewhat arbitrary and will be placed at about ~100 keV. Some techniques are limited by technical reasons to the plasma regime at < 100 keV, as we will see below. However, other techniques discussed in this section can be used both, at low and high energies.

2.1   Plasma Measurements

In the next six sections we discuss some techniques that are typically used to derive the plasma parameters density, velocity, temperature, and composition. For a recent summary of plasma measurements see, for example, Young (1998) and references therein. In section 2.7 we then discuss higher energies ranging from suprathermal particles to the MeV/nuc energy range.

2.1.1   Langmuir Probe

Langmuir probes (LP) have been used for many years on rockets and spacecraft to perform in situ measurements of plasma parameters. The LP technique is based on the measurement of the volt − ampere characteristics of metal collectors mounted on booms, sufficiently long to extend beyond the disturbed plasma close to the spacecraft. The measured parameter is the current (I) flowing when a variable voltage (V) is applied to the probe. From this I - V characteristic, the electron temperature T_e, the electron and ion densities, N_e and N_i, and the spacecraft potential V_S relative to the plasma can be derived. For a recent review on LP see Brace (1998), and references therein. A modern application is included in the SPEDE package onboard the ESA mission SMART-1 (Small Mission for Advanced Research and Technology).

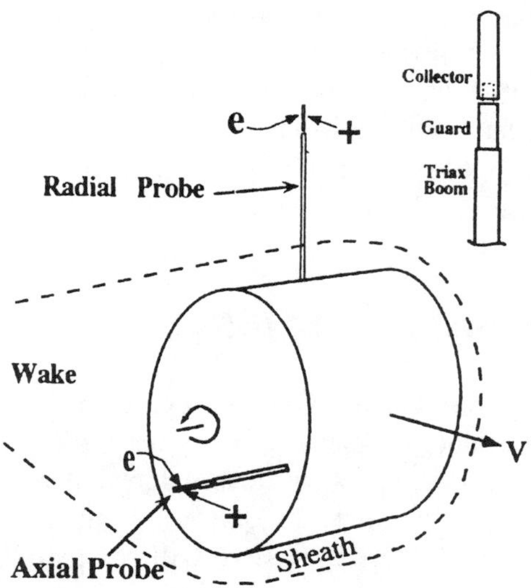

Figure 1. A typical Langmuir Probe arrangement. Two cylindrical probes are mounted on triaxial booms. The radial and axial probes are oriented perpendicular and parallel to the spin axis, respectively (Brace, 1998).

2.1.2   Retarding Potential Analyser

Another frequently used technique is the Retarding Potential Analyser (RPA), well known from laboratory plasma physics. In this technique a voltage is applied to one or several grids mounted perpendicular to the incident particle velocity so that an electric field is created retarding the incident particle motion. Directional information can be obtained by utilizing multiple detectors. Early applications of RPAs in space were in the ionosphere and high altitude plasmas on Lunik 2 (Gringauz et al., 1960) and Explorer 10 (Bridge et al., 1960). A schematic cross section of a typical RPA design is shown in Fig. 2.

In the case that the energy distribution of the ions is an isotropic Maxwellian moving supersonically with respect to the sensor, the flux of ions is given by (Heelis and Hanson, 1998, and references therein)

$$J_i\,(P) = 1/2\ N_i\ V_r\ [1+\mathrm{erf}(\beta_i\ f_i) + 1/\sqrt{}(\pi\beta_i V_r)\ \exp\ (-\beta_i^2 f_i^2)] \tag{1}$$

where N_i is the density of species i, V_r is the velocity of the ions with respect to the sensor along the sensor look direction, $1/\beta_i$ is the most probable thermal velocity $(\sqrt{}(2kT_i/m_i))$, $P = q\,(R_g+\Phi_S)$ is the potential of the retarding

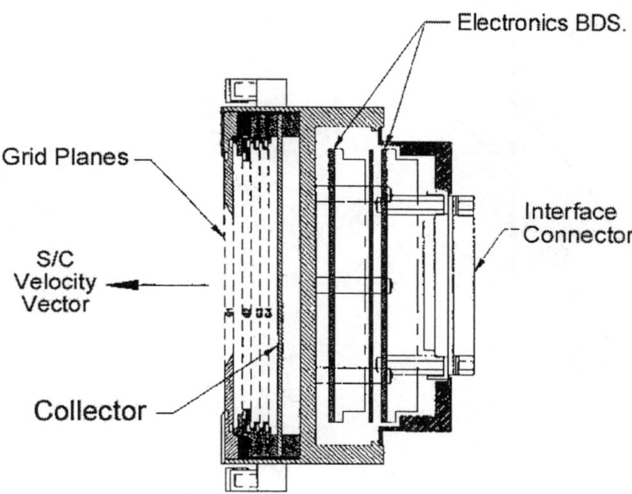

Figure 2. Schematic cross section of a planar RPA sensor (from Heelis and Hanson, 1998).

grid with respect to the plasma, $f_i = V_r - \sqrt{(2P/m_i)}$, and R_g and Φ_S are the potential of the grid and spacecraft, respectively.

The RPA can be extended to energies of several keV, making it also useful for studies of the solar wind and low energy magnetospheric plasma.

2.1.3 Energy-per-Charge Determination with Curved Plate Analysers

This type of analysers takes advantage of the central force motion of ions and electrons travelling in electric fields oriented perpendicular to the incoming particle velocity. We will discuss here analysers based on electric fields in some detail, because they are by far the most common and have a wide range of applications, in particular also for modern low-mass, miniaturized sensors. The curved plate analysers (CPA) consist of concentric electrodes (or plates), the principal configuration is shown in Fig. 3. For spherical geometry, for example, Fig. 3 shows a cross section of a quadrosphere with two concentric spherical shells and the radii R_1 and R_2 refer to the radius of the inner and outer shell, respectively.

214

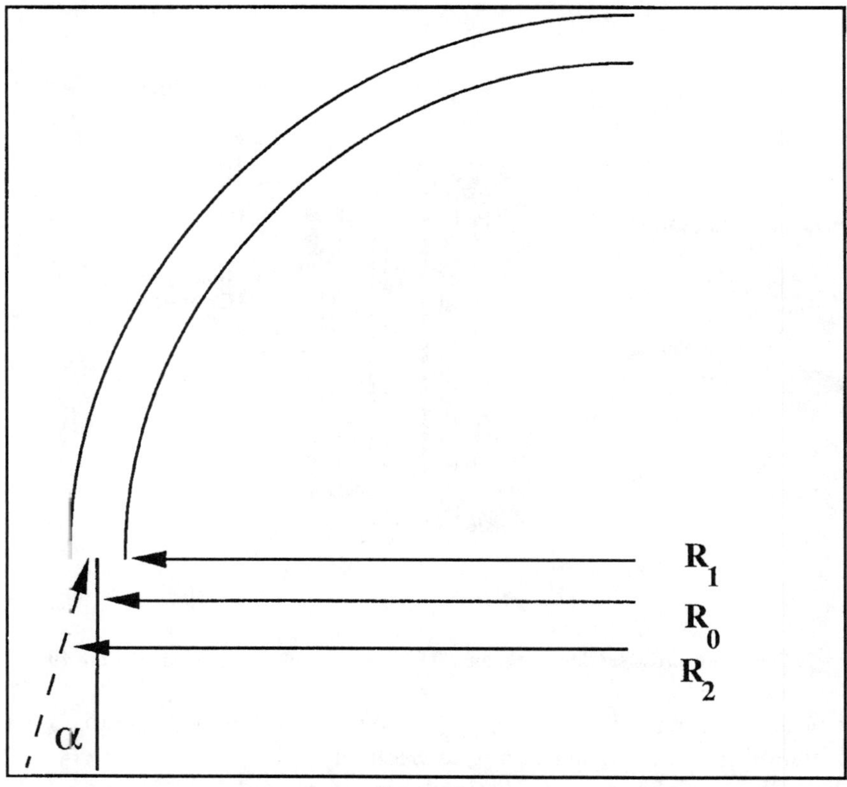

Figure 3. Schematic view of a curved plate analyser

With the definitions
T   kinetic energy of particle
q   ionic charge
α angle of incidence
$V_{1,2}$ potential of plates 1 and 2
V $= V_2\text{-}V_1$
ΔR $= R_2 - R_1$
R_c $= (R_2 + R_1) / 2$
$\phi\,(r)$ Potential between plates
E (r)   Electric field between plates

and using $V_2 = 0$, $V = \text{-}V_1$, the field E (r) between the CPA plates is given by

$$E\,(r) \;=\; V\,R_1\,R_2 / (\Delta R\,r^2) \tag{2}$$

The conditions for transmission for the special case $\alpha = 0$ are given by

$$2\,T\,/\,R_0 = q\,E(R_0), \text{ i.e.} \tag{3}$$

$$T = 0.5\,q\,V\,R_1\,R_2\,/\,(\Delta R\,R_0) \tag{4}$$

With $V_0 = V/2$ and $R_0 = 2\,R_1\,R_2\,/\,(R_1 + R_2)$, the condition for transmission can also be written as

$$T = \mathbf{k}\,q\,V_0, \tag{5}$$

where $\mathbf{k}$ is the Analyser Constant

$$k = (R_1 + R_2)\,/\,(2\Delta R) = R_c\,/\,\Delta R \tag{6}$$

Note that the Analyser Constant k depends only on the geometry of the analyser. It determines, both, the relation between the voltage of the analyser (V_0) and the transmitted energy per charge (T/q), and the geometrical factor $G = A\,TR$, where A is the acceptance area and $TR = <d\alpha\,dv/v>$ is the average transmission. For a spherical analyser with bending angle Φ the transmission TR is given by (see e.g. Paolini and Theodoridis, 1967, Gosling et al., 1978)

$$TR = <d\alpha\,dv/v> = 1/4\,k^{-2}\,\csc^3\,(\Phi/2)\,(7/8 + \cos(\Phi/2)) \tag{7}$$

In space applications spherical and cylindrical designs dominate. Cylindrical CPAs have been used, for example, for magnetospheric studies onboard OGO-3 (Frank, 1967) and for solar wind observations onboard Mariner-2 (Neugebauer and Snyder, 1962). Spherical analysers were first used for magnetosheath and bow shock studies on IMP-1 (Wolfe et al., 1966). In recent years, the design concept of a symmetric qaudrispherical analyzer (or 'top-hat' configuration) first described by Carlson et al. (1985) has been used extensively (see Fig. 4).

This configuration combines the advantages of a curved plate analyser with a 360° field of view and is therefore ideally adapted for making 4πsr particle observations from spinning spacecraft. It has been used, for example, on the AMPTE-IRM spacecraft (Paschmann et al., 1985). For an extensive review see CPAs see e.g. Young, 1998, and references therein. With these spectrometers the energy per charge, T/q, of the particles is determined by varying the electric field of the analyser. The particles are measured with channel electron multipliers (CEM) or micro-channel plates

(MCP) at the exit slit of the analyser. These detectors provide single particle counting capability at high efficiency for both, electrons and ions.

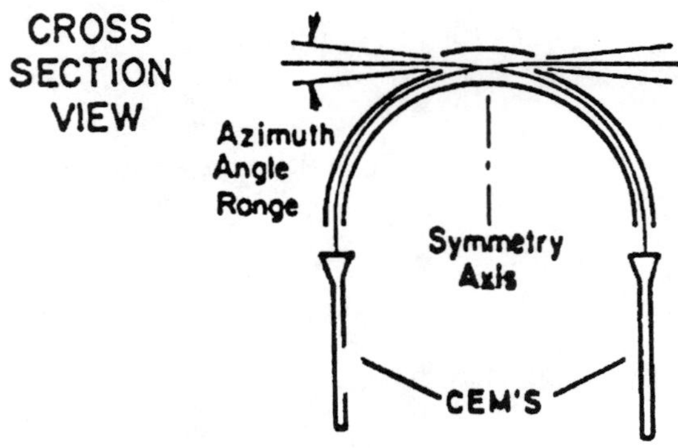

Figure 4. Schematic view of top-hat configuration

2.1.4　Velocity Determination

Although protons are usually the dominant ion in the near Earth plasma environment, the compositional information on minor ions as, for example, O^+, He^+, and He^{2+} provides important information on sources and acceleration processes. In order to obtain compositional information additional parameters of the ions have to be determined. The combination of the energy per charge measurement (T/q), and a time-of-flight measurement provide the particle parameters T/q and velocity, or T/M. Combining the determination of T/M with T/q, the mass per charge, M/q, of the ions can be identified (see Fig. 5). Various configurations are being used for the time-of flight-measurement. One example is shown in Fig. 5. In this configuration, the particles pass through a thin carbon foil. Secondary electrons from the foil, measured at the micro channel plate, provide the 'start' signal for the time-of-flight measurement. The arrival of the ions at the MCP provides the 'stop' signal. From the time difference the velocity can be determined. A typical thickness of C-foils used for this technique is about 3 $\mu g/cm^2$. Low energy ions will be stopped or significantly degraded in energy when passing through these foils. Therefore, in order to extend the energy range to low energies as needed for space plasma applications, the ions need to be accelerated before entering the TOF- section. Typical acceleration voltages applied between the CPA and the TOF-section are $V_{ACC} \sim 15 - 30$ kV.

Thus, ions of charge Q have a minimum energy of eQV_{ACC} before entering the TOF section. The energy per charge E/Q as selected by the CPA plus the energy per charge $e \cdot V_{ACC}$ gained by post-acceleration and the measured time-of-flight τ of the ions propagating through the length s of the TOF unit determine the mass per charge M/Q of the ion according to:

$$M/Q = (E/Q + e.V_{ACC}) / (E/M) = 2 (E/Q + e.V_{ACC}) / \alpha_1 (s/\tau)^2 \qquad (8)$$

The quantity α_1 takes into account the effect of energy loss in the thin carbon foil ($\approx 3\mu g/cm^2$) at the entry of the TOF section and depends on particle species and incident energy.

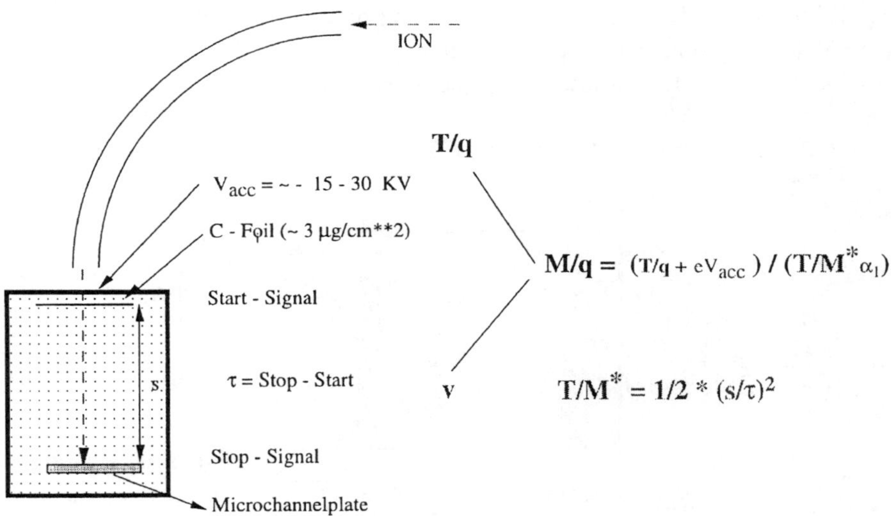

Figure 5. Schematic view of a CPA with time-of-flight measurement

2.1.5   A 3D Analyser with time-of-flight measurement: CODIF

In order to simultaneously determine the 3-dimensional distribution function separately for the main ion species observed in the near-Earth space, the concept of the fast 3D plasma instrument in top-hat configuration on AMPTE / IRM (Paschmann et al., 1985) has been combined with the TOF-technique and post-acceleration in the COmposition and DIstribution Function Analyser (CODIF) for the ESA Horizon 2000 mission CLUSTER (Rème et al., 1993, 1997). Similar instruments, based on the CLUSTER development, have been flown with TEAMS on FAST (Carlson, 1992) and with ESIC on Equator-S (Möbius et al., 1998). The major achievement of these instruments is to allow the measurement of the full velocity distribution functions for the four most abundant species (e.g., H^+, He^{2+}, He^+ and O^+) with high time resolution of typically 1 spacecraft spin. These instruments combine the selection of incoming ions according to energy per charge in a toroidal CPA with post-acceleration by up to 20 keV/e and subsequent TOF analysis. The instrument is mounted with its axis of symmetry perpendicular to the spacecraft spin axis. The front end of the sensor is protruding out of the spacecraft surface so that the 360° aperture has a free field-of view. The electrostatic analyser is of a toroidal top-hat type with a uniform response over 360° of polar angle. As illustrated in Fig. 6, the analyser consists of an inner toroid, to which a variable negative potential is applied, and an outer toroid with a cut-out at the top, and a top-cap lifted above the outer toroid. Both, the outer toroid and the top-cap are normally on ground potential, thereby exposing no high voltages to the outside. Figure 6 also shows how the start and stop signal can be derived from the same MCP by using suitable deflection of the electrons and segmented plates and grids behind the MCPs for the start and stop signals.

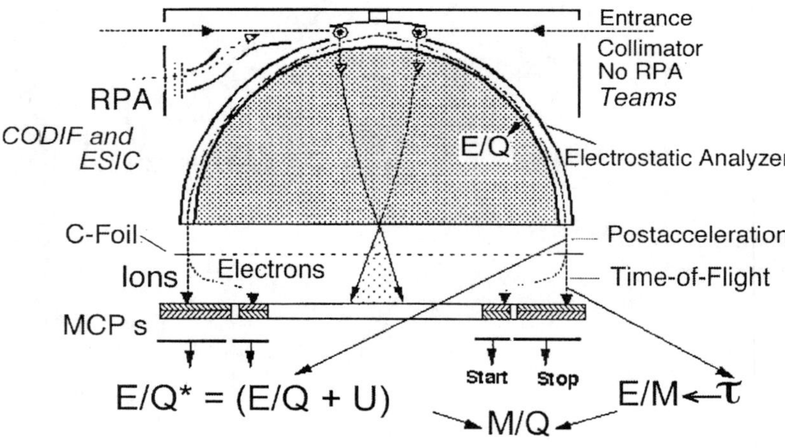

Figure 6. Schematic view of the CODIF sensor onboard CLUSTER

The MCP assemblies are ring-shaped with inner and outer radii of 6 x 9 cm and 3 x 5 cm for the stop and start detectors, respectively. The secondary electrons emitted from the carbon foil also provide the position information for the angular sectoring via segmented plates (22.5° each) behind the inner ring (start-section) of the MCPs for signal pickup.

The horizontal lines (with arrows) on top indicate the flight paths of ions through the normal aperture in a direction normal to the spin axis and in the plane of the drawing. The down pointing and crossing lines show the flight paths of 2 particles on parallel trajectories in the perpendicular direction. The cross over of these lines shows schematically that the ions are focused onto a plane close to the entrance foil of the time-of-flight section.

On Cluster and Equator-S, where regions with very low temperature plasma will be encountered, the low-energy portion of the ion distribution ($\sim$ 0 – 20 eV) is sampled by an additional retarding potential analyser (RPA) at the entrance of the CPA. In the RPA mode of the instrument the ions are collected through a separate aperture of the CPA, as indicated in Fig. 6, while the normal aperture is electrically blocked.

2.1.6   High Mass Resolution Isochronous Mass Spectrometers

High mass resolution with M/ΔM $\sim$ 100 and larger can be obtained by reflecting ions in a linearly increasing electric field (LEF). In such a field configuration, the restoring force is proportional to the penetration depth and the equation of motion of the ion is that of a simple harmonic oscillator. The period of oscillation (i.e. the flight time τ of the ion) depends only on the mass per charge of the ions and not on their energy. Thus, the mass per charge can be determined with high precision from the measured flight time τ:

$$\tau \sim \sqrt{M/Q} \tag{9}$$

At low energies, the majority of the particles leave the carbon foil neutral or singly charged (e.g. Hvelplund et al., 1970, Bürgi et al., 1990). The neutral particles are not deflected by the electric field. Since the majority of the charged particles are singly charged, the mass can be determined from (9) using Q = 1. Several 1D- and 2D-configurations with cylindrical symmetry of this sensor type have been proposed for high mass resolution analysis of the solar wind and planetary magnetospheres (Möbius et al., 1990, Hamilton et al., 1990, McComas and Nordholt, 1990, Wurz et al., 1998)). Several sensors are now in operation, for example on the WIND (Gloeckler et al., 1995), SOHO (Hovestadt et al., 1995), ACE (Gloeckler et al., 1998), and Cassini (McComas et al., 1998) spacecraft. An example of the performance of such an instrument for the measurement of solar wind isotopes is shown in Fig. 7.

2.1.7   Mass-per-Charge and Mass Determination

In order to get the full information on energetic ions, a third parameter, i.e. the energy of the ions needs to be measured. Then, with E/Q, E/M, and E, all parameters fully characterising energetic ions, i.e. E, M, and the ionic charge, Q, can be determined (e.g. Gloeckler and Hsieh, 1979). The total energy is usually measured with one or several silicon solid-state detectors (SSD), with a total thickness sufficiently large to stop the ions. Then, the total energy of an ion before entering the SSD, E_{tot}, can be determined from the measured energy E_{SSD} with

$$E_{tot} = \alpha \; E_{SSD} \tag{10}$$

where α (T, M) > 1 takes into account energy losses in the insensitive front surface of the SSD, recombination effects, and non-electronic energy losses by nuclear collisions with the target atoms in the SSD that are not contributing to the energy signal. The combined effect of non-electronic losses and recombination are usually referred to as 'pulse height defect' (PHD), the non-electronic losses due to nuclear collisions as 'nuclear defect'. The nuclear defect becomes important at low energies and can be calculated from LSS theory (Lindhard et al., 1963). In general, the parameter α depends on particle mass and energy and is determined by calibration measurements.

With E/M determined from the TOF measurement, and the total energy E_{tot}, the mass M can readily be determined from

$$M = E_{tot} \, / \, (E/M) \tag{11}$$

In space, time-of-flight mass spectrometers with energy determination have been first used successfully onboard the AMPTE mission in 1984 (Gloeckler et al., 1985; McEntire et al., 1985; Möbius et al., 1985), covering energies from ~0.3 to 300 keV/e. Since then TOF instruments have become a standard tool in space plasma physics. TOF instruments have been flown in different configurations on many missions, such as Viking, Giotto, VEGA, Phobos, Ulysses, Galileo, and the GGS spacecraft (see e.g. Wüest, 1998 for a recent review). The ion composition results from these missions have clearly demonstrated the excellent capabilities of this type of instrumentation in both the energy range of the bulk plasma as well as at energies of a few hundreds of keV/nuc.

If the ionic charge analysis by electrostatic deflection is omitted, i.e. if the measurement is limited to time-of-flight and energy, then the energy range of this type of sensor can be extended to even higher energies of ~10 MeV/nuc as has been demonstrated on the SAMPEX, WIND, and ACE spacecraft (e.g. Mason et al., 1993, 1997).

2.2   Energetic Particle Measurements at Higher Energies

The energy range of time-of-flight measurements is limited by the accurate determination of short flight times, or the constraints on the feasible length of the flight path. Note that the flight time τ is given by

$$\tau \ (\text{ns}) = 0.72 \ \text{s} / \sqrt{T/M}, \tag{12}$$

where T/M is the kinetic energy in MeV/nuc and s is the flight path in cm. Because the present technical lower limit for the accurate determination of flight times is ~1 ns and flight paths are typically limited to ~3 to 10 cm, for ion energies above ~10 MeV/nuc different techniques are needed.

2.2.1   Measurements of ions with $\Delta E - E$ particle telescopes

Ions travelling through matter loose energy continuously by coulomb collisions with the electrons and nuclei of the absorbing material. For particle velocities exceeding a critical velocity v_c the electronic energy loss is adequately described by the Bethe-Bloch formula (e.g. Anderson and Ziegler, 1977, and references therein)

$$dE/dx = \ k_1 \ Q \ (Z,E)^2 / (E_0 / M) \ f \ (E_1, k_2), \tag{13}$$

where the parameters k_1 and k_2 depend on the target material, and E_0, M, Z, and Q are energy, mass, nuclear charge, and ionic charge of the ion. At high energies, where the ions are fully stripped (several MeV/nuc), Q (Z, E) ~ Z. At low energies, however, heavy ions are only partially ionised and Q depends on the energy of the ions (e.g. Ziegler, 1980). The general concept of ΔE-E instruments is to measure the energy loss in one or several thin detectors of thickness ΔX_i and the residual energy, E_{res}, in a detector that is sufficiently thick to stop the particles (Fig. 8).

If the thickness ΔX_i is small compared to the range of the particles, the energy loss ΔE_i is given by $dE/dx \cdot \Delta X_i$. As ΔE element solid-state detectors or gas counters are being used. The elemental and isotopic resolution is provided by the dependence of dE/dx on mass and nuclear charge. At low energies, where Q^2/M becomes similar for ions of different Z and M, this method cannot be used any more for element or mass determination.

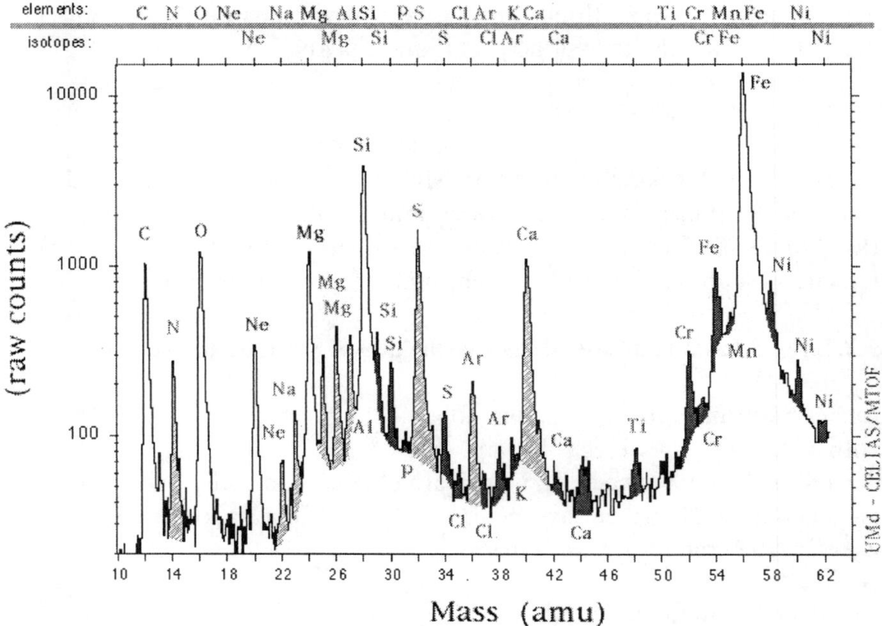

Figure 7. Measurement of solar wind isotopes with the MTOF sensor of the CELIAS experiment onboard SOHO (Lang et al., 1997).

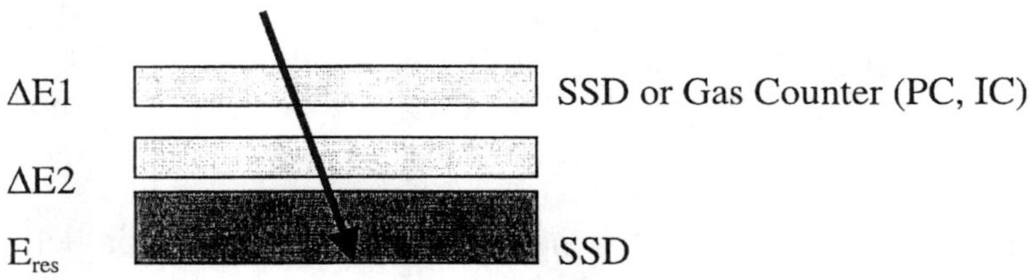

Figure 8. Schematic view of a ΔE-E particle telescope

The incident energy E_0 of the ions can be determined from

$$E_0 = \sum_i \Delta E_i + \alpha \, E_{res}, \tag{14}$$

where $\alpha \, (T, M) > 1$ is defined in equation (10).

The energy loss in the ΔE elements of the senor system will in general depend on the angle of incidence of the particles. In order to improve the mass resolution, the direction of the particles has to be measured to correct the energy loss for oblique angles of incidence. The directional information can be derived, for example, by using strip detectors (e.g. Cook et al., 1993, Stone et al., 1997), position sensitive proportional counters (Klecker et al., 1993, Möbius et al., 1998b), or drift chambers (Klecker et al., 1993) as ΔE elements. With this methods a mass resolution sufficient to resolve heavy ion isotopes can be achieved.

2.2.2   Measurements of Electrons

Similar stacks of detectors can be used for the measurement of relativistic electrons. However, because of the much increased scattering and range of electrons compared to ions of the same energy, the response of electrons is complicated and has to be evaluated by sensor simulations and calibration measurements. For a recent critical review of electron measurements see Vampola (1998).

3.   REMOTE SENSING

3.1   Ground based observations

Long before satellites could be used for near-Earth space physics, ground based instruments have been utilized to obtain information on our closest space environment. This ground based instrumentation ranges from magnetometer chains to modern radar technology operating at HF and VLF frequencies. The most powerful ground based instruments, in terms of the number of parameters measured, are probably the incoherent scatter radars. These instruments, for example, can be used to determine the ionospheric plasma parameters electron density and temperature, ion temperature, velocity and composition, and to measure ion drifts, i.e. electric fields. For a recent overview of ground based measurements see, e.g., Opgenoorth and Lockwood, 1997.

3.2   Observations from Spacecraft using Electromagnetic Radiation

In recent years, remote sensing that utilizes electromagnetic radiation in a large spectral range from visible light to UV and X-rays has been used on several missions. The advantage of this method is to obtain a more global view of the magnetosphere that cannot be obtained with single spacecraft in

224

situ measurements. For a general review of various global imaging techniques, their demonstrated feasibility and potential, see e.g., Williams et al. (1992) and references therein. This global view is particularly helpful to study global disturbances of the magnetosphere of the Earth caused by the interaction of interplanetary shocks and coronal mass ejections (CME) with the magnetosphere. Figure 9 shows an overview of one of the first events studied in detail all the way from its origin at the Sun to Earth (Fox et al., 1998). The two upper images on right of Fig. 9 show images of the aurora, recorded by the Visible Imaging System (VIS) and the Ultraviolet Imager (UVI, Elsen et al., 1998) onboard POLAR. The measured radiation allows to infer the energy input into the polar magnetosphere, and to derive the morphology, the energy spectra and time variation of precipitating electrons. The most recent mission designed, in particular, to apply various remote sensing techniques is the NASA mission IMAGE for the global exploration of the magnetosphere from the magnetopause to the aurora (Burch, 2000, and references therein).

3.3 Remote Sensing with Energetic Neutral Atoms

Energetic neutral atoms (ENA) are produced through charge exchange reactions between singly charged ions (e.g. H^+, O^+) and a neutral gas. Since ENAs are not affected by electric and magnetic fields, they travel along a straight, line-of-sight path from the point where they are created to the point of observation that could be thousands or millions of kilometres away. The flux of the ENAs observed can then be used to infer the flux and energy spectrum of energetic charged particles along the line of sight. This technique has a wide range of applications ranging from the measurement of ENAs from the outer heliosphere of the Sun (Hsieh et al., 1992, Hilchenbach et al. 1998) to the generation of images from the inner magnetosphere of the Earth (McEntire and Mitchell, 1989). In magnetospheric applications, ENA instruments are being used to generate images of the intensity and spatial distribution of energetic neutral atom (ENA) emission. Then, from these images the intensity distribution of the charged particle source population can be inferred, using charge exchange reactions with the neutral hydrogen exosphere of the Earth. Thus, this technique provides information on the global morphology of, for example, the ring current or the plasma sheet of the magnetosphere (e.g. Williams et al., 1992).

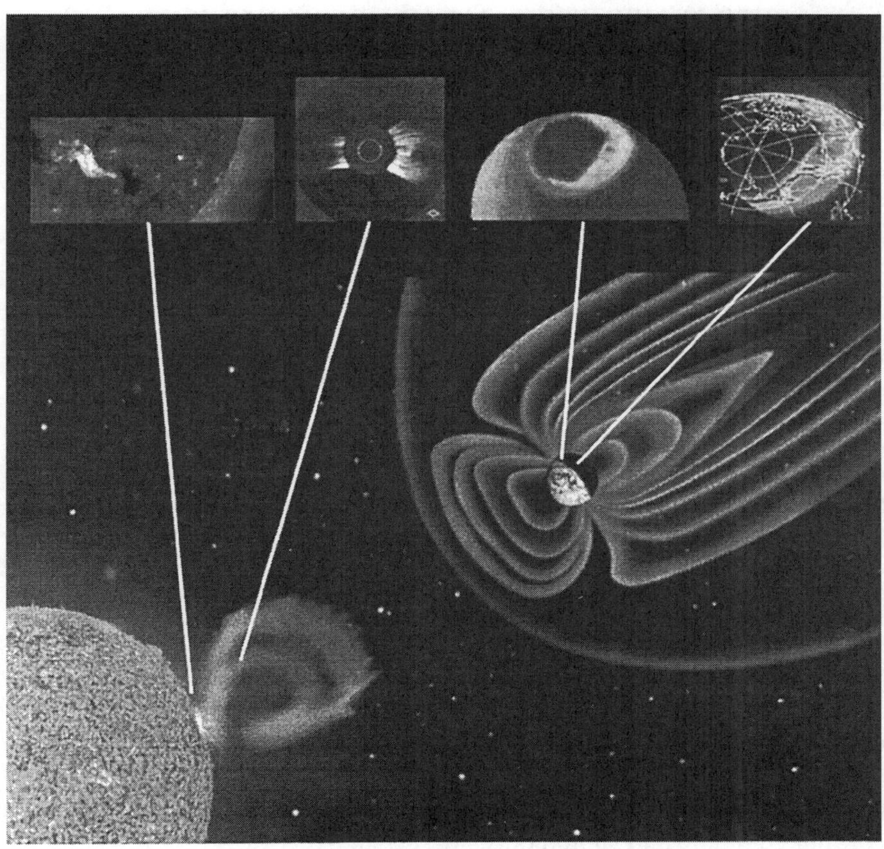

Figure 9. Global disturbances of the magnetosphere of the Earth caused by CMEs from the Sun. The upper images on right show images of the aurora, recorded by VIS and UVI onboard POLAR (from GRL, 25, Nr. 14)

The techniques for the in situ measurements of ENAs are very similar to the measurement of energetic ions. In the energy range 20 keV 500 keV/nuc, for example, electrostatic deflector, time-of-flight, and energy measurements with SSDs are used (Mitchell et al., 2000). The deflector is needed to suppress the usually much higher background of energetic ions in the near-Earth environment. Recently, the Imager for Magnetopause to Aurora Global Exploration (IMAGE) mission has been launched that is the first mission dedicated to imaging the Earth's magnetosphere (Burch, 2000).

4. THE NEXT STEP: MULTI-SPACECRAFT MISSIONS

In order to overcome the limitations inherent with single point measurements in the magnetosphere, the goals of future missions are to

- unfold temporal and spatial variations in the magnetosphere
- explore boundary structures and boundary motions in detail
- provide a dynamic picture of the magnetosphere and its interaction with the solar wind
- provide global, 3D, synoptic images of the magnetosphere

In order, for example, to unambiguously differentiate between temporal and spatial variations a minimum of four spacecraft are needed. The first mission of this type is Cluster (Escoubet et al., 1997). After the loss of all four spacecraft due to the failure of the Ariane 5 launcher in 1996 the project was repeated as Cluster-II and successfully launched in July and August 2000. Cluster-II just started routine operation on February 1, 2001.

4.1 Multi-spacecraft missions

After Cluster-II a number of missions with up to ~100 spacecraft are being explored. Table 1 summarises some of the magnetospheric missions discussed at ESA and NASA for the near and mid-term future. This table is by far not complete and should only provide an overview of the direction the research in this area is anticipated to go. Missions that could be launched within the next ~8 years are the Magnetospheric Multiscale Mission (MMS) and STORMS. MMS is proposed by NASA and includes 5 spacecraft in highly variable orbits, with apogee ranging from upstream to the distant tail STORMS, proposed by ESA, is a three spacecraft equatorial constellation for Earth magnetic storms and inner magnetosphere studies.

4.2 New Sensor Concepts

The multi-spacecraft missions, in particular those involving a large number of small satellites with small total mass (microsats and nanosats) will require the development of new technology, in particular miniaturization of the spacecraft, and development of small instruments with low mass and power for the scientific payload. Some of the technologies needed are already in development and test. The laboratory prototype of a low mass and power sensor for plasma diagnostics, for example, has been discussed by Young et al. (1998b). This sensor concept has already served as the basis of a new generation of miniaturized sensors. It has been implemented in the Plasma Experiment for Planetary Exploration (PEPE) onboard the NASA technology mission Deep Space 1, and in the Ion Electron Spectrometer (IES) onboard the ESA mission Rosetta to the comet Wirtanen (Young et al., 1998b).

Mission	Agency	Number of Spacecraft	Mass (kg)	Phase C/D	Orbit (Re)	Technology Challenge
CLUSTER-II	ESA	4	1180	Launch 2000	4 x 19.5	Variable distance
Magneto-spheric Multiscale (MMS)	NASA	5	240	2004	Apogee from 12 to 127	Large variations of apogee
STORM	ESA	3	1873	$\geq$ 2007		Variable distance
Magneto-spheric Constellation	NASA	100	10	2007	10 - 35	Dispenser Ship Miniaturi-sation
Inner Magnetosph. Constellation	NASA	42	10	2008-2014	2 – 12	Miniaturi-sation Radiation tolerance

Table 1. Active and Planned Multispacecraft Missions

5. ACKNOWLEDGEMENTS

This chapter is a summary of my invited tutorial lecture presented at the NATO Advanced Study Institute (ASI) Space Storms and Space Weather Hazards meeting (Crete, June 18-20, 2000). I would like to thank the organizers, in particular I. Daglis, for their continuous support that made this meeting possible. I also would like to thank all participants for their contributions in the discussions that helped to enhance this chapter. I also wish to thank our technical assistant at the Max-Planck-Institut für extraterrestrische Physik, Birgit Meyne, for editing this manuscript.

6. REFERENCES

Anderson, H.H. and Ziegler, J.F., Hydrogen stopping powers and ranges in all elements, Vol. 3 of the stopping and ranges of ions in matter, Pergamon press, 1977.

Brace, L.H., Langmuir probe measurements in the ionosphere, in Measurement techniques in space plasmas, AGU Geophysical Monograph 102, 23-35, 1998.

Bridge, H.S., et al., An instrument for the investigation of interplanetary plasma, J. Geophys. Res., 65, 3053-3055, 1960.

Bürgi, A., et al., Charge exchange of low energy ions in thin carbon foils, J. Appl. Phys., 68, 2547-2554, 1990.

Burch, J.L., IMAGE Mission Overview, Space Sci. Rev., 91, 1-14, 2000.

Carlson, C.W., et al., An instrument for rapidly measuring plasma distribution functions with high resolution, Adv. Space Res., 2, 67-70, 1985.

Carlson, C.W., The Fast Auroral Snapshot Explorer, EOS, 73, #23, 249, June 9, 1992.

Cook, W.R., et al., MAST: a mass spectrometer telescope for studies of the isotopic composition of solar, anomalous, and galactic cosmic ray nuclei, IEEE Trans. on Geosc. and Remote Sensing, 31, 557-564, 1993.

Elsen, R.K., et al., The auroral oval boundaries on January 10, 1997: a comparison of global magnetospheric simulations with UVI images, Geophys. Res. Lett, 25, Nr. 14, 2585-2588, 1998.

Escoubet, C.P., Schmidt, R., and Goldstein, M. L., Cluster-Science and Mission Overview, Space Sci. Rev. 79, 11-32, 1997.

Fox, N.J., Peredo, M., and Thompson, B.J., Cradle to grave tracking of the Jnanuary 6-11, 1997 Sun-Earth connection event, Geophys. Res. Lett., 25, Nr. 14, 2461-2464, 1998.

Frank, L.A., On the extraterrestrial ring current during geomagnetic storms, J. Geophys. Res. 72, 3753-3767, 1967.

Gloeckler, G., et al., The Charge-Energy-Mass Spectrometer for 0.3 − 300 keV/e ions on the AMPTE CCE, IEEE Trans. on Geosc. and Remote Sens., GE-23, 234-240, 1985.

Gloeckler, G., et al., The solar wind and suprathermal ion composition investigation of the WIND spacecraft, in The Global Geospace Mission, C.T. Russel (ed.), Kluwer Academic Publisher, 79-124, 1995.

Gloeckler, G., et al., Investigation of the composition of solar and interstellar matter using solar wind and pickup ion measurements with SWICS and SWIMS on the ACE Spacecraft, in The Advanced Composition Explorer Mission, Ed. Russell, C.T., Mewaldt, R.A., and v. Rosenvinge, T.T., 86, 497-539, 1998.

Gosling, J.T., et al., Effects of a long entrance aperture upon the azimuthal response of spherical section electrostatic analyzers., Rev. Sci. Instrum., 49, 1260-1268, 1978.

Gringauz, K.I., et al., The study of ionised gas, energetic electrons, and corpuscular emission of the sun with the aid of charged particle three-electrode traps installed on the second Soviet space rocket, Dokl. Acad. Nauk. SSR, 131 (6), 1301, 1960.

Hamilton, D.C., et al., New high-resolution electrostatic ion mass analyser using time-of-flight, Rev. Sci. Instrum., 61, 3104-3106, 1990.

Heelis, R.A., and Hanson, W.B., Measurements of thermal ion drift velocity and temperature using planar sensors, in Measurement techniques in space plasmas, AGU Geophysical Monograph 102, 61-71, 1998.

Hilchenbach, M. et al., Detection of 55-80 keV hydrogen atoms of heliospheric origin by CELIAS/HSTOF on SOHO, Ap. J., 503, 916-922, 1998.

Hovestadt, D., et al., CELIAS- Charge, element and isotope analysis system for SOHO, Solar Physics, 162, 441-481, 1995.

Hsieh, K.C., et al., Probing the heliosphere with energetic neutral hydrogen atoms, Ap. J., 393, 756-763, 1992.

Hvelplund, P.E., et al., Equilibrium charge distributions of ion beams in carbon, Nucl. Instr. Meth, 90, 315-320, 1970.

Klecker, B., et al., HILT: a heavy ion large area proportional counter telescope for solar and anomalous cosmic rays, IEEE Trans. on Geosc. and Remote Sensing, 31, 542-548, 1993.

Lang, K.R., SOHO reveals the secrets of the Sun, Scientific American, 276, Nr. 3, 32-39,1997.

Lemaire, J.F., From the discovery of radiation belts to space weather perspective (this issue), 2001.

Lindhard, J., Scharff, M., and Schiott, H.E., Mat. Fys. Medd. Dan. Vid. Selsk., 33, Nr. 14, 1963.

Mason, G.M., et al., LEICA, a low energy ion composition analyser for the study of solar and magnetospheric heavy ions, IEEE Trans. on Geosc. and Remote Sensing, 31, 549-556, 1993.

Mason, G.M., et al., The ultra-low-energy isotope spectrometer (ULEIS) for ther ACE Spacecraft, in The Advanced Composition Explorer Mission, Ed. Russell, C.T., Mewaldt, R.A., and von Rosenvinge, T.T., 86, 409 – 417, 1998.

McComas, D.J., and Nordholt, J.E., A new approach to 3D, high sensitivity, high mass resolution space plasma composition measurements, Rev. Sci. Instruments, 61, 3095-3097, 1990.

McComas, D.J., et al., The Cassini ion mass spectrometer, in Measurement Techniques in Space Plasmas, AGU Geophysical Monograph 102, 187-193, 1998.

McEntire, R.W., Keath, E.P., Fort, D.E., Lui, A.T.Y., and Krimigis, S.M., The Medium-Energy Particle Analyzer (MEPA) on the AMPTE CCE Spacecraft, IEEE Trans. on Geosc. and Remote Sens., GE-23, 230-233, 1985.

McEntire, R.W., and Mitchell, D.G., Instrumentation for global magnetospheric imaging via energetic neutral atoms, in J.H. Waite, Jr., J.L. Burch, and R.I Morre (eds), Solar System Plasma Physics, AGU Geophs. Monograph 54, 69, 1989.

Mitchell et al., High energy neutral atom (HENA) imager for the IMAGE mission, Space Sci. Rev., 91, 67-112, 2000.

Möbius, E., et al., The time-of-flight spectrometer SULEICA for ions of the energy range 5-270 keV/charge on AMPTE IRM, IEEE Trans. on Geosc. and Remote Sens., GE-23, 274-279, 1985.

Möbius, E., et al., High mass resolution isochronous time-of-flight spectrograph for three-dimensional space plasma measurements, Rev. Sci. Instrum., 61, 3609-3612, 1990.

Möbius, E., et al., The 3-D plasma distribution function analysers with time-of-flight mass determination for Cluster, FAST, and Equator-S, in Measurement Techniques in Space Plasmas, AGU Geophysical Monograph 102, 243-248, 1998a.

Möbius, E., et al., The Solar Energetic Particle Ionic Charge Analyzer (SEPICA) and the data processing unit (S3DPU) for SWICS and SWIMS on the ACE spacecraft, in The Advanced Composition Explorer Mission, Ed. Russell, C.T., Mewaldt, R.A., and von Rosenvinge, T.T., Space Sci. Rev., 86, 449-495, 1998b.

Neugebauer, M., and Snyder, C.W., Solar Plasma Experiment: Preliminary Mariner II Observations, Science 138, 1095-1097, 1962.

Opgenoorth, H.J., and Lockwood, M., Opportunities for magnetospheric reasearch with coordinated CLUSTER and ground based observations, Space Sci. Rev., 79, 599-637, 1997.

Paolini, F.R. and Theodoridis G.C., Charged particle transmission through special plate electrostatic analysers, Rev. Sci. Instrum., 38, 579-588, 1967.

Paschmann, G. et al., The plasma instrument for AMPTE/IRM, IEEE Trans. on Geosc. and Remote Sens., GE-23, 262, 1985.

Rème, H., et al., The Cluster Ion Spectrometry Eperiment, ESA SP-1159, 133-161, 1993.

Rème, H., et al., The Cluster Ion Spectrometry (CIS) experiment, Space Sci. Rev., 79, 303-350, 1997.

Stone, E. et al., The solar isotope spectrometer for the Advanced Composition Explorer, in The Advanced Composition Explorer Mission, Ed. Russell, C.T., Mewaldt, R.A., and von Rosenvinge, T.T., Space Sci. Rev, 86, 357-408, 1998.

Vampola, A.L., Measuring energetic electrons-what works and what doesn't, in Measurement Techniques in Space Plasmas, AGU Geophysical Monograph 102, 339-355, 1998.

Williams, D.J., Roelof, E.C., and Mitchell, D.G., Global Magnetospheric Imaging, Rev. of Geophys., 30, 3, 183-208, 1992.

Wolfe J.H., Silva, R.W., and Myers, M.A., Observations of the solar wind during the flight of IMP 1, J. Geophys,. Res. 71, 1319, 1966.

Wurz, P., Gubler, L., and Bochsler, P., Isochronous mass spectrometer for space applications, in Measurement Techniques in Space Plasmas, Geophysical Monograph 102, 229-235, 1998.

Wüest, M., Time-of-flight ion composition measurement technique for space plasma, in Measurement Techniques in Space Plasmas, Geophysical Monograph 102, 141-155, 1998.

Young, D.T., et al., Cassini plasma spectrometer investigation, in Measurement Techniques in Space Plasmas, Geophysical Monograph 102, 237-242, 1998a.

Young, D.T., et al., Miniaturized optimized smart sensor (MOSS) for space plasma diagnostics, in Measurement Techniques in Space Plasmas, Geophysical Monograph 102, 313-318, 1998b.

Young, D.T., Space plasma particle instrumentation and the new paradigm: faster, cheaper, better, in Measurement Techniques in Space Plasmas, Geophysical Monograph 102, 1-18, 1998.

Ziegler, J.F., Handbook of: Stopping Cross-Sections for Energetic Ions in all Elements, Ed. J.F. Ziegler, Vol. 5, Pergamon press, 1980.

Chapter 9

Solar Activity Variations and Possible Effects on Climate

Eigil Friis-Christensen
Danish Space Research Institute
2100 Copenhagen, Denmark

Abstract: The Earth's climate expresses the combined response of the atmosphere, the oceans and the continents to the energy that is received from the Sun. Any variation in the energy received at the Earth or radiated away from the Earth, and any change in the distribution of the energy over the Earth's surface may have an effect on climate. Precise measurements of the total irradiance of the Sun, the solar "constant" has indeed shown that there is a small variation of the order of 0.1 per cent during the solar cycle. Other manifestations of solar activity have even larger relative variations during the solar cycle. This applies to the UV radiation, the X-rays, the high-energy particles from the Sun, and the extension of the solar corona, the solar wind. The various manifestations of solar activity may each have an impact, small or large, on our surroundings and even on our closest surroundings, the troposphere.

Keywords Solar variability, high-energy particles, climate, solar irradiance, UV radiation, sun-climate relationship, global warming.

1. SOLAR VARIABILITY IN THE CONTEXT OF CLIMATE CHANGE

For several centuries it has been suggested that some of the variations in climate are associated with solar variability. Traditionally, solar variability has primarily been associated with the changing number of dark spots on the Sun's surface and it has been suggested that such changes might indicate variations in the irradiance emitted from the Sun. This energy is driving the

I.A. Daglis (ed.), Space Storms and Space Weather Hazards, 231–250.

dynamics of our atmosphere and any change will therefore have an effect on climate.

The really important parameter regarding climate is the amount of energy from the Sun that actually reaches the surface of the Earth and how this energy is distributed over the planet. Major climatic changes like the quasi-periodically repeating ice ages are believed to be related to the changes in the Earth's orbit around the Sun and the varying angle between the Earth's rotation axis and the direction to the Sun. These changes imply a varying distribution of solar radiation on the two hemispheres. Although the changes in the received energy may seem rather small, such changes apparently have had quite large effects on the Earth's climate. Most probably these changes have been amplified by some effective feed back mechanisms in our climate system, which we do not yet understand. This is also one of the major reasons for the current concern regarding the on-going change in climate that may be directly affected by mankind through the contribution to the changes in the composition of the atmosphere. This change is associated with a change in the radiative properties that may influence the global temperature.

Other processes may, however, also contribute to atmospheric changes. Some of these processes are related to the Sun itself. The ultraviolet radiation from the Sun has a major effect on the production of ozone, and high energy particles of different kinds may also affect the chemical composition of the atmosphere at various altitudes in a manner still not very well known. The ultraviolet radiation as well as the flux of high energetic charged particles found in the surroundings of the Earth is subject to a significant modulation by different manifestations of solar activity.

Therefore, as long as we do not have a proven physical mechanism accounting for the possible solar induced changes in climate it is difficult to define a unique solar parameter that would provide the best description of solar variability related to climate change.

1.1 Variations in total solar irradiance

The energy received at the Earth is primarily determined by the radiation from the Sun. The integrated power over all wavelengths at the position of the Earth is approximately 1368 W/m^2; this figure has traditionally been called the "solar constant".

It is not trivial to measure this figure. Due to the varying optical properties of the atmosphere it was not possible to measure the solar irradiance with sufficient precision until satellites were equipped with appropriate instruments to perform such observations outside the atmosphere. During the last twenty years or so a suite of satellites with sophisticated instruments have finally provided precise, relative measurements of the variations of the total solar irradiance. In spite of all the care that has been exercised, the absolute measurements from the various satellites are, however, not consistent. They differ by a few W/m^2, and this illustrates the difficulty of the measurements and the necessity to have overlapping satellite missions in order to perform cross calibrations so that the long-term variations in the total irradiance may be monitored and their possible climatic effects assessed.

Based on about twenty years of observations it has recently been possible to conclude that there is a distinct solar cycle variation in the total irradiance (Fröhlich et al., 2000). The solar cycle variation closely follows the variation in the sunspot number or the 10.7-cm electromagnetic radiation from the Sun. When the Sun is active and the occurrence of sunspots is high the total irradiation from the Sun is also larger. This might sound counterintuitive since the sunspots are known to be associated with areas on the solar surface that are cooler than the surroundings. It has been found, however, that during high solar activity there is also an increase in the brightness of the remaining part of the solar surface, and this more than cancels the cooling effects of the sunspots themselves (Lean et al., 1998). On top of the solar cycle variation there is a distinct and large 27-day variation, which is associated with the rotation of the Sun. The 27-day variations are considerably larger during the high activity part of the solar cycle indicating the importance of the solar surface activity for the level of solar irradiance.

Although very distinct, the magnitude of the solar cycle change in total irradiance is rather small, about 1 W/m^2 or approximately 0.1 per cent from solar maximum to solar minimum. A basic question is, of course, whether there could be larger changes over longer periods. Lean et al. (1998) used recent experimental data and empirical models of total solar irradiance to estimate that the total irradiance may have been decreased by 0.25 per cent during the lowest solar activity in recent times, the Maunder minimum at the end of the 17th century.

1.2   Variations in the ultraviolet (UV) spectral band

While the variations in the total irradiance during the solar cycle are small the relative variations in other spectral bands are quite different. The conducting part of our atmosphere, the ionosphere, is maintained by the ionisation due to the extreme ultraviolet (EUV) radiation from the Sun. From observations of the total electron density during many solar cycles we know that there is a very distinct solar cycle variation and a ratio of 2:1 from solar maximum to solar minimum in the corresponding spectral band of the UV emission from the Sun has been estimated. Most of the EUV emission is absorbed in the upper atmosphere. At stratospheric altitudes the ozone is absorbing a lower frequency part of the UV radiation from the Sun. This part of the spectrum has a solar cycle variation between 1 and 8 per cent depending on the frequency. The absorption of the UV radiation is associated with a local heating in the stratosphere, which may influence the vertical structure of the atmosphere and thereby possibly also the Hadley circulation in the troposphere.

1.3 Variations in the solar wind

The largest solar cycle changes in the various manifestations of solar activity are associated with the emission of high-energy particles and plasma from the solar surface. During high activity there is an increased number of solar flares. These are often associated with emission of X-rays and energetic protons. Another manifestation of solar surface activity is the ejection of huge clouds of plasma from the surface of the Sun, called coronal mass ejections, CMEs. The velocity of the CMEs is usually larger than the velocity of the solar wind, and both plasma regimes carry with them the frozen-in magnetic field originating from the solar surface. The complex interaction between particles and magnetic fields in the faster moving CMEs, the ambient solar wind, and finally the Earth's magnetosphere is reflected in the more than 130 year series of measurements of geomagnetic activity on various ground stations. One particular representation of geomagnetic activity, the aa-index, is plotted in Figure 1 together with the corresponding monthly average values of the sunspot number. Although there is a clear 11-year solar cycle variation in the geomagnetic activity it is evident that both the individual solar cycle variations and the long-term variations are quite different from those of the sunspot number itself. This illustrates the fact that solar activity cannot be described by one single parameter. In particular it is striking that the minimum geomagnetic activity during the recent solar cycles at the end of the century is comparable to or even larger than the geomagnetic activity measured at solar maximum during the beginning of the century. This indicates that during the last century a fundamental change of solar surface activity has taken place, which is not noticed in the sunspot number. Lockwood et al. (1999) compared the geomagnetic activity with recent satellite estimates of the open magnetic flux of the Sun. Based on this "calibration" they used the record of the global geomagnetic index, *aa*, to conclude that the open magnetic flux of the Sun has increased by 130 per cent since the beginning of the century.

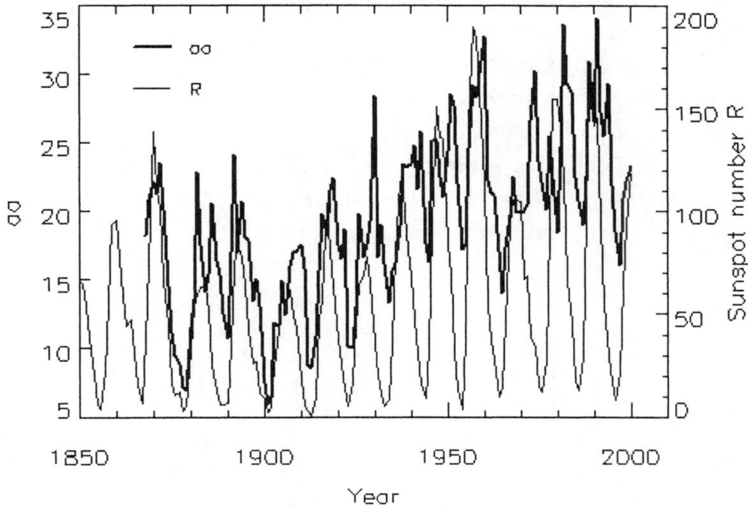

Figure 1. Yearly average values of the sunspot number, R, and the geomagnetic index, aa.

Another distinct modulation related to the heliosphere and the solar wind is observed in the intensity of the galactic cosmic ray (GCR) flux of high-energy particles that originate outside our solar system. These charged particles are affected both by the Earth's magnetic field and the interplanetary magnetic field. The impeding effect of the magnetic fields is larger for the low-energy part of the cosmic ray particles. Therefore the abundance of cosmic ray particles that reaches the top of the atmosphere at lower (geomagnetic) latitudes is considerably less than at higher latitudes. During high solar activity the cosmic ray intensity decreases and this results in an anti-correlation between solar activity and GCR flux. This anticorrelation is not perfect though. In Figure 2 we notice a varying phase difference between the sunspot number curve and the curve of the reversed cosmic ray intensity measured at the Earth.

2. CAUSES OF CLIMATE VARIATIONS

Distributed evenly over the surface of the Earth, the average power from the Sun's radiation amounts to 342 W/m^2 at the top of the atmosphere. In the steady state the Earth re-radiates the same amount of energy as a black body with a temperature corresponding to the global average, approximately 288 °K. The corresponding frequency spectrum has a peak in the infrared part. The short-wave radiation from the Sun that reaches the Earth's surface and the long-wave radiation from Earth to Space have to pass the Earth's atmosphere. Because the transparency of the atmosphere depends on wavelength, the absorption is different for the incoming and the outgoing radiation. The net effect of these processes, the Earth's radiation budget, therefore depends on the composition of the atmosphere. Today the composition of the atmosphere will keep the global temperature about 31°K higher than it would be without an atmosphere. The dominant contributors to this warming effect, the natural greenhouse effect, are water vapour and carbon dioxide (CO_2).

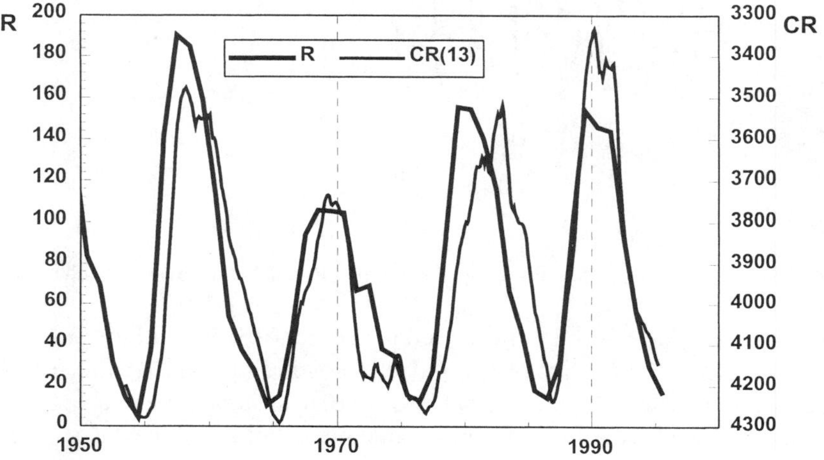

Figure 2. Yearly average values of the smoothed sunspot number, R, and the 13-month running mean of the cosmic ray flux, CR, measured at Climax station.

Any change in the balance between the incoming and the outgoing radiation may have an effect on climate. Changes in the energy received from the Sun may be associated with changes in cloudiness, in volcanic dust and aerosols, and in other aerosols, natural as well as those introduced by human activity. Changes in the energy radiated away from the Earth is associated with the atmospheric properties related to e.g. the varying amount of greenhouse gases like CO_2, Methane, CFCs, and Ozone. But also the surface itself has a significant effect on the radiation away from the Earth, because the albedo is quite different for the various kinds of surfaces like ice cover, various types of vegetation, and soil.

Today the scientific and public attention is mainly directed towards the effect of the increased emission of CO_2 in the atmosphere, which gives rise to an enhanced greenhouse effect. But it is necessary to realise that there is a number of naturally occurring processes that also affect our climate and which are difficult to quantify and understand today.

One of the least understood effects is associated with internal oscillations in the coupled atmosphere ocean system. In particular the ocean currents and their long-term variations are not very well described. Such internal variations may cause substantial year-to-year variations not only on a regional scale but also globally, in terms of the annual global average temperature. Such variations need no external forcing because the large thermal inertia of the ocean may easily account for such changes.

3.   UNDERSTANDING PAST AND PRESENT CLIMATE

Climate models constitute the most widely used tool to predict future climate change. But are today's models sufficiently advanced to be used as prediction tools? We know that the models contain elements that describe the basic physical laws, which determine the dynamics of the atmosphere. But they are also dependent on numbers that cannot be directly calculated but rely on parameterisation of various processes that occur on spatial scales much less than the grid size of the models. Never the less these parameterisations are of crucial importance like, for example, the formation of clouds. Can we be confident that such parameterisations are correct and take into account all the important physics? One obvious way to check this is, of course, to try to simulate past and present climate.

Climate describes the various aspects of the state and dynamics of the atmosphere. These may be the distribution of, say, temperature, precipitation and winds. The single most quoted parameter in the current debate is undoubtedly the global average temperature, which has been derived based on instrumental measurements during the last 140 years. A subset of the global average temperature is the global average sea surface temperature. Reid (1987) noticed a striking similarity between the variation of this temperature index and the long-term variation of the sunspot number during the same time interval and he suggested that the variation in the Sun's total irradiance might be sufficiently large to account for the observed variation in surface temperatures. Direct measurements of the variation in the total solar irradiance during the last solar cycles indicate however, that this variation is probably not sufficiently large to account for the observed change in sea surface temperatures.

Going back further in time, before systematic instrumental measures of the temperature were performed at a sufficient number of locations, our knowledge about past climate is far less quantitative. There are, however, a

238

number of records available that may be used to derive past climate. Measured temperature profiles in boreholes relate directly to past temperature changes. Other records are indirect observations of past temperature. Such proxies may include tree rings, ice core records of different isotopes, and the retreat and advance of glaciers. Proxies need to be calibrated by means of data obtained during modern times in order to provide probable estimates of past temperatures.

Even with a successful calibration of the individual series the problem about the limited spatial distribution of the data remains. This makes it difficult to derive a number that can be regarded sufficiently representative for the global average. This is one of the reasons for the significant differences between various attempts to reconstruct past temperature changes.

Two of the recent reconstructed temperature series have been extensively discussed, namely the temperature series by Jones et al. (1998) and Mann et al. (1998, 1999). These curves have been derived using different methods. The result of Jones et al. (1998) is based on a classical calibration with modern temperature measurements for each site. All sites are then used in a standard calculation of the average temperature curve. Mann et al. (1998, 1999) base their global average on data from sites that have been given varying weight factors. These weight factors reflect how well the temperature time series at a given site follows the expected trend for that site based on spatial correlation maps that have been derived by means of modern instrumental data. In this way it is possible to assign less weight to data that may be erroneous because they show an atypical behaviour. On the other hand this method is based on the assumption that the spatial climate patterns have remained unchanged during the whole period that is investigated. This assumption is questionable and is probably one of the reasons that the amplitude of the global temperature oscillations of this specific derivation are smaller, by nearly a factor of two, than observed in the classical temperature reconstruction for the same time interval by Jones et al. (1998).

The assumption of an unchanged circulation pattern during the whole period including the modern instrumental time is also inconsistent with the recent claim that a global warming caused by an enhanced greenhouse effect has a specific "fingerprint". This means that the observed regional distribution of the warming is different from the distribution that would be expected from natural variations in climate. If so, then the present climate pattern should indeed differ from previous climate patterns. This implies that temperature observations from sites that are not consistent with present day patterns will automatically receive less weight in the averaging procedure whereby the variability of the temperature curve will reduced accordingly.

One of the particular features of the temperature reconstruction of Mann et al. (1999) is the fact that the Medieval Warming around 1000 A.D. and the

Little Ice Age around 1400-1700 tend to diminish in amplitude in comparison with the recent global warming during the instrumental period. Such a levelling of past temperature changes seems to be consistent with the numerical climate models, which indicate a strong effect of the man-made greenhouse gases during the last century. The result is, however, in contradiction to direct measurements of past temperatures in Greenland based on inversion of the temperature profile in two different boreholes in the Greenland Ice Sheet (Dahl-Jensen et al., 1998). Figure 3 shows their result, which demonstrates that these observations from two sites in Greenland separated by 865 km show a very consistent maximum in the surface temperature in Greenland around 1000 A.D. A second, but smaller maximum occurs around 1930 A.D. Two cold periods are present around 1500 A.D. and 1850 A.D. When interpreting this derived temperature history it must of course be realised that the resolution of these particular data series decreases back in time. This feature is inherently associated with the heat flow process.

It may be claimed that the temperature observations on the Greenland Ice Sheet represent just some regional measurement series, which are not necessarily representing the global or even hemispheric temperature variations. But the fact that the two separate sites in Greenland do show a very consistent variation during the last 4000 years indicates that the method is robust. Furthermore, the large observed temperature variations must represent considerable and systematic changes in the circulation patterns, in particular during the last 1000 years. This makes the above-mentioned assumptions regarding the Mann et al. (1998, 1999) results quite disputable.

4. OBSERVATIONAL BASIS FOR A SUN-CLIMATE RELATIONSHIP

Taking into account the above mentioned difficulties associated with the definition of one single parameter describing solar variability, and the even more difficult task of defining past climate variations, it is obvious that a real proof of a sun-climate relationship is probably not possible by means of observations alone. Since, on the other hand, we have not yet identified a unique physical mechanism for such a relationship we need to proceed along two different paths in our search for a more complete understanding of past climate and the possible role of solar activity variations.

240

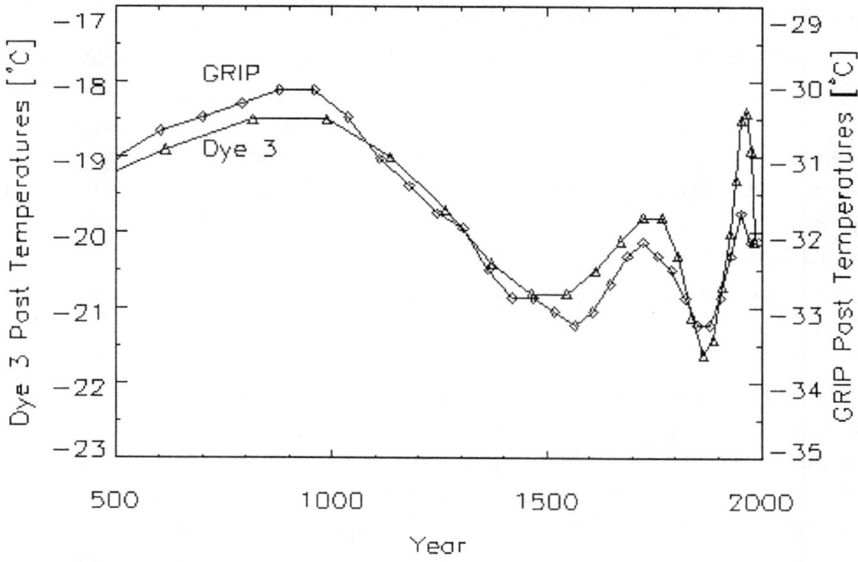

Figure 3. The reconstructed temperature histories plotted for two sites Dye 3 and GRIP on the Greenland Ice sheet (based on data presented by Dahl-Jensen et al., 1998).

One of the paths concerns the search for the consistent and systematic variations in past climate. The other path concerns the investigation of possible physical mechanisms and the magnitude of their effect. These two paths are not quite independent, though. Correlation studies may provide important hints about the most important parameters to investigate and may also be used to rule out some proposed physical mechanisms. In climate research like in most geophysical research disciplines we are faced with the fact that Nature provides the laboratory and determines the experimental set-up. Therefore the resulting observations are often due to a mixture of physical processes.

Eddy (1976) provided an important study of long-term (century scale) variations in solar activity and climate. Based on the observed correlation he hypothesized that the climate changes might be due to small changes in the solar total irradiance. Subsequently studies of paleoclimate and historical solar activity inferred by its modulation of ^{14}C in tree rings and ^{10}Be in ice cores provided evidence that long-term minima in solar activity seems to be associated with climate on Earth that is colder than average.

Inspired by the reported correlation (Reid, 1987) between the sunspot number and the global sea surface temperature during the recent century, Friis-Christensen and Lassen (1991) examined the Northern Hemisphere land temperature series. This series is supposed to be based on the most accurate measurements sufficiently well distributed to be representative of a large fraction of the globe. They concluded that if the sunspot number is a proper representation of solar activity the instrumental temperature record could not support a sun-climate relationship. The main result of their paper indicated, however, that the sunspot number is perhaps not the best parameter to be used as an indicator of solar activity. In stead they pointed at the length of the solar cycle as a more fundamental characterization, or proxy, of solar activity. This suggestion was based on a very high correlation between the filtered length of the solar cycle and the filtered Northern Hemisphere Land temperature curve from 1860 to 1990. Both the steep rise in the temperature from 1920 to 1940 and the subsequent decrease in temperature to 1970 followed the variations in the cycle length. The interpretation by Friis-Christensen and Lassen (1991) was, however, questioned.

First of all, they did not provide a physical mechanism relating the cycle length to a physical parameter that could affect climate. Secondly, because of the heavy filtering of the data, only limited statistical significance could be attributed to the observed correlation, in particular because the recent decade of increasing temperatures could not unequivocally be attributed to solar activity changes since sufficient data points for the filtering were not yet available. Figure 4 shows the filtered solar cycle length and the 11-year running mean of the Northern Hemisphere land temperature according to Jones (1988).

In a follow-up paper Lassen and Friis-Christensen (1995) investigated the significance of the found correlation by increasing the time interval by studying several independently derived sets of proxy temperature records prior to the instrumental period. For this period also the sunspot data are less reliable. Lassen and Friis-Christensen (1995) therefore included records of auroral observations in the derivation of times of solar minima and solar maxima. All of these data sets are of course associated with a higher uncertainty; never the less several independently derived proxy temperature series did show variations during the last 400-500 years that were consistent with the correlation found during the instrumental period. Figure 5 shows for example the spring temperature in the Middle and Lower Yangtze River Valley deduced from phenomenological data, and the filtered length of the solar cycle. The missing solar cycle length data during the period 1630 to 1700 is due to the fact that the very low solar activity – also called the Maunder minimum – did not permit a credible determination of solar activity extrema.

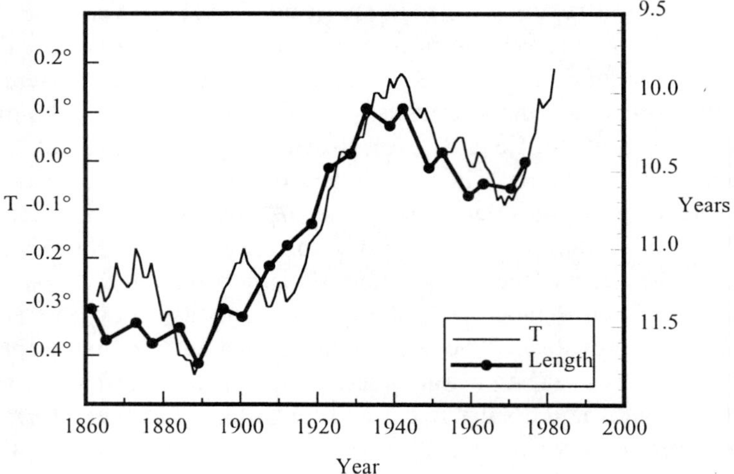

Figure 4. 11-year average values of the Northern Hemisphere Land temperature (T) and the filtered length (12221-filter) of the solar cycle (L).

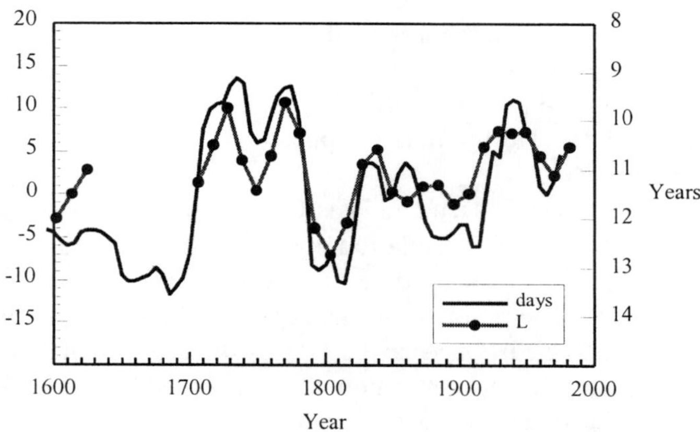

Figure 5. Five-year running average of spring temperature in the Middle and Lower Yangtze River Valley deduced from phenological data, and the filtered length of the solar cycle based on sunspot and auroral data (Lassen and Friis-Christensen, 1995).

During the latest decade the global surface temperature has continued its fast rise. It is therefore of interest to examine whether the solar activity has shown a similar increase that could support the hypothesis of a solar-climate relationship. Additional solar data have now become available that can be used to extend the time series of the derived filtered solar cycle length parameter. In Figure 6 it is demonstrated that since 1985 the Northern Hemisphere land surface temperature has continued its steep increase without a simultaneous decrease in the filtered solar cycle length. The time interval of the latest trend is still too short to allow a comparison with the long-term (1-2-2-2-1) filter used in the original paper by Friis-Christensen and Lassen (1991). Therefore Figure 6 also includes the solar cycle length parameter filtered with a shorter (1-2-1) filter. It is obvious that the surface temperature variation during latest decades is not well correlated with the solar cycle length. Thejll and Lassen (2000) estimated that whereas the solar cycle model was able to explain about two-thirds of the variance in the mean temperatures before 1988, the solar cycle model can now only explain about half of the variance in the temperature. Although this is still a large fraction, it might indicate that the emergence of the climatic effect of man-made greenhouse gasses is now observable.

Such an interpretation implies that the temperature variations during the first part of the century may primarily have been caused by the varying solar activity. Another possible interpretation is, however, that the solar cycle length – although apparently superior to the sunspot number – is not a perfect indicator of solar activity variations related to climate. A different indicator of solar activity, the geomagnetic activity caused by the interaction between the solar wind and the Earth's magnetic field, has continued to increase. Because of the complexity of the physical processes that are involved in solar climate relationship, a single and unique parameter describing solar activity does probably not exist.

Whereas solar activity may be difficult to characterise with a single parameter, the situation is not much better for the global temperature. This is illustrated by the difference in the calculated average temperature for land and sea surface, respectively (Parker et al., 2000). But the problem with the proper definition of the global temperature is further illustrated by the difference in the temperature trends measured by satellites using a microwave sounding unit, MSU. Such global observations have been performed since 1979 and are expected to give a fairly accurate measurement of the temperature in the lower troposphere from the surface to about 7 km altitude (Christy, 1995). Current climate models predict that the effect of increased greenhouse gasses should cause an increase in the lower troposphere temperature that should vary in concert with the surface temperature. However, during the latest 15 years there is no significant trend in the global MSU temperature. This discrepancy is yet not understood but reflects the difficulty in drawing conclusions based on the available data as long as our understanding of the physical processes determining climate is so relatively incomplete.

244

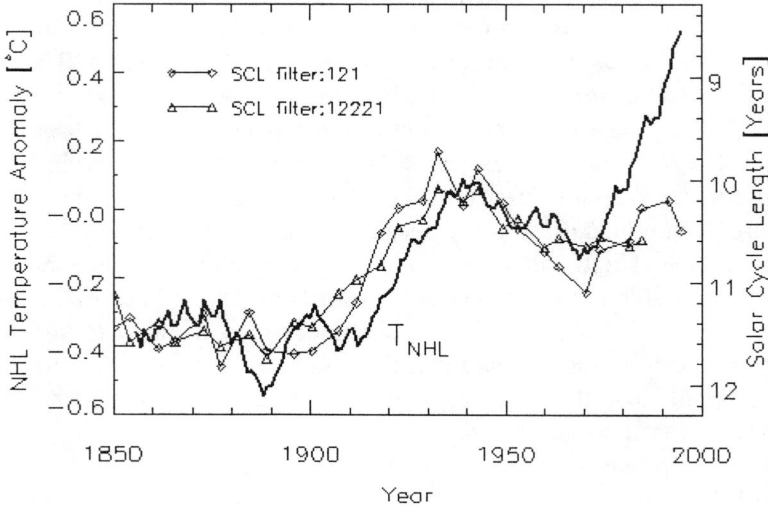

Figure 6. 11-year running average values of the Northern Hemisphere Land temperature derived from data compiled by Jones (1994) and Jones et al. (1999), plotted together with the filtered solar cycle length based on weighting factors 1-2-1 and 1-2-2-2-1 respectively.

5. PHYSICAL BASIS FOR A SUN-CLIMATE RELATIONSHIP

The simplest physical mechanism that can account for an effect of solar activity variations on climate change is the direct effect of the varying total solar irradiance (the solar "constant"). Current climate models all agree, however, that the nominal amount of such variations (~0.1 per cent during a solar cycle and perhaps amounting to 0.25 per cent over centuries) is too small to be of major importance for climate change. This is in particular the case during modern times where the nominal forcing in Wm^{-2} from the increased amount of CO_2 in the atmosphere is dominant.

If the many different reports of a correlation between solar activity variations and climate change do represent a real cause and effect relationship it is therefore necessary to look for possible indirect mechanisms by which such an effect can be manifested.

The solar activity related changes in the ultraviolet part of the solar irradiance have been proposed to be involved in a physical mechanism. Haigh (1996) and Shindell et al. (1999) have conducted model experiments based on changes in the solar spectrum and associated changes in ozone. The results indicated that the absorption of the UV-radiation by the stratospheric ozone could give rise to a stratospheric warming during high solar activity. The model simulations also indicate that the vertical structure of the atmosphere is affected and that the tropospheric circulation may be affected by a broadening of the tropical Hadley cells. Such model simulations may explain some of the observed features in the variation of the 30 mbar geopotential height (van Loon and Labitzke, 2000) but the amplitude of the simulated changes are much too small compared to the observations.

Turning to the most variable part of the solar energy output, namely the emission of particles and fields from the surface of the Sun, the first intuitively based conclusion would be that this energy is much too small. Compared with the energy emitted from the Sun in form of radiation it is quite negligible in the context of climate forcing. Never the less, distinct effects of the energetic particles are observed at all atmospheric heights, even at the bottom of the atmosphere.

In the middle atmosphere, between the stratosphere and the ionosphere, energetic particles including precipitating relativistic electrons from the magnetosphere may significantly affect the chemical processes that are involved in, for example, the formation of odd Nitrogen, NO_y, (Callis and Lambeth, 1998). These atmospheric constituents contribute to the destruction of Ozone, which, in turn, may affect the vertical structure in a way, similar to the effect of changes in UV radiation.

At ground level the major effect is due to the cosmic ray particles. These particles are the main cause of the ionisation of the atmosphere below 55-60 km, except near the ground level below 3-4 km over land. Tinsley (1996; 2000) investigated the effect of ionisation and the associated vertical electric currents on the production of large space charges, which could influence the production of ice forming nuclei and clouds in the higher troposphere. Such effects might be related to the findings by Pudovkin and Veretenenko (1995; 1996) of a local decrease in the amount of cloud cover related to short term changes in the cosmic rays due to increased solar activity (Forbush decreases). The effect, however, seemed to disappear at latitudes lower than 55°, which could be interpreted as an effect of the cosmic ray cut-off due to the increasing horizontal component of the Earth's magnetic field.

A global effect on cloud cover was, however, indicated in the studies of Svensmark and Friis-Christensen (1997) and Svensmark (1998). Figure 7 shows that the global cloud cover measured by satellites is highly correlated with the cosmic ray flux during a large part of the solar cycle. The radiative effect of the total cloud cover is rather complex, though. An increase in the high altitude clouds will imply a warming while an increase in low clouds has a cooling effect. To estimate the climatic effect it is therefore important

246

to find out which type of clouds are affected. In addition to the possibility of assessing the climatic effect, a possible distinction of the effect on different types of clouds may provide crucial information regarding the microphysical processes that may be involved.

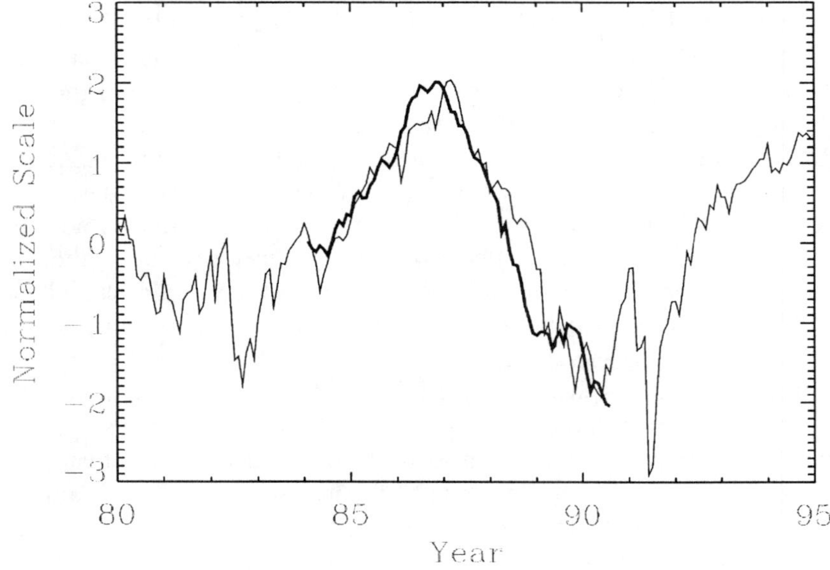

Figure 7. Variation of total cloud cover and cosmic ray flux (thin line) observed at Climax. (Svensmark and Friis-Christensen, 1997).

Kernthaler et al. (1999) did not find any indication of a cosmic ray influence on cloud cover when the cloud data were decomposed by cloud type or by height. While the C2 data set from the International Satellite Cloud Climate Project (ISCCP) used in their analysis reasonably well represents the total cloud cover, this is, however, not the case for the decomposition into individual cloud types. The reason for this is that these data have been created using an inadequate algorithm that was abandoned in 1990 due to its bad performance. A new generation of the data, named the D2 data set, was therefore established using an improved algorithm for the decomposition into the individual cloud types. Taking advantage of this improved data set Marsh and Svensmark (2000) were indeed able to demonstrate a strong correlation between the cosmic ray flux and the frequency of low clouds for an extended time period July 1983 to September 1994, as shown in Fig. 8. The fact that only the low clouds seem to be affected offers an important clue regarding the physical mechanism since low clouds consist of liquid water drops which form on cloud condensation nuclei.

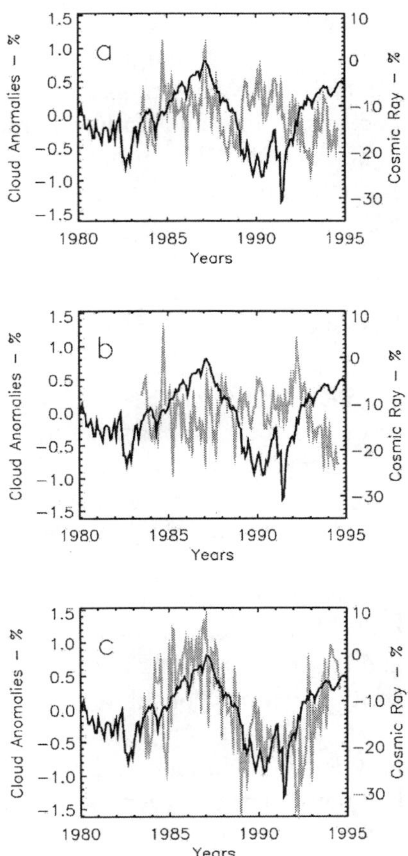

Figure 8. Monthly mean values for global anomalies of: a) high, b) middle, c) low cloud cover (gray) and galactic cosmic ray flux from Climax station (black). From Marsh and Svensmark (2000).

The abundance of cloud condensation nuclei is very dependent on the number of aerosols present in the atmosphere. Therefore an effect of cosmic rays on low clouds could be through their effect on the formation of aerosols. Marsh and Svensmark (2000) refer to observations of the effect on low clouds of ship exhaust. This demonstrates that small perturbations in the aerosol production may have a large effect on clouds. Recent computer simulations (Turco et al., 1998; Yu and Turco, 2000) of aerosol formation by ion-ion recombination indicate that some still unexplored mechanisms may contribute significantly to the enhancement of cloud condensation due to increased ionisation, in particular at low altitudes.

The result regarding the possible effect of cosmic rays on the low cloud cover is confirmed by the correlation between the cosmic ray flux and an independent parameter, the cloud top temperature. Marsh and Svensmark (2000) present a global correlation map demonstrating that this correlation is particularly high over oceans at low and middle latitudes while there is a lack of correlation at higher latitudes. This probably indicates latitudinal differences in the cloud formation processes although the exact mechanism is currently not understood.

6. CONCLUSION

During the last decade much effort has been devoted to the research on climate change. In spite of this there is still no definitive answer to the important problem: How large is the enhanced greenhouse effect due to the increase of human activity? One of the major obstacles in the search for this answer is our incomplete understanding of all the natural changes in climate.

Some of the naturally occurring changes in climate may be related to changes in solar activity. Whether such an effect exists or not is still unproven. There are a lot of indications in the observations of past climate that such effects do occur and there is a growing appreciation that during pre-industrial times, solar activity variations constituted one of the major causes of climate change.

The big question today is the magnitude of the effect and whether it is of significance compared to the enhanced greenhouse effect in modern times, and in particular in the future. This question can probably not be fully answered until a physical mechanism has been found that can explain the observations. Promising progress in this direction has recently been made, although the final answer is still not available.

Perhaps one of the reasons for this is that there is no single answer. An increasing amount of evidence indicates that we should perhaps not focus on only one single physical mechanism. There might be several possible physical mechanisms related to solar activity changes that can operate together in determining the state and dynamics of our atmosphere. Of course this does not in any way exclude the effect of mechanisms that are not related to solar variability.

7. REFERENCES

Callis, L.B. and Lambeth, J.D. (1998) NO$_y$ formed by precipitating electron events in 1991 and 1992: Descent into the stratosphere as observed by ISAMS, *Geophys. Res. Lett.* **25**, 1875.

Christy, J.R. (1995) Temperature above the surface layer, *Clim. Change.* **31**, 455-474.

Dahl-Jensen, D., Mosegaard, K., Gundestrup, N., Clow, G.D., Johnsen, S.J., Hansen, A.W. and Balling, N. (1998) Past temperatures directly from the Greenland ice sheet, *Science* **282**, 268-271.

Eddy, J.A (1976) The Maunder minimum, *Science* **192**, 1189-1202.

Friis-Christensen, E. and Lassen, K. (1991) Length of the solar cycle: An indicator of solar activity closely associated with climate, *Science* **254**, 698-700.

Fröhlich, C. (2000) Observations of irradiance variations, *Space Sci. Rev.* **94**, 15-24.

Haigh, J.D. (1996) The impact of solar variability on climate, *Science* **271**, 981-984.

Jones, P.D. (1988) Hemispheric surface air temperature variations: Recent trends and an update to 1987, *J. Climate* **1**, 654-660.

Jones, P.D. (1994) Hemispheric surface air temperature variations: a reanalysis and an update to 1993, *J. Climate* **7**, 194-1802.

Jones, P.D., Briffa, K.R., Barnett, T.P. and Tett, S.F.B. (1998) High-resolution palaeoclimatic records for the last Millennium: Interpretation, integration and comparison with general circulation model control-run temperatures, *The Holocene* **8**, 455-471.

Jones, P.D., New, M., Parker, D.E., Martin, S. and Rigor, I.G. (1999) Surface air temperature and its changes over the past 150 years, *Reviews Geophysics* **37**, 173-199.

Kernthaler, S. C., Toumi, R. and Haigh, J.D. (1999) Some doubts concerning a link between cosmic ray fluxes and global cloudiness, *Geophys. Res. Lett.* **26**, 863-865.

Lean, J., J. Beer and Bradley, R.S. (1998) Reconstruction of solar irradiance since 1610: Implications for climate change, *Geophys. Res. Lett.* **22**, 3195-3198.

Lassen, K. and Friis-Christensen, E. (1995) Variability of the solar cycle length during the past five centuries and the apparent association with terrestrial climate, *J. Atm. Solar Terr. Phys.* **57**, 835-845.

Lockwood, M., Stamper, R. and Wild, M.N. (1999) A doubling of the Sun's coronal magnetic field during the past 100 years, *Nature* **399**, 437.

Mann, M.E., Bradley, R. S. and Hughes, M. K. (1998) Global-scale temperature patterns and climate forcing over the past six centuries, *Nature* **392**, 770-787.

Mann, M.E., Bradley, R. S. and Hughes, M. K. (1999) Northern hemisphere temperatures during the past millenium: Inferences, uncertainties, and limitations, *Geophys. Res. Lett.* **26**, 759-762.

Marsh, N. and Svensmark, H. (2000) Cosmic rays, clouds, and climate, *Space Sci. Rev.*, **94**, 215-230.

Parker, D.E., Basnett, T.A., Brown, S.J., Gordon, M., Horton, E.B. and Rayner, N.A. (2000) Climate observations – the instrumental record, *Space Sci. Rev.*, **94**, 309-320.

Pudovkin, M. and Veretenenko, S. (1995) Cloudiness decreases associated with Forbush-decreases of galactic cosmic rays, *J. Atm. Terr. Phys.* **57**, 1349-1355.

Pudovkin, M. and Veretenenko, S. (1996) Variations of the cosmic rays as one of the possible links between the solar activity and the lower atmosphere, *Adv. Space Res.* **17**, No. 11, (11)161-(11)164.

Reid, G.C. (1987) Influence of solar variability on global sea surface temperatures, *Nature* **329**, 142-143.

Shindell, D., Rind, D., Balachandran, N., Lean, J. and Lonergan, J. (1999) Solar cycle variability, ozone and climate, *Science* **284**, 305-308.

Svensmark, H. (1998) Influence of cosmic rays on climate, *Phys. Rev. Lett.* **81**, 5027.

Svensmark, H. and Friis-Christensen, E. (1997) Variation in cosmic ray flux and global cloud coverage - a missing link in solar-climate relationships, *J. Atmos. Solar-Terr. Phys.* **59**, 1225-1232.

Thejll, P. and Lassen, K. (2000) Solar forcing of the Northern hemisphere land air temperature: New data, *J. Atmos. Solar-Terr. Phys.* **62**, 1207-1213.

Tinsley, B.A. (1996) Correlations of atmospheric dynamics with solar wind induced air-earth current density into cloud tops, *J. Geophys. Res.* **101**, 29701-29714.

Tinsley, B.A. (2000) Influence of solar wind on the global electric circuit, and inferred effects on cloud microphysics, temperature, and dynamics in the troposphere, *Space Sci. Rev.*, **94**, 231-258.

Turco, R. P., Zhao, J.-K. and Yo, F. (1998) A new source of tropospheric aerosols: Ion-ion recombination, *Geophys. Res. Lett.* **25**, 635.

van Loon, H. and Labitzke, K. (2000) The influence of the 11-year solar cycle on the stratosphere below 30 km: A review, *Space Sci. Rev.*, **94**, 259-278.

Yu, F. and Turco, R. (2000) Ultrafine aerosol formation via ion-mediated nucleation, *Geophys. Res. Lett.* **27**, 883-886.

Chapter 10

Cosmic Ray and Radiation Belt Hazards for Space Missions

An introductory overview

Mikhail I. Panasyuk
Skobeltsyn Institute of Nuclear Physics, Moscow State University
119899 Moscow, Russia

Abstract   Radiation hazard for space vehicles in near-Earth space is caused by a number of factors among which, besides the time of exposure to the radiation environment, the most significant are the orbital parameters of satellites, as well as the levels of solar and geomagnetic activities leading to radiation flux enhancement. The main components of the radiation environment, surrounding the Earth are: galactic cosmic rays, solar energetic particles and the radiation belts. All these components are subject to studies from the point of view of radiation impact on biological structures and spacecraft elements.

Keywords   Radiation belts, galactic cosmic rays, solar energetic particles.

1.   INTRODUCTION

The high-energy electrons, positrons, protons, nuclei and ions of different elements, encountered in space (with energies from hundreds keV to tens GeV), are one of the most important projectiles affecting spacecraft operation and constituting hazard for living organisms onboard any spaceship. This is due to the ability of these particles to penetrate the shielding material and to their interactions with spacecraft elements, electronics, and biological structures (i.e. any matter in condensed state).

I.A. Daglis (ed.), Space Storms and Space Weather Hazards, 251–284.

The radiation environment of interplanetary space is composed of particle fluxes of solar and galactic origin. Solar activity is the most critical factor, determining their characteristics (flux dynamics and energy spectra).

The radiation environment in the vicinity of the Earth, surrounded by a magnetic field, contains a larger number of different components, than the radiation in the interplanetary medium. Among them are the radiation belts (RB), a stable formation, trapped by the Earth's magnetic field, it contains radiation of both solar and terrestrial origin. The radiation environment at low altitudes is determined by the albedo of secondary particles, which are the result of cosmic ray interactions with the atmosphere, as well as particles, precipitated from the RB. The main factor, that influences the dynamics of the spatial distribution of the RB, are geomagnetic disturbances (storms and sub-storms), which as a rule, occur during solar wind plasma interactions with the interplanetary and the Earth's magnetic fields. The other components of the energetic particles in near-Earth space are galactic cosmic ray (GCR) and solar energetic particles (SEP), penetrating into the geomagnetic field.

The numerous components of the radiation environment, with their different physical characteristics, and various mechanisms of their interaction with matter, complicate modelling of the environment and predictions of their impact on objects flown in space.

This paper presents a survey of our current understanding of the basic components of the surrounding radiation environment and the mechanisms of their impact on matter. The empirical models, used to evaluate the radiation effects are also discussed.

2. SPACE RADIATION - A SHORT OVERVIEW

2.1 The Earth's Radiation Belts

The RB are fluxes of energetic particles, confined inside the geomagnetic trap, located in the Earth's inner magnetosphere at equatorial distances ranging from ~1.2 to 7 R_e (Earth radii). The RB predominantly consist of electrons and protons, however, heavy ions (at least up to Fe) are also present. The typical energies for ions are from ~100 keV/nucleon to several MeV/nucleon, and for electrons from ~100 keV to several MeV. The low energy boundary is somewhat conventional and corresponds to the highest energy of hot ring current plasma (tens keV/nucl) inside the magnetosphere, whereas, the upper boundary is determined by the value of the adiabaticity parameter (see for example (Morfill, 1973 and Ilyin et al., 1988))

$\kappa = \rho_L / \rho_m$, where ρ_L is the Larmor radius, and ρ_m is the curvature radius of the magnetic field line. The value of $\kappa \approx 0.1$ defines the maximum energy of electrons and ions for stable trapping on a given L-shell.

The charge state of trapped ions depends on their sources and charge-exchange with the background neutral exosphere.

The RB space-distribution (Figure 1) is different for electrons and protons: the electrons form an inner and outer belt, whereas the protons display a continuous increase of mean particle energy with increasing L-shell value. The most energetic protons populate the inner zone at L<2.5. Particle energy increase due to betatron acceleration occurs during their propagation across the L-shells (radial diffusion) caused by fluctuations of the magnetic field and/or electric fields in the magnetosphere (see, e.g. (Tverskoy, 1964 and Spjeldvik, 1977)). The gap between the inner and outer

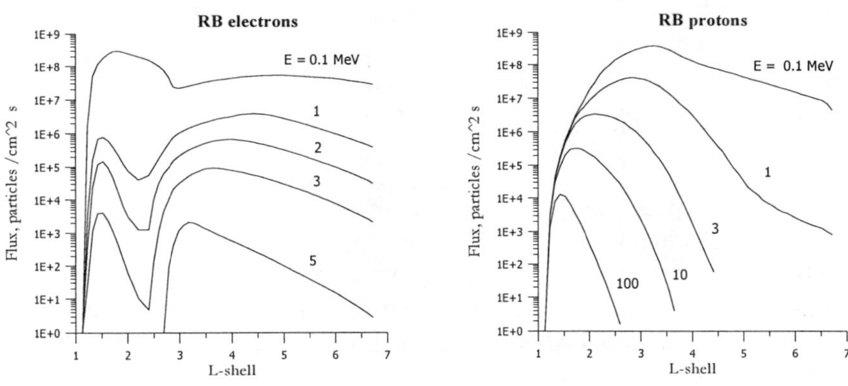

Figure 1. Radial profiles of radiation belt proton and electron intensities in the geomagnetic equator plane.

RB results from particle scattering by electromagnetic waves.

The main sources of RB particles are the solar wind, the ionosphere, and secondary (albedo) particles, which are produced as a result of cosmic ray interactions with the atmosphere. Each of these sources contributes to the population of RB particle population. Hence, the atmospheric source, unlike the solar one, mostly contributes singly-charged heavy ions (He^+, O^+) to the inner magnetosphere. Solar ions are typically multiply-charged heavy ions (He^{2+}, O^{6+}, Fe^{12+}).

Both these sources are responsible for populating the outer zone of the RB. The inner zone sources are mostly albedo particles - the result of GCR interactions with the atmosphere.

2.2   Galactic Cosmic Rays

GCR are fully ionized particles of all existing elements with relativistic energies (ranging from $\sim 10^8$ eV to at least $\sim 10^{21}$ eV). Their chemical composition approximately corresponds to the mean element abundance in the Universe (90% of protons, $\sim 7\%$ helium nuclei, etc.) GCR are of galactic and extra-galactic origin. At energies below 10^{15} eV their origin is associated with supernova blasts, at larger energies their origin is still unclear. Inside the heliosphere the energy spectrum of GCR has a distinct maximum (Figure 2) at energies of 0.3-1 GeV/nucleon. It is caused by modulation of GCR fluxes at energies below the maximum by the variable interplanetary magnetic field, which depends on the cycle of solar activity. The typical period of this modulation is about 11 years; however, periods with other duration have been identified.

The penetration of particles inside the geomagnetic field is determined by its magnetic rigidity:

$$R = \frac{A}{Q}\sqrt{E(E + 2Mc^2)}$$

where A is the mass number, Q is the charge, and M is the mass of the charged particle with energy E. The GCR radiation flux inside the magnetosphere changes due to the 'geomagnetic cut-off effect' which is determined by the deflection angle of a particle inside the geomagnetic field from its incident or initial direction. Hence, the energy spectrum of GCR is distorted due to the filtering effect of the geomagnetic field. GCR flux characteristics inside the magnetosphere are determined by the distortions of the Earth's magnetic field during its interactions with coronal ejections of solar plasma, as well as magnetic storms and sub-storms, associated with plasma injections into the inner magnetosphere.

2.3   The Anomalous Cosmic Ray Component

The anomalous component of cosmic rays (ACR) is the least energetic part of GCR their mean energies are of tens MeV. They mostly consist of O, N, Ne, Ar and Si ions. Unlike the regular GCR component, the predominant charge state of ACR ions is 1+, 2+ (see, e.g. (Blake and Friesen, 1977 and Panasyuk, 1993)). Like GCR, ACR fluxes experience an 11-year solar modulation cycle. ACR with the lowest charge states, penetrate deepest inside the Earth's magnetic field, where they interact with the atmosphere, undergoing charge-exchange. Part of these stripped ions, with higher charge states (up to +8 for oxygen), become stable trapped and form an ACR radiation belt (e.g., Grigorov et al., 1991).

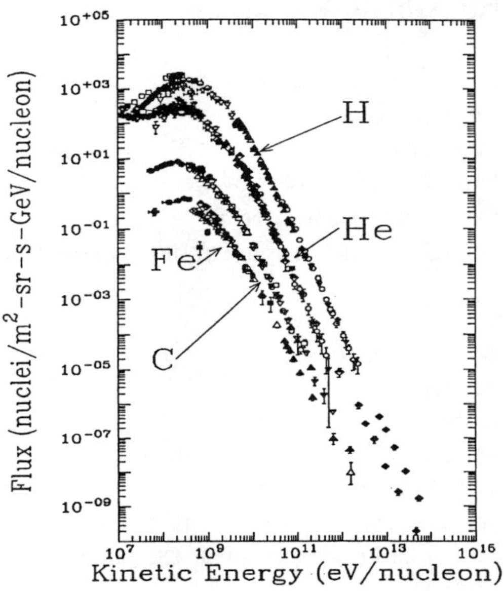

Figure 2. The energy spectrum of galactic cosmic ray nuclei.

2.4 Solar Energetic Particles

These particles are generated in solar flares and coronal mass ejections (CME). The maximum energy of these particles is ~ 100 MeV/nucleon. However sometimes, during gigantic solar events, their energies reach ~ GeV/nucleon. The ionization states of SEP can range from fully ionized for the light ions (protons, He) to partially ionized for heavier atoms (Fe), depending on the solar coronal temperature and the SEP propagation through the interplanetary medium.

SEP occurrence frequency depends on solar activity, displaying a tendency for greater occurrence frequency during maximum solar activity and during decline of the cycle. The chemical composition of SEP varies around the mean elemental composition of the Sun: protons being always predominant, He ions constitute several percent and heavier elements are much less abundant. The measured fluence of SEP events varies within a wide range from 10^5 to 10^{11} particles/cm^2. The duration of the event can range from one to several days. In comparison to GCR, SEP particle spectra are softer, and therefore subject to greater distortion due to the geomagnetic cutoff when measured within the magnetosphere.

2.5   Natural Albedo Radiation

Albedo radiation (AR) is the secondary radiation generated in the inner magnetosphere as a result of:

- nuclear reactions by cosmic ray (GCR, SEP) interactions with protons of the inner radiation belts and atoms of the atmosphere;
- secondary cosmic ray (muon) decay.

This radiation environment component mostly consists of:

neutrons;

gamma-quants;

- electrons and positrons;
- protons and nuclei.

The first measurements of relativistic electrons were made in 1966-68 onboard the Proton 3, and 4 satellites at altitudes of ~ 350 km (Grigorov, 1985). The main result of these experiments was an extremely hard spectrum ($\gamma \approx 1$) of secondary electrons, produced as a result of $\pi-\mu-e$ decay. The integral flux of these albedo particles near the equator was equal to the flux of primary GCR particles.

Detailed data on albedo protons were recently obtained in the Alpha Magnetic Spectrometer experiment (AMS collaboration, 2000). It is obvious, that such secondary particles fluxes (the albedo radiation environment) should be accounted for in radiation impact models for low earth orbiting (LEO orbits) satellite.

3.   THE MAIN RADIATION EFFECTS IN ENERGETIC PARTICLE INTERACTIONS WITH MATTER

3.1   Particle Interactions With Matter

The mechanisms of particle losses depend on their energy. For example, the main losses for *ions* with 10^4-10^8 eV energies are ionization, i.e. the particle loses energy as it ionizes and excites atoms of materials it traverses. For ions with energies exceeding 100 MeV/nucleon the major role is played by nuclear interactions. High energy *electrons* ($E_e >> m_o c^2$) lose energy mostly due to bremsstrahlung, whereas at energies $E_e << m_o c^2$ ionization losses are the most predominant loss mechanisms.

The mean free path, as well as linear energy loss per unit length (LET) in a given material, serves as a quantitative measure characteristic for interactions of penetrating radiation with matter. The electron component of radiation has greater penetrating ability than the ion component. It is to a large extent responsible for the dose effects in matter. However, ions, especially heavy ones, cause larger energy release in a given volume in comparison to light ions and electrons and, therefore, can cause more significant local radiation damage in microvolumes of matter (high-integration electronics, biological structures at the cell level).

Photon and *gamma ray* interactions with matter are also energy-dependent. At gamma-ray energies less than 1 MeV photo-effect is predominant. It leads to the production of low energy electrons, whereas at higher energies pair (electron-positron) production dominates. The Compton effect, accompanied by generation of electrons and gamma ray scattering, dominates in the MeV energy range.

There are various types of *neutron* interactions with matter. They include elastic and inelastic scattering; radiation capture, accompanied by photon emission; capture, accompanied by charged particle emission and nuclear fission. The cross-sections of these processes depend on the neutron energy and differ for all atoms. Unlike photon interactions with matter, neutron interaction displays a resonance structure. In elastic interactions the neutron changes the direction of its velocity and loses part of its energy. Elastic scattering plays a major role in the attenuation of fast neutron fluxes (with energies from 0.2 to 20 MeV). The most efficient attenuation per unit mass is observed in media containing hydrogen. In this case the neutron energy decreases rapidly; the neutrons are thermalized (to energies from $5 \cdot 10^{-3}$ to 0.5 eV) and absorbed by hydrogen nuclei.

Inelastic scattering of neutrons has a threshold energy character and is most significant for heavy nuclei. If the energy of the incident neutron exceeds the energy of the first excited state of the target-nucleus, emission of one or several photons occurs, with a spectrum, typical of the nuclei. Neutron absorption belongs to the inelastic interaction type and for most elements occurs mainly in the thermal region of neutron energies. The trapping cross-section in this case is proportional to $1/\sqrt{E}$. Photon emission occurring in radiation trapping has sufficiently high energy (6-8 MeV) and plays an important role in the formation of radiation fields behind shielding. Inelastic interactions of neutrons also include reactions involving charged particle production: (n,p), (n,α), etc.

As we will see below, besides single event upsets in microchips, the neutron component of space radiation (mainly of albedo and local origin) contributes a major dose in thick shields.

258

3.2　Radiation Effects Caused By Ionization In Solid Bodies

Numerous types of radiation transformations occur in particle interactions with solid bodies. Such interactions can be both reversible and irreversible. The main types of radiation transformation can conventionally be separated into three groups - ionization effects, charge transfer effects, and displacement effects.

3.2.1　Ionization Effects

The electron-hole pairs generated as a result of ionization energy losses lead to new reversible and irreversible properties in materials, such as radiation conductivity, radiation coloring, radioluminescence, radiation-chemical effects, etc. Radiation conductivity and radioluminescence are mainly determined by the dose rate and are reversible; they rapidly ($<$ 1 mcs) disappear when irradiation stops. Radiation-chemical transformations, and radiation coloring, on the contrary, depend on the absorbed dose, and the relaxation time for these effects can be quite long - from 1 to 10^7 seconds and even longer.

3.2.2　Charge Transfer Effects

Compton currents arising under the impact of high-energy gamma rays, or proton recoil during bombardment of light matter by neutrons, are typical examples of charge transfer effects. As a result of radiation impact electric charges can be induced inside the dielectric. The magnitude of this charge density depends on the radiation dose.

3.2.3　Displacement Effects

These effects are associated with displacement of the atoms in a solid body from the crystalline lattice nodes due to energy losses of the interacting particles. The critical energy for the displacement effects depends on the bombarded atoms in the crystal and on the nature of the incident particles: thus, in silicon it is 106 eV for protons and 160 keV for electrons. Displacement effects in a solid body lead to long and short-time defects. As a rule, irradiation by relativistic electrons and gamma-rays ($\sim$ 1 MeV), causes simple primary defects. Irradiation by neutrons, protons, and electrons of higher energies is usually accompanied by a cascade of defects.

3.3   Radiation Damage In Materials And Systems

Radiation stability of materials and systems/components is their ability to operate and preserve initial characteristics under the impact of radiation. Radiation impact causes two types of effects in the satellite's components and materials: gradual degradation of operating abilities due to absorbed dose accumulation and upsets in microelectronics caused by single ion penetration. We will now consider some aspects of both effects.

3.3.1   Dose Effects

Numerous results of laboratory testing and operation of satellite experiments provided data on the radiation stability of satellite materials and components (Akishin et al., 1983). Some of these data are reported in Table 1.

One of the most well studied examples of radiation damage in satellite systems is solar panel degradation. The main reason of their degradation is the decrease of the life-times of minority carriers in the semiconductor base region due to radiation defects. Under the impact of ionizing radiation the following relation is valid:

$$1/t = 1/t_0 + k/F$$

where t_0, t are the carrier life times, before and after irradiation by fluence F, and k is the damage coefficient (which is different for electrons and protons). For typical energies of electrons (0.5-4 MeV) and protons (30 MeV), responsible for radiation damage in solar panels, the value of k is equal to $\sim10^{-9}$ and 10^{-6} respectively. A large decrease of solar panel operation efficiency is necessarily expected in the inner zone of the radiation belts, where the high-energy proton fluxes are largest (see Figure 1.)

Table 1. Radiation stability characteristics for some satellite materials and components.

Material, or system	Irradiation dose, Gr	Impact characteristics
Semiconductor instruments	10^3-10^4	Backward current increase, amplification decrease
Solar panels	10^3-10^4	Efficiency decrease
Optical mirrors	10^3-10^4	Transparency degradation
Fiber optics	10^2-10^3	Transmitted signal damping
Polymer materials	10^4-10^6	Degradation of mechanical and insulating properties
Thermal-insulation coatings	10^5-10^7	Increase of solar radiation absorption coefficient
Metals	10^9-10^{10}	Degradation of metallic properties

Another example of space radiation impact on the satellite components are electric discharges in the irradiated dielectrics. Dielectric materials irradiated by charged particles and gamma rays can produce the accumulation of excessive charge densities. The interior electric charge creates a high electric potential difference. When the electric field intensity inside the irradiated dielectric volume induced by the implanted charge, exceeds the electric strength limit of a material (1-1.5 MV/cm) a breakthrough discharge of the dielectric to its surface will occur. Thus, a fluence of about 10^{12} particles/cm^2 is required to make a breakthrough in a Lihtenberg structure by 4 MeV electrons. The charge inside the dielectric is determined by its structure, shape, interior resistivity, and the presence of capturing centers in the forbidden zones. Electric discharges in dielectrics under the impact of protons and electrons with energies of about 100 MeV, can generate a current pulse of 1-100 A with duration from 0.1 to 1 microsecond and current density up to 10^6 A/cm^2. Dielectric discharges and plasmoid relaxations cause electromagnetic emission in the 0.1-100 MHz frequency range. Such a wide scope of physical phenomena, accompanying electric discharge processes can, of course, lead to anomalies during satellite operation inside the radiation belts. Sporadic increases of relativistic electrons in the outer radiation belt zone have caused radiation anomalies in a number of instances (see, e.g. (Baker, 1996)).

3.3.2   Dose Effect Calculation Models

Primary particles and gamma rays, as well as secondary particles, produced in primary particle interactions with satellite components, can penetrate within modules of the satellite. The absorbed dose is the universal measure of ionizing radiation impact on matter. It is defined as the amount of energy, absorbed by a unit mass of matter. Obviously, the dose load on any objects inside the spacecraft depends on the location of these objects, on the complexity of the satellite configuration and its non-uniform structure.

Radiation effect calculations are usually made in three steps:

1) Radiation environment determination, using corresponding empirical models.

2) Modelling of radiation penetration through shielding and calculations of the energy spectra of the particles, impacting the target. Here it is necessary to keep in mind the complexity of radiation propagation in a solid body. First of all, the stochastic range directions (especially significant for electrons) requires Monte-Carlo simulations. Secondly, particle propagation is accompanied by cascade processes and generation of secondary particles. The evaluation of these processes requires the use of complicated existing

transport codes (like GEANT (CERN, 1994)). These transport codes provide inputs for the third stage of calculations.

3) Estimates of the total energy and/or the energy deposition rate in a target. At this stage it is important to simulate the geometry of the object. The accuracy of the final dose effect calculations depends on how well the geometrical model describes the actual object. The SHIELDOSE (Seltser, 1979) and RDOSE (Makletsov et al., 1997) models are sophisticated mathematical models currently used for calculations of the absorbed dose in complicated satellite systems.

3.3.3 Single Event Upsets

Radiation effects in on-board microprocessor systems onboard satellites were reported for the first time in (Binder et al., 1975). These effects are caused in most cases by single event upsets in digital integral microchips. SEU are associated with deceleration of high energy nuclei in the chip materials and accumulation of excess charges on their electrodes due to charge carrier separation in the ionization track of the nucleus. These nuclei, which are present both in the radiation belts and cosmic rays, penetrate directly into the sensitive volume of the microchip and lead to upsets. The proton component can produce residual nuclei (in nuclear reactions) and nuclear recoil (in elastic collisions) which, in turn, initiate upsets. Another source of upsets are neutrons. They can form nuclei as a result of nuclear reactions in the material surrounding the microchip.

The predominant role of the heavy ions or the protons in upset occurrence depends on the relative fluxes of both components. The proton flux in GCR exceeds the heavy ion flux by 2-3 orders of magnitude, this prevents protons from being an efficient source of upsets, since the nuclear interaction probability for protons in chip materials is only $\sim 10^{-5} - 10^{-6}$. The situation is different in the inner belt, where heavy ions are practically absent: here the upsets are mostly the result of electronic system irradiation by the proton component.

Besides single event upsets in microchips multiple upsets can occur for the following reasons:

- multiple penetration of a single ion through several chips;
- charge diffusion into several sensitive regions of the microchip;
- penetration of a cosmic ray induced shower.

SEU can be regarded as recoverable microchip failures. However, besides these effects, microchips can suffer single-particle radiation effects like dielectric layer breakdown, field transistor burn-through, etc., which lead to irrecoverable failures of the whole microchip or of one of its active elements.

3.3.4 Single Event Upset Effect Models

Numerous empirical and physical models have been developed for quantitative predictions of single event upset effects. (see e.g. (Pickel, 1996; Kuznetsov and Nymmik, 1994). There are models (mostly SEU occurrence models), which describe separately the impact of the ion and the proton radiation component. Models describing both these impacting mechanisms have also been developed.

The quantitative models are based on calculations of the critical charge (which leads to SEU effects). They take into account:
- the overall area and size of the microchip surface;
- the LET spectra of the interacting particles;
- mean free path distribution in the sensitive volume of the microchip.
The numerical values of the model parameters for each microchip type are fitted to match the results of ground radiation testing.

4. EXPERIMENTAL AND MODEL CHARACTERISTICS OF THE SURROUNDING RADIATION ENVIRONMENT

4.1 Radiation Belts

4.1.1 Theoretical Considerations

The spectral and spatial distribution of RB particles is mainly determined by their radial diffusion (see e.g. (Tverskoy, 1964 and Spjeldvik, 1977) which is driven by electric and magnetic field fluctuations. Radial diffusion is accompanied by particle transfer across the drift-shells. This process is described by the Fokker-Plank equation, which includes the transport term, and the terms responsible for particle losses.

The space-energy structure of the RB in scope of the diffusion equation is determined by the balance between particle transport across drift-shells and their losses (energy degradation, pitch-angle scattering, charge-exchange of ions). The particle radial transport inside the magnetosphere leads to particle energy increase with decreasing L-shell number, in concordance with betatron acceleration $E \propto L^{-3}$. The characteristic times of particle radial transfer are determined by the diffusion coefficients D and depend on the L-shell parameter. Thus, for diffusion driven by geomagnetic field fluctuations of the sudden commencement type (SC) (Tverskoy, 1964) the radial diffusion coefficient is approximated by the empirical formula:

$$D_m = D_{0m} L^{10}$$

where D_{0m} depends on the frequency and amplitude (power spectrum) of the SC. The SC generates an induction electric field (E) which causes particles to across magnetic field lines. SCs are initiated by solar wind dynamic pressure and IMF variations. Therefore, significant correlation of long-period variations of D_m should be expected during the cycle of solar activity.

The SC occurrence frequency dependence on the sunspot number W has a high positive correlation coefficient (0,9) (Kuznetsov et al., 1999) Therefore, variations of D_m within the factor of 3-5 should be expected in the course of the solar activity cycle. On the other hand, particle (for example ion) losses are fully determined by the density of the atmosphere and the neutral exosphere, which also change during the solar activity cycle (displaying the opposite dependence on W variations). These qualitative considerations of the physical RB model indicates, that the use of empirical models, based on data sets corresponding to limited time intervals can hardly be justified for engineering calculations regarding other time intervals.

4.1.2 Empirical Models

The most extensively used empirical models of the RB are AE-8 and AP-8, for the electron and proton trapped radiation components, respectively (Vette, 1991; Sawyer and Vette, 1979). These models are based on averaged experimental data mainly obtained during the epoch of 1960-1970 (a time interval from solar activity decline to solar activity maximum). This means the use of transport parameters for the physical model, averaged over the whole interval of this solar minimum period. In other words, the AE-8 and AP-8 models (as well as similar Skobeltsyn Institute of Nuclear Physics (SINP) radiation belt model – RB/SINP (GOST, 1986) represent the averaged space-energy distribution of the radiation belts, typical for certain phases of the solar activity cycle. A comparison between the NASA and SINP radiation belt models can be found in (Beliaev and Lemaire, 1994, and Beliaev and Lemaire, 1996). In this respect, calculations, based on these models, but made for other periods should be regarded only as indicative ones: i.e. as a first approximation of the actual values. More reliable estimates can only be obtained if additional experimental data are made available. Furthermore, the above mentioned models are essentially static, though they display the correct tendency of flux variations for solar cycle minimum and maximum periods.

The recently developed NOAAPRO (Huston et al., 1996), CRESSPRO and CRESSELE (Brautigam et al., 1992) models, can be regarded as dynamic, but they can only describe the proton and electron fluxes in the framework of the corresponding experiments: extrapolation of the calculations to other time periods is obviously controversial.

Therefore, there are several types of trapped radiation variations, that impair the accuracy of model calculation of radiation doses or other radiation effects based on these static empirical or statistical models.

4.1.3   Variations of Trapped Radiation Fluxes

The time variations of the RB space-energy distributions are numerous and have different physical origins. Basically all of these variations can be subdivided into main groups - adiabatic (i.e. occurring with conservation of the first, and the second adiabatic invariants) and non-adiabatic particle motion (i.e. non-conserving the adiabatic invariants). Adiabatic variations are slow compared to the gyroperiod, the bounce period and the period of azimuthal drift of particles. They correspond to slow variations of the surrounding magnetic and electric fields. In this case the space-energy distribution of the radiation belts is restored directly after the end of the disturbance

Non-adiabatic variations can develop on different time scales. Several examples are presented and discussed in the famous paper by McIlwain (1996). Another classical example of non-adiabatic variations is the well-known 'transient' event of May 24th, 1991, when RB electrons were accelerated up to ~15 MeV within minutes and protons up to ~40 MeV, at L~2.5 (Gussenhoven et al., 1992). On the other hand, there are secular variations of the Earth's magnetic field, which lead to noticeable changes in the radiation fluxes in LEO on time scales of several decades. The trends on the distribution of RB particles resulting from these secular variations of the geomagnetic fields have been described and discussed in the final report of the TREND-2 study for ESTEC (Lemaire et al., 1995).

For the electron component it is necessary to mention the 27-day variations, associated with corotating solar wind fluxes (Williams, 1966), and seasonal variations, caused by changes of the geomagnetic dipole tilt angle relatively to the ecliptic plane (Desorger, et al., 1998), as well as non-periodic variations.

The most significant example of global variations of the outer zone electron component is radial transport of particles towards the Earth during geomagnetic disturbances (see, e.g. (Williams et al., 1968). The location of the outer radiation belt maxima (L_{max}) depends on the D_{st} variation amplitude: it is given by the empirical formula $L_{max} \propto |D_{st}|^{-4}$ (Tverskaya, 1996). Such waves diffusion are observed in a wide range of electron energies up to E>8 MeV .It is obvious that such types of variations should eventually be used in future dynamic models of the electron radiation belt component. A first electron model built along this line of thoughts has been proposed by D.J. Rodgers as part of the TREND-2 effort: his Kp - dependent electron model MEA3MSSL is based on the CRRES MEA observations for energies ranging between 153 and 1534 keV (Rodgers, 1996).

A relevant example of relativistic outer belt electron variations are long-term flux variations in geosynchronous orbit (at L~6.6). Figure 3 shows variations of electron fluxes with E_e>2 MeV and protons with E_p >1 MeV, measured on GOES during 1986-1997. It should be mentioned that in general, the electron fluxes are poorly correlated with solar wind velocity increases, as it was assumed when the studies of these types of variations began. It is more probable, that there is a tendency towards flux increase at the decline phase of the solar activity cycle (1992-1996), than during its maximum (1990-91).

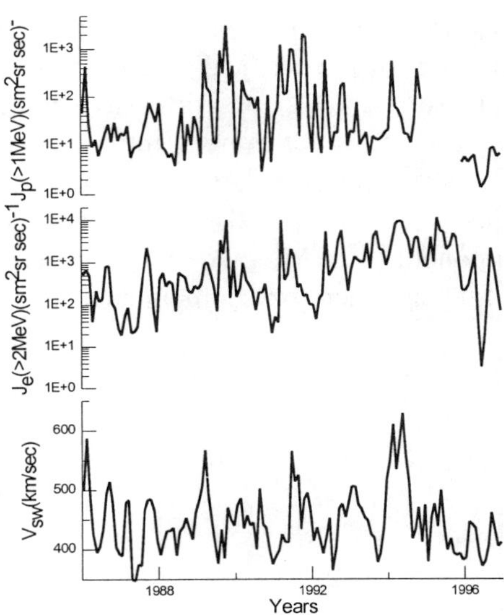

Figure 3. Variations of proton fluxes with E>1 MeV and electron fluxes with E>2 MeV (according to GOES satellite data) and solar wind velocity Vsw for the 1984-1996 time period (Dmitriev et al., 1999)

The physical mechanism of the above-mentioned variations in outer zone electron fluxes is not yet clear. However, the existence of particle diffusion transport during geomagnetic activity increases should be regarded as the basis for a comprehensive physical description of this phenomenon. Even slight variations of the steeply declining electron spectra in the outer zone during amplification of the particle diffusion transport process inside the radiation belts can cause significant changes of the fluxes in the range of relativistic energies. These types of variations are not taken into account in any of the existing radiation belt electron models. Figure 4 illustrates the difference between the actually observed variations of ~1 MeV electrons measured on the GLONASS-94/1 satellite at L≈4.0, and AE-8 as well as RB/SINP model predictions (Panasyuk et al., 1996).

The radiation effects, induced by outer zone electrons are extremely important in onboard equipment radiation stability analysis. Primarily this concerns electric discharge processes deep in dielectric materials (see 3.3). According to CRRES data (Vampola et al., 1992), the deep dielectric discharges began when the currents reached ~0.1 pA/cm^2, which corresponds to electron fluences of 10^6 electrons/cm^2 (compared to the typical fluence values of 10^{12} electrons/cm^2 for electric discharge processes, obtained from ground data). The difference between estimates of the shielding thickness, required for preventing these processes, obtained using AE-8 and CRRES MAX models is quite striking: 300 mg/cm^2 and 750 mg/cm^2, respectively (Vampola, 1996). This clearly illustrates the ambiguity existing for the determination of the 'critical' fluences inducing the electric discharge effect, and the surrounding radiation environment models. The existence of the rapidly varying outer zone electron fluxes leads to more significant long-term radiation dose variations in this region, than in the inner zone where the dose loads are mostly determined by the more stable high energy proton component (tens MeV and more). We present below the main types of variation for this inner zone component.

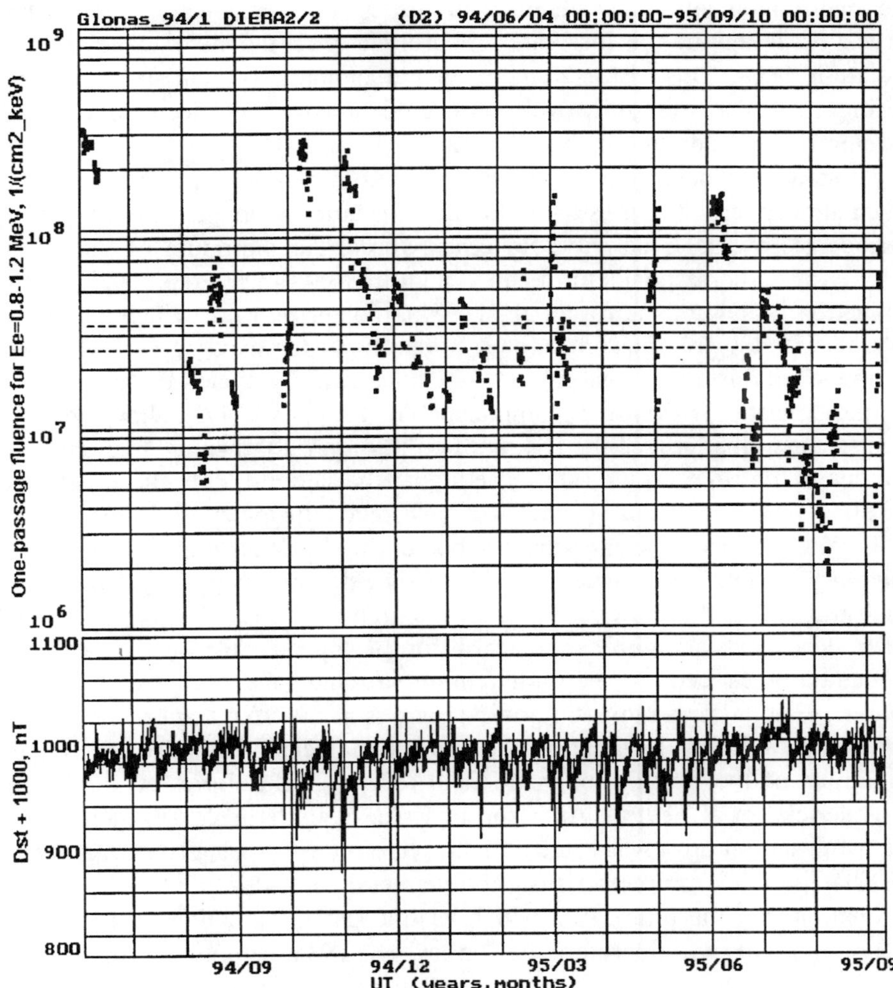

Figure 4. The 0.8-1.2 MeV electron flux as measured on the GLONASS-94/1 satellite. The data is represented in terms of fluence per passage through the belt. The corresponding model estimates are shown by the upper and lower dashed lines. The RB/SINP and AE-8 models were used. Dst index is shown at the bottom (Panasyuk, et al., 1997).

4.1.4   Inner Zone Protons

High-energy protons of the inner zone are mainly generated by the albedo mechanism - decay of neutrons produced in interactions of primary GCR with the atmosphere. Due to weak variations of the source (GCR fluxes) and the slow processes of particle radial diffusion transport ($D_m \propto L^{10}$), the particle dynamics in this case is mainly driven by losses. For protons, these are mainly ionization deceleration in the upper atmosphere. Hence, the largest variation in the loss rate is observed near the outer edge of the radiation belts. Dose measurements made onboard orbital station 'Mir' during the last solar activity cycle show evidence of a significant solar cycle variation of the radiation in the SAA region (Figure 5).

The radiation doses reach maximum values during solar cycle minimum, when atmospheric densities are minimum (the atmosphere gets cooler during periods of low solar activity). In other words, the minimum of the solar activity cycle corresponds to the maximum lifetimes of protons relatively to ionization losses (Bashkirov et al., 1998). Another process, which varies in 'phase' with the previous one, are GCR variations, the maximum of which is also reached during the minimum of the solar activity cycle. The amplitude of the variations for this component of the dose (D2), however, is significantly smaller (see the D2 variation in Figure 5), than for high-energy protons in the SAA region (D1). The higher the satellite orbit, the smaller is the above mentioned solar-cycle variation effect (Huston et al., 1996).

The radiation doses, measured onboard 'Mir' station under 2 g/cm^2 of shielding during solar cycle minimum reached a relatively large value of 2000 mRad/month; this exceeds solar maximum values by a factor of 4-5. Such doses are a significant hazard for prolonged missions since they substantially exceed the permitted limits for humans, accepted in many countries. Long-term radiation dose variations in the SAA region are driven not only by solar-cycle variations of atmospheric density, but also by secular variations of the Earth's magnetic field (see, e.g. Bashkirov et al., 1998; Heynderickx et al., 1996). As a consequence of its larger (51°) inclination the 'Mir' station orbit was more radiation hazardous (especially during years of solar activity minimum) than the standard 'Shuttle' orbit with a 28° inclination. For the future ISS, which has the same inclination as 'Mir', the radiation effects will, therefore, be similar. The ISS assembly will be carried out during the period of solar activity maximum in 2000-2002. During this time period the trapped radiation hazard will be minimum, however the SEP effect risk increases. The next increase of trapped radiation induced dose values can be expected during the next solar minimum (2004-2006).

Solar-cycle variations of high-energy radiation belt protons have been accounted for in the model based on the TIROS/NOAA data (Huston et al., 1996). This model adequately describes the radiation condition dynamics in LEO.

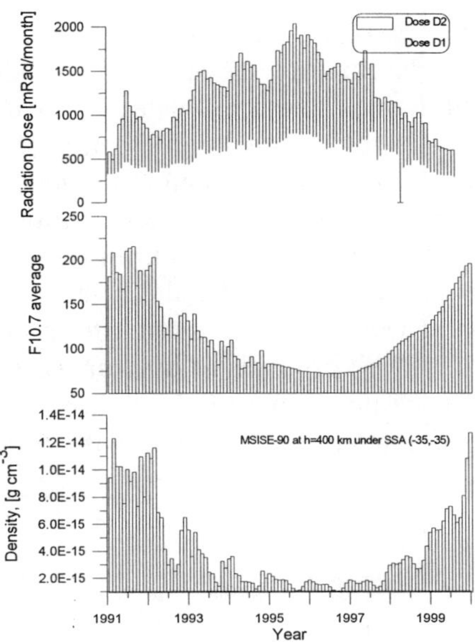

Figure 5. Radiation dose variations D1 (under 0.5 g/cm² shielding) and D2 (under 3.5 g/cm²) onboard 'Mir' station during 1991-2000 as compared to the $F_{10.7}$ solar activity index and atmospheric density (according to the MSISE-90 model) at the altitude of 400 km (Bashkirov et al., 1998).

4.2   The Neutron Environment

The neutron component in space has the following origin:

- solar neutrons, generated in the solar atmosphere as a result of nuclear reactions during active processes;

- albedo neutrons, which are a result of the interaction of primary GCR proton and SEP proton interactions with the atmosphere;

- local neutrons, generated in the satellite structure elements, as a result of GCR particle impact and energetic protons of the inner trapped radiation zone.

Neutrons, coming from the Sun, rarely reach the Earth's vicinity, since their half-decay time is short. The albedo neutron fluxes at low altitudes increase with latitude in concordance with the well-known Lingenfelter dependence, which reflects the GCR latitude effect. This dependence is in good agreement with the data of different experiments (see e.g. Lockwood, 1973).

The local neutron flux depends on both the primary particle spectra and spacecraft mass. Accordingly 'Saluyt-6' data (Yushkov, 1988) outside the SAA the efficiency of local neutron generation is greater than inside the

SAA, where the spectrum of trapped protons is softer. The dependence of neutron generation on the mass of the spacecraft is illustrated in Figure 6. With increasing satellite mass, the local neutron flux increases (Bogomolov et al., 1998). An increase of the shielding thickness of the satellites leads to an increase of the neutron component radiation doses. Thus, according to direct measurements on 'Mir' station, the radiation dose under several tens of g/cm^2 shielding is mainly induced by neutrons (see Table 2.)

Besides the dose effects, induced by the neutron component, local and albedo neutrons can cause single event upset effects in onboard electronics. This effect is associated with products of nuclear reactions, occurring in the satellite materials. It is assumed that SEU induced by background neutrons have production cross-sections comparable with proton SEUs, but with lower energy threshold (Pickel, 1996).

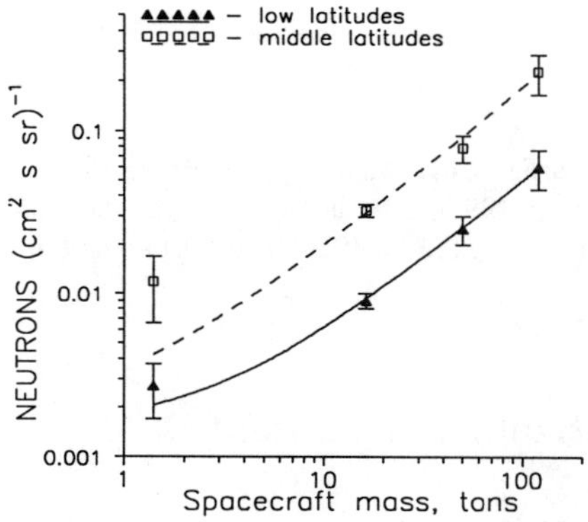

Figure 6. Mass dependence of neutron fluxes at low and middle altitudes (Bogomolov et al., 1998).

Table 2. Ionizing radiation doses (D_i) and corresponding neutron induced radiation doses (D_n) under different shielding thickness, measured on 'Mir' station (Lyagushin et al., 1997) during solar activity maximum (1990-93).

Shielding g/cm^2	Dose µSv/day	
	D_i	D_n
20	143	76
30	92	115

4.3   Galactic Cosmic Rays

4.3.1   Empirical Models

Cosmic rays of galactic origin with energies below several hundred GeV are subjected to solar modulation: with increasing/decreasing solar activity the GCR fluxes decrease/increase. The quantitative description of these variations is built in the various GCR models. At present the GCR model developed at SINP MSU most completely describes the dynamics of all the GCR components ($1\le Z \le 92$) in a wide range of energies E=1- 10^5 MeV/nucleon (Nymmik et al., 1995). The GCR/SINP models provide a more adequate description of the behavior of the main nuclear components as well as electrons, than the widely used CREME model (Adams, 1985). The dynamic GCR/SINP model establishes a direct connection between the particle fluxes and solar activity (the Wolf number - W).

Currently the GCR/SINP model is incorporated into the CREME-96 model developed in the USA (Tylka et al, 1997).

4.3.2   GCR Induced Radiation Effects

As it was already mentioned in section 2, the main radiation effect induced by GCR is single event upsets (SEU) in the memory chips of onboard electronic systems. Therefore it is interesting to compare the results of SEU model calculations and the corresponding experimental data. Currently a vast number of papers are available (see e.g. Bendel and Petersen, 1983; Campbell et al., 1992), which contain data on SEU rates in various types of memory chips, obtained in actual space flights. These data, apart from direct verification of the sensitivity of various microchips to SEU

.0occurrence in flight conditions, provide the opportunity for verification of the SEE and radiation belt environment models.

Among the numerous data on SEE, acquired in space, the TDRS-1 geostationary satellite data are very important, since they cover a large part of the 22-year solar activity cycle. Figure 7 shows simulated and experimental SEU rate values in the Fairchild 93 L422 microchip, without taking into account the large SEP events, which can also cause SEU (Adams et al., 1996). The simulated values of SEU rates were obtained using the GCR/SINP model, described above. They take into account for the distribution of shielding matter surrounding the chip. The conducted study reaches an optimistic conclusion about the capabilities of this model to predict GCR induced SEE effects.

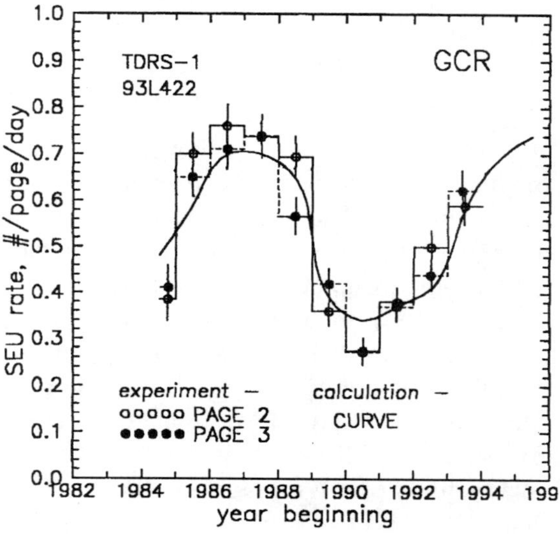

Figure 7. Calculated and experimental values of SEU rates in geostationary orbit (the TDRS satellite) (Adams et al., 1996). PAGE 2 is the working chip, PAGE 3 is the backup chip.

4.4 Anomalous Cosmic Rays

It was already mentioned in section 2.3 that ACR can penetrate inside the Earth's magnetic field and undergo charge exchange in dense layers of the atmosphere, forming a shell of multi-charged ions. The maximum particle intensity of this belt at $L \approx 2 \div 2.5$, exceeds by a factor of several hundreds the intensities of corresponding ion fluxes outside the magnetosphere (Grigorov et al., 1991). The temporal variations of the trapped ACR component are similar to interplanetary ACR flux variations, which, apparently, can be explained by the short lifetimes of trapped ACR (Grigorov et al., 1991). Therefore, the maximum fluxes of trapped ACR particles, are observed during maximum of the solar activity cycle. An analysis of the experimental data of the 'Cosmos' and 'SAMPEX' satellites was made in (Tylka et al., 1997) in order to calculate the LET spectra of trapped ACR for typical LEO orbits and their comparison of LET spectra for GCR and extra-magnetospheric ACR. The results of the analysis show that ACR cannot play a significant role in SEU generation, except for very thin shielding and small-inclination low-altitude orbits where their contribution becomes predominant.

4.5 Solar Energetic Particles

4.5.1 Empirical models

Modern solar physics is not capable of predicting SEP generation, and, thus not their dependence on solar activity. Therefore the probabilistic models of SEP events have been designed. In order to build such models is it is necessary to find the regularities in the occurrence rates of SEP events. It has been found that they depend on the following parameters:
- time;
- fluence and peak flux magnitude;
- elemental ion and isotopic composition;
- particle energy.

Most of the models developed so far are not capable of describing all these characteristics together. They are restricted to extreme cases, and, as a rule, provide only the proton flux. Hence, in the CREME (Adams, 1985) model, all the SEP events are reduced to 'ordinary', 'worst case' and 'anomalously large' types. In this model, the proton energy spectra for the first two types of events are represented by the sum of two exponential functions of energy, and the 'anomalously large' event is described by the extreme SEP of August 1972 (and subsequently by the mean of the August 1992 and February 1956 events). The CREME 96 version of the model (Tylka et al., 1997) uses the October 1998 event as the 99% 'worst case'.

The heavy ion energy spectra are separated in to two different classes - the 'ordinary' and 'heavy ion enriched' events. It is assumed that the ratios between heavy ion fluxes in SEP events do not depend on the energy of the particles. Other models (e.g., Feynman, et al., 1992) consider only the probabilities of the occurrence of SEP events of different fluences, and do not account for the existence of heavy ions in SEP events.

The recently developed SEP/SINP (Nymmik, 1999) model describes the characteristics of both fluences and peak fluxes of the solar event proton and ion events. Whereas in all the previous models it was assumed, that the probability of a SEP event over several years of active Sun is the same as over other phases of solar activity, the SEP/SINP model is based on the concept, that the mean occurrence frequency of proton fluences is indeed a function of solar activity. Thus, the mean occurrence frequency (events per year) of SEP events $\overline{v}$ with fluences $\geq 10^5$ protons/cm^2 is defined by the power law:

$$\overline{v}(t) = 0.3 \cdot \overline{\overline{W}}(t)^{0.75}$$

where $\overline{W}$ are Wolf numbers averaged over 12 months. The distribution of V as a function of fluence, as well as the differential energy spectra of both fluences and the peak flux values (for all the particles in the $E \geq 30$ MeV/nucl) are also power law functions in this model. These functional regularities were established on the basis of experimental data. The power law dependence, as opposed to the log-normal one, is the main difference between the SEP/SINP model and all other earlier alternative models. The probabilistic SEP/SINP model enables to calculate the energy spectra of all the main SEP components for any space mission duration ranging from 1 month to 11 years and to provide event probabilities.

4.5.2   SEP Induced Radiation Effects

Figure 8 shows the energy dependence of the SEP fluence energy dependence calculations (the solid lines are differential energy, the dashed lines are integral spectra) for the whole 22-nd cycle (left) and solar activity minimum periods (limited to four 4-year intervals around solar minimum). GCR calculations, employing the GCR/SINP model for the corresponding time intervals are also shown. It should be noted, that even during periods of the quiet Sun, the differential fluences of SEP protons exceed GCR fluences (for energies up to ~ 100 MeV). This result demonstrates how important it is to account for SEP when estimating the total radiation effect.

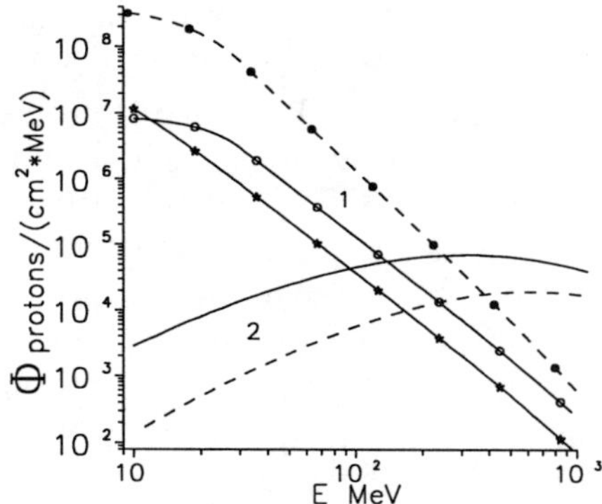

Figure 8. The differential energy spectra of annual proton fluences according to the model (Nymmik, 1999) for 'quiet' Sun periods (open circles) and time periods close to maximum (dark circles). The solid and dashed lines (1) are experimental data extrapolations to high energies for 'quiet' Sun and high solar activity, respectively. The solid and dashed lines (2) are the annual integral galactic proton spectra for the same solar activity levels.

Nevertheless, it is necessary to emphasize, that additional research is needed to verify the basic results of the SEP/SINP model and other models to achieve an adequate and more definitive description of the radiation effects with respect to SEP events.

Another important problem of SEP modelling is to determine their characteristics after penetration inside the magnetosphere. This stage of modelling implies knowledge of the cut-off rigidity values. In this case the particle fluxes depend on the geographical coordinates of the altitude, latitude, longitude of the point of observation in the magnetosphere on the particle energy E, and on its direction of arrival.

Our understanding of the cutoff rigidity has significantly changed over the past years. The transition from the first calculations using the dipole model of the terrestrial geomagnetic field (the Stormer approximation) to the multi-pole IGRF fields, accounting for various current systems inside the magnetosphere and on its surface, contributed to our current level of understanding of the rigidity cutoff. Therefore, the calculations, concerning SEP penetration inside the magnetosphere, require the use of dynamic magnetic field models, which account for geomagnetic disturbances. The widely used Tsyganenko model (Tsyganenko, 1989) satisfies these requirements.

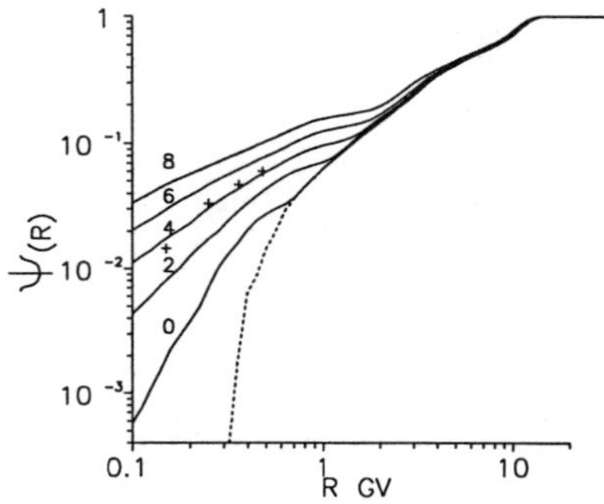

Figure 9. Penetration function (Ψ) for SEP depending on rigidity (R) for an ISS orbit at different Kp-index values. The crosses are modelling results (Aristova et al., 1991) for Kp=5.

An example of calculated penetration functions for ISS orbit depending on geomagnetic cutoff for different levels of geomagnetic activity is given in Figure 9. The calculations were made using the technique suggested in (Danilova and Tyasto, 1995), which provides penetration function calculations with accuracy better than 10%. It can be seen that the SEP particle flux in the low-energy part of the spectrum (<100 MeV) strongly depends on the level of geomagnetic activity. In reality, even in very low orbits (such as piloted orbital station orbits), the penetration of SEP, is the main radiation hazard factor during powerful events on the Sun, accompanied by magnetic storms. Thus, during the solar events of October 19th 1989 and March 24th, 1991, the integral radiation doses on 'Mir' station (induced by SEP penetration) increased by large factors of 2.5÷3 during several hours (Aristova et al., 1991). It should be noted that up to shielding thickness of several g/cm^2 the SEP radiation doses exceed the estimated dose values induced by both GCR and trapped radiation belt protons and electrons during these extreme SEP events.

Besides dose effects, protons and other SEP ions can cause SEU. Table 3 shows the experimental and calculated data of the total SEU number (over the whole SEP event) for large SEP events in 1989-1991. The experimental data are represented by the averaged annual values for control unit microchips (93L422) of the TDRS-1 geostationary satellite. The calculation was made by taking into account the distribution of shielding matter surrounding the microchips.

It is obvious that the calculated SEU rate under SEP impact is formed in approximately equal proportions by protons and ions, which is a consequence of the distribution of shielding matter around the chip. Comparison of the experimental and calculated values shows that the average discrepancy between them is ±60%, and in the worst cases it reaches a factor of ~2.5. These discrepancies are probably explained by the difference between the actual SEP spectrum and the 'mean event' spectrum used in the model.

Table 3. Experimental and calculated values of the total number of SEU on the TDRS-1 satellite (Adams et al., 1996).

SEP event	Experimental value	Calculated values		
		Protons	Ions	Total
August 1989	10±3	15	9	24
September 1989	34±6	21	17	38
October 1989	98±10	70	57	127
March 1991	8±3	8	7	15
May 1991	24±5	10	10	20

4.6 The Near-Earth 'Integral' Radiation Environment

The radiation impact of the space environment, surrounding the Earth, is formed by a combination of individual components, such as: radiation belts, albedo particles, galactic and solar cosmic rays. The actual impact of each of these factors is to a large extent determined by the orbit of the satellite, its construction (shielding of individual units and the volume of the modules), and the time of the satellites active existence in near-Earth space. The flight is conducted in different space radiation fields, with essentially different space-time characteristics.

Therefore, radiation effect characteristics should account for two cases - the maximum, peak particle flux and the averaged (daily mean, annual mean, etc.) flux, which should be recorded in orbit during prolonged missions. The first case permits to estimate the spacecraft 'survival' in extreme conditions, the second one - the overall level of serviceability. We will discuss the specific features of 'integral' radiation effects using several typical satellite orbits as an example.

4.6.1 Low Earth Orbit

The term LEO is usually used in reference to orbits with perigee altitudes below 1000 km. All piloted spacecraft, spacecraft performing Earth surface observations, etc., typically operate at these altitudes. The radiation belt particle fluxes in these orbits are extremely time-dependent, and vary significantly for certain orbit revolutions ranging from minimum levels to maximum ones (exceeding the average level by orders of magnitude) during flight intervals of several minutes. The maximum RB particle flux values are observed in the SAA region, therefore 'peak' flux values should be calculated for orbit revolutions, passing over the center of the SAA.

The mean RB particle fluxes are calculated as a result of model flux averaging for one day of the flight, during which some of the orbits pass over the SAA center. It should be mentioned, that satellites in LEO orbit spend most of the mission time at altitudes, where the particle drift shells are not distorted by geomagnetic disturbances, therefore, the calculations, are, as a rule, made for a quiet geomagnetic field. GCR particles and SEP are subject to complex time variations in LEO. First of all, these are their own variations in the interplanetary medium; secondly, these are variations associated with GCR and SEP particle penetration into different parts of the orbit under different geomagnetic activity conditions. Satellites with sufficiently large orbit inclinations at high altitudes enter regions where GCR and SEP fluxes are predominant.

When making GCR flux impact predictions it is reasonable to use their flux calculations for the epoch of solar minimum, when maximum fluxes of these particles are observed. When calculating SEP fluxes, it is necessary to account for the probabilistic nature of SEP event occurrence. Therefore, the maximum estimated SEP flux should be found according to the peak flux of an event, which is expected with a certain probability. The mean daily flux has to be substituted by the 'mean' flux value, achieved in one event with the same probability over a long (not less than 11 years) time period. The maximum estimated SEP flux should be calculated for interplanetary medium conditions, and the mean flux estimates should take into account the particle penetration function into a given orbit.

Figure 10 shows the results of annual radiation dose predictions for the ISS orbit. The calculations were made (Kuznetsov and Lobakov, 1999), using the generalized model of SEU calculations, described in (Bendel et al., 1983) and the SEP/SINP model (Nymmik, 1999), the GCR/SINP (Nymmik et al., 1995), and AP8 (Sawyer and Vette, 1979). The conclusions, concerning ISS operation, which can be made from the calculations presented above, are the following:

1) The radiation doses are mainly induced by RB particles (protons), in the whole range of reasonable shielding thickness. During minimum solar activity the radiation doses are maximum.

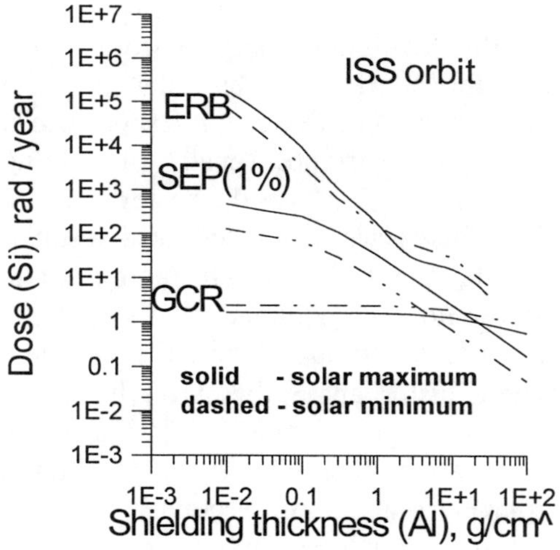

Figure 10. The radiation doses in ISS orbit under different shielding thickness, calculated using RB/SINP, GCR/SINP and SEP/SINP models for an 11-year period of active existence (Kuznetsov and Lobakov, 1999).

2) The largest SEU rates should be expected for very large SEP events, which occur with a 1% probability. Most of the SEP events give significantly lower SEU rate levels. For example, 50% of the SEP lead to SEU with rates, significantly lower, than the rates of GCR induced SEUs.

3) Simultaneously, it can be noted that under the impact of only one of the radiation components GCR or SEP, for small thickness of spherical shielding, the SEU rate grows with increasing shielding thickness. This effect is most pronounced for SEP particles.

4.6.2   Geostationary Earth Orbits - GEO

This type of orbit is characterized by the existence of rapid variations of RB particle fluxes (see 2.1.) as well as SEP and GCR variations, typical for interplanetary space. Examples of SEU calculations under the impact of GCR and SEP in geostationary orbit are given in 4.3.2 and 4.5.2. respectively. The peculiarities of dose effect calculations are given in 3.3. It is necessary to mention, that the existence of rapid variations of the electron component in the outer RB region on time scales from fractions of a minute to several days during geomagnetic storms, as well as solar cycle variations, makes it difficult to use RB models (which are essentially static) for precise calculations of the radiation load.

4.6.3   High-apogee Elliptic Orbits - HEO

Such orbits (with perigee and apogee altitudes from several hundred to several thousand kilometers) typically intersect both the inner and outer RB zones. Therefore, SEU calculations besides cosmic ray particle fluxes should account for the impact of trapped protons of the inner zone. Figure 11 shows the SEU rate calculations in microchips on the CRRES satellite, depending on the L-shell parameter for inner RB zone protons and GCR particles (for shielding thickness of $10g/cm^2$) (Bashkirov et al., 1999). The observed experimental data on L<2 can be adequately described by the impact of RB protons. The maximum number of upsets is observed in the range of L-shells, where the satellite encounters the maximum proton flux on L~1.5 (for orbit altitudes of ~2500-3000 km). However, the results of GCR induced SEU calculations in the outer zone do not coincide with the experimental results, which requires additional analysis.

5. CONCLUSIONS

Existing RB models can only be used as indicative ones. For more accurate model calculations of the radiation environment it is necessary to use available experimental data for specific orbits and time periods.

Our current understanding of the types of particle variations in the RB already permits to make estimates of the particle fluxes on time scales ranging from sub-storms to the whole solar cycle.

The existing dynamic GCR model (GCR/MSU, CREME 96) adequately describes the variations of GCR nuclei and can be utilized for accurate radiation effect calculations.

The solar energetic particle model which is essential for engineering calculations of radiation effects has not yet been completed. There are significant discrepancies between the new Solar Energetic Particle model developed at MSU (SEP/SINP) and previous models (CREME 96, JPL 91, etc.).

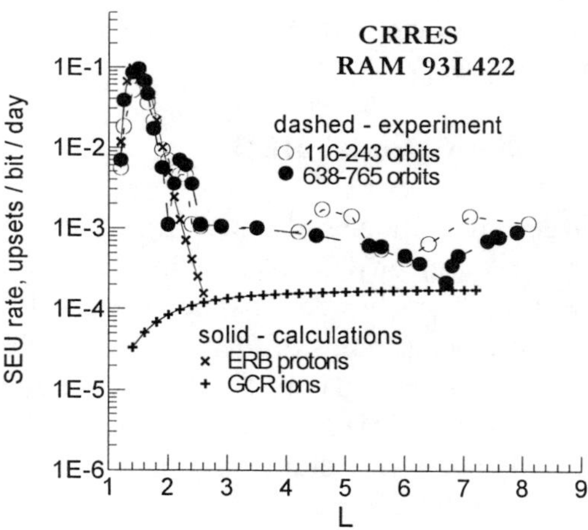

Figure 11. Experimental results from CRRES and SEU rate calculations according to the model (Bashkirov et al., 1999).

The albedo environment in LEO requires systematization, and, possibly additional physical modelling aimed at developing a comprehensive albedo radiation model, incorporating all the radiation components at low altitudes. The accuracy of secondary particle flux models depends on both the accuracy of the specific spacecraft shielding and mass distribution, and on the surrounding environment model. Utilization of modern transport codes like GEANT is required for modelling the radiation environment inside the spacecraft.

The high level of integration and miniaturization of modern electronics is the most 'sensitive' aspect of our current understanding of radiation effects on space systems. New electronics demands not only improved accuracy of radiation environment models and calculations, but also more accurate studies of the physical mechanisms of charged particle interactions with the spacecraft's material.

While accomplishing space experiments we need to keep in mind, that the space environment is aggressive towards any man-made object placed inside it, including towards man himself. There are no orbits or time intervals which would be completely radiation-safe. The question is how hazardous these effects of the aggressive environment can be, what are the most suitable and acceptable radiation thresholds to be proposed as optimized international standards.

6.   ACKNOWLEDGEMENTS

The author sincerely wishes to thank the organizers of the NATO Advanced Study Institute on Space Storms and Space Weather Hazard - the Directors of ASI Dr. Ioannis Daglis and Prof. Yurii Galperin for the opportunity to present this tutorial lecture and to publish it in these proceedings. The author is extremely grateful to Prof. Joseph Lemaire for many valuable comments. I also wish to thank Dr. Katya Tolstaya for translation and editing this manuscript.

7.   REFERENCES

Adams, J. (1985) Cosmic ray effects on microelectronics, in Naval Research Laboratory Memorandum Report, IV, Washington, USA, 1- 118.

Adams, J., Belyaev A.A., Kuznetsov, N.V. and Nymmik, R.A. (1996) Occurrence frequency of single upsets induced on synchronous orbit: model and calculations with TDRS–1 experiment, Nuclear Tracks & Radiation Measurements 22, 509- 512.

Akishin, A.I., Aleksandrov, A.P. (1983) Imitation testing space material science, in S.N.Vernov (ed), Space model (in Russian), Moscow State University Publishers, Moscow, 2, 9-28.

AMS collaboration, Protons in near earth orbit, Phys. Letters, (2000), B472, 215 – 225.

Aristova, I.N., Lyuagushin, V.I., Marin, B.V., Saraeva, M.A., Tel'tsov, M.V. (1991) Radiation environment on the orbital complex MIR during September- October 1989, Kosmicheskie Issledovaniya, 30, 794-798. (in Russian).

Baker, D.N. (1996) Solar wind-magnetosphere drivers of space weather, J. of Atmospheric and Terrestrial Physics, 58, 1509-1526.

Bashkirov,V.F., Panasyuk,M.I., Teltsov, M.V. (1998) Trapped radiation dynamical model for low altitudes in the magnetosphere, Kosmicheskie issledovaniya (in Russian) 36, 359-368.

Bashkirov, V.F., Kuznetsov N.V., Nymmik, R.A., (1999) An analysis of the SEU rate of microcircuits exposed by the various components of space radiation, Radiation. Measurements., 30 (1999), 427- 433.

Beliaev, A.A., Lemaire. J., (1994) Evaluation of the INP radiation belt models, TREND Technical note A, ESTEC contract no. 9828/92/NL/FM.

Beliaev, A.A., Lemaire J., (1996) Comparison between NASA and INP/MSU Radiation belt models, in J.Lemaire, D. Heynderickx, D. Baker, (eds.), Radiation belts: models and standards, Geophysical Monograph 97, 141-145.

Bendel, W. L., Petersen E.L. (1983) Proton upsets in orbit, IEEE Trans. on Nuclear Science NS- 30, 4481- 4485.

Binder, D., Smith, E.C., and Holman, A.B. (1975) satellite anomalies from galactic cosmic rays, IEEE Trans. on Nucl. Sci. 22, 2675-2680.

Blake, J.B., Friesen, L.M. (1977) A technique to determine the charge state of anomalous low energy cosmic rays, Proc. 15th Int. Cosmic Ray Conference., 2, 341-346.

Bogomolov, A.V., Bucik, R., Dement'ev, A.V., Denisov, Yu.I., Ryumin, S.P. (1998) Energetic neutron and gamma ray spectra under the earth radiation belts according to fluxes observed onboard Coronas-I satellite, Salut-7- Cosmos -1686 orbital complex, Adv. Space res., 21, 1801-1804.

Brautigam, D.H., Gussenhoven, M.S.,Mullen, E.G. (1992) Quasistatic model of outer zone electrons, IEEE Trans. Nucl.Sci., 39, 1797-1809.

Campbell, A., McDonald, P., Ray, K. (1992) Single event upset rates in space, IEEE Trans. on Nucl. Sci. 39, 1828-1835.

Desorger, L., Buhler, P., Zehnder,P., Daly, E., Adams,L.(1998) Outer radiation belt variations during 1995, Adv. Space Res., 22, 83-87.

Dmitriev A.V., Kalinin D.V., Kuznetsov S.N., Yushkov B.Yu. (1999) "Variations of the geomagnetic indices that control the radial diffusion of energetic particles in radiation belts".Proceeding of Space Radiation Environment Workshop (SREW), Farnborough, England. 1-3 Nov, 1999 (in print).

Feynman J., Spitale, G., Wang, J., and Gabriel, S. (1992) Interplanetary proton fluence model, J. Geophysical Research, 98, 1328-1342.

GEANT (3.16 , 3.21) Detector Description and Simulation Tool, CERN Geneva, Switzerland, 1994.

GOST 25645.138-86, GOST 25645.139-86 (1986) The natural radiation belts , Standart Publishers (in Russian).

Grigorov, N.L. (1985) High energy electrons in the vicinity of the Earth, Nauka Academic Publishers, Moscow (in Russian).

Grigorov,N.L., Kondrat'eva, M.I., Panasyuk,M.I., Tretyakova, Ch.A., Adams J., Blake, J.B., Schulz, M., Mewaldt, R.A., Tylka, A.(1991) An evidence for anomalous cosmic ray oxygen ions in the inner magnetosphere, Geophys. Res. Lett., 18, 1959-1962.

Gussenhoven, M.S., Mullen, E.G., Sperry, M., Kerns, K.J. (1992) The effect of the march 1991 storm on accumulated dose for selected orbits: CRRES dose models, IEEE Trans. Nucl. Sci., 39, 1765-1778.

Heynderickx, D., Kruglanski, Lemaire, J.F., Daly, E.L. (1996), in J.Lemaire, D. Heynderickx, D. Baker, (eds.), Radiation belts: models and standards, Geophysical Monograph ,97, Washington, 173-178.

Huston, S.L., Kuck, G.A., Pfitzer, K.A. (1996) Low- altitude trapped radiation model using TIROS/NOAA data, in J.Lemaire, D.Heynderickx, D.Baker (eds.), Radiation belts: models and standards, Geophysical Monograph, 97, Washington, 119-128.

Ilyin, V.D., Kuznetsov, S.N., Panasyuk, M.I. , and Sosnovets, E.N. (1988) Nonadiabatic effects and limit of proton capture in the Earth's radiation belts, Bull. Acad. Sci. USSR, Phys. Ser. (USA) ,48, 134-137.

Kuznetsov N.V., Lobakov A.P.(1999) An estimate of dose and single event effects on low-Earth orbit spacecraft. In: Space Radiation Environment Workshop & Workshop on Radiation Monitiring for the International Space Station. Book of Abstracts, 51.

Kuznetsov, N.V., Nymmik, R.A. (1994) Background ion fluxes as a source of single events effects of microelectronics onboard spacecrafts, Kosmicheskie issledovaniya (in Russian) 32,112-117.

Kuznetsov, S.N., Myagkova, I.N., Yushkov, B.Yu. (1999) Connection between Energetic particle Fluxes at Geostationary orbit with solar wind parameters and with solar cosmic rays. Proceeding of Space Radiation Environment Workshop SREW), Farnborough,England. 1-3 Nov.1999 (in print).

Lemaire, J., Johnstone, A.D., Heynderickx, D., Rodgers, D., Szita, S., Pierrard, V., Trapped Radiation Environment Model Development: TREND-2, Final Report, Aeronomica Acta A-393, 1995.

Lockwood, J.A. (1973) Neutron measurements in space, Space Science Rev., 14, 663-675.

Lyagushin, V.I., Sevastyanov, V.D., V.D., Tarnovsky,G.V. (1997) A measurements of energetic spectrum of neutrons on orbital station "Mir", Kosmicheskie Issledovaniya, 35, 216-225 (in Russian).

Makletsov, A.A., Mileev, V.N., Novikov, L.S., Sinolits, V.V. (1997) Radiation environment modeling onboard spacecrafts, Engineering Ecology, 1, 39-51 (in Russian).

McIlwain, C.E. (1996) Processes Acting upon outer zone electrons, in 'Radiation belts: Models and Standards', in J.Lemaire, D.Heynderickx, D.Baker (eds.), Radiation belts: models and standards, Geophysical Monograph 97, Washington.

Morfill, G.E.(1973) Guiding center approximation of trapped particles, J. Geophysical Research, 78, 588-593.

Nymmik R.A.(1999) Probabilistic Model for fluences and peak fluxes of solar energetic particles", Radiation measurements, 39, 287-296.

Nymmik,R.A., Panasyuk, M.I., and Suslov, A.A. (1995) Galactic cosmic ray flux simulation and prediction, Adv. Space Res. 17, 19-23.

Panasyuk, M. I. (1993) Anomalous cosmic ray studies on "Cosmos" satellites, Proc. 23th Int. Cosmic Ray Conference, 4, 455-463.

Panasyuk, M.I., Sosnovets,E.N., Grafodatsky,O.S., Verkhoturov,V.I., Islyaev,Sh.N. (1996) First results and perspectives of monitoring radiation belts, in J.Lemaire, D.Heynderickx, D.Baker (eds.), Radiation belts: models and standards, Geophysical Monograph 97, Washington, 211-216.

Pickel, J.C. (1996) Single event effect prediction, IEEE Trans. Nucl. Sci. 43, 483-495.

Rodgers, D.J., (1996) A New Empirical Electron Model, , in 'Radiation belts: Models and Standards', in J.Lemaire, D.Heynderickx, D.Baker (eds.), Radiation belts: models and standards, Geophysical Monograph , 97, Washington,103-107.

Sawyer, D.M., Vette, J.I. (1979) AP-8 trapped proton environment for solar maximum and solar minimum, NSSSDC/WDC-R&S.

Seltser, S.M. (1979) Electron, electron-bremsstrahlung and proton depth-dose data for space shielding applications, IEEE Trans. Nucl. Sci., NS26, 21-60.

Spejeldvik, W.N.(1979) Transport,charge exchange and loss of energetic heavy ions in the Earth's radiation belts: applicability and limitation of theory, Planetary Space Sciences, 20, 1215-1226.

Tsyganenko, N.N. (1989) A magnetospheric magnetic field model with a wrapped tail current sheet, Planetary Space Sciences 37, 5-20.

Tverskaya, L.V. (1996) Dynamics of energetic electrons in the radiation belts, in J.Lemaire, D.Heynderickx, D.Baker (eds.), Radiation belts: models and standards, Geophysical Monograph 97, Washington, 183-187.

Tverskoy, B.A.(1964) Dynamics of radiation belts, Geomgnetizm i aeronomiya, 5, 436-441 (in Russian).

Tylka, A., Adams, J., Boberg, P.R., Brownstein, B., Dietrich, W.F., Fluechiger, E.O., Petersen, E.L., Shea, M.A., Smart D.F., and Smith, E.C. (1997) CREME96: A revision of the cosmic ray effects on microelectronics code, IEEE Transactions on Nuclear Science, 44, 2150- 2160.

Vampola, A.L., Osborn, J.V., Johnson, B.M. (1992) The CRRES magnetic electron spectrometer AFGL 701-5A (MEA), J. Spacecraft & Rockets, 29, 592-594.

Vampola, A.L., (1994) Analysis of environmentally induced spacecraft anomalies, J. Spacecraft & Rockets, 31, 154-159.

Vampola, A.L. (1996) The nature of bulk charging and its mitigation in spacecraft design, In WESCON Proc, Anaheim, California, 234-241.

Vette, J.I. (1991) The AE-8 trapped electron environment, NSSSDC/WDC-R&S.

Williams, D.J., Arens, I.F., Lanzerrotti, L.T. (1968) Observations of trapped electrons at low and high altitudes, J. Geophys. Res. 73, 5673-5689.

Williams,D. (1966) 27-day periodicity of the trapped electron fluxes, J.Geophys. Res., 71, 7-21.

Yushkov., B.Yu., (1988), Neutron flux measurements onboard Salyut-6 orbital station, Kosmicheskie issledovaniya, 26, 793-796 (in Russian).

Chapter 11

Satellite Anomalies due to Space Storms

The effects of space weather on spacecraft systems and subsystems

Daniel N. Baker
Laboratory for Atmospheric and Space Physics, University of Colorado at Boulder
Boulder, CO 80309, USA

Abstract   Space weather is a term that refers to the dynamic, highly variable conditions in the geospace environment. This includes conditions on the sun, in the interplanetary medium, and in the magnetosphere-ionosphere-thermosphere system. Rapid changes in the near-Earth space environment during major geomagnetic disturbances can affect the performance and reliability of both spacecraft and ground-based systems. Modern spacecraft systems and subsystems appear to show an increasing susceptibility to effects of the space environment including communication, navigation, and reconnaissance satellite operational anomalies. This trend is probably due to "softer" designs of electronic components, reduction in subsystem sizes, and increases in performance demands. The major elements of the space environment that contribute to spacecraft anomalies can be reasonably well identified. This paper reviews operational anomaly trends and assesses the identification and potential prediction of causative space weather. We discuss the principal adverse space environmental effects presently known including cosmic rays, trapped magnetospheric radiation, and solar energetic particles. The scientific underpinnings for present prediction methods in space weather are also considered.

Keywords   Space storm, space weather, geomagnetic disturbances, substorm, surface charging, energetic particles.

1.   INTRODUCTION

As has been clearly shown in many historical instances, space storms have caused significant spacecraft operational anomalies or even complete

I.A. Daglis (ed.), Space Storms and Space Weather Hazards, 285–311.
© 2001 *Kluwer Academic Publishers. Printed in the Netherlands.*

satellite failures. We know many of the space environmental "agents" that produced such serious operational problems. Generally speaking, we can list the following as some of the main concerns:

Galactic cosmic rays: Very high energy (E > 100 MeV) ions that can cause radiation damage and "single event" upsets (SEUs);

Solar energetic particles: Moderate to high-energy (10-300 MeV) ions produced by shock waves and solar disturbances that can also cause severe radiation damage and SEUs.

Trapped energetic ions: Protons and heavier nuclei of tens to hundreds of MeV energies trapped in the Earth's inner Van Allen zone (< 2 R_E [Earth radii]) that cause radiation damage and SEUs.

Very high energy electrons: Mildly to highly relativistic (~0.2 to ~5 MeV) electrons in the Earth's radiation belts that can cause deep-dielectric charging; and

Moderate energy electrons: These are usually magnetospheric substorm-produced electrons (~10 to ~100 keV) that can cause surface charging effects on space systems throughout the magnetosphere, but mostly at the geosynchronous orbit (6.6 R_E).

Figure 1 is a schematic diagram of some of the spacecraft interactions possible with space radiation. Subpanel (a) shows the single event mechanism for penetrating ions. Panel (b) shows the deep-dielectric charging mechanism associated with penetrating (high-energy) electrons. Panel (c) illustrates the surface-charging mechanism for lower energy electrons and the current balance issues that this entails. All of these physical effects can occur any time the appropriate particle population is significantly present. However, normally only one of these space environmental factors dominates during a particular spacecraft operational problem. Combinations of effects are certainly possible in some instances.

In this review, we will first consider the space environmental factors in more detail. We will illustrate how space environmental factors change with time and location and we will illustrate with past, well-documented examples how the space environment has very likely caused spacecraft failures or serious operational anomalies. In the subsequent section of this paper, we will look at relatively recent examples of spacecraft anomalies and look at the modern spacecraft particle measurements that help assess the cause of such anomalies. In the penultimate section of this review, we will present some of our recent results in space environmental specification and forecasting (of magnetospheric energetic electrons). This will show how scientific understanding can aid in helping space system operators to protect spacecraft. Finally, we summarise and look to the future.

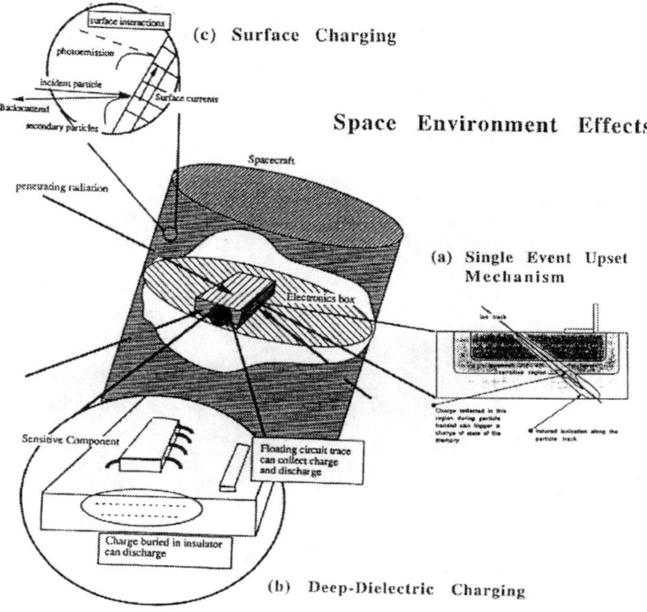

Figure 1. A schematic diagram illustrating space environmental effects due to (a) single event upsets, (b) deep-dielectric charging, and (c) surface charging [adapted from Robinson, 1989]

2. SPACE ENVIRONMENTAL FACTORS

2.1 Galactic cosmic rays

Figure 2 is a representation of several important aspects of the space environment in the vicinity of the Earth. The horizontal axis of the figure is time (in years) from 1960 to 2000. The y-axis of the diagram is the logarithm of particle energy (in MeV), while the z-axis of the three-dimensional plot is the logarithm of particle flux (particles/cm^2 s sr MeV). On the "back plane" of the diagram is the sunspot number (ranging from 0 to ~150): This gives a clear (if qualitative) indication of the 11-year solar (sunspot) cycle.

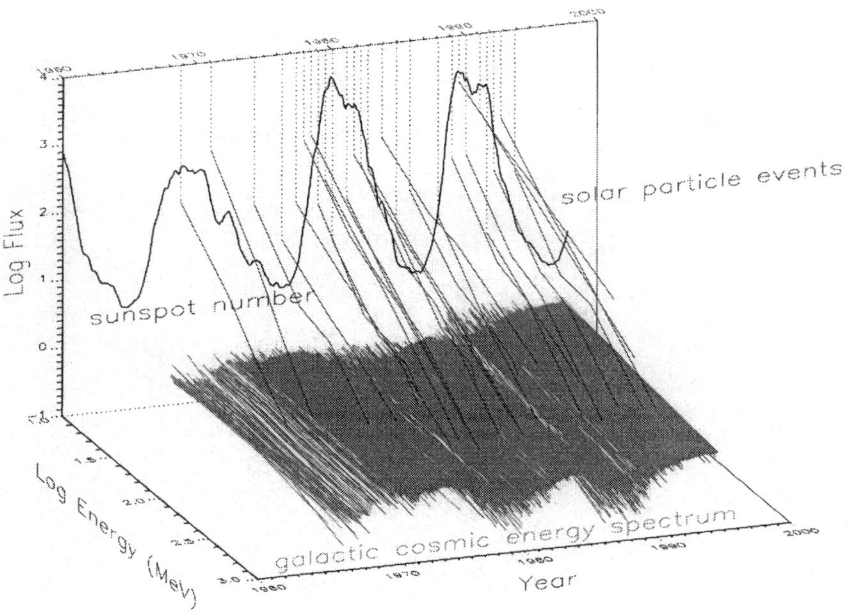

Figure 2. Illustration of fluxes of solar and galactic cosmic ray particles versus energy as a function of time (1967-1994), as described in the text. The sunspot number is shown on the "back plane" of the diagram

The grayish pattern of lines labelled "galactic cosmic energy spectrum" shows the monthly average high-energy proton spectrum measured by the IMP (Interplanetary Monitoring Platform) series of spacecraft (T.P. Armstrong, private communication). The GCR component measured near Earth extends up to (and beyond) 1000 MeV where the spacecraft measurements stop. As is evident, the galactic cosmic ray (or GCR) flux does not vary much with increasing energy. Thus, while the fluxes of GCR ions are low, the energies are extremely high: This makes the GCRs very penetrating.

Notice in Figure 2 that the GCR fluxes are modulated with time. The fluxes are highest near sunspot minimum and are lowest near sunspot maximum. This well-known GCR modulation during the solar cycle is controlled by the solar wind and interplanetary magnetic field in the heliosphere [e.g., Cummings et al., 1993]. Thus, there are more GCR effects on spacecraft near solar minimum than near solar maximum times.

2.2   Solar energetic particles

On occasion, the sun produces highly intense fluxes of relatively energetic (10 to ~300 MeV) protons and other ions. Figure 2 shows these solar energetic particle (SEP) events as the thin, slanting lines occurring at various times from ~1969 to ~1994. The spectra of these SEP events vary widely, and the hardness (flux as a function of energy) of the spectra also is highly variable. In general, SEP events can occur at any time in the solar cycle, but they tend to occur more frequently and more intensely around sunspot maximum rather than sunspot minimum. Notice that the SEP fluxes, when present, exceed the GCR fluxes up to ~100 MeV or so.

The IMP data set used to generate Figure 2 extends only from ~1967 to the early 1990s. More recent data have shown the dramatic increases in SEP events associated with the present solar maximum (expected in ~2001). Figure 3a shows some of the flux enhancements in E>100 MeV protons measured by GOES satellites that occurred during a large solar disturbance interval in April-May 1998 [see Baker et al., 1998]. In particular, there was a jump in proton intensity of about 2 orders of magnitude in one day in fluxes of very energetic protons on 2 May 1998.

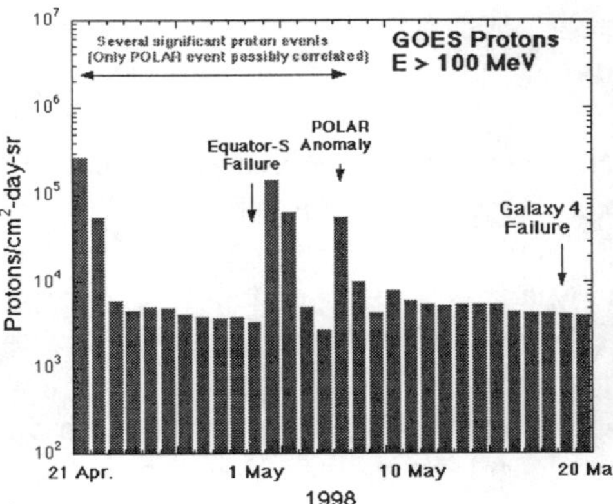

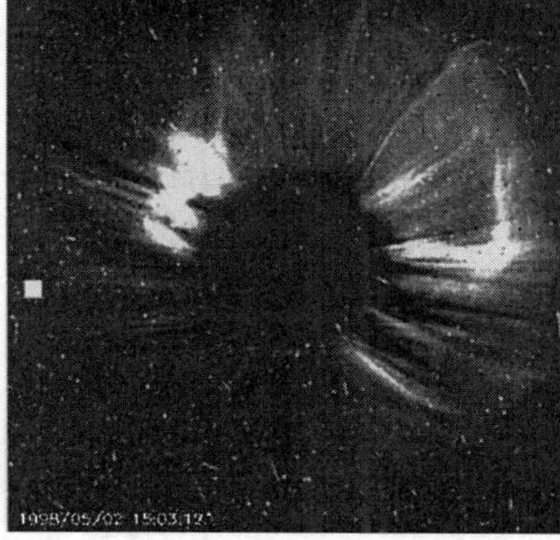

Figure 3.a. (left panel) Fluxes of protons (E>100 MeV) measured by GOES spacecraft for portions of April and May 1998. Several large events were seen in this period. b. (right panel) The "snow" in this SOHO/LASCO image of a halo coronal mass ejection on 2 May 1998 was caused by solar energetic particles (see a.).

Many spacecraft saw significant effects from the SEP events indicated in Figure 3a. For example, the charge-coupled device (CCD) cameras onboard the SOHO spacecraft were subjected to intense penetrating backgrounds as a result of the proton (ion) enhancements. The "snow" in the coronagraph picture (in Figure 3b) of a coronal mass ejection (CME) taken by the SOHO/LASCO instrument is due to such proton bombardment. While background noise is a very visible effect of SEP increases, much more significant degradation and failure of space systems can and do occur during SEP events [see Allen et al., 1989].

Single event upsets are one of the important SEP effects and these were detected in many cases in the May 1998 event (for example in the POLAR spacecraft). SEUs occur in space electronics when a charged particle, usually a heavy ion, ionizes a track along a sensitive portion of the circuit and causes that circuit to change state [Robinson, 1989]. The size of the electronics element determines the sensitivity as well as the probability that an SEU will occur. If a particle deposits sufficient charge along a sensitive path in a device, then the SEU can take place. This is illustrated in Figure 1: a heavy ion loses energy by ionizing the constituents of the material it is passing through. If the energetic particle passes through a depletion region of a transistor, for example, this can cause the device to change state. SEU effects in space memories have become more manageable in recent times by use of error-correction software.

2.3   Trapped energetic ions

Very energetic protons and heavier ions trapped in the geomagnetic field can also cause significant effects in spacecraft systems. An illustration of this is shown in Figure 4 [courtesy of D. Wilkinson of NOAA/NGDC]: The UOSAT-2 spacecraft of the European Space Agency operated at low-Earth, high-inclination orbit. The dots on the global map indicate times and geographic locations of memory upsets within the spacecraft as it cut through many different magnetic "shells". Note that these memory upsets can occur almost anywhere around the Earth, but the vast majority of them are aggregated in the southern hemisphere over South America and in the southern Atlantic Ocean. This is the so-called South Atlantic Anomaly (SAA) region where the Earth's magnetic field is weakest [see Cummings et al., 1993 and references therein]. In the SAA, trapped energetic particles have greatest access to low-altitude spacecraft and can interact most strongly with spacecraft there. The memory upsets seen by UOSAT-2 were undoubtedly caused by penetrating ions causing single event upsets in the memory chips onboard the spacecraft [J.H. Allen, private communication].

Such particle effects in the SAA have plagued virtually every low-altitude, high-inclination spacecraft, including the Hubble Space Telescope.

2.4   High energy electrons

As illustrated in Figure 1b, very high-energy electrons can penetrate through spacecraft walls and through electronics boxes to bury themselves in various dielectric materials [e.g., Robinson 1989]. This can, in turn, lead to electric potential differences in the region of the buried charge. In some instances, intense voltage breakdowns can occur leading to surges of electrical energy deep inside circuits. This can cause severe damage to various subsystems of the spacecraft.

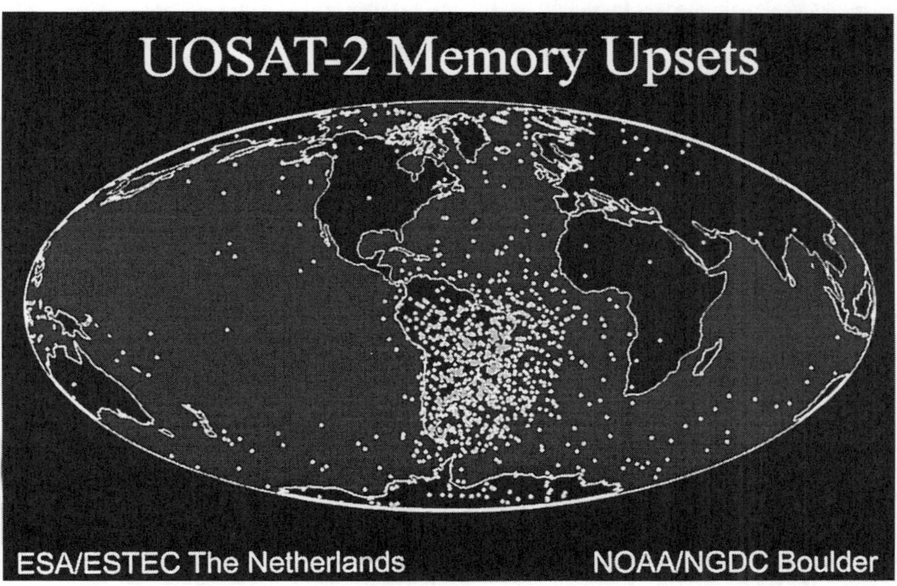

Figure 4. Locations of UOSAT-2 memory upsets in a geographic representation. The vast majority of the upsets occurred in the South Atlantic Anomaly region [figure courtesy of D. Wilkinson and J.H. Allen]

Many examples of such "deep-dielectric charging" have been presented by various authors [e.g., Vampola, 1987; Baker et al., 1987]. An interesting case study presented by Baker et al. [1987] is shown in Figure 5 using a slightly lower electron energy range. In this diagram, smoothed daily averages of E=1.4-2.0 MeV electron fluxes at geostationary orbit are plotted versus time (late 1980 through early 1982). Also shown by bold vertical arrows are some of the main occurrences of star tracker anomalies onboard

292

this military geostationary operational spacecraft. The star tracker upsets were normally associated with high intensities of relativistic electrons.

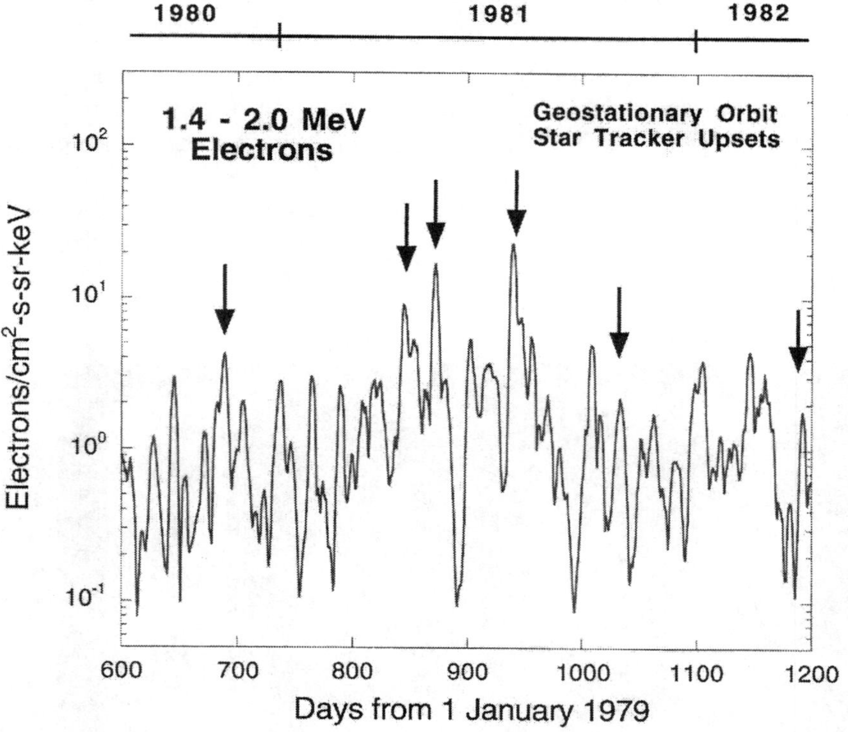

Figure 5.Fluxes of 1.4-2.0 MeV electrons at geostationary orbit from late 1980 through early 1982. High electron flux tended to be associated with star tracker anomalies (vertical arrows) on the spacecraft

However, some high intensity electron events did not produce star tracker anomalies [see Baker et al., 1987] so there are more subtle controlling factors as well. The anomalies tended to occur only during relatively long-duration events. Thus, it was not only the peak intensity of electrons, but also the duration of exposure that seemed to be important. During some intense events in late 1981, the star trackers were actually turned off and so no operational "anomalies" could be recorded.

Numerous other studies [e.g., Reagan et al., 1983; Robinson et al., 1989; Wrenn, 1995] have shown the clear role played by high energy electrons in many classes of spacecraft operational problems. Moreover, the quantitative level of radiation needed to produce deep-dielectric discharges has been rather clearly established in laboratory and spacecraft studies [e.g., Vampola, 1987].

2.5   Moderate energy electrons and surface charging

A significant effect of magnetospheric substorms (from the standpoint of space operations) is the occurrence of spacecraft surface charging [see Rosen, 1976]. During a surface charging event, insulated surfaces may charge to several kilovolts potential (usually negative relative to the ambient potential). This charging occurs because of a lack of current balance between the ambient plasma medium and the spacecraft surface. When a spacecraft is immersed in a cool, dense plasma, the incident particles (electrons and ions), as well as secondary emitted particles, photoelectrons, and backscattered electrons, all balance. This gives a low net spacecraft potential. However, in a very hot, tenuous plasma, current balance can be difficult to achieve and large potentials can build up.

Figure 1c illustrates the interaction at the surface of a spacecraft. The diagram shows that there are currents near the surface of the spacecraft due to incident, backscattered, and photoemitted particles. These various populations can be examined to calculate the charge configurations for a given spacecraft. A sheath region that forms around the spacecraft is a volume strongly affected by the spacecraft. In this region the plasma is distorted by electric fields due to the charge of the spacecraft. The sheath region can also be affected by activity on the spacecraft such as thruster firings which extend the influence of the spacecraft farther into the plasma [e.g., Robinson, 1989]. The sheath region is complex in shape and depends on the motion of the spacecraft through the plasma as well as the plasma properties and the surface materials of the spacecraft. From an operational standpoint, differential charging of spacecraft surfaces that can lead to discharges is a quite significant issue. Discharges introduce noise into the system and may interrupt normal spacecraft operation, or represent a false command. In the process of discharge breakdown, physical damage may occur. This, in turn, may change the physical characteristics (thermal properties, conductivity, optical parameters, chemical properties, etc.) of the satellite. In addition the release of material from the discharge site has been suggested as a contamination source for the remainder of the spacecraft [see Baker, 1998 and references therein].

An example of a surface charging event is presented in Figure 6 [taken from Baker, 2000]. This shows several hours of data (30-300 keV electrons in six different energy channels) for two spacecraft (LANL 1982-019 and LANL 1984-037) at geostationary orbit. Shortly after 2000 UT on 8 February 1986, the energetic electron fluxes in all energy ranges began to

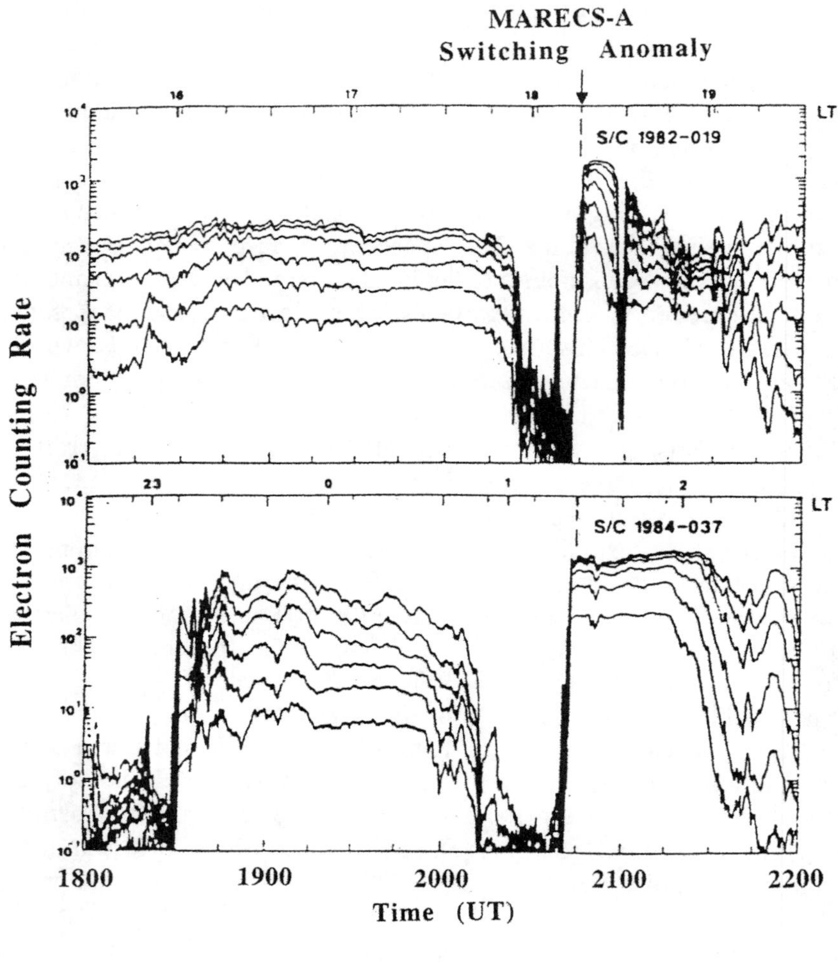

Figure 6. Electron counting rates for electrons in the energy range 30-300 keV (6 channels) for a portion of 8 February 1986. Data from two different GEO spacecraft are shown (S/C 1982-019 and S/C 1984-037). A serious switching anomaly on the MARECS-A spacecraft at 2040 UT was associated with abrupt increase of the moderate-energy electrons [Figure from Baker, 2000].

drop sharply. Then at 2040 UT sensors on both spacecraft at widely separated local times (1815 LT and 0120 LT, respectively) saw fluxes increase several orders of magnitude. At this same time, the MARECS-A spacecraft at geostationary orbit, located between 1982-019 and 1984-037, experienced an uncommanded switching anomaly. This was almost certainly caused by effects related to the abrupt enhancement of moderate energy electrons shown in Figure 6.

Figure 7, adapted from data presented in Rosen [1976], shows the number of spacecraft anomalies detected at geostationary orbit as a function of spacecraft local time. The anomalies include logic upsets as well as other significant operational problems for both military (DSP and DSCS) and commercial (Intelsat) spacecraft. As may be seen in the figure, there was a very strong local time asymmetry in the number of anomalies with the vast majority occurring roughly between local midnight and local dawn. This is where substorm-injected electrons are seen most prominently [e.g., Baker, 1998] and the LT distribution shown in Figure 7 supports the view that surface charging was (and is) a major cause of operational anomalies near geostationary orbit.

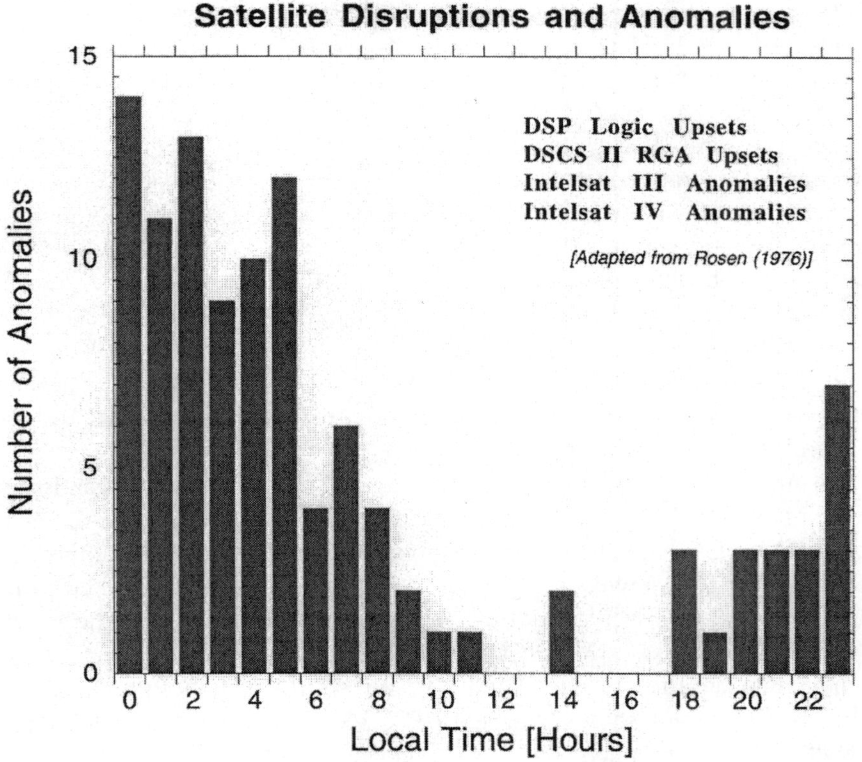

Figure 7. Local time distribution of satellite disruptions and anomalies showing a strong occurrence frequency peak in the midnight and local morning hours [data from Rosen, 1976].

3. RECENT SPACECRAFT FAILURES AND TRENDS

In the decade of the 1990s, many new observing tools became available to analyse the space environment. These include upstream spacecraft such as WIND, SOHO, and ACE as well as scientific magnetospheric spacecraft such as SAMPEX and POLAR. Figure 8 shows in a schematic way the locations of some of the many spacecraft that can be used in modern studies. In this section we illustrate the modern art of anomaly analysis by showing a few recent important spacecraft failures/anomalies.

3.1 ANIK/INTELSAT anomalies

Baker et al. [1994] showed that relativistic electron enhancements have been associated with some major spacecraft operational problems at, or near, geostationary orbit. On 20 January 1994 the momentum wheel control circuitry of the Intelsat K spacecraft at geostationary orbit suffered an operational anomaly causing a loss of attitude control. The system was switched to the backup circuitry and control was re-established. Later on 20 January, the Anik E-1 spacecraft, also at geostationary orbit, suffered the same kind of operational anomaly in the momentum wheel control circuitry.

According to newspaper accounts, Telesat Canada operators struggled for 8 hours to regain control of the Anik E-1 satellite. They were able to finally switch to the backup momentum wheel controller and resume reasonably normal operations. Unfortunately, the Anik E-2 satellite experienced a hard failure of its momentum wheel control circuitry on 21 January 1994. It was found that the backup circuitry on Anik E-2 was not functional and, therefore, it was not possible to regain full and normal control of the E-2 spacecraft until some 8 months later.

During the Anik and Intelsat anomalies, GOES-7 and other geostationary spacecraft showed large enhancements of relativistic electron (E>2 MeV) fluxes. SAMPEX data at low altitude support this observation [Baker et al., 1994]. SAMPEX showed that the Anik/Intelsat problems occurred when E>2 MeV electrons were near their historically highest values. The largest flux increases seen at SAMPEX and other spacecraft were also characterized by their long duration (10-15 days). These results are shown graphically in Figure 9a. This plot shows fluxes of E>2 MeV electrons for L=6.0 (from SAMPEX) during much of 1993 and early 1994. The Anik/Intelsat anomalies occurred during (or toward the end of) a high-intensity, long-duration electron flux enhancement.

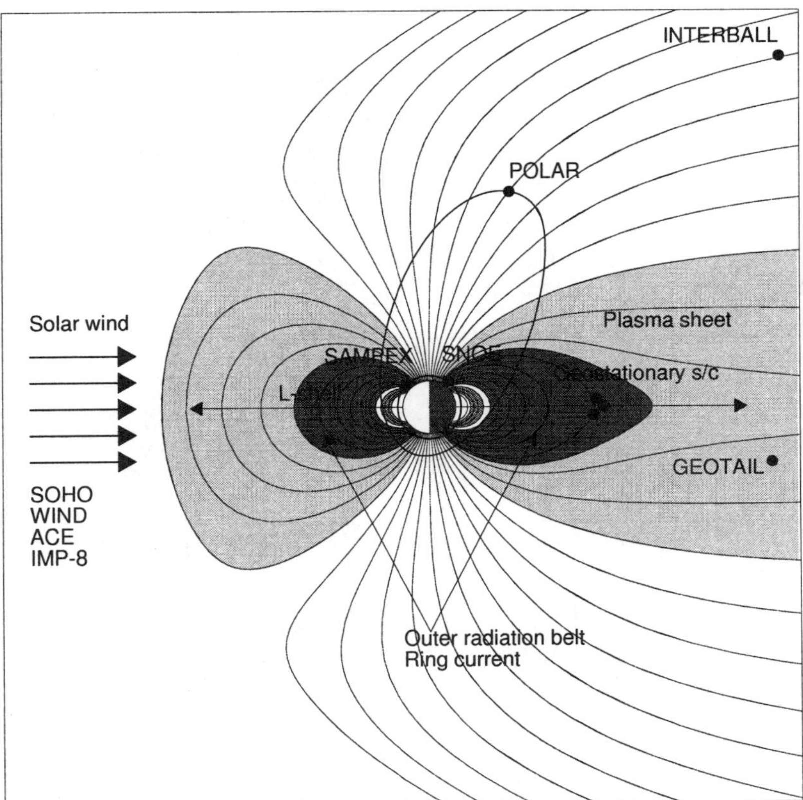

Figure 8. Schematic illustration of ISTP and other modern-day spacecraft used for anomaly analyses.

During the Anik and Intelsat anomalies, GOES-7 and other geostationary spacecraft showed large enhancements of relativistic electron (E>2 MeV) fluxes. SAMPEX data at low altitude support this observation [Baker et al., 1994]. SAMPEX showed that the Anik/Intelsat problems occurred when E>2 MeV electrons were near their historically highest values. The largest flux increases seen at SAMPEX and other spacecraft were also characterized by their long duration (10-15 days). These results are shown graphically in Figure 9a. This plot shows fluxes of E>2 MeV electrons for L=6.0 (from SAMPEX) during much of 1993 and early 1994. The Anik/Intelsat anomalies occurred during (or toward the end of) a high-intensity, long-duration electron flux enhancement.

The relationship of the electron flux increases to solar wind variations is shown in Figure 9b. The figure presents the average daily solar wind speed (from IMP-8) and the E>2 MeV electron fluxes (L=6) from SAMPEX for January 1994. The IMP-8 data show a large solar wind stream with rapidly rising speed from 10-11 January 1994, and reaching V~750 km/s. As the speed fell off, the E>2 MeV electrons rose rapidly in absolute intensity. Thus this event is similar to events observed from 1983 to 1985 in association with high-speed streams during the last solar cycle [Baker et al., 1986]. The Anik and Intelsat problems occurred after many days of high electron fluxes. Thus, it is virtually certain that the Anik anomalies were due to deep dielectric charging [Baker et al., 1994].

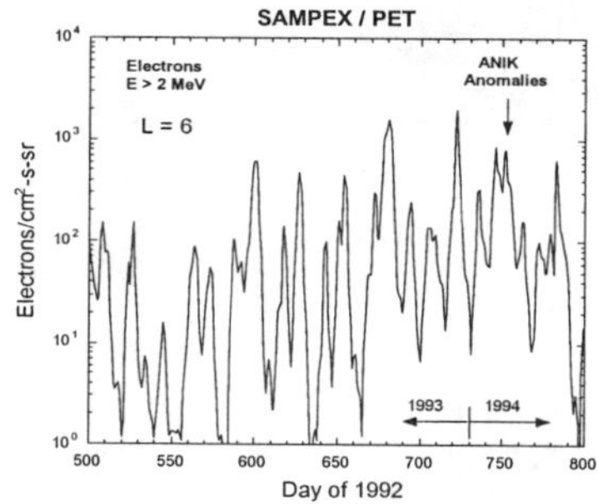

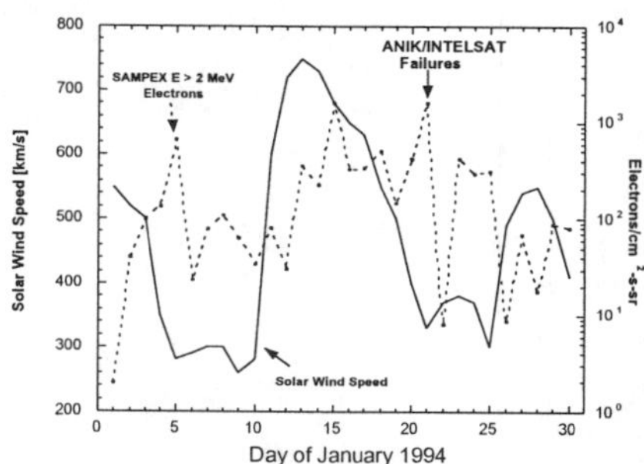

Figure 9. (a) (left panel) Data for electrons with E>2 MeV for L=6 during 1993 and early 1994 showing where Anik E-1 and E-2 anomalies occurred. (b) (right panel) An expanded view of the period in January 1994 of the Anik anomaly time as related to solar wind and electron flux data [adapted from Baker et al., 1994].

In a more recent incident, the Anik E-1 satellite suddenly lost all power from one solar panel array on 26 March 1996. The 50% power loss reduced the spacecraft's capacity from 24 C-band channels to nine channels and it reduced the Ku-band capacity from 32 channels to 10 [Baker et al., 1996]. The lost solar panel was not recovered, permanently degrading communication capability for Telesat Canada. Service to Telesat Canada customers was restored after about six hours by link switches to other spacecraft and by using backup systems such as fiber optics ground links. Figure 10, taken from Baker et al. [1996], shows E>2 MeV electron from GOES-8 at geostationary orbit covering the period from Day 1 to Day 91 of 1996. A large increase in electron fluence was seen in late March of 1996 just prior to the time of the Anik E-1 failure. This suggests a relatively hostile space environment. Several other spacecraft also suffered operational anomalies on, or near, 26 March 1996 [Baker et al., 1996]. In fact, the data from many scientific and operational spacecraft demonstrated that the high-energy electron environment was quite elevated throughout late March 1996. The satellite and ground-based data suggest that the space environment could have caused, or at least exacerbated, the conditions onboard Anik E-1 that led to the power failure that crippled the spacecraft [Baker et al., 1996], again through the mechanism of deep-dielectric charging.

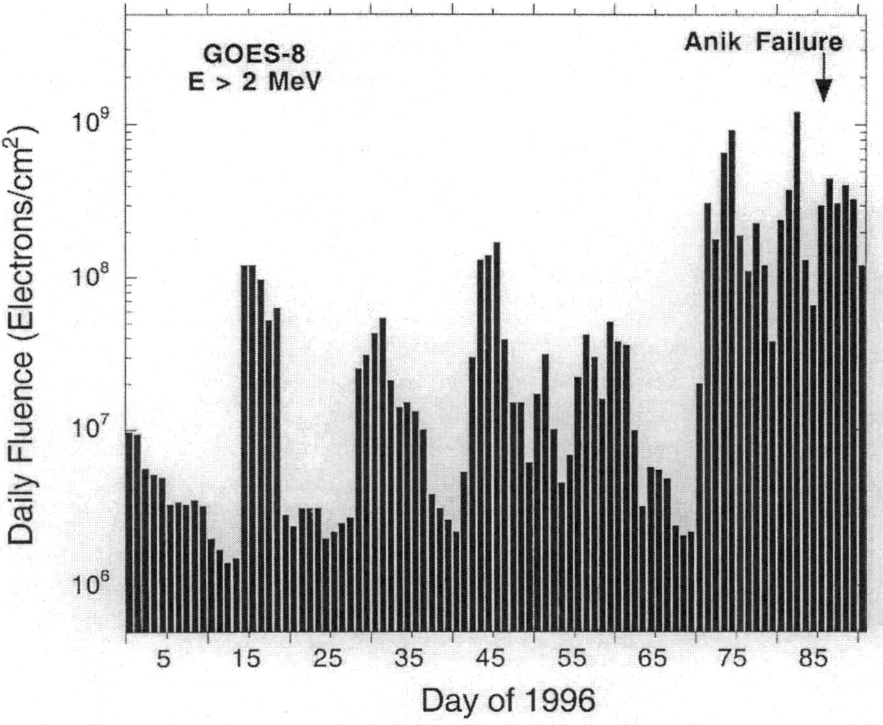

Figure 10. GOES-8 data showing the high flux of E>2 electrons during an Anik spacecraft power system failure on 26 March 1996.

3.2 Galaxy 4 failure

It is sometimes difficult to be sure about the role of the space environment in a spacecraft failure. A recent example illustrates this point. Galaxy 4 was a heavily used communication satellite at geostationary orbit. On 19 May 1998 the attitude control system on Galaxy 4 suddenly failed. A backup system on the spacecraft also failed (at the same time, or perhaps earlier) so that the PanAmSat operators were unable to maintain a stable Earth communication link. Therefore, the Galaxy 4 spacecraft was completely lost from useful service.

Galaxy 4 provided a wide range of communication services to ground users, especially in North America. There was, for example, loss of pager service to 45 million customers: Galaxy 4 was located over the central US and therefore had provided a convenient, high-throughput communication link. Its failure disrupted telephone, radio, television, and many other forms of communication. The Galaxy 4 failure was particularly notable and dangerous since many of the pagers belonged to emergency workers and medical personnel.

There is a good deal of controversy about the possible role of space weather in the Galaxy 4 case. Owners and operators of the spacecraft examined a wide range of possible causes of the failure and concluded in favor of an engineering failure. On the other hand, Baker et al. [1998] presented a case for high-energy electron (bulk charging) effects causing – or exacerbating – the situation. This is illustrated in Figure 11 from Baker et al., [1998]. The figure presents a broad history of energetic electron fluences from Los Alamos National Laboratory sensors at geostationary orbit. The data consist of 14-day running sums from January 1997 through May 1998. The data are normalized to the peak values observed in each of three energy ranges (0.7-1.8, 1.8-3.5, and 3.5-6.0 MeV). Figure 11 shows that the electron fluences in each energy range peaked quite prominently around 19 May 1998.The May 1998 event was the longest duration and spectrally-hardest electron enhancement that had occurred during 1997 and the first half of 1998.

As shown by other (GOES) data presented by Baker et al. [1998], the electron intensities at geostationary orbit were 2-3 orders of magnitude higher than the "danger" threshold for a two-week period (at least) prior to the Galaxy 4 failure. Such a long duration of elevated electron fluxes – as was discussed previously – is very conducive to deep-dielectric charging. The entire outer radiation belt electron population was highly inflated and densely populated during this complex storm interval. A combination of CMEs, solar flares, and high-speed solar wind streams led to a very intense "driving" of magnetospheric activity for an exceptionally long period of time. Despite these facts, many argue that the Galaxy 4 failure was not caused by electron charging. However, whether the space environment did, or did not, cause the Galaxy 4 problem (either directly or indirectly), the case clearly illustrates the vulnerability of modern society to service interruptions due to the loss of even a single satellite.

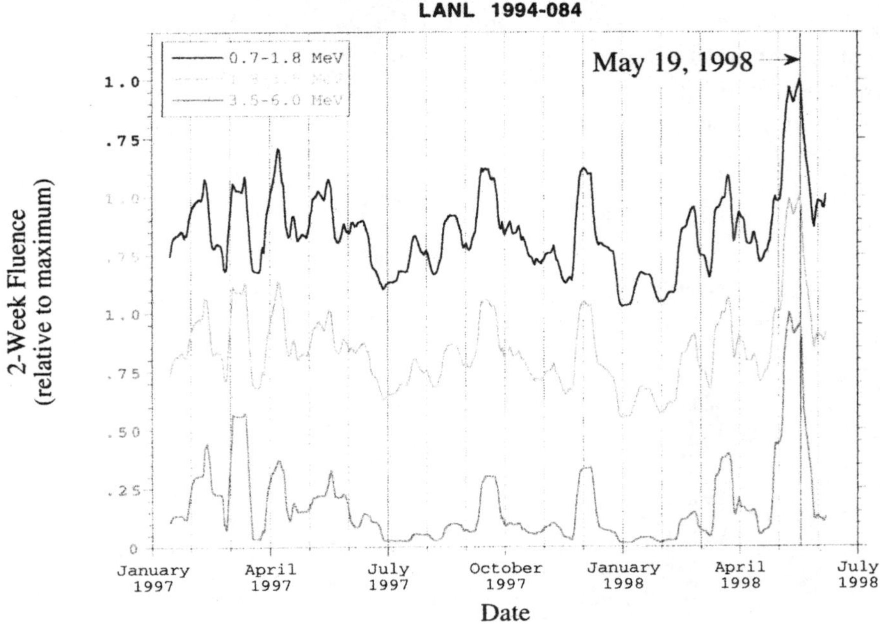

Figure 11. Two-week running fluence calculations at geostationary orbit (normalised values) from January 1997 through May 1998. A strong peak was seen in all three electron energy ranges in late May 1998 [from Baker et al., 1998].

4. SPECIFICATION AND FORECAST OF SPACE ENVIRONMENTS

The "climate" in near-Earth space is characterised from long-term observations of the space environment. However, satellite operators need to have a nowcast or even a forecast of the day-to-day space weather at a given location. This has proven to be an immense challenge. In this section, we will speak to such challenges for one aspect of the space environment, namely high-energy magnetospheric electrons, which we judge to be one of the most significant and dangerous space environmental effects.

4.1 Electron radiation belt climatology

In a recent set of papers Baker et al., [1999; 2000] examined the long-term global behavior of the Earth's outer radiation belt electron population. They

302

found that the entire outer radiation zone $(2.5 \leq L \leq 6.5)$ tends to vary in a relatively coherent way under the influence of major external drivers such as high-speed solar wind streams and CME/magnetic cloud events. Thus, it seems possible to use just a few spacecraft to measure the gross behavior of the outer zone electron population and thereby generally characterize the entire high-energy electron population on ~1-day timescales [see, also, Kanekal et al., 1999].

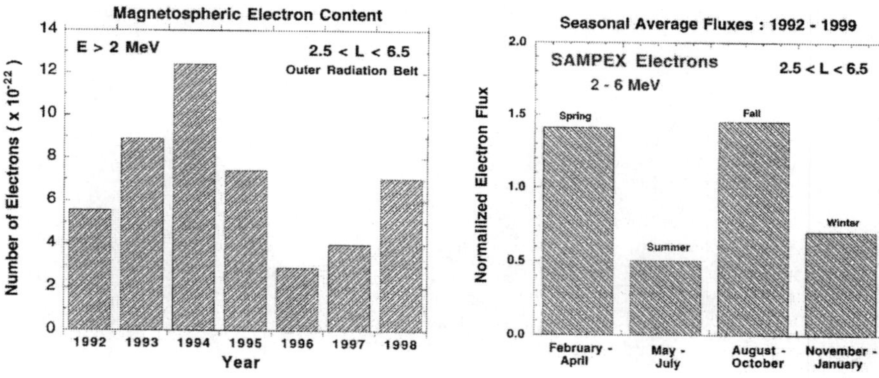

Figure 12. (a) (left panel) Annual average numbers of electrons (E≥2 MeV) trapped in the Earth's outer radiation belt (2.5<L<6.5) from 1992 through 1998. A large peak was seen in 1994 [from Baker et al., 2000]. (b) (right panel) Seasonal averages of E>2 MeV electrons (1992-1999) showing much higher average fluxes near the equinoxes than near the solstices [from Baker et al., 1999].

Baker et al. [2000] calculated, as a function of time, the total electron content for E>2 MeV electrons in the range $2.5< L <6.5$. The results of this calculation are shown in Figure 12a. As may be seen, there has been a strong long-term modulation of the electron content over recent times: The highest total number of electrons was seen in 1994 when high-speed solar wind streams were prominent. Conversely, the electron content of the outer zone was a factor of 4-5 lower in 1996 during sunspot minimum conditions when neither magnetic clouds nor solar wind streams were present to "drive" the magnetospheric accelerator very strongly. We see one important aspect of radiation belt climatology, therefore, in Figure 12a: We can expect relatively strong average changes in the typical fluxes of trapped relativistic electrons over the course of an 11-year sunspot cycle. This variation is dictated by the nature of the sun's output and by the character of the solar wind that impacts the Earth's magnetosphere during each phase of the cycle.

In a related piece of work, Baker et al. [1999] examined the seasonal dependence of high-energy electrons throughout the outer zone. SAMPEX and POLAR data were again averaged over a broad L-range (2.5-6.5). Electron fluxes with E>2 MeV were then averaged over 3-month intervals centered on the equinoxes and the solstices. As shown here in Figure 12b, a modulation was found such that the equinoctial fluxes of electrons (over a period of about 7 years) were nearly a factor of three higher than the average solstice fluxes. This appears to be a robust result.

From a space weather standpoint, the results shown in Figure 12 are quite important. All other things being equal, it seems clear that the magnetosphere is more effective in the spring and fall (or, perhaps, less effective in the summer and winter) at producing very energetic electrons. This can and should be borne in mind for space weather prediction purposes. Similarly, there is a strong solar cycle dependence of electron production in the magnetosphere apparently determined by the character of the solar wind incident on the magnetosphere. With these general "climate" aspects in mind, one can then take the next step to more detailed specification of day-to-day space weather changes.

4.2 Specification of radiation belt electrons

General properties of radiation belt electrons as presented in the last section are useful for spacecraft design: Knowledge of average flux levels, spectra, and intensity ranges can allow optimal definition of expected doses, shielding requirements, etc. However for analysis of spacecraft operational anomalies and for understanding specialised circumstances, radiation conditions must be known at particular locations in space with much higher fidelity than that which results from climatology predictions. Thus, there is a need to specify the intensity and spectrum of radiation belt electrons at a given spacecraft location at a given time. Recently, Moorer and Baker [2000] developed a specification model, which meets these requirements rather successfully.

Figure 13 shows the essential elements of the Moorer and Baker [2000] specification scheme. The model uses as its basis the CRRES electron model [CRRESELE, Brautigam and Bell, 1995]. This model uses the Ap 15 index to indicate the general level of electron response to changing geomagnetic and interplanetary conditions (Ap 15 is a running 15-day index of global geomagnetic conditions). However, this low time "resolution" of the basic model often does not capture the kind of variability (on hourly time scales) that is of concern in anomaly analyses. Thus, Moorer and Baker [2000] used the CRRESELE framework, but they developed a model with much higher intrinsic time resolution and with much closer relationship to actual data. As shown in the flow chart of Figure 13, the specification model uses all available Los Alamos geostationary, NOAA/GOES, and Global Positioning System (GPS) data to assess electron fluxes at many points throughout the radiation belts. Then a kind of four-dimensional "data assimilation" (4-DDA) approach allows the model to use these actual data to adjust and reconfigure the baseline model to a new, more realistic description of the entire radiation belt.

Many issues are dealt with systematically in the Moorer-Baker model. This includes cross-calibration of instrument data sets, diurnal (local time) variations of fluxes, and radiation belt "configurational" changes. The result is that the model produces optimised flux maps throughout the outer radiation belt (L=3-7) at all local times and over a wide energy range (0.2-2.0 MeV). As noted, the model is updated on an hourly time step. Moorer and Baker [2000] showed that the model regularly achieves accuracy of ~90% in its specification of flux levels: This assessment is made by comparing the model results with actual data. Figure 14 from Moorer and Baker shows the model "efficiency" for a two-month interval at the beginning of 1997. In this case the modeled and measured fluxes of 0.75-1.1 MeV electrons at a geostationary spacecraft (LANL 1991-080) are virtually indistinguishable (correlation coefficient r~0.95).

4.3   Forecast of radiation belt electrons

In addition to *specification* of radiation belt electron fluxes (i.e., telling what the flux is at a given point in space right now), operators of spacecraft would like to know what the flux will be one or two days in advance. Moorer and Baker [2000] also developed such a forecast tool as part of the general modelling effort discussed in the previous section. This forecast method is based on analogue methods that have been very successful in meteorology.

Figure 15 shows the forecast method flow diagram from Moorer and Baker [2000]: The method assumes that one has good knowledge (as per the previous section) of the present radiation belt (RB) conditions. Given the current "state" of the radiation belts, one then looks back through a large RB state data base to identify all prior times when fluxes and spectra were similar. The model further examines the previous history of the current state of the radiation belt changes (for the 30 hours prior to the present) and compares this 30-hour trend with the candidate nearest-analogue cases in the past (also for a 30-hour period). The closest analogues of previous RB trends then become a new subset of data. To further improve forecasting, the model then looks at the solar wind "driver" of the radiation belt response.

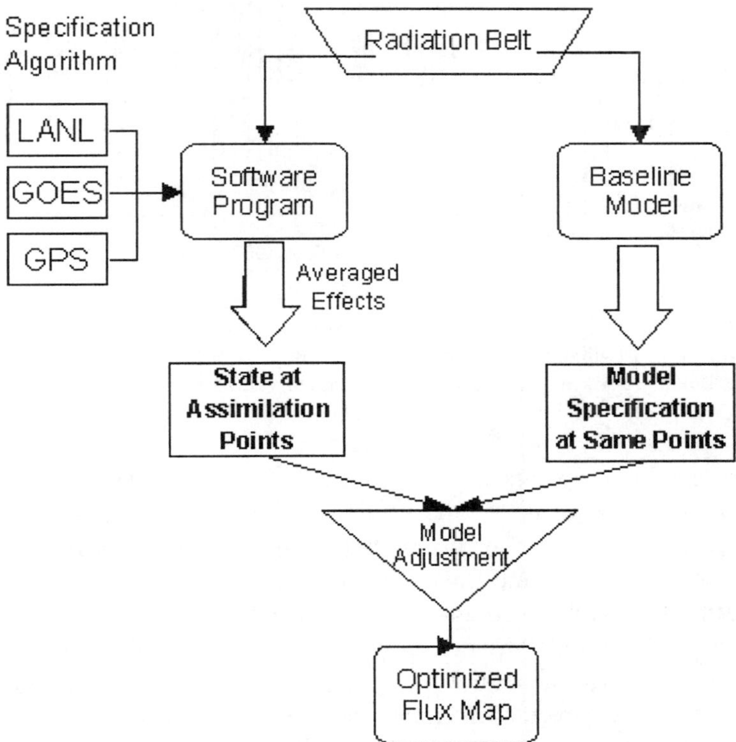

Figure 13. A schematic summary of the outer radiation belt specification modelling approach used by Moorer and Baker [2000].

For the current condition, the solar wind speed and its 5-day preceding trend is noted. Then for the subset of previous radiation belt analogue events, the 5-day solar wind trends in those cases are also examined. A cross-correlation is performed between the current case and the possible analogue

306

events. The three best analogue event intervals (i.e., with the best correlation) are then chosen as the analogue cases.

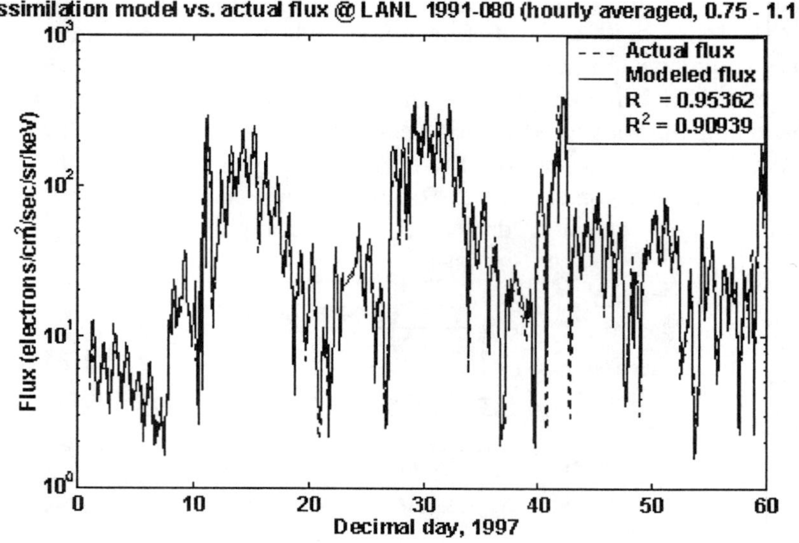

Figure 14. The predicted (solid line) and measured (dashed line) fluxes of 0.75-1.1 MeV electrons at geostationary orbit for the first 60 days of 1997. The predicted flux (in the Moorer-Baker [2000] model) shows an excellent correlation coefficient (R=0.95362) with the actual flux.

The analogue events of choice can be averaged together and the time behaviour subsequent to the "fiducial" time can be examined. This subsequent time behaviour becomes our analogue forecast of electron flux. As shown by Moorer and Baker, this method is generally quite successful in predicting behavior ahead 24-48 hours. Because both past (actual) radiation belt and solar wind "driver" information is used, this analogue forecast method is robustly successful for both quiet and disturbed conditions.

The analogue events of choice can be averaged together and the time behaviour subsequent to the "fiducial" time can be examined. This subsequent time behaviour becomes our analogue forecast of electron flux. As shown by Moorer and Baker, this method is generally quite successful in predicting behavior ahead 24-48 hours. Because both past (actual) radiation belt and solar wind "driver" information is used, this analogue forecast method is robustly successful for both quiet and disturbed conditions.

5.   SUMMARY AND FUTURE TRENDS

Recent trends suggest that there will be an increasing reliance on space systems in the future. The costs of such systems are very large, and losses of space assets have also been large, especially recently. As an example of this point, Figure 16 [from Kunstadter, 1999] shows the insurance premiums paid each year in aggregate for commercial spacecraft from 1990 to (May) 1999. Also shown for each year (by the darker bars) are the losses (i.e., the insurance claims) for each year. Obviously, both the premiums and the claims have gone up considerably in total dollar-value over the course of the 1990s. However, the matter of great alarm to the space insurance underwriters is the large excess of losses in the most recent years (1998-99) compared to premiums paid. Obviously, space (and getting into space) is dangerous and many kinds of costly losses can and do occur. As indicated in the figure, there were some $4.75B in losses from 1998 through May 1999.

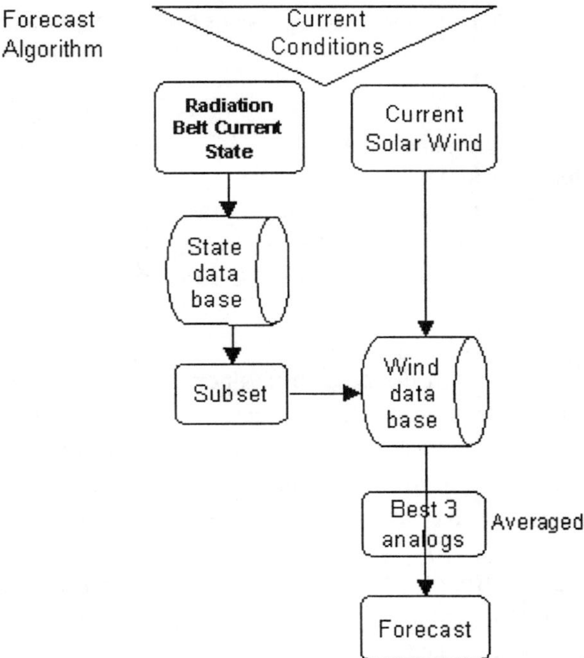

Figure 15. A schematic summary of the electron flux forecasting approach used by Moorer and Baker [2000].

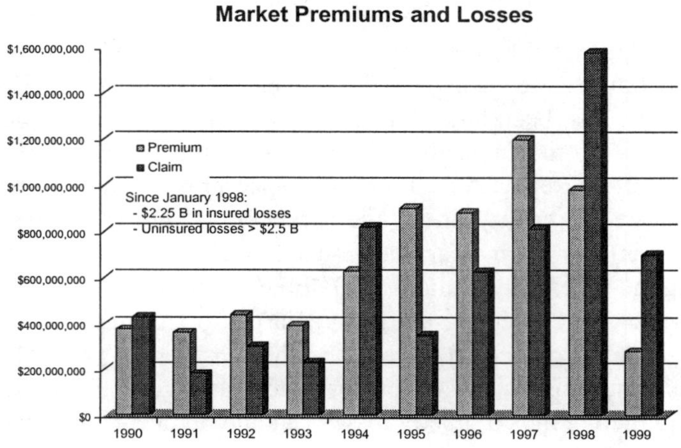

308

Figure 16. Market premiums paid to insure spacecraft from 1990 through May 1999 (light-shaded bars) and the claims (or losses) actually paid (dark-shaded bars) during the same period [from Kunstadter, 1999]

Another view of the spacecraft risks is shown in Figure 17 [also adapted from Kunstadter, 1999]. The figure shows:
The number of insured spacecraft at geostationary orbit;
The number of insured spacecraft at low-Earth orbit; and
The number of serious anomalies in these spacecraft.

Values are shown for each year from 1990 through 1998. We see that the number of GEO spacecraft doubled in this period, while the number of LEO spacecraft went up many-fold. Of considerable interest is the fact that there has been such a large increase in recent years of serious anomalies (complete loss or major loss of capability). This can be viewed with considerable alarm and may indicate that spacecraft are becoming more susceptible to on-orbit failure.

Part of the failure trend in Figure 17 may relate to the 11-year solar cycle. Indeed, we are now approaching the expected solar activity peak (in ~2001). Figure 18 shows a summary of the recent (and anticipated) annual sunspot numbers. It also tries to indicate some of the prominent solar and magnetospheric features that are or will be a risk to operating spacecraft over the course of the sunspot cycle. Although somewhat reduced at sunspot minimum, there are always worrisome features of the sun-Earth system and space system operators must be vigilant.

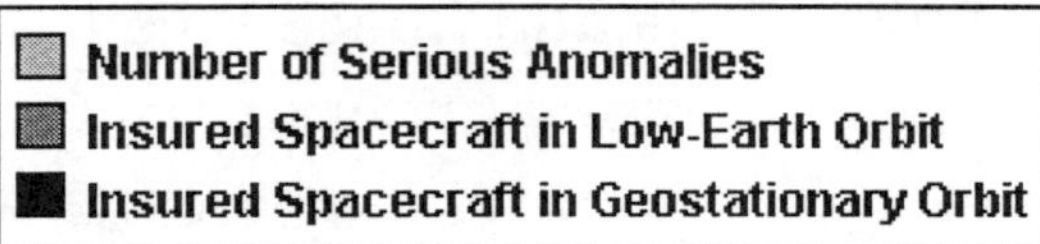

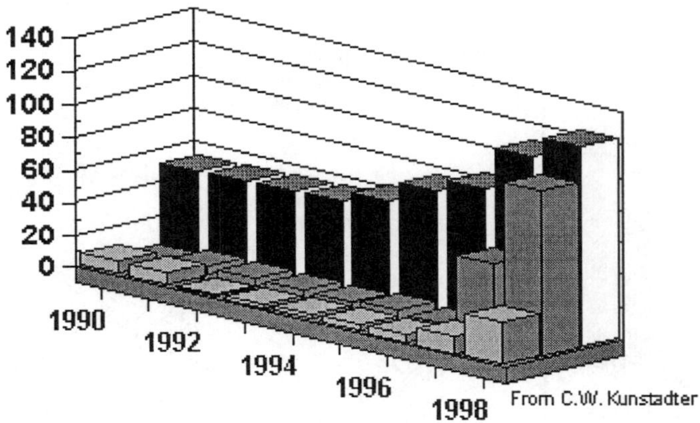

Figure 17. Insured spacecraft numbers at geostationary orbit and low-Earth orbit from 1990 through 1998. Also shown are the serious anomalies suffered during these years by insured spacecraft [from Kunstadter, 1999].

In this paper we have reviewed a number of the features of near-Earth space that represent dangers to the operation of spacecraft. We have also tried to illustrate some of the mechanisms by which space environmental agents can cause spacecraft operational anomalies or failures. We have shown how various scientific and programmatic data sets from many different satellites can be used to assess the space environment and help resolve (or at least illuminate) why anomalies occurred. We have illustrated how good scientific understanding of one aspect of the space environment (viz., radiation belt electrons) can be used to develop a powerful specification and forecasting capability. The (apparent) high reliability of this model for global radiation belt purposes may make such a model a great benefit for both spacecraft designers and spacecraft operators. Development of similarly reliable models of other aspects of the space environment would seem to be highly desirable.

310

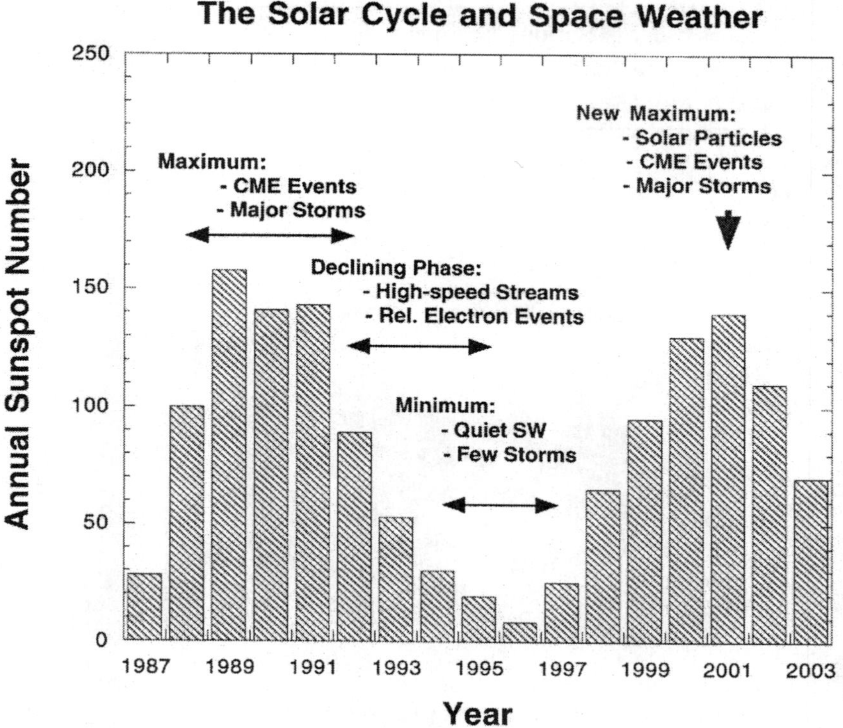

Figure 18. Sunspot and space environmental concerns during different phases of the sunspot cycle.

6. ACKNOWLEDGEMENTS

The author thanks many colleagues who have contributed to this review article. Special thanks go to D.F. Moorer, J. H. Allen, S. Kanekal, C. Connolly, G. Reeves, T. Pulkkinen, and the many members of the SAMPEX, POLAR, and other ISTP science teams. This work was supported by grants from NASA and NSF.

7.   REFERENCES

Allen, J., H. Sauer, L. Frank, and P. Reiff, Effects of the March 1989 Solar Activity, *EOS, 70*, 1479, 1989.

Baker, D.N., What is space weather?, *Adv. Space Res., 23*, 1, 7, 1998.

Baker, D.N. , The occurrence of operational anomalies in spacecraft and their relationship to space weather, *IEEE Trans. Plasma Sci.*, December issue, 2000.

Baker, D.N., J.B. Blake, R.W. Klebesadel, and P.R. Higbie, Highly relativistic electrons in the Earth's outer magnetosphere, I. Lifetimes and temporal history 1979-1984, *J. Geophys. Res., 91*, 4265, 1986.

Baker, D.N., R.D. Belian, P.R. Higbie, R.W. Klebesadel, and J.B. Blake, Deep dielectric charging effects due to high energy electrons in Earth's outer magnetosphere, *J. Electrost. 20*, 3, 1987.

Baker, D.N., J.B. Blake, S.G. Kanekal, B. Klecker, and G. Rostoker, Satellite anomalies linked to electron increase in the magnetosphere, *EOS, 75,* 401, 1994.

Baker, D.N., et al., An assessment of space environmental conditions during the recent Anik E1 spacecraft operational failure, *ISTP Newsl. 6*, 8, 1996.

Baker, D.N., J.H. Allen, S.G. Kanekal and G.D. Reeves, Disturbed space environment may have been related to pager satellite failure, *EOS, Trans. AGU, 79*, 477, 1998.

Baker, D.N., S.G. Kanekal, T.I. Pulkkinen, and J.B. Blake, Equinoctial and solstitial averages of magnetospheric relativistic electrons: A strong semiannual modulation, *Geophys. Res. Lett., 16*, 20, 3193, 1999.

Baker, D.N., S.G. Kanekal, J.B. Blake, and T.I. Pulkkinen, The global efficiency of relativistic electron production in the Earth's magnetosphere, *J. Geophys. Res.*, in press, 2000.

Brautigam, D.H., and J.T.Bell, CRRESELE Documentation, Phillips Laboratory, *Environmental Research papers, 1178*, 1, 1995.

Cummings, J.R., et al., New evidence for geomagnetically trapped anomalous cosmic rays, *Geophys. Res. Lett., 20,* 2003, 1993.

Kanekal, S.G., D.N. Baker, J.B. Blake, B. Klecker, R.A. Mewaldt, and G.M. Mason, Magnetospheric response to magnetic cloud (coronal mass ejection) events: Relativistic electron observations from SAMPEX and POLAR, *J. Geophys. Res., 104*, 24885, 1999.

Kunstadter, C.T.W., Were we crying wolf?, Insurance implications of the Leonids and other space phenomena, *AIAA Leonids Conference*, Los Angeles, CA, 1999.

Moorer, D.F., Jr., and D.N. Baker, Specification of energetic magnetospheric electrons, Space Weather, *AGU Monogr.,* in press, 2000.

Reagan, J.B., et al., Space charging currents and their effects on spacecraft systems, *ISEE Trans. Elec. Insul., E1-18*, 354, 1983.

Robinson, P.A., Jr., Spacecraft environmental anomalies handbook, *JPL Report GL-TR-89-0222*, Pasadena, CA, 1989.

Rosen, A. (editor) *Spacecraft charging by magnetospheric plasmas*, AIAA, 47, New York, 1976.

Vampola, A.L., The aerospace environment at high altitudes and its implications for spacecraft charging and communications, *J. Electrostat., 20*, 21, 1987.

Wrenn, G.L. Conclusive evidence for internal dielectric charging anomalies on geosynchronous communications spacecraft, *J. Spacecraft and Rockets, 32*, 514, 1995.

Chapter 12

Space Weather Effects on Communications

An overview of historical and contemporary impacts of solar and geospace disturbances on communications systems

Louis J. Lanzerotti
Bell Laboratories, Lucent Technologies
Murray Hill, NJ 07974, USA

Abstract   In the last century and one-half, the variety of communications technologies that are embedded in environments that can be affected by processes occurring in space have vastly increased. This paper presents some of the history of the subject of "space weather" as it affects communications, beginning with the earliest electric telegraph systems and continuing to today's wireless communications using satellites and land links. An overview is presented of the present-day communications technologies that can be affected by solar-terrestrial phenomena such as galactic cosmic rays, solar-produced plasmas, and geomagnetic disturbances in the Earth's magnetosphere.

Keywords   Solar disturbance, geomagnetic disturbance, communications technologies, cable communications, wireless communications, communication satellites, ionosphere currents, aurora, magnetosphere.

1.   INTRODUCTION AND SOME HISTORY

In 1847, during the 8th solar cycle, telegraph systems that were just beginning to be deployed in common use in Europe were found to often exhibit "anomalous currents" in their wires. W. H. Barlow [1849], a telegraph engineer with the Midland railroad in England appears to be the first to have recognized and systematically sought to understand these currents that were disturbing the operations of the railways' communications system. Making

I.A. Daglis (ed.), Space Storms and Space Weather Hazards, 313–334.
© 2001 *Kluwer Academic Publishers. Printed in the Netherlands.*

use of a spare wire that connected Derby and Birmingham, Barlow recorded during a two-week interval (with the exception of the weekend) in May 1847 the deflections in the galvanometer at the Derby station that he installed specifically for his experiment. These data (taken from a Table in his paper) are plotted in Figure 1. The galvanometer deflections obviously varied from hour to hour and from day to day by a cause (or causes) that was (were) unknown.

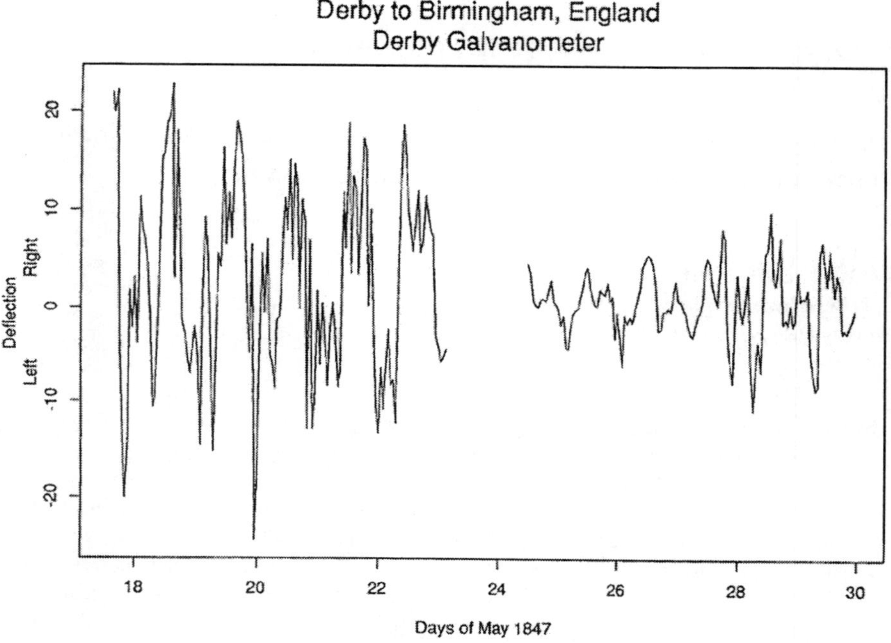

Figure 1. Hourly galvanometer recordings of voltage across a cable from Derby to Birmingham, England, May 1847.

Plotting the hourly means of the Barlow data for the Derby to Birmingham link, as well as for the measurements on the dedicated wire from Derby to Rugby, a very distinct diurnal variation in the galvanometer readings are apparent. As shown in Figure 2, the galvanometers had large right-handed swings during the local day interval, while a left-handed swing appeared during local night. Such a diurnal variation has now been recognized for many decades to be produced by solar-induced effects on the Earth's dayside ionosphere [e.g., Chapman and Bartels, 1940; Matsushita, 1967]. The systematic daily change evident in Figure 2, while not explicitly recognized by Barlow in his paper, is likely the first measurement of the diurnal component of geomagnetically induced Earth currents (often referred to in subsequent literature in the 19th and early 20th centuries as "telluric currents"). Barlow, in discussing his measurements, further noted that "… in every case which has come under [his] observation, the telegraph needles have been deflected whenever aurora has been visible".

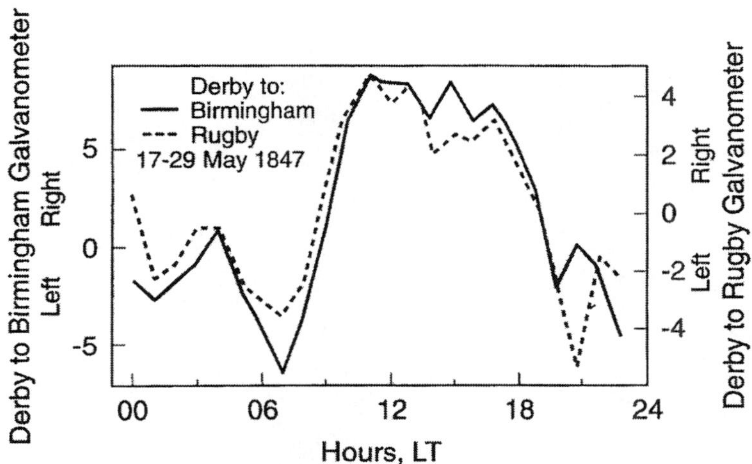

Figure 2. Hourly mean galvanometer deflections recorded on telegraph cables from Derby to Birmingham (solid line) and to Rugby (dashed line) in May 1847.

Indeed, this was certainly the case during November 1847 as the peak of the sunspot cycle approached, but after Barlow's measurements on the two dedicated Midland railway wires seem to have ceased. At that time, large auroral displays over Europe were accompanied by severe disruptions of the Midland railway telegraph lines, as well as telegraph lines in other European locations, including the line from Florence to Pisa [Prescott, 1866]

Twelve years later in the 10th solar cycle, while pursuing his very extensive and systematic program of observations and descriptions (by drawings and words) of sun spots, Richard Carrington, FRS, recorded an exceptionally large area of spots in the northern solar hemisphere at the end of August 1859. Figure 3 is a reproduction of Plate 80 from the comprehensive records of his studies, which were carried out over a more than seven year interval around the peak of that sunspot cycle [Carrington, 1863]. The large spot area at about 45° N solar latitude on August 31 is especially notable.

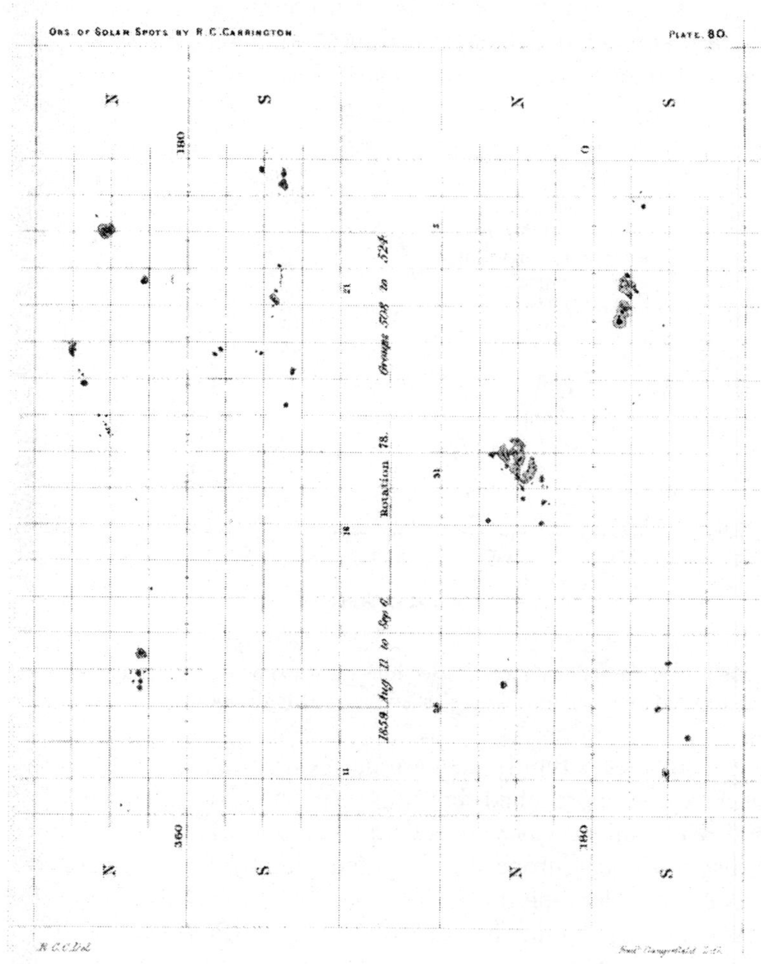

Figure 3. Plate 80 from Carrington [1863] showing his sunspot drawings for August 11 to September 6, 1859. The large spot area at about 45° N solar latitude on August 31 is especially notable.

This observation of an extensive dark region on the solar face was to ultimately prove to be more out of the ordinary than Carrington's past research would have originally suggested to him. Quoting from his description of this region, "...at [the observatory at] Redhill [I] witnessed ... a singular outbreak of light which lasted about 5 minutes, and moved sensibly over the entire contour of the spot ...." Some hours following this white light outburst from the large dark sunspot region (the first ever reported), disturbances were observed in magnetic measuring instruments on Earth, and the aurora borealis was seen as far south as Hawaii and Rome.

Although Barlow had remarked on the apparent association of auroral displays and the disturbances on his railway telegraph wires, the large and disruptive electrical disturbances that were recorded in numerous telegraph systems during the wide-spread magnetic disturbances that followed Carrington's solar event were nevertheless a great surprise. Indeed, during the same several day interval that the large auroral displays were widely seen, strange effects were measured in telegraph systems all across Europe – from Scandinavia to Tuscany. In the Eastern United States, it was reported [Prescott, 1866] that on the telegraph line from Boston to Portland (Maine) during "...Friday, September 2d, 1859 [the operators] continued to use the line [without batteries] for about two hours when, the aurora having subsided, the batteries were resumed."

In addition to the "anomalous" electrical currents flowing in the Earth, the early telegraph systems were also very vulnerable to atmospheric electrical disturbances in the form of thunderstorms. As noted by Silliman [1850], "One curious fact connected with the operation of the telegraph is the induction of atmospheric electricity upon the wires ... often to cause the machines at several stations to record the approach of a thunderstorm." While disturbances by thunderstorms on the telegraph "machines" could apparently be identified as to their source, the source(s) of the "anomalous currents" described by Barlow [1849] and as recorded following Carrington's solar event, remained largely a mystery.

The decades that followed the solar event of 1859 produced significant amounts of attention by telegraph engineers and operators to the effects on their systems of Earth electrical currents. Although little recognized for almost fifty years afterwards, the sun was indeed seriously affecting the first electrical technology that was employed for communications.

The dramatic demonstration of intercontinental wireless communications, with the long wavelength radio transmissions through the atmosphere from Poldhu Station, Cornwall, to St. John's, Newfoundland, by Marconi in December 1901, eliminated Earth currents as a source of disturbances on communications. Marconi's achievement (for which he was awarded the Nobel Prize in Physics in 1909) was only possible because of the existence of the ionosphere that reflected Marconi's signals. This reflecting layer at the top of the neutral atmosphere was definitively identified some two decades later by Briet and Tuve [1925] and by Appleton and Barnett [1925]. Because wireless remained the only method for cross-oceanic voice communications for more than five decades following Marconi's feat, any physical changes in the radio wave-reflecting layer (even before it was "discovered") were critical to the success (or failure) of reliable transmissions.

The same ionosphere electrical currents that could produce "spontaneous" electrical currents within the Earth (and thus disturb the electrical telegraph) could also affect the reception and fidelity of the transmitted long-distance wireless signals. Indeed, Marconi [1928] commented on this phenomenon when he noted that "...times of bad fading [of radio signals] practically always coincide with the appearance of large sun-spots and intense aurora-borealis usually accompanied by magnetic storms ...." These are "... the same periods when cables and land lines experience difficulties or are thrown out of action."

An example of the types of studies that were pursued in the early years of long-distance, very long wavelength, wireless is shown in Figure 4. Plotted here (reproduced from Fagen [1975], which contains historical notes on early wireless research in the old Bell Telephone System) are yearly average daylight cross-Atlantic transmission signal strengths for the years 1915 – 1932 (upper trace). The intensities in the signal strength curves were derived by averaging the values from about 10 European stations that were broadcasting in the ~15 to 23 kHz band, after reducing them to a common base (the signal from Nauen, Germany, was used as the base). Plotted in the lower trace of the Figure are the monthly average sunspot numbers per year. Clearly, an association is seen between the two physical quantities. From the perspectives of space weather predictions, this relationship of the received electrical field strengths to the yearly solar activity as represented by the number of sunspots could be used by wireless engineers at that time to provide them some expectation as to transmission quality on a year to year basis.

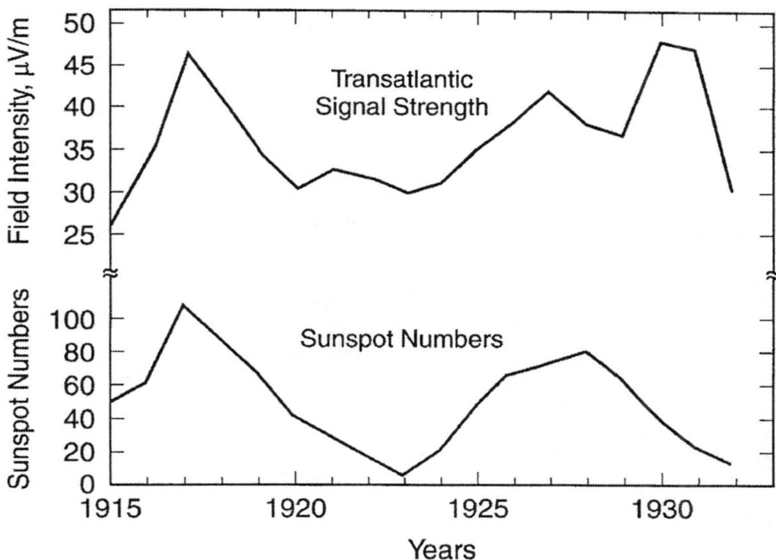

Figure 4. Yearly average daylight cross-Atlantic transmission signal strengths and monthly average sunspot numbers for the interval 1915–1932.

The relationship of "abnormal" propagation of long wavelength radio transmissions and solar disturbances was first identified in 1923 [Anderson, 1928]. The technical literature of the early wireless era showed clearly that solar-originating disturbances were serious assaults on the integrity of these communications during the first several decades of the twentieth century. Communications engineers pursued a number of methodologies to alleviate or mitigate the assaults. One of these methodologies that sought more basic understanding is illustrated in the context of Figure 4. Another methodology utilized alternative wireless communications "routes". As Figure 5 illustrates for the radio electric field strength data recorded during a solar and geomagnetic disturbance on July 8, 1928 (day 0 on the horizontal axis), the transmissions at long wave length were relatively undisturbed while those at the shorter wavelength (16m) were seriously degraded [Anderson, 1929]. Such procedures are still used today by amateur and other radio operators.

The practical effects of the technical conclusions of Figure 5 are well exemplified by a headline which appeared over a front page article in the Sunday, January 23, 1938, issue of *The New York Times*. This headline noted that "Violent magnetic storm disrupts short-wave radio communication." The subheadline related that "Transoceanic services transfer phone and other traffic to long wave lengths as sunspot disturbance strikes". The engineering work-around that shifted the cross-Atlantic wireless traffic from short to longer wavelengths prevented the complete disruption of voice messages at that time.

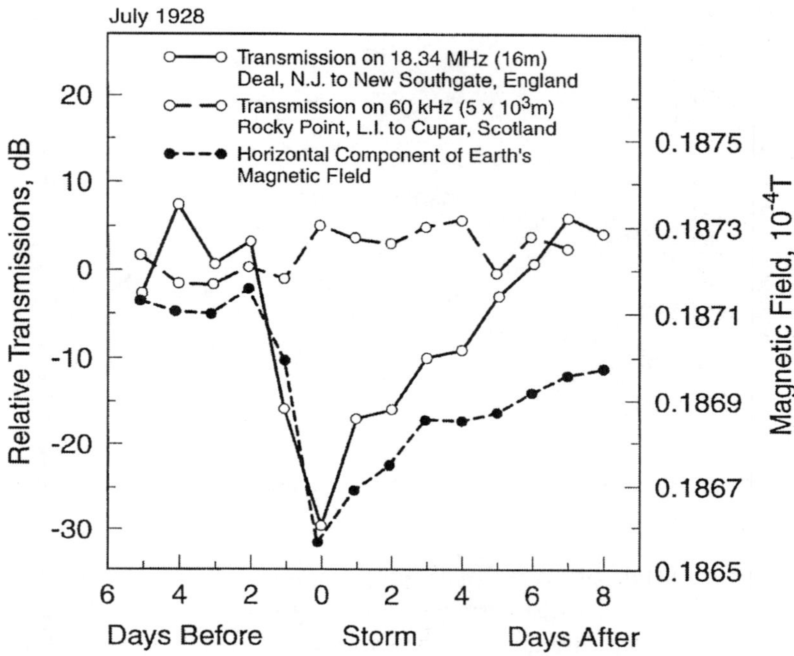

Figure 5. Trans-Atlantic wireless transmissions from the Eastern U. S. to the U. K. on two frequencies before and during a magnetic storm event in July 1928. Also shown are the values of the horizontal component of the Earth's magnetic field.

2.   THE SPACE ERA

From the earliest installations of the telegraph to the beginnings of wireless, to the extensive use of satellites, the role of the solar-terrestrial environment for successfully implementing and operating communications systems has continued to be of importance. Illustrated in Figure 6 are selected examples of the times of large solar-originating disturbances on communications. Four of the large disturbance effects indicated in Figure 6 occurred after the launch of the first Earth-encircling spacecraft, Vostok 1, in 1957. The magnetic storm of February 1958 disrupted voice communications on the first cross-Atlantic telecommunications cable, TAT-1, from Newfoundland to Scotland (and also plunged the Toronto area into darkness by the tripping of power company circuits).

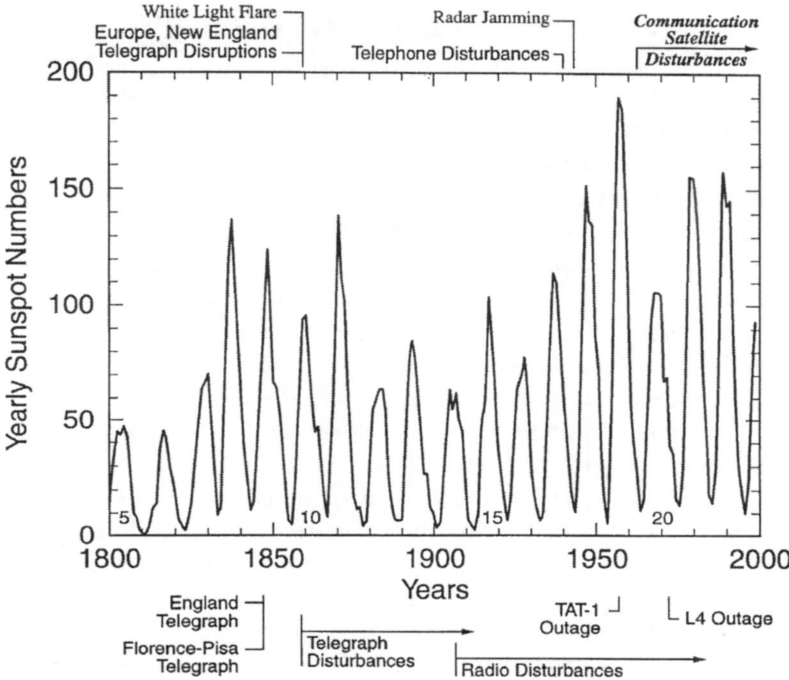

Figure 6. Yearly sunspot numbers with indicated times of selected major impacts of the solar-terrestrial environment on largely ground-based technical systems. The numbers just above the horizontal axis are the conventional numbers of the sunspot cycles.

The outage for nearly an hour of a major continental telecommunications cable that stretched from near Chicago to the west coast was disrupted between the Illinois and Iowa powering stations by the magnetic storm of August 1972 [Anderson et al., 1974; Boteler and Jansen van Beek, 1999]. In March 1989, when the entire province of Quebec suffered a power outage for nearly a day as major transformers failed under the onslaught of a large geomagnetic storm [Czech et al., 1992], the first cross-Atlantic fibre voice cable was rendered nearly inoperative by the large potential difference that was established between the cable terminals on the coasts of New Jersey and England [Medford et al., 1989].

Point-to-point high frequency (HF) wireless communications links, most often used today for some national defence communications and for civil emergency communications, continues to suffer the vagaries of the sun's interactions with the Earth's space environment. Those who are users of such systems are familiar with many anecdotes up to the present day of solar-produced effects and disruptions. For example, near the peak of the 21st solar cycle in 1979, a distress signal from a downed commuter plane was received by an Orange County, California, fire department – which responded, only to discover that the signal had originated from an accident site in West Virginia [*Los Angeles Times*, 1979].

The placing into space of ever-advancing technologies – for both civilian and national defence purposes – meant that ever more sophisticated understanding of the space environment was required in order to ensure reliability and survivability. Indeed, the first active civil telecommunications satellite, the low Earth-orbiting Telstar 1(launched June 10, 1962), carried solid state sensors to monitor the radiation environment encountered by the spacecraft [Brown et al., 1963; Buck et al., 1964]. As has been the case for many engineering investigations that have been made over the years in space-related research, these sensors also returned valuable new science information, such as measurements of the trapping lifetime of the radiation debris following high altitude nuclear explosions [Brown, 1966].

The operations and survivability of both ground- and space-based communications systems have often encountered unanticipated surprises because of natural space environmental effects. As technologies have increased in sophistication, as well as in miniaturization and in interconnectedness, more sophisticated understanding of the Earth's space environment continues to be required. In addition, the increasing diversity of communications systems that can be affected by space weather processes is accompanied by continual changes in the dominance of one technology over another for specific uses. For example, in 1988, satellites were the dominant carrier of transocean messages and data; only about two percent of this traffic was over ocean cables. By 1990, the advent of the wide bandwidths provided by fibre optics meant that 80% of the transocean traffic was now via ocean cable [Mandell, 2000]. Hence, any space weather effects on communications cables must now be considered more seriously than they may have been a decade ago.

3.   SPACE ENVIRONMENTAL EFFECTS ON COMMUNICATIONS TECHNOLOGIES

Many present-day communications technologies that must include considerations of the solar-terrestrial environment in their designs and/or operations are listed in Table 1. Figure 7 schematically illustrates these effects.

Table 1

Impacts of Solar-Terrestrial Processes on Communications

Ionosphere Variations
    Induction of electrical currents in the Earth
        Long communications cables
    Wireless signal reflection, propagation, attenuation
    Communication satellite signal interference, scintillation

Magnetic Field Variations
    Attitude control of spacecraft

Solar Radio Bursts
    Excess noise in wireless communications systems

Radiation
    Solar cell damage
    Semiconductor device damage and failure
    Failure of semiconductor devices
    Spacecraft charging, surface and interior materials

Micrometeoroids and Artificial Space Debris
    Solar cell damage
    Damage to surfaces, materials, and complete vehicles

Atmosphere
    Low altitude satellite drag
    Attenuation and scatter of wireless signals

3.1   Ionosphere and Wireless

A century after Marconi's feat, the ionosphere remains both a facilitator and an intruder in numerous communications applications. As noted above, the military, as well as police and fire emergency agencies in many nations, continue to rely on wireless links that make extensive use of frequencies from kHz to hundreds of MHz and that use the ionosphere as a reflector. Changes in the reflections produced by solar activity, both from solar UV and x-ray emissions, as well as by magnetic storms, can significantly alter the propagation of these wireless signals.

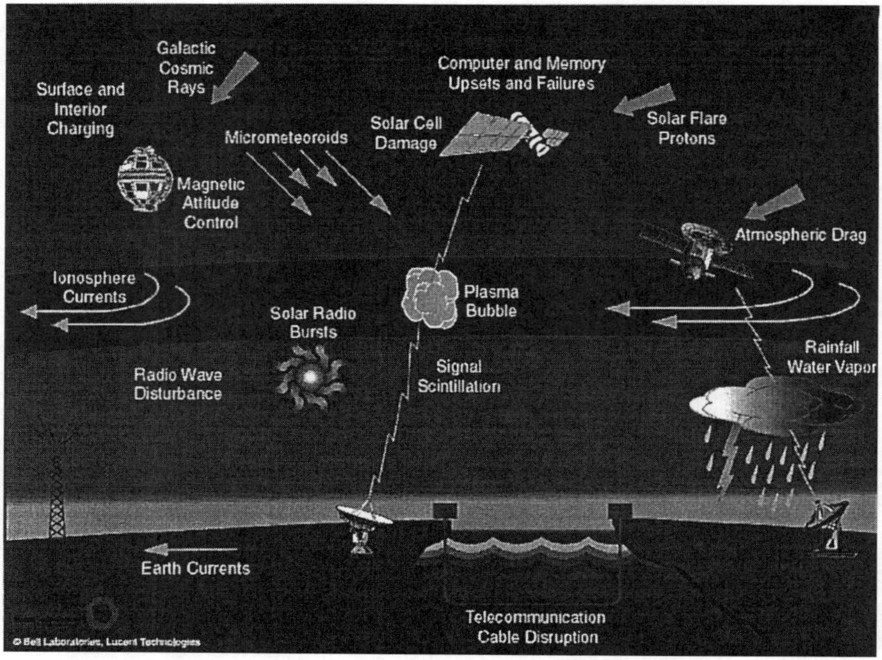

Figure 7. Some of the effects of space weather on communications systems that are deployed on the Earth's surface and in space, and/or whose signals propagate through the space environment.

At higher, few GHz, frequencies the production of "bubbles" in ionosphere densities in equatorial regions of the Earth can be a prime source of scintillations in satellite to ground signals. Engineers at the COMSAT Corporation first discovered these effects after the initial deployment of the INTELSAT network at geosynchronous orbit (GEO) [Taur, 1973]. Plasma processes in the ionosphere are also the cause of considerable problems in the use of single frequency communications signals from the Global Positioning System (GPS) for precise location determination on Earth. The intent to evolve the GPS system to a dual, and eventually a three, frequency system over the next five to ten years should eliminate this ionosphere nuisance.

There remain large uncertainties in the knowledge base of the processes that determine the initiation and scale sizes of the ionosphere irregularities that are responsible for the scintillation of radio signals that propagate through the ionosphere. Thus, it is very difficult to define diversity strategies for receivers and/or space-based transmitters that might be applicable under most conditions.

3.2   Ionosphere and Earth Currents

The basic physical chain of events behind the production of Earth potentials begins with greatly increased electrical currents flowing in the magnetosphere and the ionosphere. These increased currents then cause large variations in the time rate of change of the magnetic field as seen at the Earth's surface. The time variations in the field in turn induce potential differences across large areas of the Earth's surface which are spanned by the cable communications systems. Telecommunications cable systems use the Earth itself as a ground return for their circuits, and these cables thus provide paths for concentrating the electrical currents that flow between these newly established, but temporary, Earth "batteries" (see Kappenman, this volume, for an overview and references). The precise effects of these "anomalous" electrical currents depend upon the technical system to which the long conductors are connected. In the case of long telecommunications lines, the Earth potentials can cause overruns of the compensating voltage swings that are designed into the power supplies [e.g., Anderson et al., 1974].

There are major issues that can arise in attempting to understand in detail the effects of enhanced space electrical currents on cable systems. At present, the time variations and spatial dependencies of the space electrical currents are not at all well understood or predicable from one geomagnetic storm to the next. This is of considerable importance since the Earth potentials that are induced are very much dependent upon the conductivity structure of the Earth underlying the affected ionosphere regions. Similar electrical current variations in the space environment can produce vastly different Earth potential drops depending upon the nature and orientation of underground Earth conductivity structures in relationship to the variable overhead currents.

Modelling of these effects is becoming quite advanced in many cases. This is an area of research that involves a close interplay between space plasma geophysics and solid Earth geophysics, and is one that is not often addressed collaboratively by these two very distinct research communities (except by the somewhat limited group of researchers who pursue electromagnetic investigations of the Earth).

3.3   Solar Radio

Solar radio noise and bursts were discovered nearly six decades ago by Southworth [1945] and by Hey [1946] during the early research on radar at the time of the Second World War. Solar radio bursts produced unexpected (and unrecognised at first) jamming of this new technology that was under rapid development and deployment for wartime use [Hey, 1973]. Extensive post-war research established that solar radio emissions can exhibit a wide range of spectral shapes and intensity levels [e.g., Kundu, 1965; Castelli et al., 1973; Guidice and Castelli, 1975; Barron et al., 1985]. Research on solar radio phenomena remains an active and productive field of research today [e.g., Bastian et al., 1998].

Some analysis of local noon time solar radio noise levels that are routinely taken by the U.S. Air Force and that are made available by the NOAA World Data Center have recently been made in order to assess the noise in the context of modern communications technologies. These analyses show that during 1991 (within the sunspot maximum interval of the 22nd cycle) the average noon fluxes measured at 1.145 GHz and at 15.4 GHz were -162.5 and -156 dBW/(m^2 4kHz), respectively [Lanzerotti et al., 1999]. These values are only about 6 dB and 12 dB above the 273° K (Earth's surface temperature) thermal noise of -168.2 dBW/(m^2 4kHz). Further, these two values are only about 20 dB and 14 dB, respectively, below the maximum flux of -142 dBW/(m^2 4kHz) that is allowed for satellite downlinks by the ITU regulation RR2566.

Solar radio bursts from solar activity can have much larger intensities. As an example of an extreme event, that of May 23, 1967, produced a radio flux level (as measured at Earth) of $> 100,000$ solar flux units (1 SFU $= 10^{-22}$ W/(m^2 Hz)) at 1 GHz, and perhaps much higher [Castelli et al., 1973]. Such a sfu level corresponds to -129 dBW/(m^2 4kHz), or 13 dB above the maximum limit of -142 dBW/(m^2 4kHz) mentioned previously.

Short-term variations often occur within solar radio bursts, with time variations ranging from several milliseconds to seconds and more [e.g., Barron et al., 1980; Benz, 1986; Isliker and Benz, 1994]. Such short time variations can often be many tens of dB larger than the underlying solar burst intensities upon which they are superimposed.

As the use of wireless systems continues to grow in importance, the analysis of the possible vulnerability of these systems to solar noise must be evaluated. Further, more research is needed into the occurrence frequency of system-affecting solar bursts, including millisecond bursts. Additional fundamental research on the solar conditions for producing such bursts is also warranted.

3.4   Space Radiation Effects

The discovery by Van Allen in 1958 of the trapped radiation around Earth immediately implied that the space environment would not be benign for any communications technologies that might be placed within it. As noted by McCormac [1966], the high altitude "nuclear test Star Fish [on June 9, 1962] focused attention on the degradation of sensitive electronics by trapped electrons and protons." At that time, "…the basic processes occurring in the solar cells and transistors by the electrons and protons [were] poorly understood."

Some 200 or so in-use communications satellites now occupy the geosynchronous orbit. The charged particle radiation that permeates the Earth's space environment remains a difficult problem for the design and operations of these and other space-based systems [e.g., Shea and Smart, 1998; Koons et al., 1999]. A textbook discussion of the space environment and the implications for satellite design is contained in Tribble [1995]. The low energy (few eV to few keV) particles in the Earth's magnetosphere plasma population can produce different levels of surface charging on the materials (principally for thermal control) that encase a satellite [Garrett, 1981]. If good electrical connections are not established between the various surface materials, and between the materials and the solar arrays, differential charging on the surfaces can produce lightning-like breakdown discharges between the materials. These discharges can produce electromagnetic interference and serious damage to components and subsystems [e.g., Vampola, 1987; Koons, 1980; Gussenhoven and Mullen, 1983].

The plasma populations of the magnetosphere can be highly variable in time and in intensity levels. Under conditions of enhanced geomagnetic activity, the cross-magnetosphere electric field will convect earthward the plasma sheet in the Earth's magnetotail. When this occurs, the plasma sheet will extend earthward to within the geosynchronous (GEO) spacecraft orbit. When this occurs, on-board anomalies from surface charging effects will occur; these tend to be most prevalent in the local midnight to dawn sector of the orbit [Mizera, 1983].

While some partial records of spacecraft anomalies exist, there is not a comprehensive body of published data on the statistical characteristics of charging on spacecraft surfaces, especially from commercial satellites that are used so extensively for communications. Two surface-mounted charge plate sensors were specifically flown on the AT&T (now Loral) Telstar 4 GEO satellite to monitor surface charging effects. Figure 8 shows the statistical distributions of charging on one of the sensors in January 1997 [Lanzerotti et al., 1998]. The solid line in each panel corresponds to the charging statistics for the entire month, while the dashed lines omit data from a large magnetic storm event on January 10th (statistics shown by the solid lines). Charging voltages as large as -800 V were recorded on the charge plate sensor during the magnetic storm, an event during which a permanent failure of the Telstar 401 satellite occurred (although the failure has not been attributed specifically to the space conditions).

328

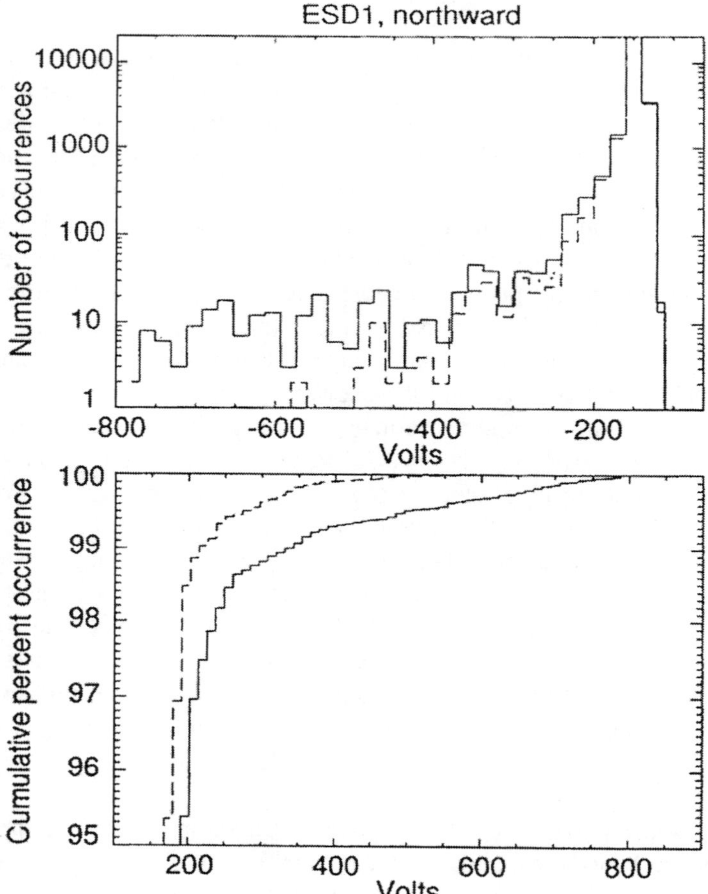

Figure 8. Statistical distribution of surface charging recorded on a charge plate sensor on the Telstar 4 spacecraft during the month of January 1997 (solid line) and for the same month with data from January 10th (the date of a large magnetic storm) removed.

The intensities of higher energy particles in the magnetosphere (MeV energy electrons to tens of MeV energy protons) can change by many orders of magnitude over the course of minutes, hours, and days. These intensity increases occur through a variety of processes, including plasma physics energisation processes in the magnetosphere and ready access of solar particles to GEO [Lanzerotti, 1968]. Generally it is prohibitively expensive to provide sufficient shielding of all interior spacecraft subsystems against high energy particles (such as including additional spacecraft mass in lieu of additional transponders or orbit control gas, for example).

The range of a 100 MeV proton in aluminum (a typical spacecraft material) is ~ 40 mm. The range of a 3 MeV electron is ~ 6 mm. These particles can therefore penetrate deeply into the interior regions of a satellite. In addition to producing transient upsets in signal and control electronics, such particles can also cause electrical charges to build up in interior insulating materials such as those used in coax cables. If the charge build-up is sufficiently large, these interior materials will eventually suffer electrical breakdowns. There will be electromagnetic interference and damage to the electronics.

A number of spacecraft anomalies, and even failures, have been identified as to having occurred following many days of significantly elevated fluxes of several MeV energy electrons [Baker et al., 1987; 1994; 1996; Reeves et al., 1998]. Plotted as histograms in Figure 9 are the daily average fluxes of > 2 MeV electrons for one month in 1998 when two spacecraft failures occurred and one anomaly was noted on a NASA satellite (all indicated in the Figure; Baker et al. [1998]). While the association of the times of occurrence of the spacecraft problems with the increased electron fluxes is evident, it is still uncertain which, if any, of the failures can be ascribed specifically to space radiation effects [Baker et al., 1998].

No realistic shielding is possible for most communications systems in space that are under bombardment by galactic cosmic rays (energies ~ 1 GeV and greater). These very energetic particles can produce upsets and errors in spacecraft electronics (as well as in computer chips that are intended for use on Earth [IBM, 1996]).

The significant uncertainties in placing, and retaining, a communications spacecraft in a revenue-returning orbital location has led to a large business in risk insurance and re-insurance for one or more of the stages in a satellite's history. The loss of a spacecraft, or one or more transponders, from adverse space weather conditions is only one of many contingencies that can be insured against. In some years the space insurance industry is quite profitable, and in some years there are serious losses in net revenue after paying claims [e.g., Todd, 2000]. For example, Todd [2000] states that in 1998 there were claims totalling more than $1.71 billion after salvage, an amount just less than about twice that received in premiums. These numbers vary by large amounts from year to year.

3.5   Magnetic Field Variations

The designs of those GEO communications spacecraft that use the Earth's magnetic field for attitude control must take into account the high probability that the satellite will on occasion, during a large magnetic disturbance, find itself outside the magnetosphere on the sunward side of the Earth. Enhanced

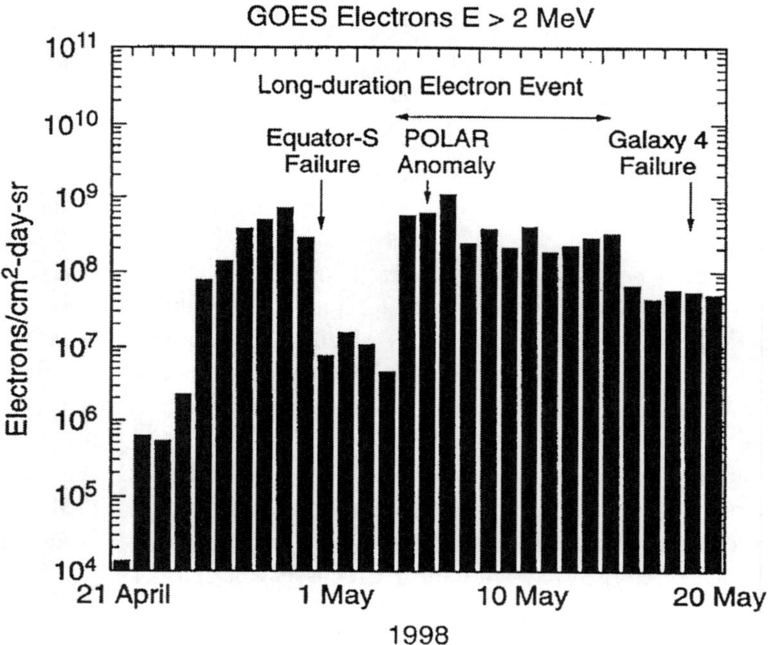

Figure 9. Daily flux values of electrons (> 2 MeV) measured on the NOAA geosynchronous GOES satellite for a one month interval, April – May, 1998. Dates of operational problems on three spacecraft are indicated.

solar wind flow velocities and densities, such as those that can occur in a coronal mass ejection event, can easily distort the dayside magnetopause and push it inside the geosynchronous orbit. The highly varying, in space and time, magnetic fields that occur at the boundary and outside the magnetosphere can seriously disrupt the satellite stabilization if appropriate precautions have not been incorporated into the design of the control system. The magnetic field outside the magnetopause will have a polarity that is predominantly opposite to that in which the spacecraft is normally oriented, so a complete "flip" of the orientation could occur when the magnetopause is crossed.

3.6 Micrometeoroids and Space Debris

The impacts on communications spacecraft of solid objects, such as from micrometeoroids and from debris left in orbit from space launches and from satellites that break up for whatever the reason, can seriously disorient a satellite and even cause a total loss [e.g., see Beech et al., 1995; 1997;

McBride, 1997]. The U.S. Air Force systematically tracks thousands of space debris items that are circling the Earth, most of which are in low altitude orbits.

3.7 Atmosphere: Low Altitude Spacecraft Drag

The ultraviolet emissions from the sun change by more than a factor of two at wavelengths ≤ 170 nm during a solar cycle [Hunten et al, 1991]. This is significantly more than the ~ 0.1 % changes that are typical of the visible radiation. The heating of the atmosphere by the increased solar UV radiation causes the atmosphere to expand. The heating is sufficient to raise the "top" of the atmosphere by several hundred km during solar maximum. The greater densities at the higher altitudes result in increased drag on both space debris and on communications spacecraft in low Earth orbits (LEO). Telecommunications spacecraft that fly in LEO have to plan to use some amount of their orbit control fuel to maintain orbit altitude during the build-up to, and in, solar maximum conditions [e.g., Picholtz, 1996].

3.8 Atmosphere Water Vapour

At frequencies in the Ka (18–31 GHz) band that are planned for high bandwidth space-to-ground applications (as well as for point-to-point communications between ground terminals), water vapour in the neutral atmosphere is the most significant natural phenomenon that can seriously affect the signals [Gordon and Morgan, 1993]. It would appear that, in general, the space environment can reasonably be ignored when designing around the limitations imposed by rain and water vapour in the atmosphere.

A caveat to this claim would arise if it were definitely to be proven that there are effects of magnetosphere and ionosphere processes (and thus the interplanetary medium) on terrestrial weather. It is well recognized that even at GHz frequencies the ionised channels caused by lightning strokes, and possibly even charge separations in clouds, can reflect radar signals. Lightning and cloud charging phenomena may produce as yet unrecognised noise sources for low-level wireless signals. Thus, if it were to be learned that ionosphere electrical fields influenced the production of weather disturbances in the troposphere, the space environment could be said to effect even those wireless signals that might be disturbed by lightning. Much further research is required in this area of speculation.

4.   SUMMARY

In the last 150 years, the diversity of communications technologies that are embedded within space-affected environments have vastly increased. The increasing sophistication of technologies, and how they relate to the environments in which they are embedded, means that ever more sophisticated understanding of the physical phenomena is needed. At the same time, most present-day communications technologies that are affected by space phenomena are very dynamic. These technologies can not wait for optimum knowledge to be acquired before new embodiments are created, implemented, and marketed. Indeed, those companies that might seek perfectionist understanding can be left behind by the marketplace. A balance is needed between seeking deeper understanding of physical phenomena and implementing "engineering" solutions to current crises. The research community must try to understand, and operate in, this dynamic environment.

5.   ACKNOWLEDGEMENTS

I thank numerous colleagues for their discussions on this topic over many years, including C. G. Maclennan, D. J. Thomson, G. Siscoe, J. B. Blake, G. Paulikas, A. Vampola, H. C. Koons, and L. J. Zanetti.

6.   REFERENCES

Anderson, C. N., Correlation of radio transmission and solar activity, *Proc. I. R. E.*, **16**, 297, 1928.

Anderson, C. N., Notes on the effects of solar disturbances on transatlantic radio transmissions, *Proc. I. R. E.*, **17**, 1528, 1929.

Anderson, C. W., III, L. J. Lanzerotti, and C. G. Maclennan, Outage of the L4 system and the geomagnetic disturbance of 4 August 1972, *The Bell Sys. Tech. J.*, **53**, 1817, 1974.

Appleton, E. V., and M. A. F. Barnett, Local reflection of wireless waves from the upper atmosphere, *Nature*, **115**, 333, 1925.

Baker, D. N., R. D. Balian, P. R. Higbie, et al., Deep dielectric charging effects due to high energy electrons in Earth's outer magnetosphere, *J. Electrost.*, **20**, 3, 1987.

Baker, D. N., S. Kanekal, J. B. Blake, et al., Satellite anomaly linked to electron increase in the magnetosphere, *Eos Trans. Am. Geophys. Union*, **75**, 401, 1994.

Baker, D. N., An assessment of space environment conditions during the recent Anik E1 spacecraft operational failure, *ISTP Newsletter*, **6**, 8, 1996.

Baker, D. N., J. H. Allen, S. G. Kanekal, and G. D. Reeves, Disturbed space environment may have been related to pager satellite failure, *Eos Trans. Am. Geophys. Union*, **79**, 477, 1998.

Barlow, W. H., On the spontaneous electrical currents observed in the wires of the electric telegraph, *Phil. Trans. R. Soc.*, **61A**, 61, 1849.

Barron, W. R., E. W. Cliver, D. A. Guidice, and V. L. Badillo, *An Atlas of Selected Multi-Frequency Radio Bursts from the Twentieth Solar Cycle*, Air Force Geophysics Laboratory, Space Physics Division, Project 4643, Hanscom AFB, MA, 1980.

Barron, W. R., E. W. Cliver, J. P. Cronin, and D. A. Guidice, Solar radio emission, in *Handbook of Geophysics and the Space Environment*, ed. A. S. Jura, Chap. 11, AFGL, USAF, 1985.

Bastian, T. S., A. O. Benz, and D. E. Gary, Radio emission from solar flares, *Ann. Rev. Astron. Astrophys.*, **36**, 131, 1998.

Beech, M., P. Brown, and J. Jones, The potential danger to satellites from meteor storm activity, *Q.J. R. Astr. Soc.*, **36**, 127, 1995.

Benz, A. O., Millisecond radio spikes, *Solar Phys.*, **104**, 99, 1986.

Beech, M., P. Brown, J. Jones, and A. R. Webster, The danger to satellites from meteor storms, *Adv. Space Res.*, **20**, 1509, 1997.

Boteler, D. H., and G. Jansen van Beek, August 4, 1972 revisited: A new look at the geomagnetic disturbance that caused the L4 cable system outage, *Geophys. Res. Lett.*, **26**, 577, 1999.

Breit, M. A., and M. A. Tuve, A test of the existence of the conducting layer, *Nature*, **116**, 357, 1925.

Brown, W. L., Observations of the transient behavior of electrons in the artificial radiation belts, in *Radiation Trapped in the Earth's Magnetic Field*, ed. B. M. McCormac, D. Reidel Pub. Co., Dordrecht, Holland, 610, 1966.

Brown, W. L., T. M. Buck, L. V. Medford, E. W. Thomas, H. K. Gummel, G. L. Miller, and F. M. Smith, *Bell Sys. Tech. J.*, **42**, 899, 1963.

Buck, T. M., H. G. Wheatley, and J. W. Rogers, *IEEE Trans. Nucl. Sci.*, **11**, 294, 1964.

Carrington, R. C., *Observation of the Spots on the Sun from November 9, 1853, to March 24, 1863, Made at Redhill*, William and Norgate, London and Edinburgh, 167, 1863.

Castelli, J. P., J. Aarons, D. A.. Guidice, and R. M. Straka, The solar radio patrol network of the USAF and its application, *Proc. IEEE*, **61**, 1307, 1973.

Chapman, S., and J. Bartels, *Geomagnetism*, 2 vols, Oxford Univ. Press, 1940.

Czech, P., S. Chano, H. Huynh, and A. Dutil, The Hydro-Quebec system blackout of 13 March 1989: System response to geomagnetic disturbance, *Proc. EPRI Conf. Geomagnetically Induced Currents*, EPRI TR-100450, Burlingame, CA, 19, 1992.

Fagen, M. D., *A History of Science and Engineering in the Bell System*, Bell Tel. Labs., Inc., Murray Hill, NJ, 1975.

Garrett, H. B., The charging of spacecraft surfaces, *Revs. Geophys.*, **19**, 577, 1981.

Gordon, G. D., and W. L. Morgan, *Principals of Communications Satellites*, John Wiley, New York, 178-192, 1993.

Guidice, D. A., and J. P. Castelli, Spectral characteristics of microwave bursts, in *Proc. NASA Symp. High Energy Phenomena on the Sun*, Goddard Space Flight Center, Greenbelt, MD, 1972.

Gussenhoven, M. S., and E. G. Mullen, Geosynchronous environment for severe spacecraft charging, *J. Spacecraft Rockets*, **20**, 26, 1983.

Hey, J. S., Solar radiations in the $4-6$ metre radio wavelength band, *Nature*, **158**, 234, 1946.

Hey, J. S., *The Evolution of Radio Astronomy*, Neale Watson Academic Pub. Inc., New York, 1973.

Hunten, D. M., J.-C. Gerard, and L. M. Francois, The atmosphere's response to solar irradiation, in *The Sun in Time*, ed. C. P. Sonett, M. S. Giampapa, and M. S. Matthews, Univ. Arizona Press, Tucson, 463, 1991.

IBM Journal of Research and Development, **40**, 1-136, 1996.

Isliker, H., and A. O. Benz, Catalogue of 1 – 3 GHz solar flare radio emission, *Astron. Astrophys. Suppl. Ser.*, **104**, 145, 1994.

Koons, H. C., Characteristics of electrical discharges on the P78-2 satellite (SCATHA), *18th Aerospace Sciences Meeting*, AIAA 80-0334, Pasadena, CA, 1980.

Koons, H., C., J. E. Mazur, R. S. Selesnick, J. B. Blake, J. F. Fennel, J. L. Roeder, and P. C. Anderson, *The Impact of the Space Environment on Space Systems*, Engineering and Technology Group, The Aerospace Corp., Report TR-99(1670), El Segundo, CA, 1999.

Kundu, M. R., *Solar Radio Astronomy*, Interscience, New York, 1965.

Lanzerotti, L. J., Penetration of solar protons and alphas to the geomagnetic equator, *Phys. Rev. Lett.*, **21**, 929, 1968.

Lanzerotti, L. J., C. Breglia, D. W. Maurer, and C. G. Maclennan, Studies of spacecraft charging on a geosynchronous telecommunications satellite, *Adv. Space Res.*, **22**, 79, 1998.

Lanzerotti, L. J., C. G. Maclennan, and D. J. Thomson, Engineering issues in space weather, in *Modern Radio Science*, ed. M. A. Stuchly, Oxford, 25, 1999.

Los Angeles Times, Sunspots playing tricks with radios, Metro Section, pg. 1, Febr. 13, 1979.

Mandell, M., 120000 leagues under the sea, *IEEE Spectrum*, 50, April 2000.

Marconi, G., Radio communication, *Proc. IRE*, **16**, 40, 1928.

Matsushita, S. Solar quiet and lunar daily variation fields, In *Physics of Geomagnetic Phenomena*, S. Matsushita and W. H. Campbell, eds., Academic Press, New Yourk, pg. 301, 1967.

McBride, N., The importance of the annual meteoroid streams to spacecraft and their detectors, *Adv. Space Res.*, **20**, 1513, 1997.

McCormac, B. M., Summary, in *Radiation Trapped in the Earth's Magnetic Field*, ed. B. M. McCormac, D. Reidel Pub. Co., Dordrecht, Holland, 887, 1966.

Medford, L. V., L. J. Lanzerotti, J. S. Kraus, and C. G. Maclennan, Trans-Atlantic earth potential variations during the March 1989 magnetic storms, *Geophys. Res. Lett.*, **16**, 1145, 1989.

Mizera, P. F., A summary of spacecraft charging results, *J. Spacecraft Rockets*, **20**, 438, 1983.

Pickholtz, R. L., Communications by means of low Earth orbiting satellites, *in Modern Radio Science 1996*, ed. J. Hamlin, Oxford U. Press, 133, 1996.

Prescott, G. B., *Theory and Practice of the Electric Telegraph*, IV ed., Tichnor and Fields, Boston, 1860.

Reeves, G. D., The relativistic electron response at geosynchronous orbit during January 1997 magnetic storm, *J. Geophys., Res.*, **103**, 17559, 1998.

Shea, M. A., and D. F. Smart, Space weather: The effects on operations in space, *Adv. Space Res.*, 22, 29, 1998.

Silliman, Jr., B., *First Principals of Chemistry*, Peck and Bliss, Philadelphia, 1850.

Southworth, G. C., Microwave radiation from the sun, *J. Franklin Inst.*, **239**, 285, 1945.

Taur, R. R., Ionospheric scintillation at 4 and 6 GHz, *COMSAT Technical Review*, **3**, 145, 1973.

Todd, D., Letter to *Space News*, pg. 12, March 6, 2000.

Tribble, A. C., *The Space Environment, Implications for Spacecraft Design*, Princeton Univ. Press, Princeton, NJ, 1995.

Vampola, A., The aerospace environment at high altitudes and its implications for spacecraft charging and communications, *J. Electrost.*, **20**, 21, 1987.

Chapter 13

An Introduction to Power Grid Impacts and Vulnerabilities from Space Weather

A review of geomagnetic storms, impacts to ground-based technology sytems, and the role of forecasting in risk management of critical systems

John G. Kappenman
Metatech Corporation
Duluth, MN 55802, USA

Abstract   Geomagnetic disturbances (i.e. space storms) can impact the operational reliability of electric power systems. Solar Cycle 22 (the most recent solar cycle extending from 1986-1996) demonstrated to the electric power industry the need to take into consideration the potential impacts of geomagnetic storms. Experience gained from the unprecedented scale of these recent storm events provides compelling evidence of a general increase in electric power system susceptibility. Important infrastructure advances have recently been put in place that provides solar wind data. This new data source along with numeric model advances allows the capability for predictive forecasts of severe storm conditions, which can be used by impacted power system operators to better prepare for and manage storm impacts.

Keywords   Geomagnetic disturbance, geomagnetically-induced currents (GIC), reactive power demand, half-cycle saturation, transformer, electrojet currents, geo-electric field, forecast, nowcast, voltage regulation.

1.   INTRODUCTION

Reliance of society on electricity for meeting essential needs has steadily increased for many years. This unique energy service requires coordination of electrical supply, demand, and delivery—all occurring at the same instant.

I.A. Daglis (ed.), Space Storms and Space Weather Hazards, 335–361.
© 2001 *Kluwer Academic Publishers. Printed in the Netherlands.*

Geomagnetic disturbances can disrupt these complex power grids when Geomagnetically-Induced Currents (GICs) flow through the power system, entering and exiting the many grounding points on a transmission network. GICs are produced when shocks resulting from sudden and severe magnetic storms subject portions of the Earth's surface to fluctuations in the planet's normally quiescent magnetic field. These fluctuations induce electric fields across the Earth's surface—which causes GICs to flow through transformers, power system lines, and grounding points. Only a few amperes (amps) are needed to disrupt transformer operation, but over 300 amps have been measured in the grounding connections of transformers in affected areas. Unlike threats due to ordinary weather, Space Weather can readily create large-scale problems because the footprint of a storm can extend across a continent. As a result, simultaneous widespread stress occurs across a power grid to the point where widespread failures and even regional blackouts may occur. Systems in the upper latitudes of the Northern Hemisphere are at increased risk because auroral activity and its effects centre on the magnetic poles. North America is particularly exposed to these storm events because the Earth's magnetic north pole tilts toward this region and therefore brings it closer to the dense critical power grid infrastructure across the mid-latitude regions of the continent.

Geomagnetic disturbances will cause the simultaneous flow of GICs over large portions of the interconnected high-voltage transmission network, which now span most developed regions of the world. As the GIC enters and exits the thousands of ground points on the high voltage network, the flow path takes this current through the windings of large high-voltage transformers. GIC, when present in transformers on the system will produce half-cycle saturation of these transformers, the root-cause of all related power system problems. Since this GIC flow is driven by large geographic scale magnetic field disturbances, the impacts to power system operation of these transformers will be occurring simultaneously throughout large portions of the interconnected network. This saturation produces voltage regulation and harmonic distortion effects in each transformer in quantities that build cumulatively over the network. The result can be sufficient to overwhelm the voltage regulation capability and the protection margins of equipment over large regions of the network. Combinations of events such as these can rapidly lead to systemic failures of the network.

Power system designers and operators expect these systems to be challenged by the elements, and where those challenges were fully understood in the past, the system design has worked extraordinarily well. Most of these challenges have largely been terrestrial weather related and therefore confined to small regions. The primary design approach undertaken by the industry for decades has been to weave together a tight network, which pools resources and provides redundancy to reduce failures. In essence, an unaffected neighbour helps out the temporarily weakened neighbour.

Ironically, the designs, that have worked to make the electric power industry strong for ordinary weather, introduce key vulnerabilities to the electromagnetic coupling phenomena of space weather. The large continental grids have become in effect a large antenna to these storms. Further, Space Weather has a planetary footprint, such that the concept of unaffected neighbouring system and sharing the burden is not always realizable. To add to the degree of difficulty, the evolution of threatening space weather conditions are amazingly fast. Unlike ordinary weather patterns, the electromagnetic interactions of space weather are inherently instantaneous. Therefore large geomagnetic field disturbances can erupt on a planetary-scale within the span of a few minutes.

2. LESSONS LEARNED FROM SOLAR CYCLE 22

The events that led to the collapse of the Hydro Quebec system in the early morning hours of March 13, 1989 illustrate the challenges that lie ahead in managing this risk [*NERC*, 1990]. At 2:42am LT (7:42 UT), all operations were normal; at that time a large impulse in the Earth's magnetic field erupted along the US/Canada border (see Figure 1 and animations on Kluwer WWW site of March 13, 1989 storm). Within a matter of seconds, the voltage on the network began to sag as the storm developed; automatic voltage compensating devices rapidly turned "ON" to correct this voltage problem. However, these automatic devices were themselves vulnerable to the storm and all 7 of these critical compensators failed within less than a minute. The failure of the compensators led to a voltage collapse and complete blackout of the second largest utility grid in North America. All together, the chain of events from start to complete province-wide blackout took an elapsed time of only 90 seconds. The rapid manifestation of storm events and their impacts on the Hydro Quebec system allowed no time to even assess, let alone provide time for meaningful human intervention.

The rest of the North American system also reeled from this Great Geomagnetic Storm. Over the course of the next 24 hours, additional large disturbances propagated across the continent, the only difference being that they extended much further south and very nearly toppled power systems from the Midwest to the Mid-Atlantic Regions of the US. The NERC (North American Electric Reliability Council) in their post analysis attributed over 200 significant anomalies in the power grids across the continent to this one storm [*NERC,* 1990]. In spite of the large number of significant events that

338

were observed, it is now recognized that North America did not experience the largest and most intense geomagnetic disturbances associated with this storm. This same storm produced dB/dt fluctuations twice as intense over the lower Baltic, than any that were experienced in North America. (dB/dt is the rate-of-change of the geomagnetic field as measured at ground level, the usual concern for GIC is rate-of-change in the horizontal component).

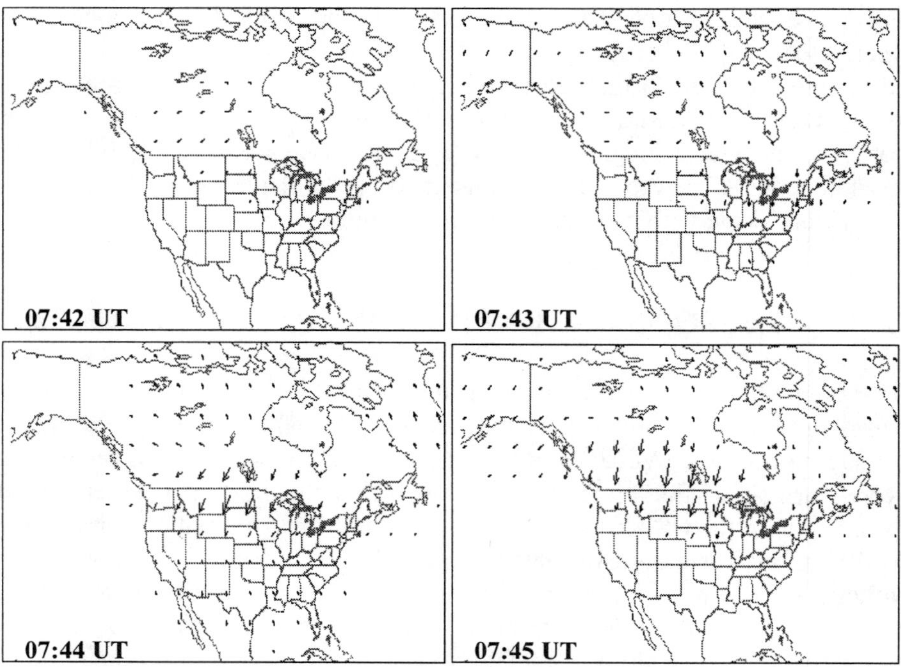

Figure 1. Four minutes of a superstorm. Space weather conditions capable of threatening power system reliability can rapidly evolve. The system operators at Hydro Quebec and other power system operators across North America faced such conditions during the March 13, 1989 super storm. The above slides show the rapid development and movement of a large geomagnetic field disturbance between the times 7:42-7:45 UT (2:42-2:45 EST) on March 13, 1989. The disturbance of the magnetic field began intensifying over the eastern US-Canada border and then rapidly intensified while moving to the west across North America over the span of a few minutes. With this rapid geomagnetic field disturbance onset, the Hydro Quebec system went from normal operating conditions to complete collapse in a span of just 90 seconds due to resulting GIC impacts on the grid. The magnetic field disturbances observed at the ground are caused by large electrojet current variations that interact with the geomagnetic field. The dB/dt intensities ranged from 400 nT/min at Ottawa at 7:44UT to over 892 nT/min at Glen Lea. Large-scale rapid movement of this disturbance was evident. The large magnetic field disturbances transited from eastern Canada to Alaska in less than 8 minutes, a velocity of approximately 10 to 15km/sec.

Over the next 5 years, smaller storms demonstrated time and again for the power industry that significant impacts could be triggered at even lower storm levels. The anecdotal notion that storms only occur late at night and during the peak of the sunspot cycles has also been proven to be dangerously mistaken; experience has demonstrated that geomagnetic storms are instead an "ever-present risk". A storm in February 1986 nearly caused a voltage collapse across New England and mid-Atlantic regions of the US and simultaneously shook many European locations. This occurred not only in the afternoon sunshine in the eastern US (2:45pm LT) but at the absolute minimums between solar cycles 21 and 22. Similar storm scenarios produced problems continent wide during 1991 through 1994 across North America to as far south as Los Angeles, and New Mexico in the US [*NERC,* 1990, *NERC,* 1992, *NERC,* 1986].

For perspective, the limited climatology data available suggests that storms of even larger intensity and with a larger planetary footprint are possible than the one that occurred in March 1989. Also, the power industry realizes that its vulnerability continues to incrementally grow over time [*IEEE,* 1996]. As a result, the challenges of future solar cycles may be even greater than those posed by the storms of solar cycle 22.

3. AN OVERVIEW OF POWER SYSTEM RELIABILITY AND RELATED SPACE WEATHER CLIMATOLOGY

The maintenance of the functional integrity of the bulk electric systems (i.e. Power Systems Reliability) at all times is a very high priority for the planning and operation of power systems worldwide. Power systems are too large and critical in their operation to easily perform physical tests of their reliability performance for various contingencies. The ability of power systems to meet these requirements is commonly measured by deterministic study methods to test the system's ability to withstand probable disturbances through computer simulations. Traditionally, the design criterion consists of multiple outage and disturbance contingencies typical of what may be created from relatively localized terrestrial weather impacts. These stress tests are then applied against the network model under critical load or system transfer conditions to define important system design and operating constraints in the network.

System impact studies for geomagnetic storm scenarios can now be readily performed on large complex power systems [*Kappenman,* 1998]. For cases in which utilities have performed such analysis, the impact results indicate that a severe geomagnetic storm event may pose an equal or greater stress on the network than most of the classic deterministic design criteria now in use. Further, by the very nature that these storms impact simultaneously over large regions of the network, they arguably pose a greater degree of threat for precipitating a system-wide collapse than more traditional threat scenarios.

The evaluation of power system vulnerability to geomagnetic storms is, of necessity, a two-stage process. The first stage is one of assessing the exposure to the network posed by the climatology. In other words, how large and how frequent can the storm driver be in a particular region? The second stage is one of assessment of the stress that probable and extreme climatology events may pose to reliable operation of the impacted network. This is measured through estimates of levels of GIC flow across the network and the manifestation of impacts such as sudden and dramatic increases in reactive power demands and implications on voltage regulation in the network. The essential aspects of risk management become the weighing of probabilities of storm events against the potential consequential impacts produced by a storm. From this analysis effort meaningful operational procedures can be further identified and refined to better manage the risks resulting from storms of various intensities [*Kappenman, 1998*].

Consistent advances have been made in the ability to undertake detailed modelling of geomagnetic storm impacts upon terrestrial infrastructures. The scale of the problem is enormous, the physical processes entail vast volumes of the magnetosphere, ionosphere, and the interplanetary magnetic field conditions that trigger and sustain storm conditions. In addition, it is recognized that important aspects and uncertainties of the solid-earth geophysics need to be fully addressed in solving these modelling problems. Further, the effects to ground-based systems are essentially contiguous to the dynamics of the space environment. Therefore, the electromagnetic coupling and resulting impacts of the environment on ground-based systems requires models of the complex network topologies overlaid on a complex geological base that can exhibit variation of conductivities that can span 5 orders of magnitude.

These subtle variations in the ground conductivity play an important role in determining the efficiency of coupling between disturbances of the local geomagnetic field caused by space environment influences and the resulting impact to ground based systems that can be vulnerable to GIC. Lacking full understanding of this important coupling parameter hinders the ability to better classify the climatology of space weather on ground-based infrastructures.

This coupling efficiency can best be illustrated through an example. Table 1 provides a conductivity profile versus depth for two regions of the earth. In both cases the conductivity profile is given to depths of 700km.

Consideration to this depth is necessary as the frequency of the geomagnetic field disturbances are typically less than .01 Hertz, which will result in a large skin depth for diffusion of the fields. In this table, Region 1 displays a conductivity profile versus depth in which 4 orders of magnitude variation of conductivity occurs dependent upon the depth. Further, as shown, the high conductivity deep earth region does not begin until a depth of approximately 600 km. Region 2 exhibits a similar 4 orders of magnitude variation of conductivity over depth, but an important difference is the high conductivity deep earth layer begins at a depth of only 200 km. These differences will produce significant differences in the electromagnetic coupling and resultant geo-electric field.

Table 1. Deep Earth conductivity depth profiles for two different regions

Depth (km)	Region 1 Profile (S/m)	Depth (km)	Region 2 Profile (S/m)
0-15	0.0001	0-10	0.00001
15-25	0.0001	10-30	0.00067
25-35	0.0001	30-70	0.01
35-400	0.00017	70-200	0.01
400-500	0.0017	200-300	0.1
500-600	0.017	300-400	0.1
600-700	0.1	400-700	0.1

Testing the response of the earth profiles to a standard geomagnetic disturbance impulse can readily compare these differences. Figure 2 provides a waveform of a large geomagnetic field impulse typical in characteristics to that caused by a geomagnetic storm. This waveform has a 2000 nT total change in geomagnetic field with a rapid rise resulting in a rate of change of 2400 nT/min. Using this magnetic field impulse, the ground coupling of the two ground conductivity profiles was simulated to determine the resulting geo-electric fields. Figure 3 provides the voltage versus time waveforms of the two profiles. Region 1 with the deeper high conductivity layer reacts much more readily with the magnetic field disturbance and produces an approximate factor of five larger electric field magnitudes than Region 2. Each region will also produce subtle variations in their respective frequency responses as well. The variations that result in the geo-electric field as illustrated in Figure 3 will have a proportionate role in the magnitude of GIC in a regional power grid as well and the magnitude of the GIC largely is an important determinant in the level of impacts that a storm will produce on the infrastructure.

It is difficult to directly assess impacts from variations in local geomagnetic fields as considerable variations are possible in resulting geo-electric field and GICs given a known level of geomagnetic field variation.

342

Further, ground conductivity profiles throughout the world are not well established, which further hinders an ability to more precisely classify the impact of magnetic field environment in terms of resulting geo-electric fields.

Even with these aforementioned uncertainties, an approach towards meaningful climatology classifications can still be achieved through the use of rate-of-change of geomagnetic field as a proxy for resulting geo-electric fields. Figure 4 shows plots of the dB/dt for the magnetic field disturbance of Figure 2 with the resultant electric field for Region 1; this illustrates that the geo-electric field in this case is primarily driven by large dB/dt variations in the geomagnetic field. Therefore dB/dt can be used as a first order proxy for classification of disturbances of importance to GIC levels. While the exact magnitude in any region will not be known without full details on ground conductivity profiles, this classification approach more accurately characterizes magneto-telluric processes than does the K and A index classification approach.

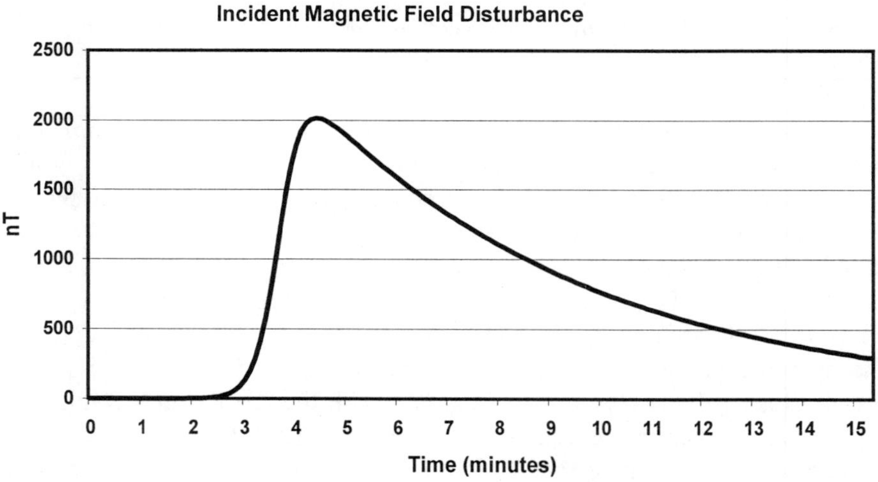

Figure 2. Magnetic field impulse waveform with dB/dt of 2400 nT/min on the rise time, typical of a large geomagnetic storm caused disturbance to the horizontal geomagnetic field.

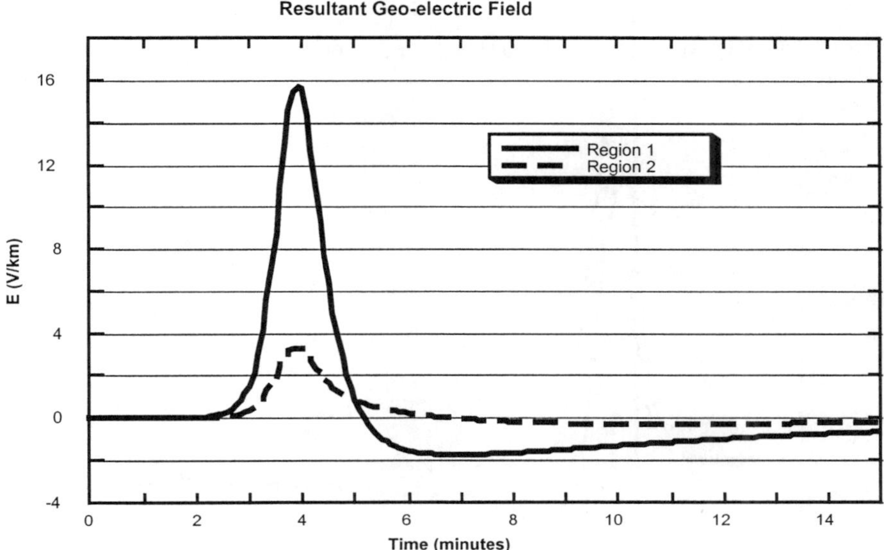

Figure 3. The resultant geo-electric fields from two different ground conductivity profiles (Regions 1 and 2) for the magnetic field impulse shown in Figure 2.

A large dB/dt observed in one location is due to coincidental aspects of the timing of substorm events, therefore these events can have equal probability of producing the same intensity disturbances at other locations around the world at equivalent geomagnetic latitudes. In assessing threat potentials for geomagnetic storm activity, it is useful to review peak dB/dt intensities from prior storms and the equatorward expansion of these disturbances to characterize threat potentials at equivalent worldwide locations.

Large rate-of-change impulses in the geomagnetic field (dB/dt in nT/min) observed during the March 13, 1989 storm were extensive and illustrative of a large storm (see Figure 5). The largest dB/dt observed, occurred at a Danish magnetic observatory Brorfelde (BFE) with a 2000 nT/min rate of change of local magnetic field. This Iso-Telluric chart describes equivalent locations worldwide had the same 2000 nT/min impulsive disturbance occurred at other local times (the top iso-level). The locations depicted would follow the geomagnetic latitudes rather than geographic latitudes, hence the large low-latitude excursion over North America. For perspective the level of dB/dt that precipitated the Hydro Quebec collapse was only 400 nT/min. Levels this high were observed at very low latitudes (for instance Bay St. Louis, BSL on the Gulf of Mexico), which would encompass most of North America and Europe (the mid iso-level). Levels of 200 nT/min can

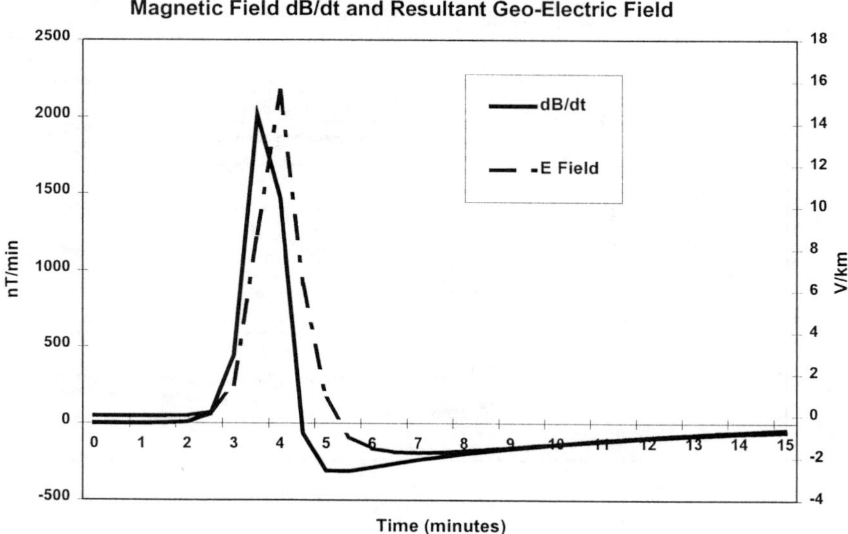

Figure 4. A waveform comparison is provided between the incident magnetic field rate of change and the resultant geo-electric field for a region. Lacking details on the deep-earth conductivity, the dB/dt can provide a first-order proxy for the magneto-telluric climatology.

easily produce significant power system impacts, which for this storm extended to even lower latitudes (as shown in lowest iso-level). It is likely even more severe power system disruptions would have resulted had the 2000 nT/min impulse (or approximately 5 times larger than levels responsible for triggering the Quebec system collapse) occurred over North America (Figures 6 and 7).

Long-term, detailed climatology data on geomagnetic field disturbances is not available. Therefore, the extremes of dB/dt impulses and the resulting equatorward boundaries are not well known. Other limited data (such as K and Ap indices) cannot be used to extrapolate dB/dt, but they suggest higher levels of dB/dt than those experienced during the March 1989 storm are possible. For example, a storm in August 1972 (with an Ap index of 223) produced dB/dt's of approximately 2200nT/min over major portions of North America [*Anderson,* 1974, *Boteler,* 1992]. Anecdotal information of other storms over the past few decades suggests even larger dB/dt disturbances may have occurred. The largest storm on record in the limited 67 years of geomagnetic storm climate data occurred in 1941. This storm had an Ap index of 312 (for comparison the March 1989 superstorm had an Ap of 285). Data on dB/dt for this storm is not available. However, this storm and a similar sized storm in 1940 resulted in power system problems as far south as southern Georgia. In events like this, the above-depicted boundaries could have extended even further equator-ward [*Albertson,* 1972, *McNish,* 1940, *Davidson,* 1940].

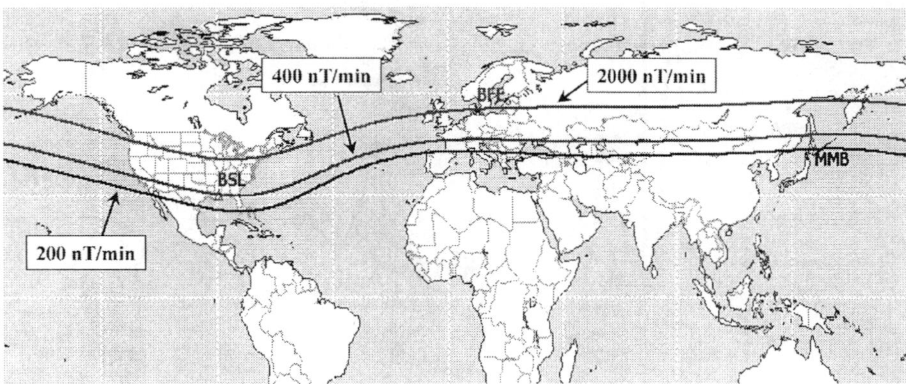

Figure 5. The equator-ward boundaries of large dB/dt impulses observed during the March 13, 1989 superstorm. The above map provides the equator-ward boundaries of large geomagnetic field impulses that can threaten power systems in the Northern Hemisphere. This climatology experience also suggests extensive regions of North America and Europe can be at risk from future large storms.

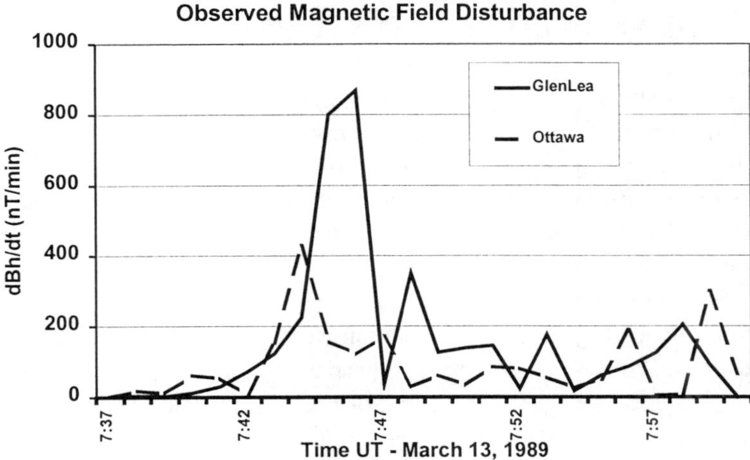

Figure 6. Observed dBh/dt (nT/min) at two magnetic observatories near the US-Canadian border during the time of the Hydro Quebec blackout and other noteworthy power system impacts. Of particular importance is the sudden impulsive behaviour of the observed storm conditions.

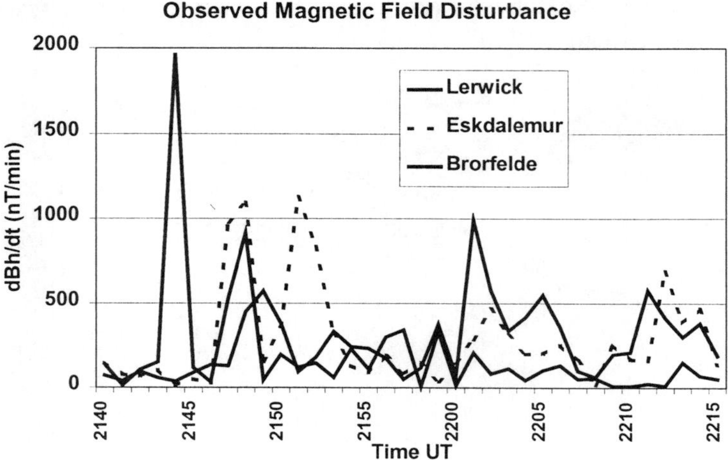

Figure 7. At a later time during the March 89 superstorm, dB/dt levels more than twice as large as any observed over North America were observed at the Danish observatory Brorfelde). The two British observatories (Lerwick and Eskdalemur) also recorded large disturbance levels.

4. POWER GRIDS AND INCREASING VULNERABILITY TO STORMS

Geomagnetic disturbances have proven to be an exceedingly unique, large scale, and important threat to the reliable operation of power systems. GIC when it flows through power system transformers is the root-cause of all power system problems. Because a storm can have such a large footprint, nearly all large-high-voltage transformers across a power grid will experience simultaneous exposure to GICs. The combined stress of what may be thousands of transformers saturating simultaneously can cause impacts such as voltage sag/collapse, distortions of the power system sine-wave that can confuse and disrupt protective systems and in extreme cases lead to permanent damage to the transformer itself. Therefore, rather than the transmission network experiencing a double or triple contingency event which the system is designed to withstand, system-wide stress and dozens of simultaneous contingency equipment failures can occur. For perspective, the Great Geomagnetic Storm of March 13, 1989 triggered over 200 significant equipment failures and deviations from acceptable operating limits across North America [*NERC,* 1990]. Figures 6 and 7 are examples of large dB/dt variations that were observed during this storm. These sudden and dramatic dB/dt variations and onsets are the important drivers for GIC-caused impacts to power grids. Figure 6 provides the observations at the Ottawa and Glen Lea observatories (along USA-Canada border) in the time range coincident with the Hydro Quebec collapse at 7:45 UT. Figure 7 provides a plot of the largest dB/dt's observed during this great storm which happened to be over northern Europe at 21:44 UT. Figure 8 illustrates the wide spread geographic scale of simultaneous impacts observed across the North American power grids during these same two intervals of intense storm activity during the March 13, 1989 superstorm.

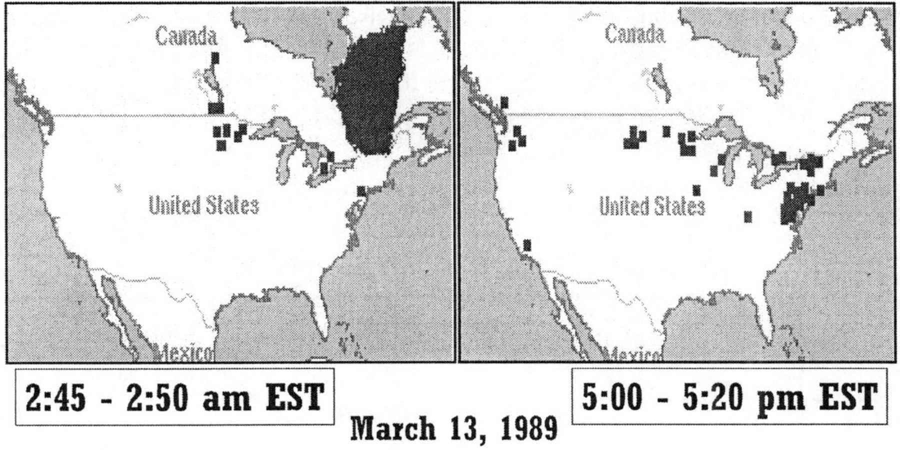

Figure 8. Widespread power system failures/problems occurred across North America during the course of the storm. In addition to the Quebec blackout, many other regions across North America were severely stressed and came close at times to similar voltage collapse conditions. Several large surges of geomagnetic field disturbances occurred over the two days of storm activity. The maps above show locations and time windows of some of the many reported power system problems.

The likelihood of severe storms is quite high. A storm of Ap 150 or greater has occurred at a rate of one or more per year over the climatology records available. While the dB/dt extremes of large important geomagnetic storms remains somewhat unanswered, it is important to recognise that the problem of power system impacts is compounded by growing vulnerability of this infrastructure to space weather disturbances. The extent of the change or growth in vulnerability over time can be due to a number of factors stemming from either growth in the infrastructure base or technology changes within the existing base that introduce new impact problems. In the case of the electric power industry, aspects of both have been occurring. Figure 9 shows the growth of the US high voltage transmission grid over the

348

last 50 years. The high voltage transmission grid is the portion of the power network that spans long distances. This geographically widespread infrastructure readily couples through multiple ground points to the geo-electric field produced by disturbances in the geomagnetic field. As shown in Figure 9, from cycle 19 (late 1950's), through solar cycle 22 (early 1980's), the high voltage transmission grid has grown nearly tenfold. In essence the antenna that is sensitive to disturbances has grown dramatically over time. Similar development rates of transmission infrastructure have occurred simultaneously in other developed regions of the world.

As this network has grown in size, it has also grown in complexity. Some of the more important changes in technology base that can increase impacts from a geomagnetic storm include higher design voltages, changes in transformer design and other related apparatus. The operating levels of high-voltage networks have increased from the 100-200kV thresholds of the 1950's to 400 to 765kV levels of today's networks.   The large high voltage

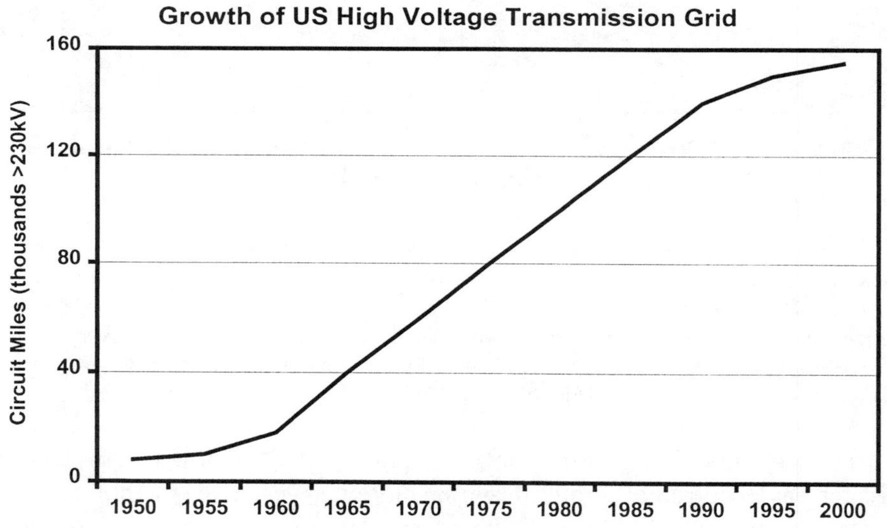

Figure 9. The US high voltage transmission network has grown significantly from Solar Cycle 19 to Solar Cycle 23, making the system a larger antenna to geomagnetic field disturbances.

transformers associated with modern networks only require a few amps of GIC for the transformer to be driven into half cycle saturation. The higher operating voltages of the network further amplify the impact of this transformer saturation. Figure 10 illustrates that a 500kV transformer, which is saturated by GIC, will produce a reactive demand that is more than twice as large as that in a similar 230 kV transformer that is saturated by the same level of GIC. Further, the GIC flows tend to concentrate in the highest voltage portion of the system, because this portion of the network has the lowest per unit resistances and spans the largest distances, which creates a larger geo-electric source voltage.

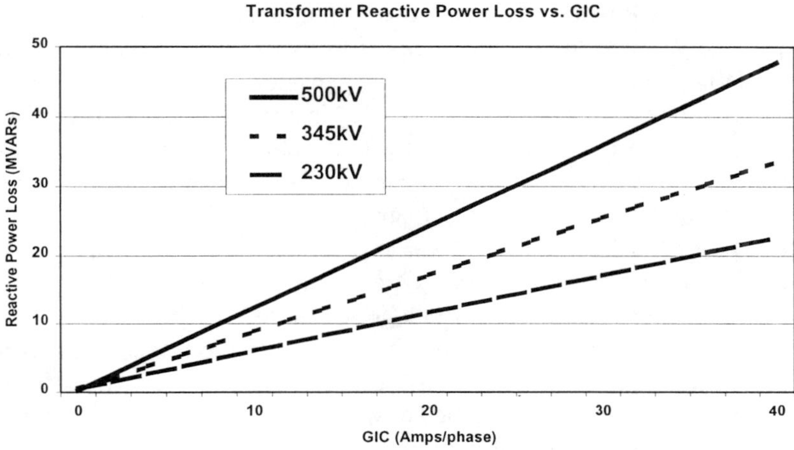

Figure 10. Reactive power losses in a transformer that is being saturated by GIC increase not only as a function of GIC levels, but also with increased operating voltages on the network.

Since the 1980's, the growth of the transmission grid infrastructure has slowed considerably. This has come about primarily due to environmental opposition to the siting and construction of large high voltage transmission lines. Based upon this trend, it would be reasonable to conclude that increasing vulnerability to space weather has also reached a plateau. Ironically, the opposite is occurring. While transmission infrastructure growth has stopped, the growth of electricity usage has increased almost unabated in the ever more technologically sophisticated societies around the world. This increased electric demand has placed an added burden that acts to strain the existing infrastructure of the power grids more than originally designed.

Deregulation of the power industry is also placing new burdens, in particular, on the transmission infrastructure. The industry is changing from a mode of vertical integration to one in which customers are allowed to choose different suppliers of power. The transmission network is now required to be accessible to allow for more competitive opportunities for energy delivery from remote supply locations. This has the affect of also placing substantial congestion onto networks to accommodate the new mode of economic operation. As such, the industry is beginning to acknowledge that operation margins are uncomfortably low at many times. Space weather storms pose a risk because they place additional burdens on this already heavily burdened infrastructure in a rapid, widespread, and heretofore unpredictable fashion.

New voltage regulation devices have also introduced new network reliability vulnerabilities related to widespread common-mode failures of apparatus across the network. The collapse of the Hydro Quebec system in 1989 was primarily due to the simultaneous failure of seven large high-speed voltage regulation devices (static var compensators or SVCs) on the transmission network that were precipitated by harmonic distortion caused by the geomagnetic storm. Any failure of a key asset in isolation would not jeopardise network integrity, but common-mode failures on a simultaneous basis pose substantial risks to network integrity. The simultaneous loss of all seven of the Hydro Quebec voltage-regulating devices was never considered as a design possibility for the network. When it did occur during this storm, it led to the instantaneous collapse of voltage regulation on the network. As power companies begin to assess vulnerability of this type, the previously mentioned change in design over the past 10 years has in some cases been substantial. For example, Hydro Quebec now has 11 SVCs on their network. National Grid in England had only four shunt devices on the network in 1989; today they have 35 of these voltage regulation devices.

Figure 11 illustrates the role of shunt compensation voltage regulation and the contributing factors that increased network-loading cause in increasing vulnerability to geomagnetic storm impacts. The important interrelationship between voltage regulation and the loading capacity is shown in this simple graphic for networks with and without shunt capacitive voltage regulation (either shunt capacitors or SVCs). In networks without SVCs (a typical design in previous solar cycles), transmission networks had to be operated with ample margins of loading versus capability in order to keep steady-state voltages within operating limits. The relationship between voltage regulation and load on the network for systems of previous solar cycles was generally a passive and very linear response. As energy demand increased on the network, voltage regulation of the network would be governed by the inductive load loss (line current times the reactive impedance of the network) caused voltage drop brought about from the increased current loading across the transmission networks. In other words as load gradually increase, voltage would gradually decrease at a predominantly linear rate. This gradual response of the network would allow ample time for system operators to recognise and correct for any threatening voltage regulation problems. Also, the design approach with networks in this era was simply to construct additional transmission circuits as load on the network increased. This would preserve the well-behaved linear relationship between load and voltage regulation on the networks.

Since the early 1980's, the ability to add additional transmission circuits with increased load has not been an available option in many cases. In order to gain additional load serving capability from the existing network infrastructure, power system designers have developed and installed fast responding automatic voltage compensation devices, principally SVCs. The devices allow greater loading of the network by adding capacitive reactive power to compensate for the inductive reactive loss increase due to higher current loading on the network. This has the important benefit of allowing loading to increase virtually to the thermal current ratings of the network, as long as enough compensating devices are added to maintain proper voltage regulation in steady state conditions. This approach towards improving the network load serving capability without adding extensive new transmission circuits has produced some undesirable trade-offs. The normal linear relationship between load and voltage levels on the network no longer exists. These voltage compensation devices are able to support the network voltage regulation needs up to the operating range of the devices. However, once this limit is reached, the behaviour between load and voltage regulation becomes highly non-linear. The shunt capacitors decrease their compensating reactive power output as a function of voltage squared. When the capacity levels of the network are exceeded and as a result voltage begins to drop, a more precipitous exponential collapse in voltage regulation can occur due to the reduced compensation capabilities. These changes can occur rapidly and therefore may not allow adequate time for operator recognition or intervention. When a geomagnetic storm occurs, a system-wide increase in reactive power demand can be instantaneously placed upon the network. This threatens to push the system beyond safe loading limits and into the region of exponential voltage collapse. This risk is further heightened due to load demand increases upon the network over the years while limited transmission circuit additions have occurred. Under normal load conditions that occur today, these systems are typically operated closer to the collapse thresholds. As a result, power networks under these operational constraints are inherently less robust and are less able to absorb the added stresses when triggered by increased reactive demands caused by a geomagnetic storm.

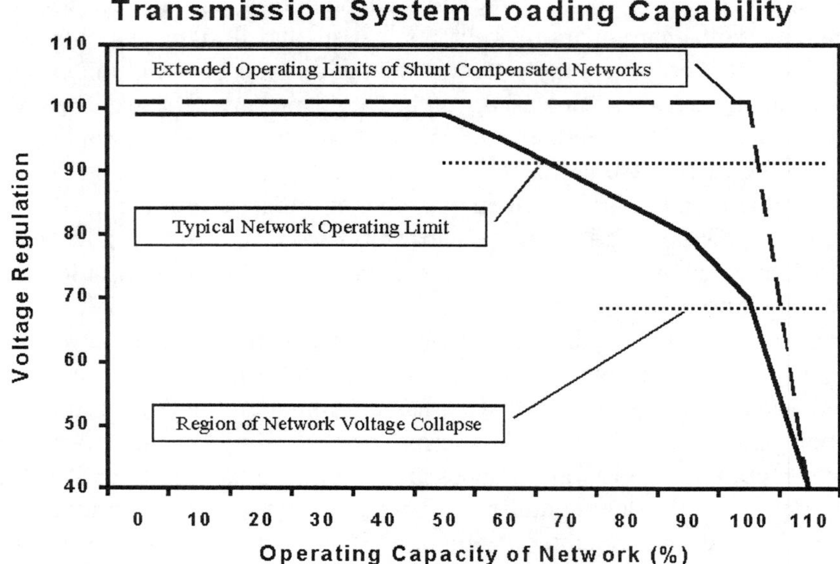

Figure 11. Shunt compensated power networks use SVCs and capacitive devices to allow higher loading levels on existing transmission networks. This makes these networks more at risk for voltage collapse due to geomagnetic storms.

Further compounding the risks to voltage regulation is that the SVCs and shunt capacitor devices necessary on the network for primary voltage regulation have been highly prone to failure during a geomagnetic storm. These failures stem from harmonic waveform distortions due to transformer saturation. The capacitive devices act as a sink to harmonics. This increase in harmonic currents can be sensed as an overload of the device, which can trigger protective device operation to remove the equipment from the network. The saturation affects can also interact with protective relays to cause mis-operations, due to the highly distorted waveforms that may result, or even due to DC saturation of the current transformers, which supply relays with sensory information from the network. This can lead to multiple or common-mode failure of these and other critical devices just when they are most needed for network reliability. The Hydro Quebec collapse followed within a matter of seconds the simultaneous failure of all seven large high-speed voltage-regulating devices on their network. Other power companies throughout North America and other places in the world have experienced similar bulk device failure problems associated with geomagnetic storm conditions. As these voltage regulation devices become more critical to the reliable operation of the network, the network therefore becomes more vulnerable to these geomagnetic storm impacts. As a result, many networks now have less inherent ability to absorb geomagnetic storm related impacts and as a result are more vulnerable to storm related problems of this type.

5.   GIC CAUSED TRANSFORMER FAILURES

Experience from recent solar cycles has provided evidence about transformer failures due to geomagnetic storms. Transformer saturation caused by the presence of GIC is a mode of operation for modern transformers that is not intended for long durations. The complex behaviour of the internal magnetic field patterns under saturation is a source of internal heating that is seldom considered in the apparatus design. A large high-voltage transformer can cost several million dollars and may take up to a year to remanufacture and replace if damaged. Subtle and hard to anticipate design variations, which are common in transformers of this design class, make the estimation of device vulnerability to this mode of failure difficult to estimate until a failure has occurred.

While the complexities of this process are poorly understood, better understandings of contributing geomagnetic storm conditions are being developed. Both large storms as well as weaker but repetitive storms can contribute to transformer failure problems. Large storms can cause internal heating damage in a very brief period of time. Measurements conducted on several transformers indicated large temperature excursions in hotspots within a matter of a few minutes. For example, during the March 1989 storm, several incidents of transformer heating problems were reported. The most significant failure occurred at a GSU (generation step-up) transformer at a nuclear plant in New Jersey in which a 1,200 MVA, 500kV transformer was damaged beyond repair. The suspected failure linkage is stray flux impinging on external core structures in concentrations intense enough to develop hot-spots. In a subsequent storm event in 1992, Allegheny Power captured on a well-monitored transformer (one that had misbehaved in previous storms) a rapid rise in internal heating due to GIC in a 350 MVA, 500/138 kV transformer. In this case, in as few as 10 minutes of GIC exposure, a hot-spot on the monitored tank surface increased in temperature from 60°C to over 175°C. Weak but long-duration storms can also cause transformer-heating damage. These extended duration heating insults raise the likelihood of loss-of-life to transformer insulation. This damage can be cumulative and acquired over repeated exposures. Most electric power companies have very limited awareness of GIC-related problems of this type in transformers. When failure eventually occurs, the cause of failure may be in combination with and usually attributed to other none Space Weather factors.

Devices to block the flow of GIC into the transformer neutral where it attaches to the ground have been investigated and are feasible. The device design employs a capacitor that blocks DC while allowing limited AC current flows. The device, however, needs to be able to rapidly bypass this capacitor under AC system disturbance conditions to provide a more solid ground connection for large transient AC current flows. This requirement raises the costs of the device considerably. Since numerous ground points exist on a typical power grid, it would be complex and expensive to blanket such blocking devices across a network. It is feasible to selectively deploy these devices on key or vulnerable individual transformers, with the recognition that the remainder of the network remains vulnerable.

6. FORECASTING FOR RISK MANAGEMENT

Since preventing the flow of GIC in power systems is usually not a viable threat mitigation strategy; a management plan to prepare the system for the stress imposed by a resulting geomagnetic storm is the most prudent course of action. Decisions on when to implement operational measures have been problematic in the past because of the inherent low quality of forecasts that had been provided. Further, storm onsets can develop suddenly and as a result some operational changes cannot be implemented in time to address the paramount priority of system reliability. Obtaining reliable advanced warnings is the first step toward preparing for disturbances and appropriate preparation of operational strategies to preserve network reliability.

In Cycle 22, in particular, utilities employed storm operational management strategies that would be triggered when local monitoring detected the manifestation of storm conditions such as GIC above some threshold. This essentially produced two important and highly undesirable drawbacks; it could not guarantee sufficient lead-time to prepare for sudden severe impulses and it created extended periods of unnecessarily cautionary and restrictive constraints on the transmission market. Hydro Quebec maintained such an operating posture for 854 hours in 1991 alone (approximately 10% of the calendar year) and as a result had to unnecessarily forgo substantial feasible energy transactions [*IEEE,* 1996].

Current space weather forecast products are primarily designed to provide broad "Environmental Assessments" of storm conditions typically in an "index" style of severity classification. However, power-industry users of forecasts who are responsible for important operational functions during storm events need to have a timely "System Impact Assessments" of the storm potential. The primary focus of System Impact analysis is the desire to "Quantify" the region and system specific severity and impact of a storm in terms meaningful to power system operators such as magnitudes and distributions of GIC in their respective grid.

7. DEFINING NEW STANDARDS FOR STORM FORECAST CAPABILITIES

Methods of classifying geomagnetic storm activity in the past have typically used two letter indices, (for example K1 to K9 for the smallest to largest geomagnetic storm) to rank the severity of the storm over broad three-hour time windows and planetary or large region locations. Space Weather is a very complex, detailed, and dynamic process that is ever changing over the course of the storm. Just as the diversity of terrestrial weather impacts to critical operational infrastructure (such as rain/snow, thunderstorms, heat/cold, hurricanes, etc.) cannot be adequately classified by 3-hour, 2-letter, planetary indices; neither should the inherently dynamic impacts of space weather remain with this outdated classification approach.

Geomagnetic storm forecasting is difficult because the storm processes can be extremely dynamic. Unlike the thermodynamic processes that largely govern the behaviour and rules of terrestrial weather forecasting, the plasma and electromagnetic coupling processes of space weather are inherently instantaneous. The operation of critical infrastructures such as power grids is a continuous minute-by-minute coordinated and supervised operation. Thus, the forecasting capability for geomagnetic activity also needs to provide continuous updates of the rapidly changing space environment conditions to best meet the operational needs of power systems in managing this storm risk.

Also, unlike the terrestrial weather conditions that are monitored routinely at thousands of locations worldwide, the conditions in space are much more difficult to monitor. As a result only a handful of space-based and ground-based monitoring stations are available. In early 1998, a NASA satellite called ACE "Advanced Composition Explorer" began continuous real-time monitoring and transmission of the Solar Wind conditions at a point in space upwind from the Earth's magnetic field. Monitoring from this point in space (an orbital position called L1, about 1.6 million km toward the Sun from Earth) provides data fundamental to enabling the formulation of highly accurate forecast techniques and the subsequent issuance of alerts and warnings of impending major geomagnetic disturbances. Because it takes a disturbance in the solar wind about an hour to travel from where ACE is to

356

Earth, telemetry from ACE will allow alerts of imminent, severe geomagnetic storms to be issued nominally an hour in advance of their onset.

With this continuous data available and the previously discussed forecast needs of this power industry, predictive electrojet models driven by solar wind conditions have been an important focus. These advances would provide the resolution specification of the geomagnetic storm environment with a typical lead-time in the range of 45 minutes.

In addition to a detailed specification or forecast of the environment, it is necessary to provide a detailed assessment of the potential impact upon operations of a power network due to a storm. This requires the needed utilization of end-to-end set of models to discretely extend from solar wind or locally observed geomagnetic field conditions to derive a forecast or nowcast set of impacts upon the operation of the client infrastructure. These measurements and models increase the capability to predict not only on a global scale, but also more importantly for concerned transmission grid operators, to provide a projection of region and time-specific meso-scale processes of concern.

In predictive forecasts of impending geomagnetic storm conditions, solar wind conditions are used to derive expected electrojet current conditions as depicted in Figure 12. The electrojet model determines the characteristics of the current flowing in the electrojet approximately 100 km above the Earth's surface. This model predicts the ionospheric Hall and Pedersen currents flowing in the auroral zone approximately 30 to 45 minutes ahead in time using the real-time solar wind data. The model also needs to take into consideration ionospheric composition conditions (adjustments in conductivity, current intensity, and equator-ward boundaries of the current systems) during the course of the storm as well in order to provide sufficiently accurate estimates of eastward and westward current conditions. To compute the coupling of the electromagnetic fields produced by the electrojet to a fixed power grid, it is necessary to translate the fields produced by these moving currents to calculate the local magnetic field in the regions of interest to the client. The derivation of the geomagnetic field disturbances is performed by two-dimensional calculations of a gaussian spatial profile for an electrojet over a layered (with depth) ground conductivity profile. Figure 13 illustrates a comparison of a predicted versus an observed magnetic field disturbance [*Kappenman,* 2000].

Given either the predicted (forecast) or observed (nowcast) magnetic fields in the region of interest, it is necessary to compute the electric fields at the same locations. These electric fields couple to the high-voltage power lines and thereby induce the quasi-dc currents that flow in the network. As previously discussed, the conductivity of the Earth itself is crucial in determining the electric fields produced by a given magnetic field.

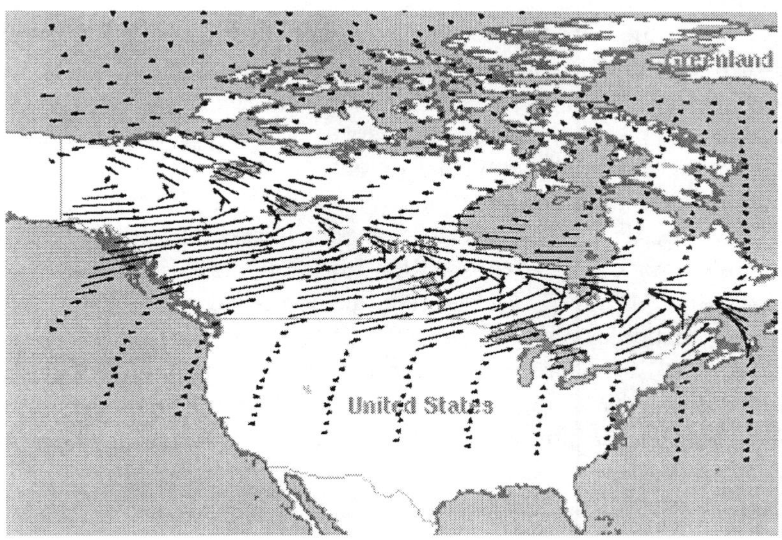

Figure 12. Electrojet forecast example for May 4, 1998 at 05:00 UT. This forecast provides the electrojet current intensity, direction, and location with up to 45-minute lead-time. This provides the first step in the calculation of estimated GIC and power system impact potential due to a geomagnetic storm.

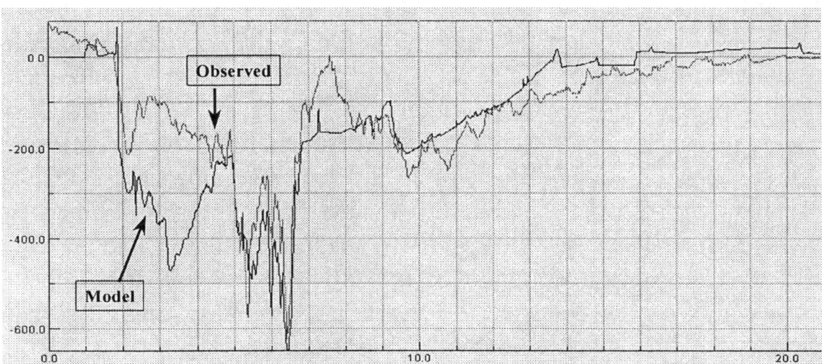

Figure 13. Predicted and observed magnetic field disturbance comparisons for Sept 25, 1998 at St. John's observatory. Major dB/dt excursions are accurately captured.

It is straightforward to compute the coupling to each line in the complex power grid. In this calculation, the orientation of the electric field vectors is applied throughout the client network. This is done by interpolation and analytic integration, so that the net voltage (electric field times distance) induced in each line segment can be determined rapidly. This allows the computation of the GIC flow in every line and transformer in the power network. Once the GIC is determined throughout the network and in each transformer, additional calculations can be made to provide power system operating staff with precise system impact estimates such as system and regional reactive power demands, numbers of transformers in saturation, and other important system impact visualisations.

While the set of approximations made in these solutions may seem substantial, experience with end-to-end benchmarks has been very good. For example, modelling of a specific storm in the Minnesota region is presented below in Figure 14. Note that the computed and measured current flowing at a particular point in the network compare very well. Agreement on the order of a factor of 2 is certainly reasonable given the goal of providing advanced warnings to utilities of an impending geomagnetic storm that could impact the normal operation of their power grid [*Kappenman, 2000*].

8. USER DISPLAYS OF FORECAST INFORMATION

One of the most important challenges is to present forecast disturbances in a clear and descriptive manner to impacted users of the data. The presentation of the information must not only be accurate, but also to-the-point. Therefore tailoring the data in a manner in which the operator is provided a clear picture of expected impacts is important. Figure 15 provides an example of presenting the conditions of exposure of a bulk power system in a way that is readily intuitive without inundating the operator with superfluous details. In this example the storm conditions are displayed over England and Scotland. The 400kV and 275kV transmission system is displayed with small circles indicating the magnitude (circle size varies) and polarity (circle shading changes) of the GIC at each transformer. Also shown are the vector icons of the magnitude and direction of the local magnetic field during the storm, which is responsible for the GIC flows. Text and graphic summaries can also be provided on System or Region reactive power demands, numbers of transformers in saturation and other important system impact details [*Kappenman, 1998*].

These data visualisations mimic the familiar terrestrial weather projections that most power system operators currently use and are quite familiar with in the management of operation of their networks. The primary difference amounts to supplanting the ordinary weather imagery with the pertinent Space Weather equivalent of a weather-radar tracking system.

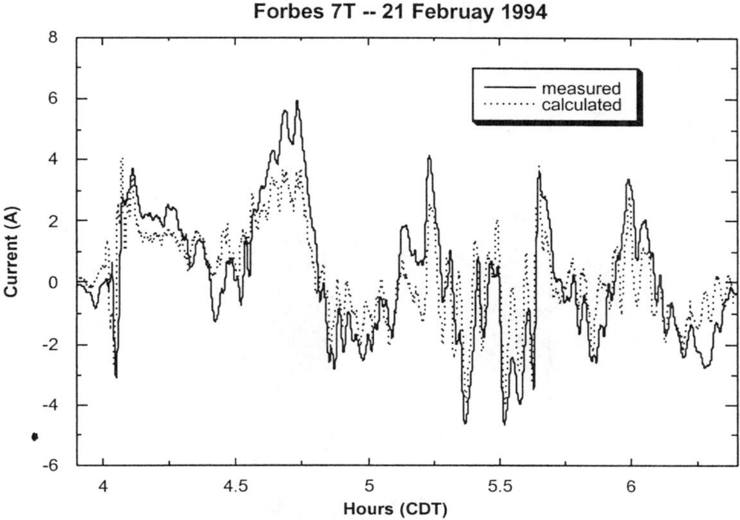

Figure 14. Validation of the ground-induction current modelling in a transformer neutral (amperes) at the Forbes substation in Minnesota on 21 February 1994. This example shows the ability to replicate GIC flow through transmission networks for storm events with reasonable accuracy over extended time, even when driven by local geomagnetic field observations as in the case shown (add 5 hours to obtain time in UT).

9.   CONCLUSIONS

Power system operators and designers will have to acquire new skills and employ new tools to successfully manage the ever-present risk posed by the space environment. Nature has presented a number of difficult challenges with Space Weather to impacted systems. In the long run, improvements in forecasts will be needed to allow better assessment of situations and management of storm-related risks. Using real-time solar wind monitoring, the ability exists to accurately predict the occurrence of large threatening storms with enough lead-time to take meaningful actions. Detailed models can further extend to first order impact estimates of the storm severity on a regional basis, allowing operators of critical systems to anticipate storm onsets and better prepare the operation of the system for storm impacts.

No large power system control centre in the world would do without continuous high quality weather data in managing the operation of their systems. The same paradigm arguably needs to be developed for the power industry for the threats posed by Space Weather.

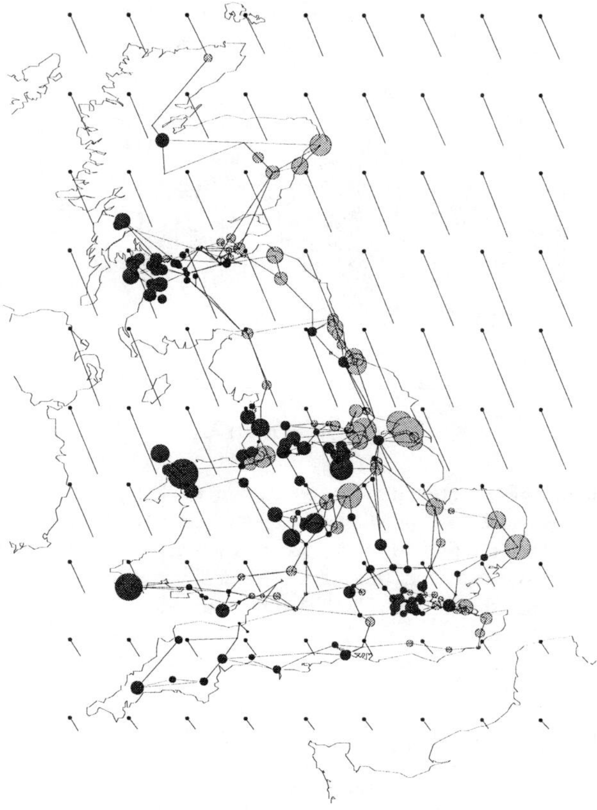

Figure 15. The storm visualisation shown above is designed to provide a clear and concise picture of the location and intensity of storm impacts across the transmission network. The circles indicate magnitudes of GIC flow at each transformer in the network and the vector icons depict the intensity and direction of changes in the horizontal magnetic field. In this example the storm conditions are displayed over England, Wales and Scotland.

10. REFERENCES

Albertson, V.D., J.M. Thorson, R.E. Clayton, and S.C. Tripathy, Solar Induced Currents in Power Systems: Causes and Effects, *IEEE Transactions Paper T72416-6*, presented at IEEE PES Summer Meeting, San Francisco, CA, July 9-14, 1972.

Anderson, C.W., L.J. Lanzerotti, and C.G. MacLennan, Outage of the L-4 system and the geomagnetic disturbances of August 4, 1972, *Bell Syst. Tech J.,* 53, 1817, 1974.

Boteler, D.H., Jansen Van Beek, G., Mapping the March 13, 1989, Magnetic Disturbance and its Consequences across North America, *Solar Terrestrial Predictions IV*, Proceedings of a Workshop at Ottawa, Canada May 18-22, 1992, Volume 3, pages 57-70.

Davidson, W.F., The Magnetic Storm of March 24, 1940 – Effects in Power Systems, EEI Bulletin, May 7, 1940.

IEEE Working Group K-11 Report , The Effects of GIC on Protective Relaying, IEEE Transactions on Power Delivery, Vol. 11, No.2, April 1996, pages 725-739.

Kappenman, J.G., Geomagnetic Storm Forecasting Mitigates Power System Impacts, IEEE Power Engineering Review, November 1998, pages 4-7.

Kappenman J. G., W. A. Radasky, J. L. Gilbert, I. A. Erinmez, Advanced Geomagnetic Storm Forecasting: A Risk Management Tool for Electric Power Operations, Submitted for IEEE Plasma Society Special Publication on Space Plasmas and Space Weather, March 2000.

McNish, A.G., Magnetic Storms EEI Bulletin, May 7, 1940.

NERC Disturbance Analysis Working Group Report, The 1986 System Disturbances: Solar Magnetic Disturbance, February 7-8, 1986, page 32, 1987.

NERC Disturbance Analysis Working Group Report, The 1989 System Disturbances: March 13, 1989 Geomagnetic Disturbance, pages 8-9, 36-60, 1990.

NERC Disturbance Analysis Working Group Report, The 1991 System Disturbances: Geomagnetic Disturbances in 1991, pages 50-57, 1992.

Chapter 14

Global Magnetospheric Modelling

Methods, results, and open questions

Manfred Scholer
*Centre for Interdisciplinary Plasma Science, Max-Planck-Institut für extraterrestrische Physik
85740 Garching bei München, Germany*

Abstract   Over the last two decades considerable progress has been achieved in modelling the response of the magnetosphere to changing interplanetary conditions by solving the time-dependent magnetohydrodynamic equations in three dimensions. We will briefly review the different formulations of the MHD equations, the numerical methods, which are used in the global modelling codes, and the boundary conditions at the ionosphere, where field-aligned currents are closed. For due southward interplanetary magnetic field (IMF) the models predict reconnection at the magnetopause, the occurrence of a near-Earth neutral line, and the tailward ejection of a plasmoid. There is considerable debate about the state of the magnetosphere for due northward IMF, and various codes give conflicting answers. An important question is how well the global models are able to predict current and potential patterns in the polar ionosphere. For selected time periods the codes have been run with measured interplanetary conditions as input and comparison was made with polar cap observations. We discuss in detail the comparisons between global modelling results and results using the Assimilative Mapping of Electrodynamics (AMIE) procedure. Open problems, as the nonideality (resistivity) in the MHD formulation and the importance of kinetic effects, are discussed.

Keywords   Magnetosphere, global modelling, magnetohydrodynamics, substorms, ionospheric currents, ionospheric potentials.

I.A. Daglis (ed.), Space Storms and Space Weather Hazards, 363–388.

1. INTRODUCTION

Understanding how energy, mass, and momentum is transported from the solar wind through the magnetosphere, and how the energy is finally deposited into the ionosphere, is pivotal in magnetospheric physics. This becomes increasingly also important from a practical point of view as we depend more and more on sophisticated and sensitive land- and space-based systems. One of the means to study the response of the magnetosphere to interplanetary conditions are global magnetohydrodynamic (MHD) simulations. In global MHD simulations a steady solar wind usually enters initially from one side into a three-dimensional simulation box, which contains a dipolar-like magnetic field. The dipolar-like field is rooted at an inner sphere, where boundary conditions appropriate for the ionosphere have to be assumed. The interaction of the solar wind with the dipolar-like field generates a magnetosphere, and the response of the magnetosphere to changing interplanetary conditions can subsequently be investigated.

The history of global MHD simulations dates back more than two decades ago: Leboeuf et al. (1978) were first to perform two-dimensional (2-D) MHD simulations of the interaction of a 2-D dipole with a magnetized solar wind. They recovered the classic Dungey (1961) picture for southward interplanetary magnetic field (IMF). Looking today at these early simulations one has the impression that they were not much better than the cartoons, may be worse, since the field lines had a little bit more wiggles. A few years later several groups started with three-dimensional (3-D) simulations (Leboeuf et al., 1981; Lyon et al. 1981; Wu et al., 1981; Brecht et al., 1982; Ogino and Walker, 1984). These codes were subsequently improved, not only by increasing the resolution and abandoning certain spatial symmetries, but most importantly by the implementation of realistic ionospheric boundary conditions. In recent years the models have been developed to the point that the results can be quantitatively compared with in situ space observations and ground based observations.

In the next section we will briefly describe the basic equations, the boundary conditions, and various numerical methods. The applications of global magnetospheric codes are twofold: First, the codes can be applied in order to investigate fundamental problems of the interaction of the solar wind-magnetosphere-ionosphere system (S-M-I coupling), and second, they can be used to predict the response of the magnetosphere and ionosphere to a given solar wind input as measured in situ by spacecraft outside of the magnetosphere. Thus we will first summarize what has been learned from global simulations for different solar wind conditions (IMF directions), and then discuss the various attempts to compare results from global simulations with in situ spacecraft and ground based measurements.

2.   BASIC EQUATIONS, MODEL ASSUMPTIONS, AND NUMERICS

2.1   Magnetohydrodynamic Equations

In global MHD simulations the interaction of the solar wind with the magnetosphere and the processes occurring within the magnetosphere are described by the time-dependent MHD equations which are solved by finite difference methods on a suitable spatial grid. The MHD equations can be written down in various forms. In the simplest case the equations for the density ρ, velocity $\mathbf{v}$, pressure p, magnetic field $\mathbf{B}$ and electric field $\mathbf{E}$ can be written in the non-conservative form

$$\frac{\partial \rho}{\partial t} = -\nabla \cdot (\rho \mathbf{v})$$

$$\rho \frac{\partial \mathbf{v}}{\partial t} = -(\mathbf{v} \cdot \nabla)\mathbf{v} - \nabla p + \mathbf{j} \mathbf{x} \mathbf{B}$$

$$\frac{\partial p}{\partial t} = -(\mathbf{v} \cdot \nabla)p - \gamma p \nabla \cdot \mathbf{v}$$

$$\frac{\partial \mathbf{B}}{\partial t} = -\nabla \mathbf{x} \mathbf{E}$$

$$\mathbf{E} = -\mathbf{v} \mathbf{x} \mathbf{B} + \eta \mathbf{j}$$

where $\mathbf{j} = \nabla \cdot \mathbf{B}$ is the electrical current, η a resistive term, and γ the ratio of specific heats. Solving these equations by a finite difference method does not guarantee numerical conservation of momentum and energy. Furthermore, there are numerical difficulties with the convective derivatives. The MHD equations can be written in a fully conservative form by rewriting the momentum equation in the form

$$\frac{\partial \rho \mathbf{v}}{\partial t} = -\nabla \cdot \left\{ \rho \mathbf{v}\mathbf{v} + p\mathbf{I} - \left(\mathbf{B}\mathbf{B} - \frac{1}{2}\mathbf{B}^2\mathbf{I} \right) \right\}$$

where $\mathbf{I}$ is a unit tensor, and by introducing an equation for the energy density U:

$$\frac{\partial U}{\partial t} = -\nabla\{(U + p)\mathbf{v} + \mathbf{E} \mathbf{x} \mathbf{B}\}$$

366

where the pressure is given by $p = (\gamma - 1)(U - \rho v^2/2 - B^2/2)$. This system allows strict numerical conservation of mass, momentum, and energy. The conservative set of equation is used in the BATS-R-US code (e.g., Gombosi et al., 1996). Since the pressure is determined by the difference between the total energy density and the magnetic field energy density, it may become negative in low beta (β = particle pressure to magnetic field pressure) regions of the magnetosphere (difference of two large numbers). A possible remedy is to write just the gas dynamic part in conservative form by introducing the plasma energy density $e = p / (\gamma-1)+(1/2)\rho v^2$:

$$\frac{\partial e}{\partial t} = \nabla \cdot \{(e + p)v\} + j \cdot E$$

This scheme allows strict numerical conservation of mass, momentum, and plasma energy, but not strict conservation of total energy, and has been used by J. Raeder (e.g., Raeder, 1999) and in the Lyon/Fedder code. In the fully conservative scheme the problem occurring in regions where the magnetic field is very strong can be avoided by solving only for the deviation of the magnetic field from the intrinsic magnetic field. This method was introduced by Tanaka (1994), and has also been used in the BATS-R-US code (Powell et al., 1999) and in the Lyon/Fedder code. In addition to the above equations the condition $\nabla \cdot \mathbf{B} = 0$ has to be satisfied throughout the numerical system.

In most simulations a three-dimensional dipole is initially placed into the numerical box. In order to counterbalance the dipole field in the solar wind a mirror dipole is placed at some distance on the sunward side of the Earth's dipole. This creates a surface sunward of the Earth with $B_x = 0$ (GSE coordinates). The region sunward of this surface is then replaced with a solar wind, i.e. at this boundary the solar wind parameters are prescribed. Since the solar wind parameters are usually assumed to be independent of y and z, the normal magnetic field component B_x has to stay constant. On all other outer boundaries the plasma is allowed to flow freely in and out, i.e., the normal derivatives of the plasma parameters are assumed to be zero.

2.2   Numerical Grids

The MHD equations are solved by finite difference methods on a 3-D spatial grid. The grid structure determines the programming overhead, the computing overhead and the memory overhead. A uniform Cartesian grid has the lowest programming and computing overheads and no memory overhead; on the other hand, the desired resolution in regions with large gradients determines the mesh size everywhere, and the simulation becomes rather computing time intensive. A uniform Cartesian grid with a mesh size from 0.5 R_E down to 0.15 R_E has been used by T. Ogino (e.g., Ogino et al., 1994; Walker et al., 1998). Similar low overheads have stretched Cartesian grids, where the highest resolution is assumed to be near Earth and the resolution diminishes in the tailward direction. Such a non-uniform grid with near-Earth resolution of 0.5 R_E is used by Raeder (1999). Adaptive grid schemes partition space into regions according to the gradients in solutions of physical quantities: in a region where large gradients exist, the resolution will be refined. In the scheme used by Gombosi et al. (1996) a Cartesian block can be replaced by eight child sub-blocks, one for each octant of the parent block. Likewise, if coarsing is needed, the eight children are replaced by a parent block (Powell et al., 1999).

Non-Cartesian distorted meshes have been designed to afford maximal resolution near critical regions, as the magnetopause, the ionosphere, and the central geomagnetic tail, and poorer resolution in the magnetosheath and the upstream solar wind. Fedder and Lyon (e.g., Fedder and Lyon, 1995) use a computer-generated distorted spherical coordinate system with its axis aligned with the solar wind flow. Tanaka (2000) uses a distorted grid with a dense allocation at the inner boundary (3 R_E) and in the plasma sheet

2.3   Non-Ideality in MHD Simulations

In an ideal MHD plasma with infinite conductivity the magnetic field is frozen in, and magnetic reconnection is not possible. In order to allow entry of solar wind energy, mass, and momentum into the magnetosphere reconnection between the interplanetary magnetic field and the magnetospheric field has to occur at the magnetopause. In the global simulations the non-ideality of the plasma is either due to numerical diffusion or can be explicitly introduced via a resistivity η in Ohm's law.

According to Fedder et al. (1995), in the Fedder/Lyon model numerical magnetic reconnection occurs because of the averaging error within localized regions when oppositely directed magnetic flux convects into a single computational cell. The averaging then results in a certain amount of flux annihilation. This type of numerical resistivity is only significant when oppositely directed magnetic fields are being forced into a cell from opposite directions. Fedder et al. (1995) point out that in their simulation the reconnection rate is not determined by the cell size or the strength of the numerical dissipation, but that it "is controlled by the global character of the solutions and the physical boundary conditions". Similarly, the Gombosi model solves the ideal MHD equations and dissipation is purely numerical. In addition to the forced numerical reconnection scenario described by Fedder et al. (1995) discretization of the ideal MHD equations leads to errors proportional to second derivatives of the magnetic field components. Thus these errors have similar effects as a finite resistivity. Such errors depend on local plasma parameters as well as on the local mesh resolution. Magnetic reconnection can be controlled by introducing an 'anomalous' resistivity, which is taken to be larger than the numerical resistivity. Raeder (1995) used a resistivity which is proportional to the square of the local current density, η ~ $j^2\Delta/(|B|+\varepsilon)$ with Δ = grid spacing and ε a very small number to avoid dividing by zero. This resistivity is only switched on when it exceeds some critical value and is zero otherwise. A similar anomalous resistivity proportional to $j^2/|B|^2$ was used by Tanaka (2000). However, no local threshold mechanism is introduced. Furthermore, in the Tanaka model the assumption is made that the magnetotail becomes more diffusive with downtail distance: the $j^2/|B|^2$ term is multiplied by a function $f(x)$ that is small in the near-Earth region, increases linearly between x = -20 R_E and x = -60 R_E, and saturates beyond x = -60 R_E.

2.4   Numerical Methods

We will only very briefly describe some of the numerical methods in use, and refer the interested reader to the original literature. Let us consider an equation of the form

$$\frac{\partial U}{\partial t} = -\nabla \cdot F$$

where $F(U)$ is some function of U (see for instance the energy equation given above in conservative form). The time differencing is usually done in explicit form by a predictor-corrector scheme, by a leap-fog scheme, or by an Adams-Bashfors method, which are all second order accurate. 'Explicit' means that when determining U^{n+1} at time step n+1 the function $\nabla \cdot F$ is evaluated at the previous time steps only, $F(U) = F(U^n, U^{n-1},...)$. For explicit time differencing the so-called Courant-Friedrichs-Levy condition holds, i.e., the time step Δt has to be smaller than the smallest grid size divided by the largest possible velocity. The latter is the absolute value of the bulk velocity plus the fast magnetosonic velocity. Implicit time differencing schemes are unconditionally stable, but require the solution of a large system of equations and are rather difficult to handle.

The divergence of the physical flux can be approximated by differences of numerical fluxes

$$\nabla \cdot F = \left(f_{i+1/2,j} - f_{i+1/2,j}\right)/\Delta x + \left(f_{i,j+1/2,j} - f_{,ji+1/2,j}\right)/\Delta y$$

A variety of numerical fluxes are possible. A method that has been used in MHD quite extensively is the two step Lax-Wendroff method. A Lax flux $f_{i+1/2} = (F(U_i)+F(U_{i+1}))/2-(U_{i+1}-U_i)/2$ is used for the predictor time step and subsequently the second order central numerical flux $f_{i+1/2} = (F(U_i)+F(U_{i+1}))/2$ is used for the corrector time step. High order schemes, like a fourth order central scheme, work well in regions where the solution is smooth, but have problems in regions where new extrema might develop, for instance at shocks. A so-called flux limiter can be introduced, which results in high order fluxes in regions with a small gradient and a low order flux in regions of large gradients. J. Raeder uses a scheme that switches between fourth order fluxes and first-order (Rusanov) fluxes. Details can be found in Harten and Zwas (1972) and Hirsch (1990). In the Fedder/Lyon code a switch is used which is based on the partial donor method (Hain, 1987). In smooth regions the scheme reduces to a centred eighth order spatial differencing. When the solution varies rapidly over 1 - 2 cells a beam scheme (Sanders and Prendergast, 1974) is used which allows smooth transitions, i.e. no overshoots occur at shocks or other discontinuities. The Gombosi model is described in detail by Powell et al. (1999). The code is based on an approximate solution of the Riemann MHD problem. The Riemann solver is a Roe scheme (Roe, 1981). The Tanaka code is based on a Total-Variance-Diminishing scheme which works due to formulation by the finite volume method on an unstructured grid (Tanaka, 1994). A combination of the leapfrog scheme with the two-step Lax-Wendroff scheme is used by T. Ogino (e.g., Ogino et al., 1994) in his global code based on the non-conservative MHD equations.

2.5   Ionospheric Boundary Conditions

The coupling of the magnetosphere to the ionosphere (M-I coupling) occurs through field-aligned currents, which flow into and out of the ionosphere and close in the ionosphere by horizontal Hall- and Pedersen currents. The inner boundary, where the MHD quantities are connected to the ionosphere, is usually assumed to be a shell with a radius R_S of a few Earth's radii centred at Earth. It is not feasible to place this shell at low altitudes since the Alfvén speed increases drastically with lower altitude and large magnetic field gradients occur close to the Earth. Within this shell the magnetic field is assumed to be a static dipole. The field-aligned current j_n obtained from the MHD simulation at R_S is mapped along the dipole field lines onto the polar cap by taking into account flux tube convergence, and is then used as input for the determination of the ionospheric potential. It is assumed that the normal component of the magnetospheric current j_n at the interface between the magnetosphere and the ionosphere is diverted into horizontal currents in a height-integrated thin ionosphere. With the help of Ohm's law, applied to the ionosphere with an electrostatic potential ϕ and a height integrated conductivity tensor Σ, this can be written as

$$\nabla_t \cdot \Sigma \cdot \nabla \phi = - j_\parallel \sin \varphi$$

where the symbol ∇_t denotes the tangential component of the gradient operator in the ionosphere surface, and $j_n = j_\parallel \sin \varphi$ with φ as the inclination of the dipole field at the ionosphere. Once the ionospheric potential ϕ is derived, it is mapped along the magnetic field lines onto the inner shell assuming the field lines to be equipotentials. The electric field $\nabla\phi$ then determines the convection velocity at the inner shell: $\mathbf{v}_c = \mathbf{B}_c \cdot \nabla\phi / B_c^2$ where the subscript c denotes the values at the inner boundary.

The ionospheric conductance tensor consists of two terms: the ionospheric Hall and Pedersen conductances, Σ_H and Σ_P. In most simulations these conductances have two contributions: a solar EUV conductance and an auroral-precipitation enhanced conductance. The solar EUV ionisation depends mainly on the solar 10.7-cm radio flux ($F_{10.7}$) and the solar zenith angle χ. The auroral precipitation enhanced conductance should be found in regions of upward field-aligned currents: an effective resistivity in the region of field-aligned currents results in a field-aligned potential drop $\Delta\phi_\parallel$, so that the mean energy E_o and the energy flux F_E of precipitating electrons that are accelerated by the potential drop is $E_o = e\Delta\phi_\parallel$ and $F_E = \Delta\phi_\parallel | j_\parallel |$, respectively (e.g., Raeder et al., 1998). Reader et al. (1998) assume a potential drop given by the density n_e and temperature T_e of the plasma at the inner shell:

$$\Delta\phi_\parallel = \frac{e^2 n_e}{\sqrt{2\pi m_e k T_e}} \max\left(0, -j_\parallel\right)$$

Because single fluid MHD does not provide an electron temperature the temperature of the MHD fluid has to be taken, possibly modified by an empirical factor (Raeder et al., 2000). Similarly, Fedder et al. (1995) define the field-aligned potential energy $E_\parallel$ between the ionosphere and the inner shell:

$$E_\parallel = R \frac{j_\parallel}{p} c_s$$

where c_s is the sound speed at the inner shell and R is an adjustable scaling factor which is taken to be 5 times larger for currents out of the ionosphere as opposed to currents into the ionosphere. Fedder et al. (1995) then adjust the precipitating flux, which is at the inner shell proportional to density times sound speed, for the effects of the field-aligned potential in a geomagnetic mirror field. A second auroral precipitation enhanced conductance is caused by diffuse electron precipitation. Raeder et al. (1998) assume complete pitch angle scattering of thermal electrons at the inner boundary, which fills the loss cone with thermal electrons of a mean energy kT_e. In order to compute from the precipitation parameters the Pedersen and Hall conductances empirical relations given by Robinson et al. (1987) can be used.

3. GLOBAL SIMULATIONS: SOUTHWARD IMF

Under southward IMF reconnection is expected to occur at the dayside magnetopause. This allows entry of solar wind mass, momentum, and energy into the magnetosphere, which subsequently undergoes substantial alterations. One of the most dramatic changes of the magnetosphere in response to a prolonged southward turning of the IMF is the expansion phase of the geomagnetic substorm. Although there is wide disagreement as to the cause-effect relationship in the substorm expansion phase, there is little disagreement that the substorm expansion phase incorporates dipolarization in the near-Earth region (or equivalently current sheet disruption), dispersionless injection of energetic plasma at synchronous orbit, and the occurrence of a neutral line at a distance of $\approx 20 - 25$ R_E leading to plasmoid formation. In early global simulations a southward IMF was imposed at time $t = 0$ at the upstream edge of the simulation box (e.g., Walker et al, 1993). These simulation resulted in dayside reconnection, plasma sheet thinning, and the occurrence of a neutral line in the near-Earth tail at about 40 min after imposing the southward IMF. Figure 1, taken from Walker et al. (1993), shows magnetic field lines at $t = 47$ min; in the upper panel field lines in the tail lobes have been removed and magnetosheath field lines have been added, in the lower part field lines are shown which are attached at least with one end to the Earth. Here the resistivity was taken to be constant throughout the simulation box and line tying was assumed at the ionospheric boundary.

A neutral line has formed in the near Earth region at about 14 R_E, and reconnection has progressed to lobe field lines. Near midnight the plasmoid has started to move tailward. A quasi-steady state convection pattern is established where reconnection transports flux toward the dayside as fast as it is being transported again into the tail by dayside reconnection. This establishes a two cell convection pattern over the polar cap. Walker et al. argue that the location of the neutral line is determined by the energy flux following dayside reconnection: the Poynting flux is concentrated in the

372

region where the tail attaches onto the dipole-dominated inner magnetosphere. Since in these simulations the time between occurrence of the near-Earth neutral line, high speed flows, and dipolarization on one hand, and detachment and tailward motion of the plasmoid on the other hand, is rather large, it is not clear whether substorm expansion is defined by the beginning of plasma sheet reconnection with accompanied strong flows, or later by the onset of lobe field line reconnection.

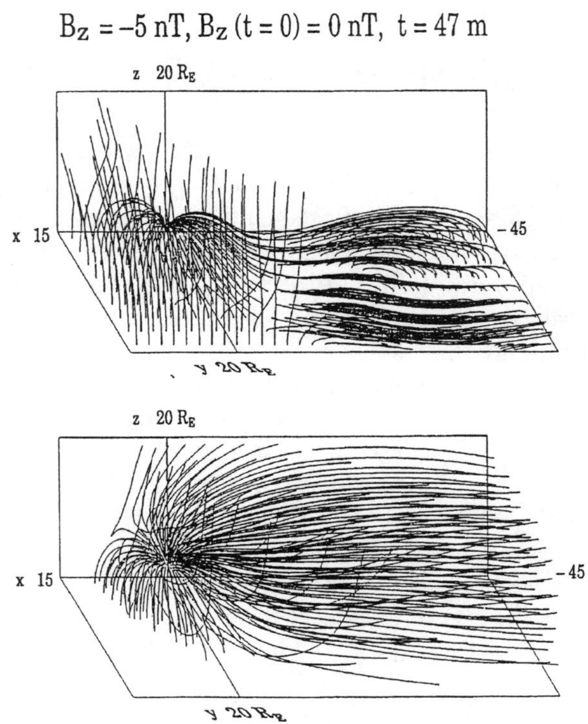

Figure 1. Magnetic field lines 47 min after imposing a southward IMF. Top: closed magnetospheric and magnetosheath field lines. Bottom: field lines with at least one end attached to the earth (from Walker et al., 1993)

In other global simulations, which investigate the response of the magnetosphere to prolonged southward IMF, first a quasi-steady configuration for northward IMF is produced and then the IMF is switched to southward. Under a northward IMF the polar convection can be described by a four cell convection pattern (see following section). Ogino and Walker (1998) have shown that switching the IMF from north to south leads to a convection front over the polar cap which changes the four cell pattern to a two cell pattern in about 60 min. This convection front corresponds to a discontinuity surface with reconnected dayside field lines as they move across the polar cap. When the antisunward flow over the polar cap reaches a low latitude midnight stagnation point, tail reconnection sets in, and subsequent Earthward convection of reconnected flow accomplishes the two cell pattern. The near-Earth neutral line appears about 60 min after the north-south IMF switching in the centre of the tail and expands then toward the dusk and dawn magnetopause. Figure 2 is a schematic of the time evolution of the polar convection and the reconnected field lines as obtained from the Ogino and Walker (1998) simulation. Dipolarization is here also due to the strong Earthward flow from the tail reconnection site.

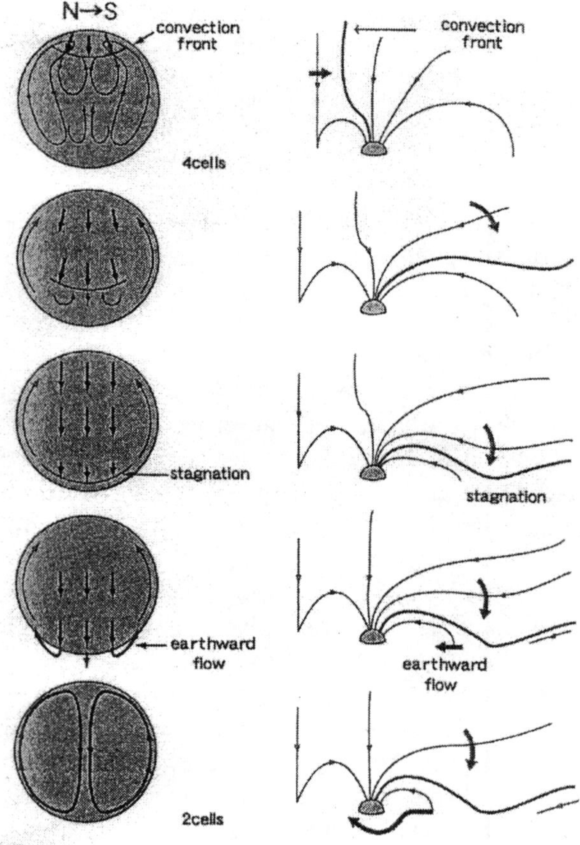

Figure 2. Schematic of the time evolution of the polar cap convection and the reconnected field lines when the IMF turns from northward to southward (from Ogino and Walker, 1998).

The Ogino/Walker model does not allow for closure of parallel currents in the ionosphere as described above. Rather the ionospheric boundary

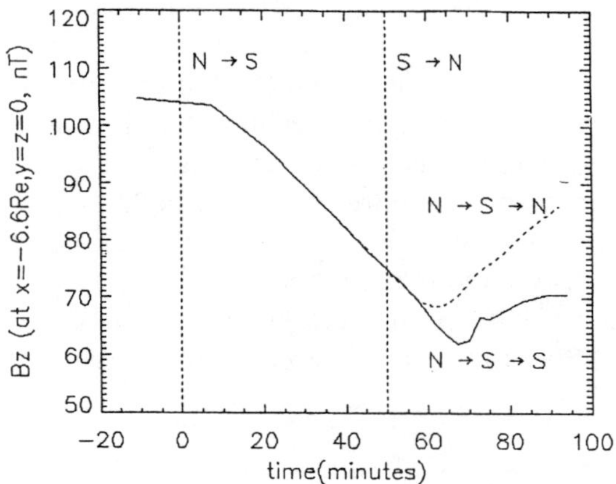

Figure 3. Development of B_z at the midnight synchronous orbit position. Solid curve for a north-south turning of the IMF; dashed curve for a north-south-north turning of the IMF (from Tanaka, 2000)

conditions imposed at the inner shell are determined by requiring a static equilibrium, and field-aligned currents partly close in a smoothing region just above the inner shell. Tanaka (2000) has recently investigated the magnetospheric dynamics following a southward turning of the IMF by using the magnetosphere-ionosphere coupling model described above, where the field-aligned currents close in the ionosphere by Pedersen and Hall currents. Tanaka (2000) starts from a stationary solution with a northward IMF, which is inclined by 30° to the northward direction and then switches the IMF to an inclination of 150°. In the growth phase southward turning of the IMF results in enhanced polar cap convection. The coupling to the inner magnetosphere leads in turn due to diversion of the flow to both sides of the Earth to rapid removal of plasma and closed magnetic flux from the inner edge of the plasma sheet. The small supply of closed flux from the distant tail (by reconnection at a distant neutral line) leads to a configuration change, i.e., a plasma sheet thinning. According to Tanaka's simulation it is this imbalance of flux transport near the inner edge of the plasma sheet, which brings about plasma sheet thinning and not the change in the attack angle of the solar wind at the lobe magnetopause due to addition of reconnected flux to the lobes. Figure 3 shows the development of the magnetic field B_z component at 6.6 R_E near midnight. Time $t = 0$ is the IMF southward turning. The solid curve is a run where the IMF is southward all the time, the dashed curve is a run where the IMF was switched to northward again at t=50 min. Dipolarization at synchronous orbit sets in about 1 hour after the southward turning. However, dipolarization sets in earlier when the IMF is switched again northward. Since northward turning is expected to

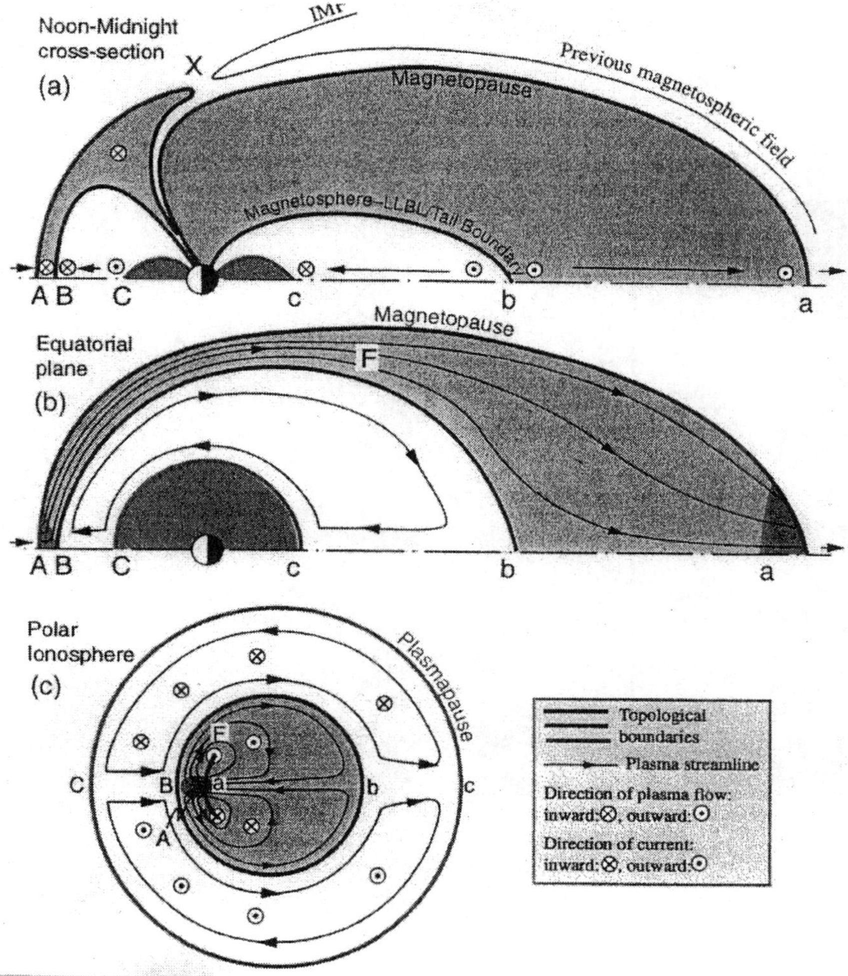

Figure 4. Schematic of results from global simulations under northward IMF. Top: Noon-midnight cross section. Middle: Mapping into the equatorial plane as seen from north. Bottom: The northern ionosphere as seen from top (from Song et al., 1999).

cause a reduction of the thinning rate in the plasma sheet, Tanaka (2000) concludes that the near-Earth neutral line is not the primary cause of dipolarization. Dipolarization may be due to enhanced transport of flux from a neutral line in the midtail, but can also be due to reduced transport to the dayside. The reduced transport to the dayside is caused by reduced polar cap convection under northward IMF, helped by an increase in the ionospheric conductivity. Such dynamic processes can only be investigated by taking into account the global convection pattern, and can by design not be obtained when starting from static tail equilibria. Tanaka (2000) points out that the latter studies necessarily always have to come to the conclusion that the near-Earth neutral line is the trigger for substorm onset.

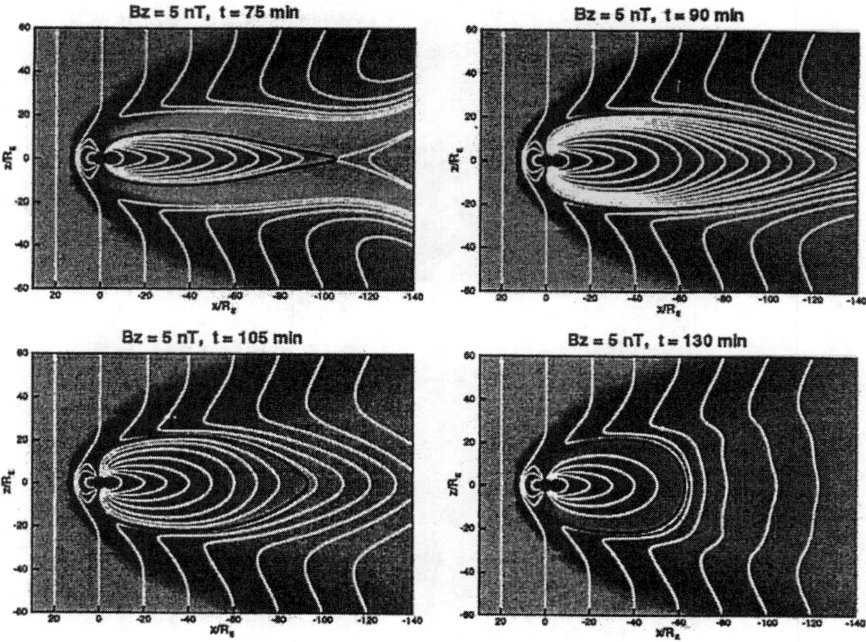

Figure 5. Time evolution of the magnetospheric configuration following northward turning of the IMF. The simulation started with steady state southward IMF (from Gombosi et al., 2000).

4. GLOBAL SIMULATIONS: NORTHWARD IMF

Global simulations for prolonged times of due northward IMF by different groups give conflicting answers, and there is some debate going on in the literature with respect to the causes. One of the key questions is whether the magnetosphere for strictly northward IMF is closed or open. In the seminal paper by Dungey (1961) the magnetosphere for northward IMF is closed, except for the two reconnection points at the nightside magnetopause. On the other hand, later models for northward IMF assume an open magnetosphere. In all global models for northward IMF plasma entry is due to merging of lobe field lines with IMF field lines tailward of the cusp. The closed models require that reconnection occurs simultaneously and symmetrically both at the northern and southern lobes. Global simulations for strictly northward IMF and no dipole tilt have resulted in most cases indeed in a closed magnetosphere. The simultaneous reconnection of IMF field lines in the northern and southern lobe erodes the tail lobes, but creates at the same time newly closed dayside magnetospheric field lines filled with magnetosheath plasma, which move down the flanks of the tail and are again added to the lobes. In the simulations by, e.g., Usadi et al., (1993), Fedder and Lyon (1995), Gombosi et al. (1998), and Song et al. (1999), the magnetosphere closes under northward IMF for more than one hour and reaches a steady state. Fedder and Lyon (1995) demonstrated that the large magnetic shear of the newly closed field lines downstream of the cusp results in the so-called NBZ (northward B_z) current system at high latitudes in the polar cap, where the field-aligned current is downward on the dawn side of the northern polar cap and upward on the dusk side, respectively. The tailward convection of the flux tubes filled with magnetosheath plasma leeds to two convection cells over the polar cap with an antisolar flow diverging from the cusp footprints, and sunward convection along the midnight-noon meridian line.

Figure 4 from Song et al. (1999) shows schematically (from top to bottom) the topology in the noon-midnight cross-section, in the equatorial plane, and in the polar ionosphere. The bright grey area corresponds to the low-latitude boundary layer (LLBL) of closed flux tubes produced by cusp reconnection. The white part represents the inner magnetosphere. The thin lines with arrows in the LLBL in the middle panel are streamlines. In the lower panel of the northern ionosphere downward field-aligned currents are shown by circles with crosses, upward currents are shown by circles with dots. The closure of these field-aligned currents in the polar ionosphere by Hall and Pedersen currents leads to a second field-aligned currents system at lower latitudes, a region I system. Connected with the region I system is a second pair of polar cap convection cells, which maps into the inner magnetosphere (white area in the middle panel of Figure 4). Currents are driven in the ionosphere by electric fields, which map back into the magnetosphere and cause **E** x **B** drift motion of the magnetospheric plasma. There is a considerable pressure decrease from the solar magnetopause to the last closed field line at midnight. This pressure gradient is the driving force for the antisunward convection in the LLBL (Song et al., 1999).

Raeder et al. (1995) start with southward IMF and turn the IMF northward after 2 hours. During the southward IMF a near-Earth neutral line develops. In contrast to the work by Usadi et al. (1993), Fedder and Lyon (1995), Gombosi et al.(1998), and Song et al. (1999), Raeder et al. (1995) obtain even after more than 3 hours following the northward turning a tail with a well defined structure at least up to 400 R_E downstream. There exists still an X line at -20 to -40 R_E in the centre of the tail, but lobes with open

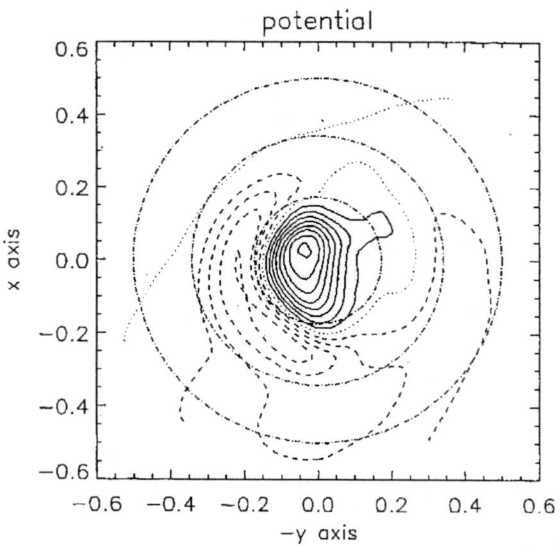

Figure 6. Ionospheric convection potential in the northern polar cap for a simulation where the IMF was tilted by 45 degrees with respect to the dipole axis (from Tanaka, 1999).

magnetic flux also still persist. They also obtain high latitude magnetopause reconnection, leading to closed flux tubes filled with magnetosheath plasma which are convected downstream and constitute a flank boundary layer. The reason for the open magnetosphere is according to Raeder et al. (1995) a rather slow reconnection rate at the lobe reconnection sites. Models which lead rapidly to a short and closed magnetosphere under northward IMF may simply exhibit higher reconnection rates at the lobe reconnection sites. However, recently Raeder (1999) has suggested that numerical diffusion may be another factor leading to rapid closing of the magnetosphere under northward IMF. To prove this Raeder (1999) has run his code with different values of the (spatially constant) resistivity. By increasing the resistivity he eventually obtains a magnetosphere without open lobe field lines, and the tail length decreases with increasing resistivity, reaching a tail length as short as 50 R_E in the highest resistivity case. First, the increased diffusion annihilates lobe flux, thus leading to disappearance of the tail lobes. Second, the tail ends of the newly produced flux tubes constituting the flank boundary layer eventually stop to convect tailward since diffusion violates the frozen-in condition: convection of the magnetic flux becomes decoupled from the plasma convection.

Gombosi et al. (2000) disagree with Raeder's (1999) statement about the numerical resistivity being the cause of a closed magnetosphere for northward IMF. Since the different contributions to numerical errors are

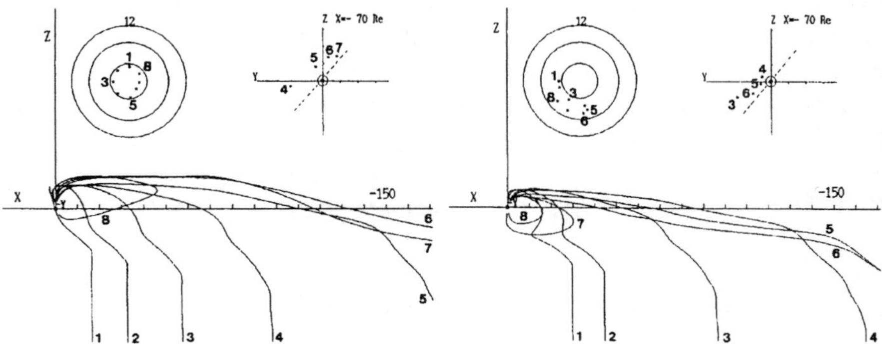

Figure 7. Configuration of magnetic field lines as viewed from dusk. Left: Field lines traced from the outer part of the round cell. Right: Field lines traced from the crescent cell (from Tanaka, 1999).

difficult to evaluate Gombosi et al. (2000) feel that the size of numerical errors (including numerical resistivity) is best measured by a mesh-convergence study: the same case is run with a series of sequentially finer meshes. When the difference between two runs is small enough, the calculation is called mesh-converged. Gombosi et al. (2000) have presented a mesh-convergence study for the northward IMF case and have shown that the magnetosphere becomes closed with a fixed tail length independent of the mesh size. They have also excluded the possibility that the due northward IMF case might be different when the simulation starts from an initial magnetosphere under southward IMF. Figure 5 shows the time evolution of the magnetospheric configuration after northward turning of the IMF at $t = 0$, when starting with a steady state magnetosphere under southward IMF. As a counterargument against numerical diffusion in the BATS-R-US code being responsible for the closed magnetosphere Gombosi et al. (2000) point out the high resistivity case of Raeder (1999) leads to a highly diffused bow shock, whereas in the Gombosi et al./Song et al. simulations the bow shock is very well resolved. In his reply Raeder (2000) stresses first that none of the codes listed by Gombosi et al. (2000), which result in a closed magnetosphere, really quantify the diffusivity in their codes. The mesh-convergence test performed by Gombosi et al. (2000) is critisized as a necessary, but not sufficient condition, and secondly, mesh-convergence does not necessarily lead to a solution of the ideal MHD equations since there is still (numerical) reconnection going on. It is extremely important for the community that this problem is solved. But the opponents even disagree about possible means to resolve the problem: Gombosi et al. (2000) feel that careful comparisons of the various models on several simple benchmark cases should be carried out. Raeder (2000), on the other hand, believes that the complex magnetosphere itself is the only relevant benchmark case, and that model event studies, i.e., comparing simulation results for given solar wind input with in situ measurements, are much more worth pursuing. The letter is definitively important, in particular under the space weather aspect. However, it should be possible to determine for each code the amount of diffusivity, eventually just by adding a resistive term in Ohm's law and comparing results for different resistivity values.

5. GLOBAL SIMULATIONS: FINITE B_Y

The case of a strictly northward IMF is a rather singular case and only of academic interest. In reality, a finite IMF B_y exists which destroys in the case of northward IMF the reconnection symmetry between the northern and southern lobes. This leads to newly reconnected open field lines which are added to the tail lobes. Figure 6, taken from Tanaka (1999), shows the ionospheric convection pattern for a case where the IMF is inclined 45° from the due-northward IMF case. Solid, dotted, and dashed contours show positive, zero, and negative potentials, contour spacing is 4 kV. The circles show 60°, 70°, and 80° latitudes. The convection consists of two cells: a round cell over the center of the polar cap and a crescent cell, which is for negative B_y on the evening side. The field lines connected to the core of the round cell are on open lobe field lines. Field lines connecting to the boundary of the lobe cell are field lines which result from reconnection between IMF field lines and lobe field lines. The reconnection site and the footpoint of the field lines are in the same hemisphere. Field lines connected to the crescent cell are open field lines which are due to reconnection of the IMF with closed magnetospheric field lines. The reconnection point in the lobe and the footpoint of these field lines are in different hemispheres. Figure 7a shows field lines traced from the outer part of the round cell and Figure 7b shows field lines traced from the crescent cell. The two inserts shows the footpoints of the field lines in the polar cap and the positions of the same field lines on the z-y plane in the distant tail. The dashed line in the latter insert shows the position of the twisted neutral sheet. These simulations by Tanaka (1999) have revealed the origin of the round cell and the crescent cell in the polar cap and have shown that the merging-cell convection consists of the outer part of the round cell and the crescent cell.

Global simulations with a large IMF B_y by White et al. (1998) have revealed a new feature of the magnetospheric structure: the so-called magnetospheric sash. The sash is a band of weak magnetic field which emanates from the cusps, spreads fan-like tailward along the high-latitude magnetopause flanks, and closes via the cross-tail neutral sheet. On a cross-sectional plane in the tail the sash has the form of a cross-tail S. The low field strength region coincides with the separator line on the magnetopause. Thus, the sash bears a strong resemblance to an antiparallel merging situation. The cross-tail S is according to Tanaka (1999) due to secondary reconnection of field lines at the flanks of the magnetotail, which is unavoidable in the reclosure of the polar cap merging cell convection.

6.   COMPARISON OF GLOBAL SIMULATIONS WITH DATA

Important for an assessment of the predictive capability of global magnetospheric codes is a comparison of modelling results with measured data. In order to investigate to what extent and accuracy models can predict the ionosphere's response to given solar wind conditions, interplanetary plasma and magnetic field data have been used as input for the global codes. The codes then predict ionospheric currents, potentials, and precipitation patterns. For the same periods synoptic maps of the polar cap were derived from all available data by using the Assimilative Mapping of Electrodynamics (AMIE) procedure. The AMIE technique is based on the mapping procedure described by Richmond and Kamide (1988): maps of the high latitude electric fields and currents are reproduced from localised sets of observational data. These data may be ground-based magnetometer data alone, or in combination with other data, like magnetic perturbations at satellite altitudes, electric fields obtained by Radar measurements or satellite measurements, electric currents from Radar measurements, as well as conductivities derived from particle fluxes and/or photometric images. Certain periods have been selected as the Geospace Environment Modelling (GEM) Challenge events. Raeder et al. (1998) and Fedder et al. (1998) have used their global models in order to simulate the response of the magnetosphere-ionosphere system to solar wind input parameters for the time period January 27-28, 1992, and have compared the results with the synoptic maps obtained by Lyons et al. (1996) for the same time period by the AMIE technique. Both groups found very good agreement of the polar cap potential pattern predicted by the global models with those obtained by AMIE. However, the simulation models predicted cross polar cap potential drops, which are a factor of 2 larger than those predicted by AMIE. This discrepancy could not been resolved. The disagreement points to a serious deficiency, since the cross polar cap potential is a measure for the global magnetic merging rate between the IMF and the magnetospheric magnetic field at the magnetopause. The January 27-28, 1992 time period was some-what unusual, in that the IMF had a rather large magnitude of about 20 nT.

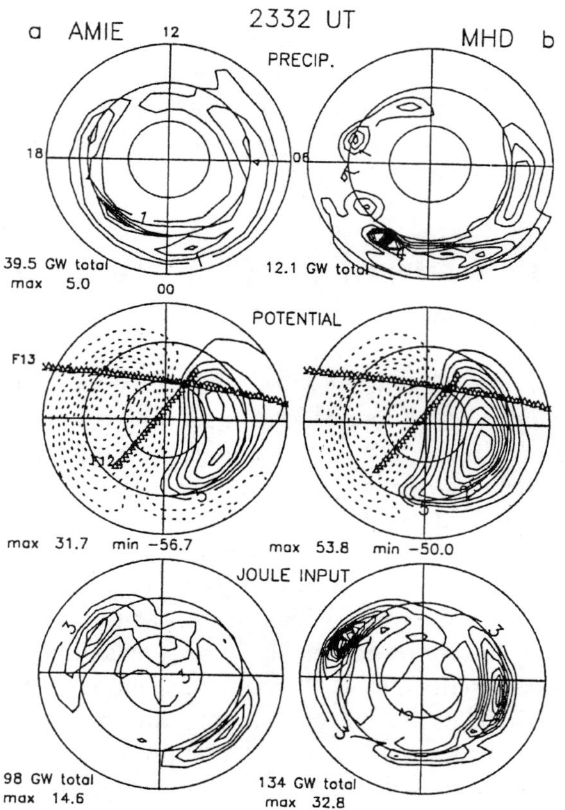

Figure 8. Comparison of northern polar cap quantities between the AMIE model and the Fedder/Lyon MHD model. Contours of the precipitating electron energy flux, of the polar cap potential, and of the Joule energy deposition rate (after Slinker et al., 1999)

A time period with considerable smaller IMF (May 19-20, 1996) has been used by Slinker et al. (1999) in order to perform a comparison between the results from the Fedder/Lyon model and the results from the AMIE analysis. Figure 8 shows a comparison of AMIE and MHD results during a time period where the results are rather close. Shown are northern polar cap quantities; noon is at the top and dawn is to the right, and latitude circles are drawn every 10°. The first panel shows contours of the precipitating electron energy flux. The maximum AMIE value is 40 GW, while the maximum MHD value is about a factor 4 smaller, about 12 GW. This could, in principle, be accommodated by a fetch factor. However, as pointed out by Slinker et al. (1999), the MHD code does not describe the motion of the electrons to the dawn dayside by drift, so that the morning precipitation will still be too low. The next panel shows contours of the polar cap potential with negative potential dashed. Also indicated is the path of two DMSP spacecraft, F12 and F13. The AMIE potential pattern is surprisingly well reproduced by the MHD simulation; the maximum potential drop in the MHD simulation is slightly larger than in the AMIE results but the difference is not as large as in the high IMF GEM event of January 27-28, 1992. The last panel shows the Joule input, which is somewhat larger in the MHD case. Figure 9, from Slinker et al. (1999), shows a comparison of polar cap potential, Joule heating, and auroral heating by precipitating electrons for the AMIE (dashed line) and the MHD (solid line) simulation. The bottom panel shows auroral indices (AE, AU, AL). During quiet times potential and Joule heating compare favourably well; deviations occur during periods of higher activity. The same is true for the polar cap potential. MHD greatly underestimates polar cap precipitating electrons. The dotted line in the panel exhibiting auroral heating is the result estimated from precipitation measurements on two NOAA and three DMSP spacecraft.

Raeder et al. (2001) recently compared results from a global simulation with data obtained during a geomagnetic substorm. Input into the MHD model were again solar wind parameters as measured upstream of the bow shock. The result of the global model were compared with AMIE synoptic maps, with ground magnetometer data from the IMAGE magnetometers, with magnetic field from the synchronous spacecraft GOES 8, with magnetic field data from the mid-tail spacecraft IMP 8, and with magnetic field and plasma data from Geotail. As in the former modelling attempts the MHD model predicts too large polar cap potentials when compared with AMIE, although the polar cap potential pattern was well represented. We note that the MHD model predicts strong localised field-aligned currents in the polar cap, which are absent in the AMIE results. Similar strong current concentrations have been reported by Slinker et al. (1999) for their model event study. Also, the electrojet during the substorm growth phase is by a factor 5 stronger than the observed ground signature. The model predicts the occurrence of a neutral line at about 25 R_E in association with substorm onset. However, it is not possible to delineate the cause-effect relationship between auroral signatures and the occurrence of the neutral line. Raeder et al. (2001) performed an interesting, although disturbing experiment. They use in the model an anomalous resistivity proportional to the square of the current density above a threshold. By changing the threshold or the absolute value of the resistivity above the threshold they obtain model solutions for the same solar wind parameters without the occurrence of a substorm. The tail rather settles into a state of steady convection in which dayside reconnection is balanced by reconnection in the tail. Tanaka (2000) has criticised the threshold model for an anomalous resistivity: a local threshold mechanism, like a step function for the resistivity, does not allow a global state transition where the substorm sequence is controlled by the global development of the topology, rather than by a local instability which results in an anomalous resistivity. The Tanaka (2000) model is based on the assumption that the tail becomes more diffusive with increasing distance.

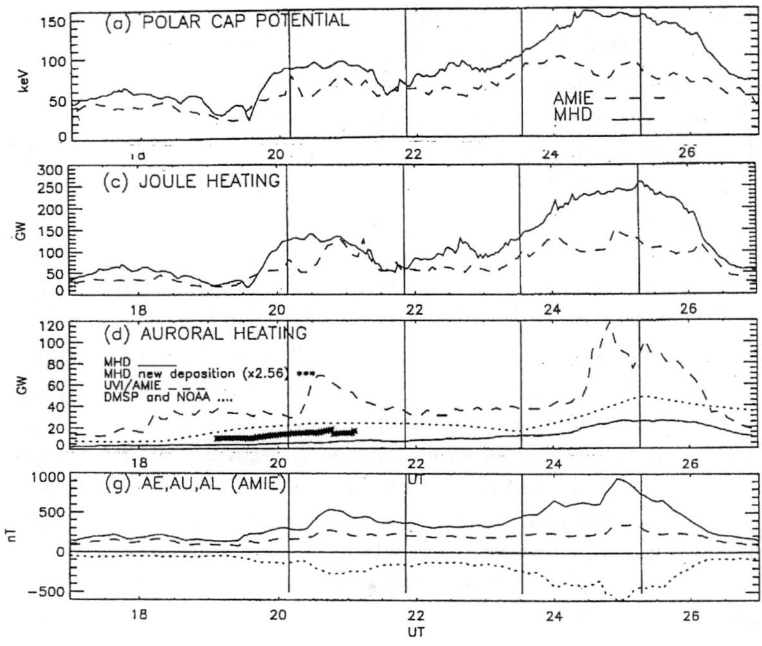

Figure 9. From top to bottom: Total potential drop across the polar cap, Joule heating, auroral heating, and geomagnetic indices. Solid lines from MHD simulation, dashed line from AMIE technique (after Slinker et al., 1999).

7.   OPEN PROBLEMS

As pointed out above, the nonideality in the MHD equations is essential for the interaction of the solar wind with the magnetosphere. The reason for such a nonideality, as well as its form, have to be carefully evaluated and considered. This concerns the question of numerical diffusivity in the codes as compared to a finite resistivity in Ohm's law, as well as the form of such a finite resistivity. The parameter variation exercise by Raeder (2000) and the work by Tanaka (2000) are here important contributions.

The difference between polar cap potentials obtained from global simulations and between the potentials obtained by AMIE may not be too much of a problem: AMIE is also a model and depends on a number of assumptions, in particular about the Hall and Pedersen conductivities. A more direct methods is a comparison between the polar cap potentials and drift measurements obtained by the DMSP satellites, which is under way.

All the modellers point out that there are deficiencies in the inner magnetosphere, since the MHD models do not include a ring current. The MHD formalism does not allow for gradient and curvature drifts. As pointed out by Raeder (2000) the plasma cannot be captured on drift shells, which leads to a pressure in the MHD models which is far too low. Higher pressure in the inner magnetosphere may have an important effect as to where the neutral line forms. Without a ring current the region II current and the associated convection cells in the ionosphere can not be adequately described. The absence of correct region II currents may influence the current closure by field-aligned currents, so that one important contribution to the global current system is missing.

Application of single fluid MHD may have also it's problems in the outer magnetosphere and tail. The magnetotail current sheet prior to substorm onset may become so thin, that the ion inertial length is not negligibly small and the Hall term in Ohm's law can no longer be neglected. 3-D Hall MHD simulations of reconnection have recently been performed by Yokokawa et al. (2001). These simulations have shown that the Hall term leads to a modification of the field-aligned current system. The existence of a small current sheet thickness is not a necessary condition for these Hall field-aligned currents to occur; rather the aspect ratio of the length of the reconnection line to the current sheet thickness is the important parameter. Based on a parameter study Yokokawa et al. (2000) conclude that the Hall field-aligned currents dominate during reconnection as long as the aspect ration is larger than about 7, i.e., when the reconnection topology is more two-dimensional as opposed to three-dimensional with a short cross-tail neutral line.

Kinetic effects can become important on large scales in the reconnection layer following reconnection in the magnetotail. In low beta plasmas, like in the lobe, a reconnection layer can develop which is not bounded by slow mode shocks, but a thin current sheet exists in the center. Such a reconnection layer has been proposed by Hill (1975) and has been seen in 2-D hybrid simulations of reconnection (Lottermoser et al., 1998; Nakamura et al., 1998). Arzner and Scholer (2001) find in 2-D hybrid simulations of magnetotail reconnection that such a current sheet exists up to 30 R_E away from the neutral line. The current sheet then breaks up and becomes turbulent. This indicates that kinetic effects can be rather important for the large-scale structure.

386

8.   SUMMARY

The global models have come a long way from the early simulations by Leboeuf et al. (1978) to the sophisticated simulations, which include a realistic ionospheric boundary where the field-aligned currents into and out of the magnetosphere are closed. The codes have come to the point where they actually are able to make quantitative predictions for ionospheric parameters. Worrying for the non-specialists are conflicting answers of different codes for the same solar wind input. What is the effect of the resistivity, numerical or assumed, in the various codes? How are reconnection rates and ultimately polar cap potentials determined by the resistivities? The effect of the numerical diffusivity compared to an imposed finite resistivity in Ohm's law has to be delineated. Is there anything as an anomalous resistivity in the magnetosphere? If not, can collective processes in a collisionless plasma be approximated in MHD simulations by a certain form of the resistivity in Ohm's law? There is ample room for further improvement. Apart from the quest for better and better resolution, there is the problem of the inner magnetosphere with the ring current which is so far not adequately addressed. Furthermore, more realistic ionospheric conditions require an ionosphere of finite thickness, as well as the coupling of the ionosphere to the thermospheric circulation.

9.   ACKNOWLEDGEMENTS

I am grateful to J. Raeder and T. Tanaka for helpful comments on an earlier version of the manuscript. I wish to thank E. Friis-Christensen for careful reading of the manuscript.

10.   REFERENCES

Arzner, K., and Scholer, M. (2001) Kinetic structure of the post-plasmoid plasma sheet during magnetotail reconnection, *J. Geophys. Res.,* in press.

Brecht, S. H., Lyon, J. G., Fedder, J. A., and Hain, K. (1982) A time dependent three dimensional simulation of the Earth's magnetosphere: Reconnection events, *J. Geophys. Res.,* **87**, 6098.

Dungey, J. (1961) Interplanetary magnetic field and the auroral zones, *Phys. Rev. Lett.,* **6**, 47.

Fedder, J. A., and Lyon, J. G. (1995) The Earth's magnetosphere is 165 R_E long: Self-consistent currents, convection, magnetospheric structure, and processes for northward interplanetary magnetic field, *J. Geophys. Res.,* **100**, 3623.

Fedder, J. A., Slinker, S. P., Lyon, J. G., and Mobarry, C. M. (1995) Topological structure of the magnetotail as a function of interplanetary magnetic field direction, *J. Geophys. Res.,* **100**, 3613.

Fedder, J. A., Slinker, S. P.,and Lyon, J. G. (1998) A comparison of global numerical simulation results to data for the January 27-28, 1992, Geospace Environment Modelling challenge event, *J. Geophys. Res.,* **103**, 14799.

Gombosi, T. I., DeZeeuw, D. L., Häberli, R. M., and Powell, K. G. (1996) Three-dimensional multiscale MHD model of cometary plasma environments, *J. Geophys. Res.,* **101**, 15233.

Gombosi, T. I., Powell, K. G., and van Leer, B. (2000) Comment on "Modelling the magnetosphere for northward interplanetary magnetic field: Effects of electrical resistivity" by Joachim Raeder, *J. Geophys. Res.,* **105**, 13141.

Hain, K. (1987) The partial donor method, *J. Comp. Phys.,* **73**, 131.

Harten, A., and Zwas, G. (1972) Self-adjusting hybrid schemes for shock computations, *J. Comput. Phys.,* **9**, 568.

Hill, T. W. (1975) Magnetic merging in a collisionless plasma, *J. Geophys. Res.,* **80**, 4689.

Hirsch, C. (1990) *Numerical Computation of External and Internal Flow,* **II**, JohnWiley, New York.

Leboeuf, J. N., Tajima, T., Kennel, C. F., Dawson, J. M. (1978) Global simulations of the time-dependent magnetosphere, *Geophys. Res. Lett.,* **5**, 609.

Leboeuf, J. N., Tajima, T., Kennel, C. F., Dawson, J. M. (1981) Global simulations of the three-dimensional magnetosphere, *Geophys. Res. Lett.,* **8**, 257.

Lottermoser, R.-F., Scholer, M., and Matthews, A. P. (1998) Ion kinetic effects in magnetic reconnection: Hybrid simulations, *J. Geophys. Res.,* **103**, 4547.

Lyon, J. G., Brecht, S. H., Huba, J. D., Fedder, J. A., and Palmadesso, P. J. (1981) Computer simulation of a geomagnetic substorm, *Phys. Rev. Lett.,* **46**, 1038.

Lyons, L. R., Lu, G., de la Beaujardiere, O., Rich, F. J. (1996) Synoptic maps of solar caps for stable interplanetary magnetic field intervals during January 1992 Geopspace Environment Modelling campaign, *J. Geophys. Res.,***101**, 27283.

Nakamura M. S., Fujimoto, M., and Maezawa, K. (1998) Ion dynamics and resultant velocity space distributions in the course of magnetotail reconnection, *J. Geophys. Res.,* **103**, 4531.

Ogino, T., and Walker, R. J. (1984) A magnetohydrodynamic simulation of the bifurcation of the tail lobes during intervals with a northward interplanetary magnetic fields, *Geophys. Res. Lett.,* **11**, 1018.

Ogino, T., Walker, R. J., Ashour-Abdalla, M. (1994) A global magnetohydrodynamic simulation of the response of the magnetosphere to a northward turning of the interplanetary magnetic field, *J. Geophys. Res.,* **99**, 11027.

Ogino, T., and Walker, R. J. (1998) Response of the magnetosphere to a southward turning of the IMF: Energy flow and near Earth tail dynamics, in *Substorms-4,* Ed. S. Kokobun and Y. Kamide, Terra Scientific, p. 635.

Powell, K. G., Roe, P. L., Linde, T. J., Gombosi, T. I., and De Zeeuw, D. L. (1999) A solution-adaptive upwind scheme for ideal magnetohydro-dynamics, *J. Comput. Phys.,* **154**, 284.

Raeder, J. (1999) Modelling the magnetosphere for northward interplanetary magnetic field, *J. Geophys. Res.,* **104**, 17357.

Raeder, J. (2000) Reply, *J. Geophys. Res.,* **105**, 13149.

Raeder, J., Berchem, J., and Ashour-Abdalla, M. (1998) The Geospace Environment Modelling Grand Challenge: Results from a Global Geospace Circulation Model, *J. Geophys. Res.,* **103**, 14787.

Raeder, J., McPherron, R. L., Frank, L. A., Kokobun, S., Lu, G., Mukai, T., Paterson, W. R., Sigwarth, J. B., Singer, H. J., and Slavin, J. A. (2001) Global simulation of the geospace environment modelling substorm challenge event, *J. Geophys. Res.,* in press.

Richmond, A. D., and Kamide, Y. (1988) mapping electrodynamic features of the high-latitude ionosphere from localized observations: Technique, *J. Geophys. Res.,* **93**, 5741.

Robinson, R. M., Vondrak, R. R., Miller, K., Dabbs, T., and Hardy, D. (1987) On calculating ionospheric conductances from the flux and energy of precipitating electrons, *J. Geophys. Res.*, **92**, 2565.

Roe, P.L. (1981) Approximate Riemann solvers, parameter vectors and difference schemes, *J. Comput. Phys.*, **43**.

Sanders, R. H., and Prendergast, K. H. (1974) The possible relation of the 3-kiloparsec arm to explosions in the galactic nucleus, *Astrophys. J.*, **188**, 489.

Slinker, S. P., Fedder, J. A., Emery, B. A., Baker, K. B., Lummerzheim, D., Lyon, J. G., and Rich, F. J. (1999) Comparison of global MHD simulations with AMIIE simulations for the event of May 19-20, 1996, *J. Geophys. Res.*, **104**, 28379.

Song, P., DeZeeuw, D. L., Gombosi, T. I., Groth, C. P. T., and Powell, K. G. (1999) A numerical study of solar wind- magnetosphere interaction for northward interplanetary magnetic field, *J. Geophys. Res.*, **104**, 28361.

Tanaka, T. (1994) Finite volume TVD scheme on an unstructured grid system for three-dimensional MHD simulations of inhomogeneous systems including strong background potential fields, *J. Comput. Phys.*, **111**, 381.

Tanaka, T. (1999) Configuration of the magnetosphere-ionosphere convection system under northward IMF conditions with nonzero IMF B_y, *J. Geophys. Res.*, **104**, 14683.

Tanaka, T. (2000) The state transition model of the substorm onset, *J. Geophys. Res.*, **105**, 21081.

Usadi, A. Kageyama, A., Watanabe, K., and Sato, T (1993) A global simulation of the magnetosphere with a long tail: Southward and northward interplanetary magnetic field, *J. Geophys. Res.*, **98**, 7503.

Walker, R., Ogino, T., Raeder, J., and Ashour-Abdalla, M. (1993) A global magnetohydrodynamic simulation of the magnetosphere when the interplanetary magnetic field is southward: The onset of magnetotail reconnection, *J. Geophys. Res.*, **98**, 17235.

Walker, R. J., Ogino, T., and Ashour-Abdalla, M. (1998) The response of the magnetosphere to a solar wind density pulse, in *Substorms-4*, p. 527, Terra Scientific Publ./Kluwer Academic Publ.

White W. W., Siscoe, G. L., Erickson, G. M., Kaymaz, Z., Maynard, N. C., Siebert, K. D., Sonnerup, B. U. Ö., and Weimer, D. R. (1998) The magnetospheric sash and the cross-tail S, *Geophys. Res. Lett.*, **25**, 1605.

Wu, C. C., Walker, R. J., and Dawson, J. M. (1981) A three dimensional MHD model for the Earth's magnetosphere, *Geophys. Res. Lett.*, **8**, 523.

Yokokawa, N., Fujimoto, M., Yamade, Y., and Mukai, T. (2001) Hall effects on field-aligned current generation in three-dimensional reconnection, *Earth Planets and Space,* in press.

Chapter 15

The Role and Form of Modelling in Space Weather

Konstantinos Papadopoulos
Departments of Physics and Astronomy, University of Maryland
College Park, MD 20782, USA

Abstract   A critical element of the space weather research and forecasting is the development and use of models. Models vary in form, content, context and sophistication. On one extreme are the contemporary forecasting models, based on heuristic algorithms, statistical relationships, and human estimates currently in use in space weather forecasting and system engineering design. On the other extreme are physics based, dynamic, quantitative models, critical in understanding the fundamental chain of physical processes that affect the state of the sun, magnetosphere, ionosphere, and atmosphere, and their couplings and feedback; the elements that constitute and control the space weather. It is clear that physics based models, such as MHD, hybrid and particle codes, have played and will continue to play a critical role in answering research questions. The important issue at hand is to what extent physics based models can, by themselves or in combination with statistical or non-linear dynamics models, become real time forecasting tools. This tutorial will first review the types of models currently in use in space weather, with emphasis on their strengths and limitations. This will be followed by the presentation of several magnetospheric storms and substorms recorded by ground and space based instruments and compared with data driven simulations using the Lyon-Fedder-Mobarry model. These examples will be used as aides in the assessment of the utility and limitations of physics based models in space weather and in addressing issues such as validation, testing and implementation of such models as research and forecasting tools.

Keywords   Space weather, modelling, MHD, substorm, magnetic storm, LFM model, magnetosphere, ionosphere, magnetosphere-ionosphere coupling.

I.A. Daglis (ed.), Space Storms and Space Weather Hazards, 389–400.

390

1.   INTRODUCTION

The ultimate objective of space weather research is the development of quantitative forecasting models. There are two generic classes of forecasting models, system models and first principles models. System models – empirical, statistical or non-linear dynamics – are the only ones currently used in forecasting. Their main advantage is their minimal algorithmic complexity, which permits real time implementation. Their biggest problem is reliability, a generic issue to all empirical modelling. First principle models - models that rely on well founded closed sets of mathematical equations derived from well-founded physical principles - have always been important in addressing research questions. Their biggest problem is the enormous computer power required to achieve the required spatial and temporal resolution.

The critical current question in modelling is the feasibility and extent to which first principles models, by them selves or in some combination with system models, can be used as real time forecasting tools. As is obvious from previous talks in the school and the workshop, this is an actively debated issue. This issue will not be explicitly addressed in this review, because I believe that the answer depends on the specific requirements and reliability. Instead, I will present the background information that, in conjunction of the other workshop presentations, will allow you to reach your own conclusions. Time constraints force me to be selective on the issues and examples I will present, as well as parochial since for my personal convenience and availability of graphics I selected mostly examples from the University of Maryland group.

2.   ELEMENTS OF SPACE WEATHER

The critical elements of space weather are shown in Figure 1.

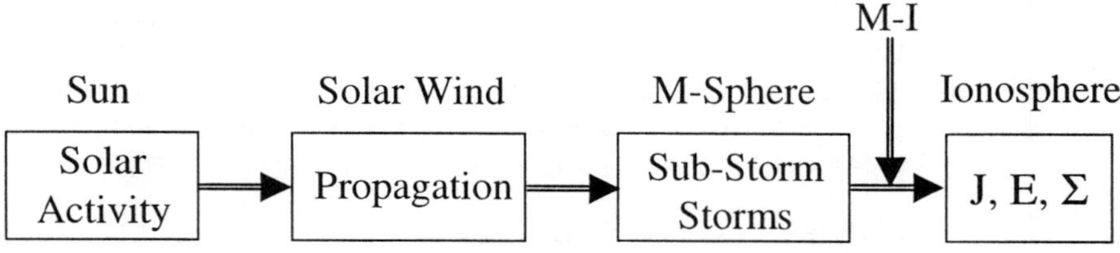

Figure 1. Critical elements of Space Weather

The cause of space weather is the active Sun that generates disturbances in the form of Solar Flares and Coronal Mass Ejections (CME). The disturbances propagate towards the earth where they interact with the earth's magnetosphere and ionosphere. The effects of space weather are manifested mostly in the radiation belts, in the form of enhanced particle fluxes, the ionosphere, in the form of strong current systems and perturbations in the electron density, and on the ground in the form of strong and dynamic electromagnetic fields. All of the components of space weather shown in Figure 1 have intellectually challenging research and forecasting questions. Previous talks at the conference addressed the first two components. In this talk I will emphasize the last two components. The strongly non-linear response of the magnetosphere to the solar wind input and its coupling and feedback with the ionosphere are perhaps the most critical elements determining the geo-effectiveness of the solar disturbance.

3.   GLOBAL GEO-SPACE MODELS – THE LFM MODEL

The development of space weather models can be guided to a great extent by the modelling effort of the International Solar Terrestrial Program (ISTP). The objective of ISTP was the quantitative understanding of the flow of energy, mass and momentum through geo-space. The Space and Plasma Physics (SPP) group at the University of Maryland was tasked with the development of the first three-dimensional (3D), global, dynamic "cause and effect" models that utilize dynamic input from the ISTP satellites. A methodology was developed that:

- Was compatible with computer resources and state of the art physics.
- Utilized as best as possible the space and ground based ISTP measurements; first as input to the code and then as test-bed of the results.

This led to the development of a 3D MHD code interactively coupled to a 2D electrostatic, height-integrated ionospheric model. The model, known as the LFM (Lyon-Fedder-Mobarry) model from the initial of its authors, is driven by dynamic input upstream provided by satellites. Subsidiary models received input from LFM to address local problems. Extensive diagnostic and visualization tools were developed to complement the effort.

The magnetospheric part of the LFM model was based on the solution of the 3D ideal MHD equations in a conservative form given below:

392

$$\frac{\partial \rho}{\partial t} + \nabla \cdot (\rho \mathbf{u}) = 0$$

$$\frac{\partial}{\partial t}(\rho \mathbf{u}) + \nabla \cdot \left[\rho \mathbf{u}\mathbf{u} + \left(p + \frac{B^2}{8\pi} \right) \mathbf{I} - \frac{\mathbf{B}\mathbf{B}}{4\pi} \right] = 0$$

$$\frac{\partial \mathbf{B}}{\partial t} + \nabla \cdot (\mathbf{u}\mathbf{B} - \mathbf{B}\mathbf{u}) = 0$$

$$\frac{\partial}{\partial t}(\rho E) + \nabla \cdot \left[\mathbf{u} \left(\rho E + p + \frac{B^2}{8\pi} \right) - \mathbf{B}(\mathbf{u} \cdot \mathbf{B}) \right] = 0$$

with

$$\rho E = \frac{1}{2}\rho u^2 + \frac{p}{\gamma + 1} + \frac{B^2}{8\pi}$$

$$\nabla \cdot \mathbf{B} = 0$$

These equations are discretized and solved on a cylindrical staggered mesh, typically 60 R_E in radius and 330 R_E long, containing the solar wind and the magnetosphere. A spider web type of computational grid places maximal resolution on critical locations. The code uses diffuse solar wind matching conditions along the outer edges of the computational domain. This allows use of time dependent solar wind parameters as input conditions. A simple supersonic outflow condition is used at the far boundary. The inner boundary condition is located on a geocentric sphere of radius 2-3 R_E, where the magnetospheric solution is matched to an ionospheric model. The ionospheric model solves the 2D height integrated electrostatic potential equation driven by the field-aligned currents within the magnetosphere,

$$\nabla \cdot \Sigma \cdot \nabla \Phi = \mathbf{J} \cdot \mathbf{B}$$

where Φ is the ionospheric potential and Σ the ionospheric conduction tensor. The $\mathbf{J} \cdot \mathbf{B}$ term represents the Magnetosphere-Ionosphere (MI) coupling. Solar UV and auroral conductance models were used for Σ. To include dipole rotation, the simulations are usually performed in SM coordinates. Details of the code and of the numerical algorithms can be found in Wiltberger (1998). Before implementation the code was successfully tested against several standard numerical problems. The code initialization is shown graphically in Visualization #1. We start with the earth's dipole magnetic field and allow the solar wind to flow and form its steady state configuration.

The LFM diagnostics and subsidiary models were tailored to the determination of the solar terrestrial energy chain shown in Figure 2.

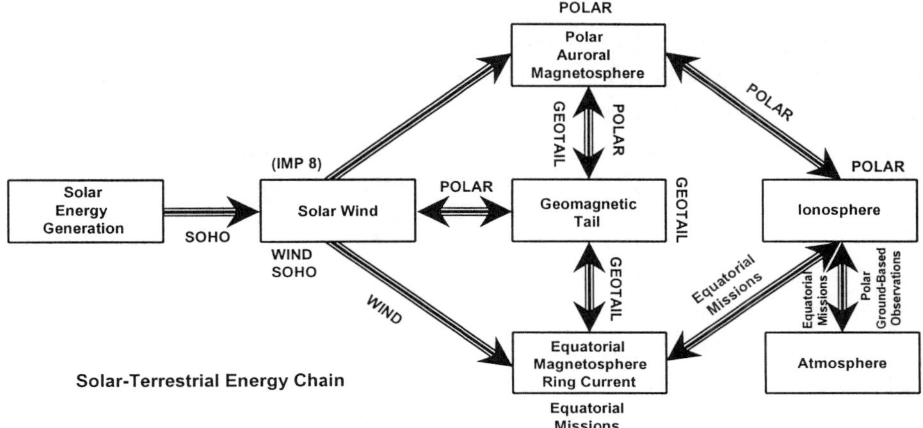

Figure 2. The solar-terrestrial energy chain: the sun's energy flows from the interior through the photosphere, corona, and interplanetary medium to the vicinity of the earth where it interacts with the geomagnetic field and atmosphere.

The main LFM code was complemented with subsidiary models that describe the radiation belts, and extensive output modules that allow comparison with measurements and assessment of metrics and extensive visualization tools. The structure of the complete model is shown in Fig. 3.

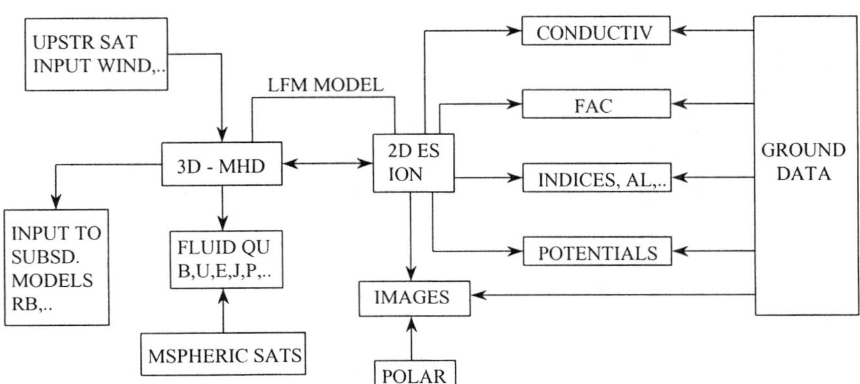

Figure 3. LFM model including subsidiary codes and diagnostic output

4.　CASE STUDIES

The first serious case study using real data was the substorm that occurred on March 9, 1995. This event occurred before the launch of the Polar satellite so the main comparison with data was through the ground magnetometer response. What was remarkable in this event was the timing accuracy and similarity between the CANOPUS magnetometer measurements and the simulations.

The event started with a long, longer than 8 hours, northward IMF, which allowed the magnetosphere to approach a "ground state" configuration. At approximately 0330 UT a rotational discontinuity reached the earth, imposing a southward IMF on the magnetospheric boundary. Shortly afterward ground magnetometers recorded signatures consistent with growth phase.

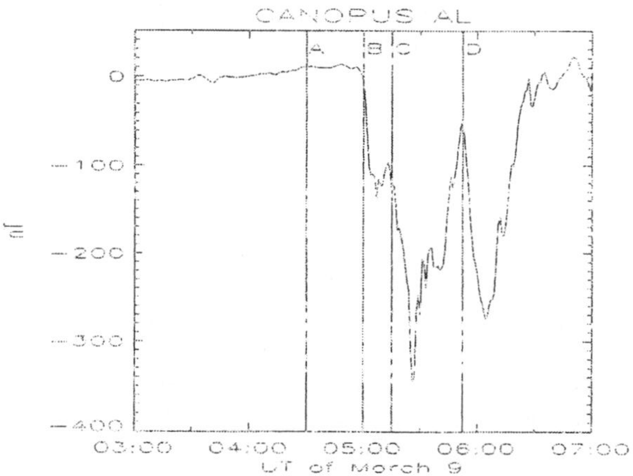

Figure 4a. CL index using CANOPUS the magnetometer array (March 9, 1995)

Figure 4a shows the CL index constructed from the CANOPUS array of magnetometers. The index shows a weak eastward electrojet associated with the substorm phase (A) at 0430 UT, onset at 0500 (B), intensification at 0520 (C), recovery from 0530 to 0552, and a second onset (D). The same features and timing are apparent in the simulation data (Figure 4b).

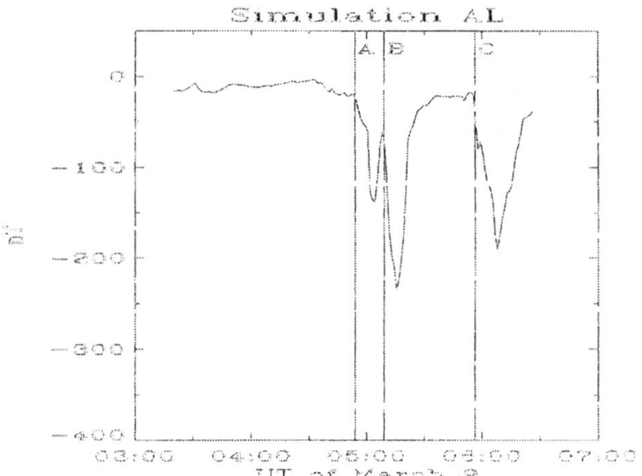

Figure 4b. CL constructed from the LFM simulations

We present next a series of visualizations (Visualizations #2—6 in the accompanying CD-ROM) from the output of the LFM simulations pertaining to this case. It is important that the visualizations be viewed in conjunction with the timing of the phases A, B, and C, seen in Figures 4 and described above. For a more detailed description of this case we refer the reader to Lopez et al. (1998) and Papadopoulos et al. (1999). Visualization #2 shows simulations describing the dynamic evolution of the surface of the last closed magnetic field lines shown as a translucent surface. On the left of each frame is the clock angle of the IMF and the arrow is proportional to the log of the ram pressure. The density is colour coded, while the arrows are flow vectors. In visualization #3 the initial state occurs at 0300 UT and is consistent with northward IMF configuration. Substorm signatures appear at 0456 and 0511. Notice in particular the appearance of hot plasma (orange surface) in the vicinity of 10 R_E. This substorm behaviour is consistent with the CL behaviour shown above and satellite measurements. The various phases of the substorm are also in evidence in the Visualization #3. This Visualization is similar to Visualization #2 but the colour plots represent the value of the electric field instead of the density. It is important to note the low value of the electric field early on. At 0456 a reconnection region forms near 30 R_E, while an enhanced electric field appears in the inner tail region (<10 R_E). This field is far from the reconnection region that is far down the tail. The Visualization shows that after the electric field impulse reached the centre of the plasma sheet it launched a tail-ward propagating signal. When the signal reaches the reconnection region enhances its rate significantly as seen from the associated flows: close to 10 R_E at substorm onset. Visualization #4 shows in a dramatic fashion the reconfiguration of the tail.

The final two Visualizations #5 and #6 show the evolution of the field aligned current and plasma flows in the equatorial plane. Notice the formation of current filaments and the current diversion towards the aurora.

A real test of the model and its capabilities occurred in January 1997. For the first time ever the ISTP satellites tracked a solar eruption, from the CME expelled from the Sun, through interplanetary space, until it encountered the earths magnetosphere, causing violent disturbances and spectacular auroral displays. The explosion started on the Sun January 6, 1997, and the magnetic cloud hit the earth on January 10. Following its arrival there were many communications disruptions and the loss of a TELSTAR satellite. The simulation of the event was a grand challenge problem and a computational coup-de-force. The simulation covered 42 hours of real time, required 160 C90 CPU hours, and produced over 6 GB of data. Details of the case are discussed in Goodrich et al. (1998) and shown graphically in Visualization #7. The format is similar to Visualization #2, except that in the lower right side the UV emission in the northern hemisphere is shown as computed by the code. There was good agreement with Polar observations. The period can be divided into three case segments. Between 0100-0300 UT the magnetosphere was in a closed tadpole like configuration. The first significant activity started at 0300 UT with a modest substorm onset at 0315 UT. The magnetosphere expanded to a more relaxed state between 0340 and 0410 UT. The arrival of the magnetic cloud marked the beginning of the next period, 0500-1500 UT, characterized by a large reduction in density and temperature. The decrease in ram pressure caused bow shock motion outwards. Strong activity occurs till 0600 UT. An approximate equilibrium occurs between 0600-0700 UT. Sudden increase in ram pressure at 0700 UT leads to increased emission. Ionospheric activity with significant intensifications correlates well with density pulses at 0730, and 0840. The last period occurred between 1500 January 10 and 0300 January 11 and was characterized by steady northward IMF. However, beginning at 2200 UT the density increased gradually reaching values between 100 and 180 $\#/cm^3$. The enormous increase in pressure, as seen clearly in the simulations, resulted in moving the dayside magnetopause well within the geo-synchronous orbit. This period is not included in the CD-ROM, but it can be obtained at the Space Plasma Physics group's website: www.spp.astro.umd.edu/Research/Mhd/mhd.htm. Notice that unlike earlier pulses this more massive pulse resulted in minimal ionospheric activity due to the northward IMF orientation.

5. CODE VALIDATION – IMPROVEMENTS

In testing the code against ground-based observations, we found that while there was relatively good agreement with the CANOPUS magnetometer chain, the results of the code were totally inconsistent with riometer data. This is evident from Figure 5a, which shows the VHF (38 MHz) absorption measured by the riometers at three different locations during 10 January 1997 (solid trace) along with the model results (dashed trace). The riometers were located in Sondersrom, Ingaluit and Gakona, Alaska. This led to further research to identify the causes of the discrepancy and improve the appropriate part of the code. In examining the times of the major discrepancies we realized that they coincided with the times that the electrojet current was very large. This led us to hypothesize that instabilities connected with the electrojet might result in turbulent electron heating caused by electrojet instabilities. The non-linear theory of the modified two-stream instability, also known as the Farley-Buneman instability, was developed by Ossakow et al (1974). Later on Schlegel and St-Maurice (1981) using radar observations confirmed that the electron temperature rises significantly in the polar electrojet during substorms. In view of this we developed a model that describes the temporally and spatially averaged electron temperature, based on non-linear physics considerations and comparison with available radar observations of the electron heating. The model gives good agreement with the observations. When anomalous heating is taken into consideration the agreement between the computed VHF absorption and observations improves considerably. This is shown in Figure 5b, where the VHF absorption measured by the riometers at the above three locations is shown by solid trace while, the model results are shown by dashed trace.

We expect that inclusion of anomalous collision frequency in the ionospheric conductance could improve the MHD model by providing feedback between the model ionosphere, serving as a dynamic boundary condition, and the 3-D global magnetospheric simulations.

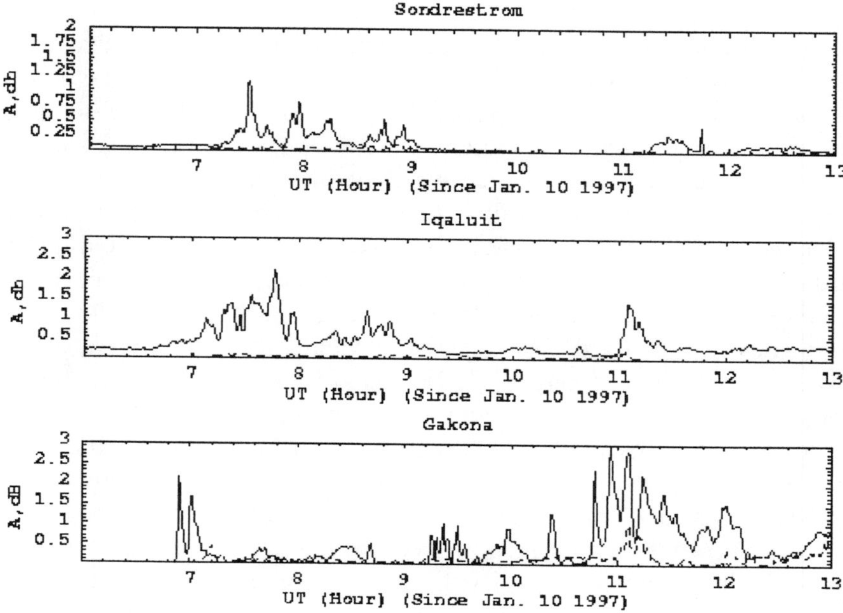

Figure 5a. Comparison of the code results with data, without turbulent heating

6. CONCLUDING REMARKS

The purpose of this tutorial was to review the progress in first-principles magnetospheric modelling during the last 20 years, using case studies. There is no question that enormous progress has been achieved from the cartoons and the steady state models of the seventies to today's data driven dynamic models. We have learned that geo-space is dynamic, and static equilibrium seldom applies. Average and statistical models are good for classification but not for forecasting. Are then first principles models ready for implementation in space weather forecasting? I hope I gave you enough information to make up your own mind. My answer is a qualified may be. It clearly depends on the objective and the required reliability. A major issue is reliability and lack of automation. It reminds me of a new chemical analysis tool, measuring blood count for example, that requires several PhD scientists to analyse and interpret. While the overall approach is promising we have not yet analysed sufficient cases to develop the proper sixth sense. Let us not forget that the art of modelling complex systems is simplification.

Getting rid of all the little details while keeping the essence. We are still keeping too many details and under-resolving important areas. We can reach this stage only by using our current tools to examine as many cases as possible.

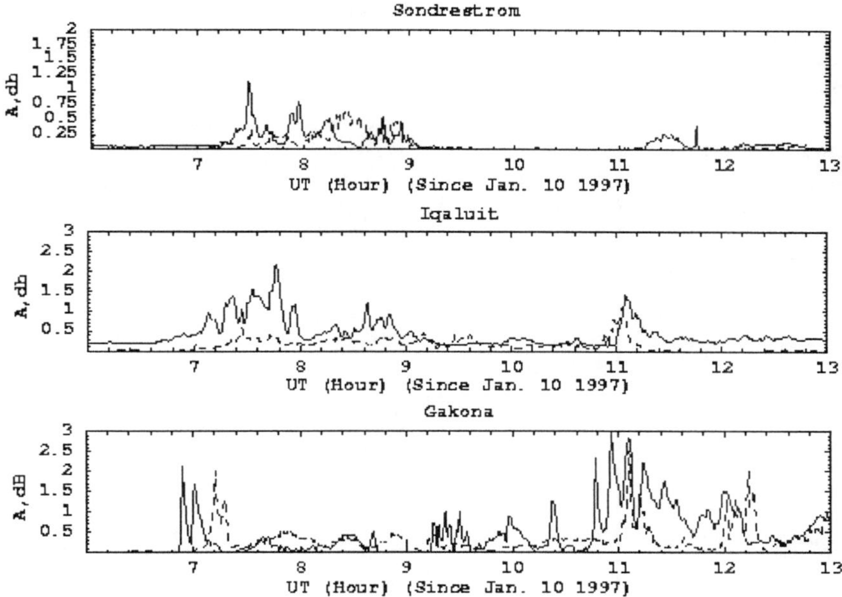

Figure 5b. Comparison of the code results with data, *with* turbulent heating

7.   ACKNOWLEDGEMENTS

While I presented the tutorial, the credit should go to the Space and Plasma Physics group at the University of Maryland that formed the core of the ISTP modelling work. I would like to thank in particular Drs C. Goodrich, M. Wiltberger, J. Lyon (Dartmouth College), X. Shao, G. Milikh, P. Guzdar, R. Lopez and S. Sharma for providing with the material of the presentation. They are the true authors of the paper.

400

8.   REFERENCES

Goodrich, C. C., W. Wiltberger, R. E. Lopez and K. Papadopoulos, An Overview of the Impact of the January 10-11, 1997 Magnetic Cloud on the Magnetosphere via Global MHD Simulations, *Geophys. Res. Lett.*, **25**, 2537-2540, 1998.

Lopez, R., C. Goodrich, M. Wiltberger, K. Papadopoulos and J. L. Lyon, Substorm Onset and Evolution: Coupling Between Tail Regions in MHD Simulations, *Physics of Space Plasmas*, 1998.

Ossakow, S., K. Papadopoulos, J. Orens, and T. Coffey, Parallel Propagation Effects on the Type I Electrojet Instability, *J. Geophys. Res.*, *80*, 141, 1975.

Papadopoulos, K., C. C. Goodrich, M. Wiltberger, R. E. Lopez and J. G. Lyon, The Physics of Substorms as Revealed by the ISTP, *Physics and Chemistry of the Earth,* **24**, 1-3, 189-202, 1999.

Schlegel, K. and J.P. St-Maurice, Anomalous Heating of the Polar E region by Unstable Plasma Waves -1. Observations, *J. Geophys. Res.*, **86A**, 1447-1452, 1981.

Wiltberger, M. S., Global Magnetohydrodynamic Simulations of Magnetospheric Substorms, Ph. D. Dissertation, University of Maryland, College Park, 1998.

Chapter 16

MHD Modelling of Space Weather Drivers

Kanaris Tsinganos
Department of Physics, University of Crete
710 03 Heraklion, Greece

Abstract   The heliolatitudinal dependence of the solar wind quantities is discussed by comparing observations with analytical and numerical modelling. First, an analytical MHD model is outlined which is obtained via a non-linear separation of the variables in the governing full set of the steady MHD equations describing the axisymmetric helicoidal magnetized outflow of the solar wind. This analysis yields three parameters, which measure the anisotropy in the latitudinal distribution of various flow quantities. The parameters are estimated by comparing observations by SoHO/CDS and *Ulysses* to the analytical solution. The solution is also checked by means of a numerical solution of the full, time dependent MHD equations which yields a steady state and in which the out-flow slightly deviates from radiality at large distances from the Sun. The calculated proton flux is compared with SOHO Lyman α observations and it is analytically explained why the proton flux of the solar wind decreases with in-creasing latitude, despite magnetic focusing which would favour the opposite trend.

Keywords   Solar wind, interplanetary magnetic field, magnetohydrodynamics, plasma physics, Ulysses, SOHO, numerical simulations.

1.   INTRODUCTION

The classical theory of the solar wind was originally based on spherically symmetric (1-D) hydrodynamic and polytropic models (Parker 1958, 1963). Magnetic effects were first introduced by Weber and Davis (1967) and 2-D effects as in coronal streamers by Pneumann and Kopp (1971). At the same time, *non-polytropic* solar wind modelling with energy and momentum addition and finite thermal conductivity has been increasingly used because observations have highlighted the fact that the acceleration of the solar wind in high-speed streams (Feldman et al. 1996) does

I.A. Daglis (ed.), Space Storms and Space Weather Hazards, 401–417.
© 2001 *Kluwer Academic Publishers. Printed in the Netherlands.*

imply energy and momentum addition in the solar corona (Leer and Holzer 1980, Steinolfson 1988, Suess et al. 1996, Hansteen, Leer and Holzer 1997). In the numerical simulation of Wang et al. (1998) a heliolatitudinally-dependent momentum addition was also included together with heat conduction and energy heating and they obtained the observed bimodal structure of the solar wind with high-speed streams above the lower-density/temperature coronal holes and low-speed streams above the higher-density/temperature helmet streamers. For a review of the dynamics of the interplanetary magnetic field see Ness (2001).

A rather different approach for analysing steady 2-D hydromagnetic wind-type outflows with energy addition in open magnetic fields was introduced by Low and Tsinganos (1986). In this study, the novel approach was to *deduce* the heating and the corresponding variable polytropic index γ which is consistent with a specific angular and radial distribution of the solar wind flow, rather than to adopt *a priori* a polytropic law with a constant γ. The density was taken as spherically symmetric and the resulting solutions showed either zero or a low terminal speed. This was also the case when this treatment was improved with the inclusion of static zones (Tsinganos and Low 1989). The explanation of this result is as follows. A dipolar magnetic field needs to be kept open by a pressure that must decrease towards the pole. If the density does not vary with latitude, there is a smaller pressure gradient to drive the flow near the pole, exactly where the magnetic field is open to allow the wind to escape. The resulting acceleration is too low, since gravity dominates, and the flow does not reach a high enough terminal speed. The only way out is to allow the density to increase with latitude, faster than the pressure does (Hu and Low 1989).

Solutions of the MHD equations with a latitudinally-dependent density and a helicoidal geometry of the streamlines were analysed in Tsinganos and Trussoni (1991). The latitudinal dependence of the different quantities was assumed *a priori*. The increase of the radial speed from equator to poles (as in the SW), and the increase of the density towards the equator (as in coronal streamers), was controlled by a single parameter. Such a choice for the density distribution was in fact motivated by earlier observations of a coronal hole at sunspot minimum (Munro and Jackson 1977). In the next section we shall *deduce* from the governing MHD equations via a nonlinear separation of the variables, the latitudinal variation of the density and other relevant physical quantities of the wind, instead of adopting *a priori* for them a specific form, as it was done in Tsinganos and Trussoni (1991).

At larger heliocentric distances, early results obtained from experiments on board Mariner 10 in 1974 (Kumar and Broadfoot 1979), Prognoz 5-6 in 1976-77 (Bertaux et al. 1985), and also by Voyager 1-2 and Pioneer-Venus, have all shown that there is more emission of Lyman α UV light of interplanetary H atoms near the ecliptic poles than predicted by an isotropic solar wind (Bertaux et al. 1997). The increase of the total ionization from the ecliptic towards the solar poles is of the order of $(20 - 40)\%$ and suggests that the solar wind proton flux increases from the solar poles towards the ecliptic, a fact fully confirmed by *in situ* observations of *Ulysses* in 1994 (Goldstein et al. 1996) and recently by the SWAN instrument on board SoHO (Kyrola et al. 1998).

The organisation of this paper is as follows. In Sect. 3 we briefly outline the MHD model of the solar wind and then in Sections 4 and 5 SoHO/CDS and *Ulysses* observations are used to constraint the model parameters. Finally in Sect. 6 we outline a time-dependent modelling of the solar wind and compare its prediction for the heliolatitudinal distribution of the proton flux with the SoHO/SWAN observations.

2.   THE MHD MODEL

In the simplest modelling, the solar wind can be regarded as an inviscid, compressible and highly conducting plasma with an axially symmetric magnetic field; thus, its dynamics can be described by the classical set of the MHD equations (Parker 1958, Weber and Davis, 1967; Sakurai, 1985; Tsinganos and Low, 1989). Since a numerical simulation of the time-dependent problem yields a steady state solution where the flow and magnetic field lines are roughly radial, at least above coronal holes at distances of several solar radii (Tsinganos and Bogovalov 2000), a reason-able assumption is to take the geometry of the stream/fieldlines helicoidal on cones of fixed opening angle. Mathematically this implies that $V_\theta = B_\theta = 0$, when using spherical polar coordinates (r, θ, φ), with θ as the co-latitude. Then, to find an analytical solution of the highly intractable set of the coupled MHD equations, we will limit ourselves to solutions that are separable in r and θ.

The technique to find a general separable solution is as follows (see, e.g., Lima and Priest 1993, Lima et al 2001). We take the r- and θ-components of the momentum equation and differentiate the first with respect to θ and the second with respect to r. By eliminating the pressure terms between these two equations we arrive at a differential equation involving functions of r and θ. Under the assumption of separation of variables, this equation can be transformed into an ordinary differential equation involving functions of r alone. For that purpose, the functions of θ in each term must be proportional to one another. The resulting solution has finally the following form

$$\rho(R,\theta) = \frac{\rho_0}{YR^2}(1 + \delta \sin^{2\varepsilon}\theta) \tag{1}$$

$$V_r(R,\theta) = V_0 Y(R) v_r(\theta) = V_0 Y(R)\sqrt{\frac{1 + \mu \sin^{2\varepsilon}\theta}{1 + \delta \sin^{2\varepsilon}\theta}} \tag{2}$$

$$V_\phi(R,\theta) = \lambda V_0 \frac{R \sin^\varepsilon \theta}{\sqrt{1 + \delta \sin^{2\varepsilon}\theta}}\left(\frac{Y_* - Y}{1 - M_A^2}\right) \tag{3}$$

$$B_r(R,\theta) = \frac{B_0}{R^2} b_r(\theta) = \frac{B_0}{R^2}\sqrt{1 + \mu \sin^2\theta} \tag{4}$$

$$B_\phi(R,\theta) = \lambda B_0 \frac{\sin^\varepsilon \theta}{R}\left(\frac{R^2/R_*^2 - 1}{1 - M_A^2}\right) \tag{5}$$

404

where R=r/r$_0$ is the radial distance normalised to the base of the wind, V$_0$, B$_0$ and ρ_0 correspond to reference values at this level for the radial velocity, radial magnetic field and density, respectively, λ is the ratio of the azimuthal and radial velocities at the base, $M_A^2=(V_r/V_A)^2$ is the Alfvénic Mach number and $V_A = B_r / \sqrt{4\pi\rho}$ is the radial Alfvénic speed.

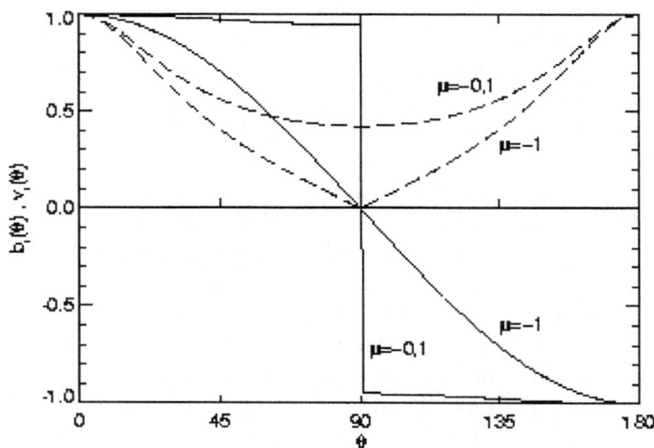

Figure 1. Co-latitudinal dependence of the radial magnetic field b$_r$(θ) (solid line) and the radial velocity (dashed line) for ε =1 and δ=4. Reproduced from Lima et al. (2001).

The dimensionless pressure Q(R, θ)=2P(R,θ)/ρ_0V$_0^2$ has the form,

$$Q(R,\theta) = Q_0(R) + Q_1(R)\sin^{2\varepsilon}\theta \tag{6}$$

where Q$_0$ represents the spherically symmetric part, while Q$_1$ includes the effects of the anisotropy.

Momentum balance in the radial and meridional directions yields then a single expression for Y(R),

$$\frac{\mathrm{d}Y}{\mathrm{d}R} = \frac{F(R)}{G(R)} \tag{7}$$

where

$$F(R) = \frac{\delta v^2}{YR^4} + \frac{4\mu}{M_{A0}^2 R^5} + \frac{2\lambda^2 Y}{\varepsilon R M_A^2(1-M_A^2)^2}x$$

$$\left[\frac{(1+\varepsilon)M_A^2 - \varepsilon}{M_A^2} \frac{R^4}{R_*^2} - (2+\varepsilon)M_A^2 + \varepsilon + 1 \right]$$

$$G(R) = -\frac{2\mu M_{A0}^2 Y}{M_A^2} - \frac{\lambda^2}{\varepsilon(1-M_A^2)^2}\left[\frac{2M_A^2-1}{M_A^4}\frac{R^4}{R_*^4}-1\right] \qquad (9)$$

v is the ratio of the escape speed to the radial speed at the base and $M_{A0}=V_0/V_0^A$ is the radial Alfvén number at the polar base.

Equation (7) yields the most general solution for the density and hydromagnetic field, under the assumptions of this work. This solution depends on three parameters that are common to magnetized wind models: (λ, v, M_{A0}). In addition, we have the three-anisotropy parameters: (δ, μ, ε). The parameter δ

$$\delta = \frac{\rho(r_0,\pi/2)}{\rho(r_0,0)} - 1 \qquad (10)$$

is related to the ratio of the density at the equator ($\theta = \pi/2$) to the density at the pole ($\theta = 0$) and so the higher it is, the more the density distribution deviates from the spherically symmetric case ($\delta = 0$). The lowest value of δ is -1, while as we shall see later δ is positive in solar polar coronal holes.

The parameter μ is related to the ratio of the kinetic energy density in the radial direction at the equator, to that at the pole,

$$\mu = \frac{\rho(r_0,\pi/2)V_r(r_0,\pi/2)^2}{\rho(r_0,0)V_r(r_0,0)^2} - 1 \qquad (11)$$

In order that the radial component of the magnetic field, B_r, decreases from the pole to the equator, μ must be negative. If $\mu = -1$, the radial component of the magnetic field and the outflow are zero at the equator. Thus the range of variation for μ is

$-1 < \mu < 0$. It is interesting to note also that the mass efflux $\dot{m}$ (θ) and the angular momentum efflux $\dot{\ell}$ (θ) increase from the pole to the equator if $\mu \neq -1$ and they are zero at the equator only for $\mu = -1$.

Finally, the parameter ε controls the width of the speed and density profile of the outflow for some fixed variation between pole equator (Lima et al., 2001). For large values of ε, the density ρ (θ) and velocity $V_r(\theta)$ have narrow profiles. Thus, the density increases rapidly only very close to the equator remaining practically constant for all other values of θ. On the other hand, $V_r(\theta)$ is uniform for all θ except very close to the equator, where it drops rapidly. In the following we shall see that the value of ε varies with the phase of the solar cycle.

A quantity of interest is the mass efflux ($\rho V_r r^2$) or, equivalently, the mass loss rate per infinitesimal solid angle $d\Sigma$ at the angle θ. From the equation of

406

conservation of mass this has to be a function of θ alone, which we shall denote by $\dot{m}(\theta)$:

$$\dot{m}(\theta) = \rho_0 r_0^2 \sqrt{(1 + \mu \sin^{2\varepsilon}\theta)(1 + \delta \sin^{2\varepsilon}\theta}\qquad(12)$$

Figure 2 shows the variation of the mass efflux with latitude. It vanishes at the equator only for $\mu=-1$.

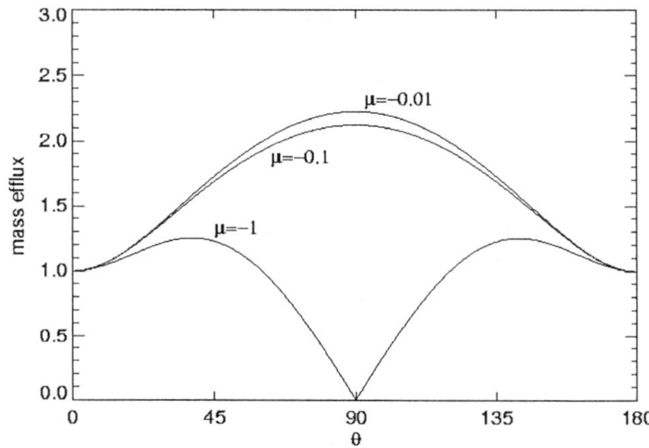

Figure 2. Co-latitudinal dependence of the mass efflux $\dot{m}(\theta)$, for $\delta=4$ for $\varepsilon=1$. Reproduced from Lima et al. (2001).

For a particular δ, if $|\mu| < \delta/(2\delta+1)$ the maximum of $\dot{m}(\theta)$ occurs for $\theta=90°$, while if $|\mu| > \delta/(2\delta+1)$ it occurs for $0 < \theta < 90°$. This is shown in Fig. 2 for $\delta=4$. In the following section 5 we shall see that a time-dependent modelling of the solar wind yields a mass efflux which increases towards the equator.

Equation (7) has a singularity at $R=R_*$, $Y=Y_*$, where $M_A=1$. In other words, a singularity appears at the spherical distance where the radial velocity equals the radial Alfvénic velocity $V_A=B_r/[4\pi\rho]^{1/2}$. This critical point corresponds to the familiar Alfvénic transition. There is a second singular point found by satisfying simultaneously $F=G=0$ in Eq. (7). A first-order analysis around this point shows that only two slopes are allowed, one positive and one negative, giving an X-type point (see also Tsinganos and Trussoni 1991). It can be shown that it corresponds to the point where the r-component of the flow speed equals the fast MHD mode wave speed in that direction, which is perpendicular to the direction θ of self-similarity (Tsinganos et al. 1996).

3. SOHO OBSERVATIONS AT THE BASE OF THE SOLAR WIND, $R \approx R_0$

In this section we will try to constrain the values of the parameters of the previous model with observations made by the SOHO/CDS instrument. These observations correspond to the base of the solar wind. In the next section we shall see how the values of the same parameters may fit observations at several AU by *Ulysses.* Finally, in section 5 we shall use the obtained values of the model parameters in a numerical simulation of the solar wind. Furthermore we shall compare the proton flux obtained from the simulation with Lyman α observations by the SWAN instrument onboard SOHO.

Fig. 3 shows a full-Sun SOHO EUV imaging Telescope Image (EIT) in the region of Fe XII $\lambda 195$, on 1998 February 26 (Gallagher et al. 1999). On the other hand, the SOHO/CDS instrument can form solar images at some particular spectral lines by moving an image of the Sun across the entrance slit with a scan mirror. Thus, using the density-sensitive line ratios Si IX $\lambda 349.9/\lambda 341.9$ and Si X $\lambda 356.0/\lambda 347.7$, we may derive the density as a function of the position angle for the south-west quadrant of the off-limb corona, over the ranges $1.00\, R_0 < R < 1.20\, R_0$.

In Fig. 4 we show three density profiles ranging from $180°$ to $270°$ over three radial intervals. These density profiles were fitted with Eq. (1) of the model discussed in Sec. 2. The optimum values of the parameters δ and ε have been evaluated by fitting the selected quiet coronal density profiles with Eq. (1). Thus $\delta \approx 1.70$ and $\varepsilon \approx 1.2$ for this particular period of the solar cycle (1998). In the next section we shall com-pare these values of δ and ε with data obtained by *Ulysses.* The *Ulysses* observations were obtained at a different part of the solar cycle, namely around solar minimum (1994 - 1995).

4. ULYSSES OBSERVATIONS AT $R > 1$ AU

To illustrate the effects of the anisotropy parameters, δ, μ, ε, and check how these parameters vary with the solar cycle, in Fig. 5 we show hourly averaged data for the radial proton speed by the *SWOOPS* ion experiment on *Ulysses* during the fast scan between September 1994 and July 1995, Goldstein et al. (1996). At this period the spacecraft was at a distance 1.3 to 2.3 AU from the Sun, the poloidal lines were radial and the radial speed was constant with radial distance over the pole, 77 km/sec while the proton density was $n_0 = 2.48\ \mathrm{cm}^{-3}$.

A least mean squares fit to these data gives for $\delta = 1.70$, $\mu = -0.50$, the high value of $\varepsilon = 9.9$, relatively higher than the value we obtained in the previous section from the SOHO/CDS data. Such a change of the parameter ε across the solar cycle can be also checked with interplanetary scintillation (IPS) data which give the solar wind speed at various latitudes (Rickett and Coles 1993). These IPS observations show that the velocity profiles at times of solar maxima are much flatter than the U-shaped curves which are characteristic of minimum activity (1976, 1987, 1997). This variation is well reproduced by the parameter ε of our model where it is found that during solar minimum conditions the value of ε is considerably higher than during the period around solar maximum (Lima and Tsinganos, 1996). The other two

408

anisotropy parameters (δ, μ) are maintained almost constant throughout the period of a solar cycle.

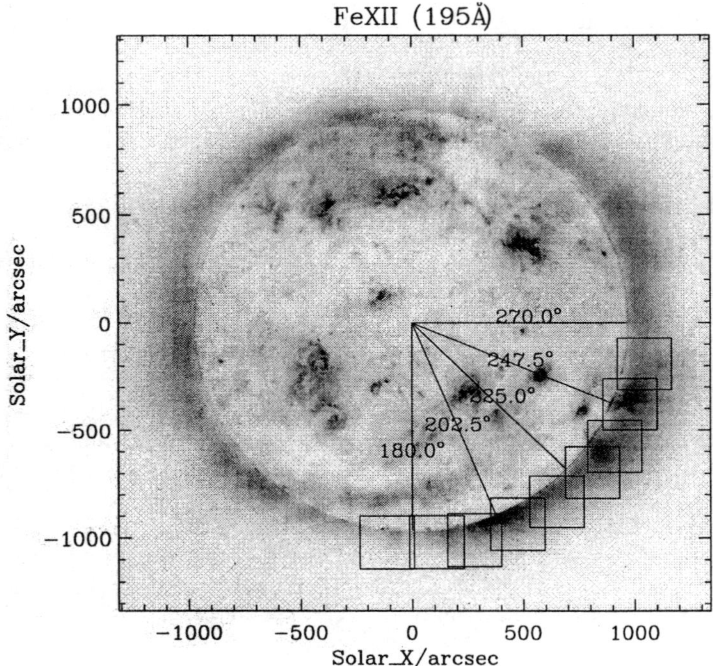

Figure 3. Full Sun *EIT* image in Fe XII (195 Å) taken on 1998 February 26, together with the 9 *CDS* raster regions studied. Reproduced from Gallagher et al. 1999.

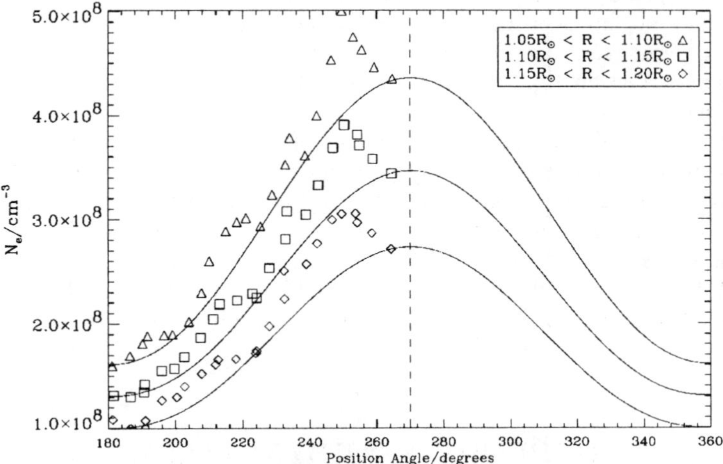

Figure 4. Comparison of electron densities at the base of the solar wind, with the model of Lima and Priest (1993), Lima et al (2000). In each case, the parameters δ and ε from Equ. (1) were evaluated using a χ^2 minimization routine. The values of δ and ε for each density profile are: δ = 1.74, ε = 1.20 for 1.05 R_0 < R < 1.10 R_0, δ = 1.65, ε = 1.18 for 1.10 R_0 < R < 1.15 R_0, and δ = 1.70, ε = 1.09 for 1.15 R_0 < R < 1.20 R_0. Reproduced from Gallagher et al. (1999).

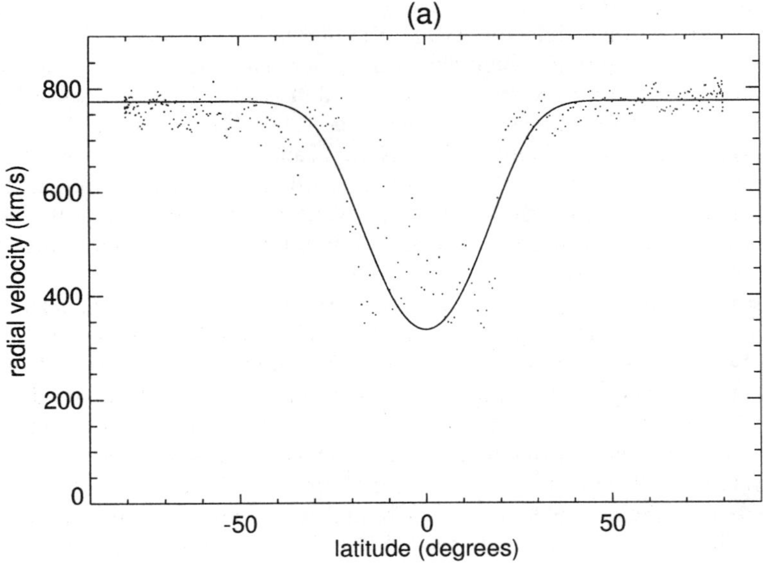

Figure 5. A fitting of the hourly averaged solar wind speed from *Ulysses* with the model of Lima et al (2000). The parameters δ, μ and ε from Eq. (2) were evaluated using a χ^2 minimisation routine.

The model presented in Sec. 2 assumes that Alfvénic surfaces of constant M_A are spherical and do not depend on the latitude. This is a somewhat restrictive assumption. However, data given by *Ulysses* show that this may be a good approximation even far from the Sun (see also Tsinganos and Bogovalov 2000 and next section).

Figure 6 shows the latitudinal dependence of the scaled M_A for radial poloidal field lines and constant velocity. This figure indicates that the assumption of M_A not de-pending on colatitude is a good one. The large scattering of the points around the equator is only due to the reversal of magnetic polarity where B goes through zero.

The above estimated values of the parameters for this model also yield a latitudinal dependence of the radial magnetic field which reproduces the right trends in *Ulysses* observations for the magnetic field (Smith at al. 1995, Forsyth et al 1996).

5. TIME-DEPENDENT MHD MODELLING

Observationally, information on the degree of latitudinal anisotropy in the solar wind at the large distances of several AU can be inferred from anisotropies in the Lyman α emission. These solar UV photons are scattered by neutral H-atoms of interstellar origin and where the SW mass flux is increased the neutral H atoms are destroyed and thus the Lyman α emission is reduced. Early observations by the Mariner 10 (Kumar and Broadfoot 1979) and Prognoz (Bertaux et al. 1985) satellites have shown that there is less Lyman α emission near the equator in comparison to the ecliptic poles, than predicted by an isotropic SW (Bertaux et al. 1997). Therefore, these Lyman alpha observations imply that the SW mass efflux should be maximum at the equator and minimum at the poles. The same trend is confirmed by *in situ* observations of *Ulysses* (Goldstein et al. 1996) and the SWAN instrument onboard of the SOHO spacecraft (Kyrola et al. 1998). However, the effect of SW collimation around the ecliptic poles (Nerney and Suess 1975, Bogovalov and Tsinganos 1999) would cause the opposite effect on Lyman α observations. In other words, although UV observations infer a SW mass efflux peaked at the equator, magnetic collimation would cause a SW mass efflux peaked at the poles, for an isotropic at the base wind.

To resolve this *paradox* we numerically simulated the time-dependent MHD equations describing the axisymmetric outflow of plasma from the magnetised and rotating Sun, with attention directed towards the collimation properties of the solar out-flow at large heliocentric distances. We have chosen to simulate the t-dependent problem in order to avoid the singularities of the steady state system at the critical surfaces. A Lax-Wendroff method on a lattice with dimension 1000 has been used to integrate the full set of the t-dependent polytropic MHD equations in the nearest zone containing the critical surfaces. Then, the steady state reached is used in order to supply the boundary conditions needed in order to solve the hyperbolic problem posed by the steady and superfast SW outflow to very large distances from the base of the outflow (Tsinganos and Bogovalov 2000).

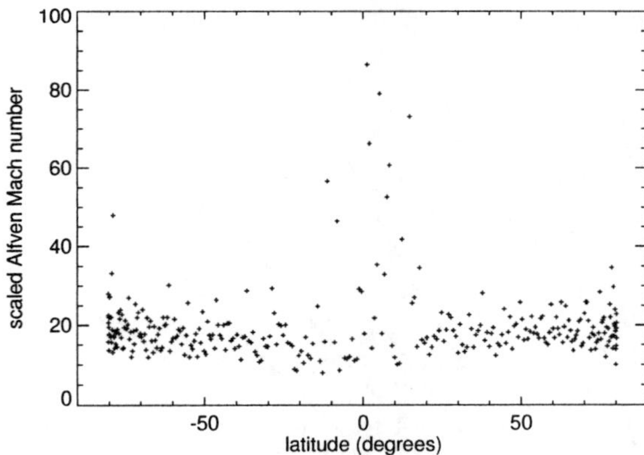

Figure 6. Scaled Alfvén Mach number as a function of latitude. The points represent daily averaged data from the *Swoops Ions* experiment on *Ulysses*, during the fast-scan (between September 1994 and July 1995). Reproduced from Lima et al. (2000).

Thus, for an isotropic at the base solar wind it is found that the poloidal streamlines and fieldlines are slightly focused toward the solar poles. However, even such a modest compression of the flow by the azimuthal magnetic field would lead to an increase of the mass flux at the polar axis by about 20% at 1 AU, relatively to its value at the equator, contrary to older and recent (Prognoz, Ulysses, SOHO) observations. But, for an anisotropic in heliolatitude wind with parameters at the base inferred from *in situ* observations by *Ulysses/Swoops* and SOHO/CDS, the effect of collimation is almost totally compensated by the initial velocity and density anisotropy of the wind.

In Fig. 7 the poloidal field lines of the SW are plotted in a logarithmic scale, to magnify their slight bending towards the axis. This logarithmic scale extends to the huge distance of $10^{10}R_{slow}$, i.e, about 4.10^8 AU $\approx$ 60 light years. This figure shows clearly that the isotropic solar wind is indeed collimated toward the axis of rotation, although this collimation is indeed very weak.

Despite the lack of the formation of a jet core in the isotropic solar wind, the lines of the plasma flow are certainly bent to the axis of rotation. And some observable effects might arise due to this bending. If the base density is isotropic, the SW mass efflux increases with the latitude θ because of the magnetic focusing by about 20% from the equator to the pole (Fig. 8a) at a distance from the Sun of about 5.8 AU.

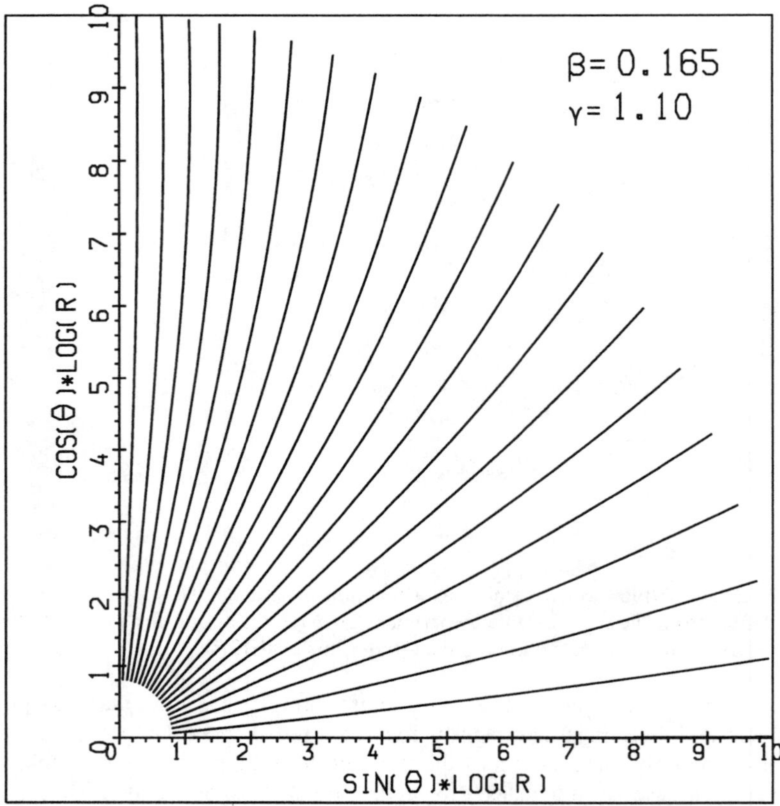

Figure 7. Shape of poloidal magnetic field lines in the far zone of an *isotropic* SW. The poloidal field lines are plotted in a logarithmic scale, which magnifies their slight bending towards the axis. Reproduced from Tsinganos and Bogovalov 2000.

Approximately at this distance the interplanetary Ly α emission is formed, as observed by the SWAN instrument on board of SOHO. This theoretical anisotropy is in contradiction with the measurements of the anisotropy of the SW at large distance from the Sun by SWAN. This discrepancy however, can be eliminated if we take into account some initial *anisotropy* in the SW.

To study in more detail the effect of the focusing of the SW, we performed calculations for a more realistic model of the SW including some initial anisotropy of the wind at its base.

For this simulation we used the latitudinal dependence of our exact MHD solution discussed in Sect. 1 with $\rho(\theta)$ and $V_r(\theta)$ given by Eqs. (1) and (2). The values of ε, δ and μ have been chosen as in Sect. 4 to best reproduce the observed values of the wind speed at various latitudes, as obtained recently by Ulysses (Goldstein et al. 1996). Their deduced values which we adopted in this study are $\delta = 1.17$, $\mu = -0.38$ and $\varepsilon = 8.6$. A similar density enhancement about the ecliptic and an associated increase of the wind speed around the poles is also found in recent MHD simulations as well (Keppens and Goedbloed 1999), while for the latitudinal variation of radiatively driven winds see Owocki et al (1996).

The distribution of the mass efflux at the distance of 5.8 AU for the anisotropic at the base SW is shown in Fig. 8b. The effect of the magnetic focusing almost totally disappears at high latitudes. Near the equator the excess of the mass efflux remains remarkable in the region below 30 degrees, although evidently the mass efflux decreases below 15 degrees. Therefore these results are in reasonable agreement with the distribution of the mass efflux of the SW which is deduced by SWAN at distances 5-7 AU.

The drastical decrease of the effect of the focusing of the solar wind in the anisotropic case can be naturally explained by the larger velocity of the solar wind at high latitudes. According to observations by *Ulysses,* the velocity at high latitudes is almost twice higher than the velocity at the equator (Feldman et al. 1996). In the following we give a brief physical explanation of how this works.

Let $(\hat{n}, \hat{p}, \hat{\phi})$ represent three unit vectors in the perpendicular directions of the normal and tangent to a poloidal fieldline and in the azimuthal direction around the symmetry axis. These unit vectors form a right-handed curvilinear orthogonal co-ordinate system and let g_n, g_p, g_ϕ denote the corresponding metric coefficients.

Evidently $g_\phi = \varpi$ where ϖ is the cylindrical distance from the axis. To calculate g_n note that the poloidal magnetic field can be defined in terms of the magnetic flux function A as

$$\vec{B}_p = \frac{\vec{\nabla} A \times \hat{\phi}}{\varpi} = \vec{\nabla} \times \left(\frac{A \hat{\phi}}{\varpi} \right) \tag{13}$$

It follows that $\varpi B_p = |\vec{\nabla} A|$ and since $g_n = |\partial \vec{r} / \partial A| = 1/|\vec{\nabla} A|$, we have $g_n \varpi B_p = 1$. In terms of the components of the current $\vec{J}$ in the system $(\hat{n}, \hat{p}, \hat{\phi})$, the Lorentz force is

$$\vec{F}_L = \frac{(J_p B_\phi - J_\phi B_p)}{c} \hat{n} - \frac{J_n B_\phi}{c} \hat{p} + \frac{J_n B_p}{c} \hat{\phi} \tag{14}$$

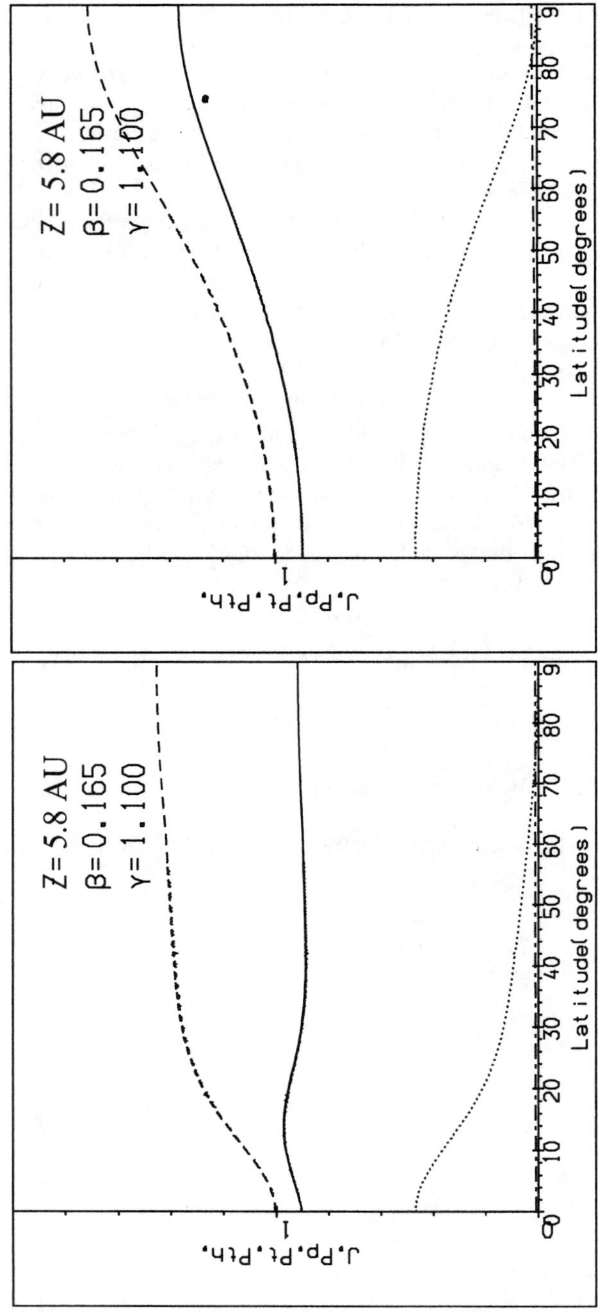

Figure 8. Distribution with latitude of characteristics of isotropic (upper panel) and anisotropic (lower panel) solar wind at 5.8 AU for $\beta = 0.165$ and $\gamma = 1.1$. Dashed lines indicate thermal pressure $P_{th}(\theta)/P_{th}$ $\theta =0$), solid lines mass flux J (θ) $=\rho$ v R^2, dotted lines pressure of toroidal magnetic field $P_t(\theta)/P_{th}(\theta =0)$ and dotted-dashed lines pressure of poloidal magnetic field $P_p(\backslash\theta)/P_{th}(\theta =0)$. Reproduced from Tsinganos and Bogovalov 2000.

The current parallel to a poloidal field line is

$$\frac{J_p}{c} = \frac{(\vec{\nabla} \times \vec{B}) \cdot \vec{B}_p}{4\varpi B_p} \tag{15}$$

or,

$$\frac{J_p}{c} = -\frac{1}{4\pi g_n \varpi}\frac{\partial}{\partial n}(\varpi B_\phi) = \frac{B_p}{4\pi}\frac{\partial}{\partial A}(\varpi B_\phi) \tag{16}$$

and

$$\frac{J_p B_\phi}{c} = \frac{B_p}{8\pi\varpi}\frac{\partial}{\partial A}(\varpi B_\phi)^2. \tag{17}$$

The second term $-J_\varphi B_p$ in the Lorentz force F_n is proportional to B_p. In the asymptotic domain and far from the Alfvén surface, this term is negligible in comparison to the $J_p B_\varphi$ term, which alone balances the inertial term of the poloidal flow, $\rho V_p^2/R_c$, where R_c is the curvature radius of the streamlines in the poloidal plane at large distances. Then we have,

$$\frac{B_p}{\varpi}\frac{\partial}{\partial A}\frac{\varpi^2 B_\phi^2}{8\pi} \sim \frac{\rho V_p^2}{R_c}. \tag{18}$$

On the other hand, the toroidal magnetic field at large distances can be estimated from the frozen in condition, i.e., $B_\varphi \sim -(\varpi\,\Omega/V_p)\,B_p$. Substituting this expression of B_φ in the previous force balance equation we obtain,

$$\frac{1}{R_c} \sim \frac{1}{8\pi\rho V_p^2}\frac{B_p}{\varpi}\frac{\partial}{\partial A}\left(\frac{\varpi\Omega}{V_p}\right)^2 B_p^2\varpi^2 \tag{19}$$

The latitudinal distribution of ρV_p is approximately constant with a ~15% increase only near the equator. Therefore the curvature radius R_c depends on V_p as

$$\frac{1}{R_c} \sim \frac{1}{V_p^3} \tag{20}$$

provided that the mass flux density is fixed. Due to this strong dependence of the effect of collimation on the velocity of the plasma, the focusing practically disappears at high latitudes where the velocity is almost twice the velocity near the equator and the distribution of the mass efflux is in pretty good agreement with the observed one. The decrease of the mass efflux below $15°$ cannot be found by the equator and the distribution of the mass efflux is in pretty good agreement with the observed one. The decrease of the mass efflux below $15°$ cannot be found by the present SWAN data analysis since it was assumed in this analysis that the mass flux can only monotonically increase with decreasing latitude.

6. ACKNOWLEDGEMENTS

This research has been supported in part by a NATO collaborative research grant CRG 972857 and a Research and Training Network grant (RTN) funded by the EEC. I am grateful to Drs. J. Lima, M. Mathioudakis, P. Gallagher, F. Keenan, K. Phillips and S. Bogovalov since many of the results discussed in this review grew out of collaborations with them. I also thank Dr N. F. Ness for a critical reading of the manuscript.

7. REFERENCES

Bertaux J.L., Lallement R., Kurt V.Z., 1985, J. Geophys. Res. 90, 1413

Bertaux J.L., Quemerais E., Lallement, R., Kyrola, E., Schmidt, W., Summanen, T., Makinnen, T., Holzer, T., 1997, The Corona and Solar Wind near Minimum Activity, ESA SP-404, 29

Bogovalov S.V., Tsinganos K., 1999, MNRAS 305, 211 (Paper I)

Feldman W.C., Phillips J.L., Barraclough B.L., Hammond C.M., 1996, in *Solar and Astrophysical MHD Flows*, K. Tsinganos (ed.), Kluwer Academic Publishers, 265

Forsyth R.J., Balogh A., Horbury T.S., Erdos G., Smith E.J., Burton M.E., 1996, A&A, 316, 287

Gallagher P.T., Mathioudakis M., Keenan F.P., Phillips K.J.H., Tsinganos, K., 1999, ApJ Lett., 524, L133

Goldstein B.E., Neugebauer M.N., Phillips, J.L., Bame, S., Gosling, J.T., McComas, D.J., Wang, Y.M., Sheeley, N.R., Suess, S.T., 1996, A&A 316, 296

Hansteen V.H., Leer E., Holzer T.E., 1997, ApJ 482, 498

Hu Y.Q., Low B.C. 1989, ApJ 342, 1049

Keppens R., Goedbloed J.P., 1999, A&A 343, 251

Kumar S., Broadfoot A.L., 1979, ApJ 228, 302.

Kyrola E., Summanen T., Schmidt W., Makinnen, T., Quemerais, E., Bertaux, J.-L., Lallement, R., Costa, J., 1998, J. Geophys. Res. 103(A7), 14523

Leer E., Holzer T.E. 1980, J. Geophys. Res. 85, 4681

Low, B.C., Tsinganos, K. 1986, ApJ, 302, 163

Lima, J.J.G., Priest, E.R. 1993, A&A, 268, 641

Lima, J. and Tsinganos, K. 1996, Geophys. Res. Letters, 23(2), 117

Lima, J.J.G., Priest, E.R., Tsinganos K. 2001, A&A, in press

Munro R.H., Jackson B.V. 1977, ApJ 213, 874

Nerney S.F., Suess J., 1975, ApJ 196, 837

Ness, N.F., 2001 (this volume)

Owocki, S.P, Crammer, S.R. & Gayley, K. ApJ Letters, 472, L115 - L118

Parker E.N. 1958, ApJ 128, 664

Parker E.N., 1963, *Interplanetary Dynamical Processes*, Interscience Publishers, New York

Pneumann G.W., Kopp R.A. 1971, Solar Phys. 18, 258

Rickett B.J., Coles W.A. 1991, J. Geophys. Res. 96, 1717

Sakurai T. 1985, A&A 152, 121

Smith E.J., Neugebauer M., Balogh A., Bame S.J., Lepping R.P., and Tsurutani B.T., 1995, Space Science Reviews 72, 165

Steinolfson R.S. 1988, J. Geophys. Res. 93, 14261

Suess J., Wang A.H., Wu S.T. 1996, J. Geophys. Res. 101, 19957

Tsinganos K., Low, B.C., 1989, ApJ 342, 1028

Tsinganos, K., Trussoni, E., 1991, A\&A, 249, 156

Tsinganos, K., Sauty, C., Surlantzis, G., Trussoni, E., Contopoulos, J., 1996, MNRAS, 283, 811

Tsinganos, K., Bogovalov S.V., 2000, A&A 356, 989

Wang A.H., Wu S.T., Suess S.T., Poletto G., 1998, J. Geophys. Res. 103, 1913

Weber E.J., Davis L.J. 1967, ApJ 148, 217.

Chapter 17

State of the Art in Space Weather Services and Forecasting

An introduction to space weather operations in the U.S.A.

Jo Ann Joselyn
Cooperative Institutes for Research in Environmental Sciences
Boulder, Colorado 80309, USA

Abstract   Space weather services are an international endeavour co-ordinated by the International Space Environment Service (ISES), which sanctions nine Regional Warning Centers positioned around the globe. In the USA, the NOAA Space Environment Center's Space Weather Operations (SWO) in Boulder plays a special role as "World Warning Agency", acting as a hub for communication. The SWO is a joint operation of the National Oceanic and Atmospheric Administration (NOAA) and the US Air Force, that is staffed 24 hours/day, 7 days/week. The Sun is a focus of attention and is monitored in a number of wavelengths (optical and radio) by ground-based observatories around the world and instruments in space. In particular, solar monitors include the X-ray sensors on all NOAA Geostationary Operational Environmental Satellites (GOES), and optical instruments onboard interplanetary spacecraft, most notably the Japanese YOHKOH and the National Aeronautics and Space Administration/European Space Agency (NASA/ESA) SOHO spacecraft. Also of primary importance, the interplanetary medium is monitored by instruments on NASA's ACE and WIND spacecraft. Energetic solar particles are observed on GOES, ACE and WIND. The geomagnetic field is observed by a network of ground-based observatories, and by magnetometers onboard GOES. All in all, over 2000 data streams received daily contribute to the assessment of the space environment. If the space environment is disturbed beyond preset thresholds, alerts are issued. There are alerts for solar X-ray events and energetic particle events, extraordinary radio sweeps and bursts, and geomagnetic storms. In 1999, SWO began issuing warnings (short-term, high-confidence predictions) of imminent geomagnetic activity a few minutes to a few hours in advance. Forecasts are made daily for each of the next three days for the probability of energetic flares, proton events, and geomagnetic storms, the expected value of the solar flux at a wavelength of 10.7 cm and two geomagnetic indices.

I.A. Daglis (ed.), Space Storms and Space Weather Hazards, 419–436.

transition of new models and data into operations. In addition, SEC works with value-added vendors who use their data and products to develop commercial space weather products.

Keywords   Forecasts, space weather services, space environment, energetic particle events, event probabilities, geomagnetic storms, verification, space environment, geomagnetic activity probabilities, solar X-ray events.

1.   INTRODUCTION TO SPACE WEATHER SERVICES

The need for space environment services became apparent during World War II, when battlefield communications using high frequencies (HF) were compromised by a natural phenomenon that was determined to be short wave fading caused by the effects of solar x-ray emission on the dayside ionosphere. Since then space weather services have developed and have steadily grown for both military and civilian systems in concert with increasing reliance on satellites and other technological systems. Because space weather is inherently global, there has always been a need for international cooperation to acquire the necessary data and expertise needed to keep pace with technical advances. This tutorial summarizes the present services offered by the Space Environment Center (SEC), which is a United States government agency within the Department of Commerce under the auspices of the National Oceanic and Atmospheric Administration (NOAA). SEC is the official U.S. source of space weather alerts, warnings, and forecasts. These SEC services are concentrated within a division named Space Weather Operations (SWO), a joint operation of NOAA and the 55th Space Weather Squadron of the US Air Force. The Space Environment Center began this daily service in 1965 by building on the foundation originally provided by the Central Radio Propagation Laboratory under the National Bureau of Standards. As the need for space weather products and alerts grew, the service became 24/7 (around the clock, every day) in 1978. In 1998 the former Space Environment Services Center (SESC) became Space Weather Operations (SWO) and in 1999 the Center moved into a new building with new state-of-the-art data display and communication facilities.

The origin of the international component of space weather services can be traced to the International Commission on Scientific Radiotelegraphy, active in 1913-14. This commission evolved into the International Union of Radio Science (URSI) by 1919. By the 1930's, "URSIgrams" were in use as a means of rapid data exchange. General studies of the relationships between solar and geophysical phenomena were undertaken in conjunction with the

International Geophysical Year, 1957-58, and by 1959 the International World Days Service was initiated to continue various aspects of the IGY World Days program. Various countries unilaterally began programs to monitor and even predict solar/geophysical events at about that time, during the maximum of the largest observed sunspot cycle - Cycle 19. By 1962, the International Union of Radio Science, the International Astronomical Union and the International Union of Geodesy and Geophysics formed a "permanent service" known as the International URSIgram and World Days Service (IUWDS) under the International Council for Science (ICSU). In 1996, the IUWDS evolved into the International Space Environment Service (ISES; http://www.sec.noaa.gov/ises). In addition to its data exchange and forecasting role, ISES provides several other services to the world scientific community including the annual International Geophysical Calendar and the monthly Spacewarn Bulletin, produced by ICSU's Committee on Space Research (COSPAR). Figure 1 is a map of ISES Regional Warning Centers. The RWC in Boulder presently functions as the World Warning Center, coordinating the daily messages from the regional centers and issuing the daily *GEOALERT* message.

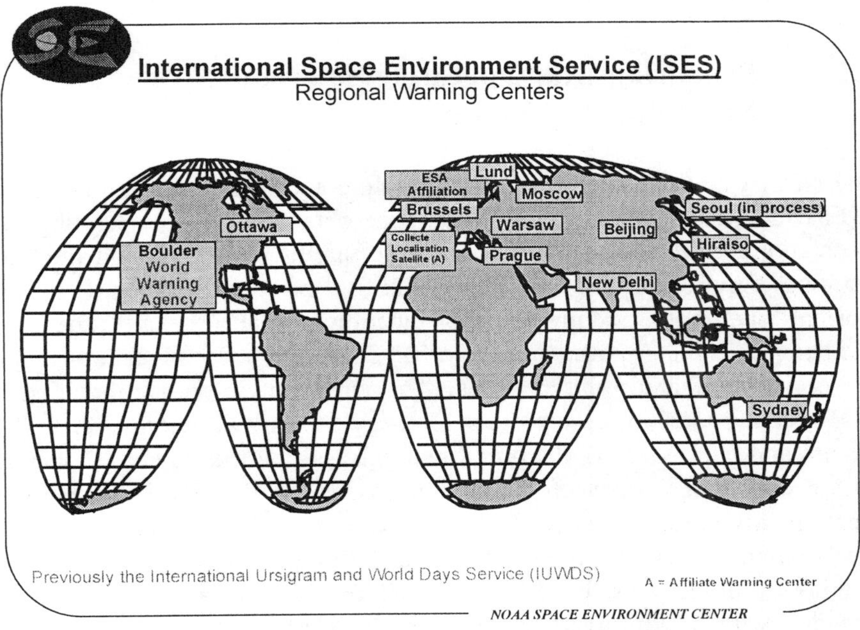

Figure 1. The Regional Warning Centers of the International Space Environment Service. Each center provides and relays data to the other centers. The center in Boulder plays a special role as "World Warning Agency", acting as a hub for data exchange and forecasts.

2. SERVICES PROVIDED BY THE NOAA SPACE ENVIRONMENT CENTER

The activities of SEC/SWO are summarized in the following list; each is described in more detail below:
- procuring and operating solar and space environment monitors
- ingesting solar and geophysical data from other national and international agencies
- interpreting and analyzing data
- alerting users of extreme/hazardous conditions
- summarizing space environment conditions
- forecasting space environment conditions
- warning of expected hazardous conditions
- conducting research into the causes and consequences of fluctuations in the space environment
- transitioning research into operational algorithms and guidelines
- educating users and the general public about the consequences of variations in the space environment
- facilitating a space weather "vendor" community.

2.1 Procuring And Operating Solar and Space Environment Monitors

The NOAA National Environmental Satellite and Data Information Service (NESDIS) has the responsibility for obtaining and operating the Geostationary Operational Environmental Satellites (GOES) and the Polar Orbiting Environmental Satellites (POES) and their on-board sensors. Normal operations require 2 fully-functional GOES at all times, one stationed over the Atlantic and the other over the Pacific, to meet the need for ordinary weather observations. The space environment monitors on GOES continuously measure the solar x-ray flux in two energy bands (0.1-0.8 nm and 0.05-4.0 nm), the omnidirectional proton flux in eleven channels with energy ranges between 0.6 and approximately 850 MeV, the energetic electron flux in three energy channels (greater than 0.6, 2, and 4 MeV), and the ambient magnetic field. Data from the GOES space environment monitors (currently GOES 8 and 10) are received on antennas located in Boulder, with backup available from the NESDIS facilities on the US east coast.

In like manner, two fully functional POES are flown, one well positioned over the USA in the morning and the other in the evening. The space environment monitors onboard POES measure energy input into the upper atmosphere by incoming charged particles. The POES program is being merged with similar Department of Defense and European satellites for cost-saving, joint operation.

In the near future, NOAA will launch and operate a Solar X-ray Imager (SXI) as an addition to the GOES space environment monitors. These images, similar to the images now being provided by the Lockheed instrument onboard the Japanese YOHKOH satellite, will arrive in Boulder at the rate of one every 10 minutes with the option of higher cadences under flare conditions. NOAA management is seeking funding for an operational solar wind monitor, GEOSTORMS, in order to assure the acquisition of real-time, upstream solar wind data now being received from the NASA Advanced Composition Explorer (ACE) mission. Because the solar wind is sampled near the "L1" position approximately 0.01 AU in front of Earth (about 4 times the distance of the Moon), international cooperation is required to receive and relay the data to Boulder for analysis. At the present time, the Communications Research Laboratory of Japan, the Rutherford Appleton Laboratory in England, and the Indian Space Research Organization in Bangalore, India are complementing the capabilities of the USA (NOAA, NASA, and the USAF) in receiving this data and sharing in the analysis.

2.2 Ingesting Solar And Geophysical Data From Other National And International Agencies

Data critical to SWO operations are provided by a number of agencies. These data vary from single numbers (e.g., the daily 10.7-cm solar flux measurement provided by the Solar Radio Monitoring Programme of the National Research Council of Canada), to 1-minute resolution vector measurements of geomagnetic fields via the US Geological Survey (http://geomag.usgs.gov) and the INTERMAGNET consortium of geomagnetic observatories (http://www.intermagnet.org). Images and analysis are provided by the USAF Solar Optical Observing Network (SOON) and cooperating solar observatories such as the National Solar Observatory (http://argo.tuc.noao.edu). Images from instruments onboard the NASA/ESA Solar and Heliospheric Observatory mission (SOHO) are relatively new and have proved to be a valuable resource (http://sohowww.nascom.nasa.gov). In particular, the coronagraph images from the Large Angle and Spectrometric Coronagraph experiment (LASCO; http://lasco-www.nrl.navy.mil/lasco.html) and the analysis provided by the NASA Goddard Space Flight Center offer useful information about potentially geoeffective solar eruptions. In all, approximately 2000 different data files (data and data-derived products) are updated each day at the SEC.

The SEC Web site documents these data and their sources, and identifies links to related space weather sites. The observed data are subsequently archived by the NOAA National Geophysical Data Center, which is also a World Data Center, coordinated under the ICSU.

2.3   Interpreting and Analyzing Data
2.4   Alerting Users of Extreme/Hazardous Conditions
2.5   Summarizing Space Environment Conditions
2.6   Forecasting Space Environment Conditions
2.7   Warning of Expected Hazardous Conditions

These five functions are performed continuously at the SWO by 9 forecasters and 7 technicians who work rotating schedules to meet requirements. Products vary from immediate telephone alerts and warnings of forecast conditions to daily summaries. Management personnel are available to assist during periods of extreme solar or geophysical activity. An example of one of the primary products, the Report and Forecast of Solar and Geophysical Activity, is shown in Figure 2. This report, issued daily at 2200 UTC (denoted as 2200 Z) combines analysis of the solar and geophysical activity with detailed evaluation of expected conditions for each of the next three days. An explanation of the forecast parts of this report is given later.

Alerts are provided directly to customers who can demonstrate an urgent need for such service. These alerts take the form of phone calls, faxes, and e-mail messages. Thresholds for alerts include:

- X-ray events exceeding M5 [5 x 10^{-5} Watts m^{-2}] and X1 [1 x 10^{-4} Wm^{-2}] levels as measured in the 0.1-0.8 nm wavelength band by instruments on GOES;
- Solar Energetic Proton levels exceeding 10 proton flux units (pfu; protons cm^{-2} sec^{-1} sr^{-1}) at greater than 10 MeV energies and exceeding 1 pfu at greater than 100 MeV energies;
- Geomagnetic K index levels[*] exceeding 4, 5, and 6 (measured every 3 hours on a US Geological Survey magnetometer located at Boulder); rapid alerts are issued for observed K's of 6 and above;

---

[*] The geomagnetic K index is a quasi-logarithmic number between 0 and 9, which is assigned at the end of specified 3-hour periods during a Universal Time day (0000-0300, 0300-0600, etc.) by measuring the maximum deviation in nanoteslas beyond expected quiet field conditions in either component of the observed horizontal magnetic field. The observed deviation is converted to a K-index value by using a table appropriate to the observing site, based primarily on the geomagnetic latitude of the site. At Boulder (and Fredericksburg, which is at the same geomagnetic latitude), measured values of 4 or less are coded as a K of 0,

- Geomagnetic A index levels[*] exceeding 20, 30, and 50;
- Solar radio bursts and noise storms reported by any observatory in a chain of solar radio observatories operated by the USAF exceeding 100 flux units [10^{-22} Watts m^{-2} Hz^{-1}] at 245 MHz, five times the background flux at 2695 MHz (10 cm), and/or Type II and Type IV radio sweep signatures;
- Magnetospheric energetic electron levels exceeding 1000 flux units at greater than 2 MeV energies;

In addition to 'alerts', 'warnings' and 'watches' are also issued as needed. 'Watches' are used to bring forecasts of hazardous conditions to the attention of vulnerable customers. Typical 'watches' are:
- forecasts of an imminent proton event;
- forecasts of geomagnetic A-index alert levels (A exceeding 20, 30, or 50).

'Warnings' are relatively new products. The intent is to provide high confidence forecasts of imminent alert conditions. An example of a 'warning' is the forecast of a K index of 5 or greater that will occur in the next tens of minutes based on solar wind data from ACE. An algorithm is presently under evaluation (see below) that assists the forecaster with the decision to issue a warning.

Recently, the SEC has initiated a new strategy to inform some customers and especially the general public about space weather conditions. The strategy is to assign a level of importance from 1 (low) to 5 (extreme) to disturbed conditions based on the anticipated severity of the effects. The detailed explanation for the NOAA Scales can be found on the SEC Web site (http://www.sec.noaa.gov). These scales are analogous to the scales used by the National Weather Service to describe tornadoes and hurricanes. Space Weather Bulletins, issued as press releases at the discretion of the forecaster, use these scales. An example of a recent Bulletin is shown in Figure 3.

Other summary products that are issued daily include the Solar Region Summary, the Solar and Geophysical Activity Summary and the Solar Coronal Disturbance Report. Once a week, a consolidated summary of activity and a general outlook for expected space weather conditions is published electronically. All of these products are described and are available for download from the SEC Web site.

---

whereas deviations of 500 nT or greater are coded as a 9. An excellent review and description of the K index and related global indices is given by *Menvielle and Berthelier* [1991].

[*] The A index is an average of the 8 daily K indices, after they have been linearized using another conversion table. The maximum value of A is 400, but storm conditions begin at a threshold of 30. Values greater than 100 for a given day are indicative of severe storm conditions.

Product: Report of Solar-Geophysical Activity
:Issued: 2000 Jun 07 2210 UT
Prepared jointly by the U.S. Dept. of Commerce, NOAA,
#Space Environment Center and the U.S. Air Force.
JOINT USAF/NOAA REPORT OF SOLAR AND GEOPHYSICAL ACTIVITY
SDF NUMBER 159 ISSUED AT 2200Z ON 07 JUN 2000
IA. ANALYSIS OF SOLAR ACTIVE REGIONS AND ACTIVITY FROM 06/2100Z TO
07/2100Z: SOLAR ACTIVITY WAS HIGH. REGION 9026 (N20W03) PRODUCED AN
X1/3B EVENT AT 07/1553Z. THIS EVENT HAD AN ASSOCIATED 200 SFU
TENFLARE, A TYPE II SWEEP WITH A SPEED OF 826 KM/S, AND A WEAK TYPE IV
SWEEP. LASCO/EIT IMAGERY ALSO OBSERVED A FAINT HALO CME WITH THIS
EVENT. EARLIER, REGION 9031 (S31W77) PRODUCED AN M2/1B EVENT AT
07/0444Z. REGION 9026 RETAINED ITS BETA-GAMMA-DELTA MAGNETIC
CONFIGURATION WITH 27 SPOTS, BUT HAS DECREASED SOME IN OVERALL
AREA. THE REGION AREA DECREASED FROM APPROXIMATELY 800
MILLIONTHS YESTERDAY TO 590 MILLIONTHS TODAY. NO NEW REGIONS
WERE NUMBERED TODAY.
IB. SOLAR ACTIVITY FORECAST: SOLAR ACTIVITY IS EXPECTED TO BE AT
MODERATE TO HIGH LEVELS. REGION 9026 IS CAPABLE OF PRODUCING MORE
M-CLASS AND ISOLATED X-CLASS EVENTS.
IIA. GEOPHYSICAL ACTIVITY SUMMARY FROM 06/2100Z TO 07/2100Z:
THE GEOMAGNETIC FIELD WAS QUIET TO UNSETTLED. THE GREATER THAN 10
MEV PROTON FLUX AT GEOSYNCHRONOUS ORBIT CROSSED THE 10 PFU EVENT
THRESHOLD AT 07/1335Z AND HAS NOT YET PEAKED (07/2100Z FLUX AT 25
PFU).
IIB. GEOPHYSICAL ACTIVITY FORECAST: DUE TO THE EXPECTED ARRIVAL OF
AN EARTH-DIRECTED FULL HALO CME FROM THE X2/3B EVENT ON 06 JUNE,
THE GEOMAGNETIC FIELD IS PREDICTED TO INCREASE DURING THE FIRST
TWO DAYS OF THE PERIOD TO ACTIVE TO MAJOR STORM LEVELS. THE THIRD
DAY MAY FIRST SEE A SHORT DECLINE IN ACTIVITY FOLLOWED BY A RETURN
TO STORMING LEVEL CONDITIONS DUE TO THE POSSIBLE EFFECTS FROM THE
X1/3B EVENT TODAY.
III. EVENT PROBABILITIES 08 JUN-10 JUN
CLASS M   70/70/70
CLASS X   20/15/15
PROTON   15/15/10
PCAF RED
IV. PENTICTON 10.7 CM FLUX
OBSERVED   07 JUN 180
PREDICTED   08 JUN-10 JUN 190/200/210
90 DAY MEAN   07 JUN 189

Figure 2a. An example of the "Report and Forecast of Solar and Geomagnetic Activity, a
plain language message issued daily at 2200 UTC by the Space Environment Center. This
product is continued in Figure 2b.

> *JOINT USAF/NOAA REPORT OF SOLAR AND GEOPHYSICAL ACTIVITY*
> *SDF NUMBER 159 ISSUED AT 2200Z ON 07 JUN 2000*
>
> V. GEOMAGNETIC A INDICES
> OBSERVED AFR/AP 06 JUN 013/016
> ESTIMATED AFR/AP 07 JUN 012/015
> PREDICTED AFR/AP 08 JUN-10 JUN 040/045-060/075-025/040
> VI. GEOMAGNETIC ACTIVITY PROBABILITIES 08 JUN-10 JUN
> A. MIDDLE LATITUDES
> ACTIVE   50/30/40
> MINOR STORM   25/40/20
> MAJOR-SEVERE STORM   20/25/15
> B. HIGH LATITUDES
> ACTIVE   30/10/30
> MINOR STORM   40/60/30
> MAJOR-SEVERE STORM   25/30/20

Figure 2b. The continuation of Figure 2a: an example of the "Report and Forecast of Solar and Geomagnetic Activity, a plain language message issued daily at 2200 UTC by the Space Environment Center.

> Official Space Weather Advisory issued by NOAA Space Environment Center
> Boulder, Colorado, USA
>
> **SPACE WEATHER BULLETIN #00- 6**
>
> 2000 June 06 at 12:28 p.m. MDT (2000 June 06 1828 UT)
>
> **** STRONG GEOMAGNETIC STORM EXPECTED ****
>
> Strong geomagnetic storm levels (G3) are expected during June 8-9. The storming is predicted due to a large coronal mass ejection (CME) that followed a major solar flare which occurred earlier today. This storm may cause some or all effects on the following: power system grids may require voltage corrections, false alarms may be triggered on protection devices, and high "gas-in-oil" transformer readings may occur; spacecraft may experience surface charging, increased drag, and orientation problems may need corrections; HF (high-frequency) radio propagation may be intermittent; intermittent low-frequency radio navigation and satellite navigation problems may occur; and the aurora may be seen as low as 50 degrees.
>
> Data used to provide space weather services are contributed by NOAA, USAF, NASA, NSF, USGS, the International Space Environment Services and other observatories, universities, and institutions. More information is available at SEC's Web site http://sec.noaa.gov or (303) 497-5127. The NOAA Public Affairs contact is Barbara McGehan at bmcgehan@boulder.noaa.gov or (303) 497-6288.

Figure 3. An example of a Space Weather Bulletin

2.8   Conducting Research Into the Causes and Consequences of Fluctuations in the Space Environment

The Research and Development Division of NOAA's Space Environment Center directly supports space weather services by conducting research into solar and interplanetary disturbances and the physical processes and effects of those disturbances in the near-Earth space environment and atmosphere. Joint projects with the University of Colorado through the Cooperative Institute for Research in the Environmental Sciences, and with the National Research Council through their Research Associateship program, also focus attention on scientific questions associated with space weather operations.

2.9   Transitioning Research Into Operational Algorithms And Guidelines

The Space Environment Center's mission expressly includes the application of research to operational needs, including both nowcasting and forecasting. SEC personnel collaborate with colleagues in universities and other government agencies to conduct research campaigns, and implement research results including new data streams. Recent test products now released for public availability (but which are not yet certified as operational) include the following:

- the "Costello" Kp index predictor that uses solar wind data and the estimated transit time from L1 to Earth to provide advice useful for issuing warnings of global geomagnetic conditions within the next few hours (accessible through links on the SEC web site or directly at http://www.sec.noaa.gov/rpc/costello/index.html).
- a predicted Magnetospheric Specification Model that calculated the expected energetic electron environment at geosynchronous orbit (accessible through links on the SEC web site or directly at http://www.sec.noaa.gov/rpc/msm/index.html).
- a predicted background solar wind speed and interplanetary magnetic field polarity at Earth using the Wang-Sheeley Model [*Arge and Pizzo*, 1999] (accessible through links on the SEC web site or directly at http://solar.sec.noaa.gov/%7Enarge/).

A product now being tested internally (not yet available on the web site) is a forecast of the probability of energetic protons (MeV energies) from solar flares. This algorithm uses the ratio of the fluxes in the two X-ray bands measured by GOES to estimate the likelihood of prompt arrival of energetic protons [*Garcia et al.*, 1999].

2.10 Educating Users And The General Public About The Consequences Of Fluctuations In The Space Environment

The Space Environment Center has prepared teaching guides and student worksheets, and provides speakers for classroom lectures. Public tours of the SWO are routine, and groups may schedule specialized tours. SEC also actively seeks feedback from customers at User Conferences [*Space Weather Week*, usually held in May] and conducts internal and external customer-satisfaction surveys.

2.11 Facilitating A Space Weather "Vendor" Community

Because SEC is not able to provide all the services that users may want, value-added vendors are encouraged to use SEC data and models to develop products that would be commercially available. Information for potential vendors is available on the SEC web site under Customer Services, and there are plans to add a directory of known vendors and their products.

3. THE STATE OF SPACE WEATHER FORECASTING

In their primary forecast product (Figure 2), SEC's Space Weather Operations describes the expected state of the space environment during each of the next three days by forecasting the following quantities in four sections (Parts III. - VI.), described below:

Part III. event probabilities
    the probability of M-class event(s)
    the probability of X-class event(s)
    the probability of a proton event (more than
        10 protons/cm^2-s-sr of greater than 10 MeV energies)
    the likelihood of a Polar Cap Absorption (PCA) event

As explained above under 'alerts', M and X-class events are determined by measuring the whole-sun radiation in the wave-length band from 0.1 to 0.8 nm using monitors on geosynchronous satellites (GOES). The Solar X-ray flux plotted on the SEC "Today's Space Weather" web page illustrates

430

this measurement as a function of time. If the peak flux exceeds 1×10^{-6} Wm^{-2} (but less than 10^{-5}) then a C-class event has occurred. An event with a peak flux 10 times larger is M-class, and 10 times larger still is denoted as an X-class event. Roughly speaking, there are about 20,000 C-class events in the period of a solar cycle, 2000 M-class events and 200 X-class events. This system of quantitatively classifying solar flare output was initiated in 1969, and was defined on the basis of radio propagation effects on High Frequency (HF; 3-30 MHz) broadcasts. X-ray radiation increases the ionization in the dayside upper atmosphere that then absorbs HF radio frequencies. Effects begin to become noticeable at C1 levels and become dramatic at M5 and higher levels by ultimately stopping all communication on those frequencies until the event subsides (minutes to hours). M and X class events are forecast by analyzing solar active region (sunspot group) characteristics. The larger and more complex a region, the more likely it is to produce the bigger events. Persistence is a factor in the forecasts - if a region is producing M or X-class activity, it is expected that it will persist in that behavior. The forecasts are probabilistic forecasts ranging from 0.01 to 0.99. Corresponding observations are either 1 (an M or X-event occurred), or 0 (no event occurred). How successful are these forecasts? Over the course of the 6-year period of 1989-1994, for M-class events the rate of occurrence of events was well forecast for probabilities of less than 30%. Probability forecasts greater than 30 % were associated with a somewhat smaller occurrence rate indicating a tendency to over-forecast. There were too few X-class events in this time period to produce a meaningful result. The most recent verification statistics are posted on the SEC Web site.

Like solar X-rays, energetic protons at geosynchronous orbit are counted onboard GOES and a 3-day plot, updated at 15-minute intervals, is available on the SEC "Today's Space Weather" web page. Proton events tend to originate with activity in the most complex and flare-rich sunspot groups and massive, fast coronal mass ejections. Roughly speaking, there are fewer that 100 proton events in a solar cycle and big ones (pfu $\geq 10,000$) are rare indeed. However, the significance of the radiation effects (radiation dose) of proton events on high-altitude aircraft, satellites, and manned missions to orbital altitudes and beyond (especially beyond the magnetosphere) mandates that forecasters consider and forecast the likelihood of these severe conditions. A contributing factor to the possibility of an energetic proton event is the physical location of the active region on the solar surface relative to Earth and the characteristics of the energy release in an accompanying solar flare. Further, it is known that a proton event or an enhancement of a proton event in progress may accompany the arrival of an interplanetary shock wave. Precursors to the arrival of the shock can be seen in the energetic electron and proton channels onboard ACE that were selected for ingest into Space Weather Operations (SWO). Forecasters are learning how to use this information to anticipate both interplanetary shocks (that subsequently become geomagnetic sudden impulses) and increased levels of energetic protons near Earth.

PCAF is an acronym for Polar Cap Absorption Forecast. There are four possible forecast conditions: green, yellow, red, and "in progress." Polar cap absorption again refers to high frequency radio communications, which are compromised on the daylit side of Earth by solar X-radiation and within the polar cap by solar energetic protons (regardless of the time of day).

Part IV. Penticton 10.7 cm flux

The 10.7 cm (2800 MHz) full Sun background radio flux is measured and reported daily by the Dominion Radio Astrophysical Observatory of the Canadian National Research Council at Penticton, British Columbia, Canada. Measurements are made at local noon, approximately 2000 UTC. Values are reported in units of 10^{-22} Watts m^{-2} Hz^{-1} and are not corrected for the variable Sun-Earth distance resulting from the eccentric orbit of Earth around the Sun. Smoothed monthly and annual averages of the 10.7 cm solar flux correlate very well with sunspot number (e.g., *Smerd,* [1969]) and this readily available and objective daily number is a well-known, albeit imperfect, proxy for overall solar output in ultraviolet wavelengths. Thus it is useful in algorithms that estimate radio propagation performance and satellite drag characteristics. Because the whole-sun 10.7-cm flux does not usually change rapidly and shows a high level of recurrence with solar rotation, it is relatively easy to forecast. Verification statistics for 1994 show that the correlation coefficient between forecast and observed daily values, one day in advance, was 0.97 [*Doggett,* 1995]. Even three days in advance the correlation coefficient was 0.88. The most salient feature of the 3-day forecast is the estimate of the trend of values – up, down, or steady.

Part V. Geomagnetic A indices

There are two geomagnetic A indices that are forecast: the A index (AFR) for the geomagnetic observations at Fredericksburg, Virginia, and the globally averaged A_p index (AP) that is reported by the Adolf-Schmidt-Observatorium für Geomagnetismus, GeoForschungsZentrum Potsdam, Germany. The first line of this section reports the indices for yesterday. The value for Fredericksburg is calculated from the 8, 3-hourly K indices provided at the end of the day by the USGS, and the value for AP is calculated from 8, estimated K indices that are determined from a set of proxy geomagnetic observatories located near geomagnetic latitudes near 50 degrees with data that are accessible in near-real-time. Again, it must be emphasized that the forecasts are for the Potsdam A_p index, but the "observed" AP reported in this product is not the actual one, but only an estimate because of the necessity for immediate access to a global measure of geomagnetic activity. Because the product is issued before the end of the UTC day, the next line of this section reports estimated values for the AFR and the AP indices. Finally, the predicted values for AFR/AP are given sequentially for each of the next 3 days.

Because the AP (estimated Potsdam A_p) reports and forecasts are the purview of the USAF, SEC/SWO depends on the Air Force Weather Agency (AFWA) to take the responsibility to validate these parameters. However, SEC/SWO does keep verification statistics on the Fredericksburg predictions. The latest reported statistics are soon to be updated on the SEC Web site, but the following table from Doggett [5] offers a general picture of the state of the art of geomagnetic forecasting, a subject that was examined in detail in *Joselyn* [1995]. For the purposes of verification, a "storm" is defined as a value of 30 or greater on a UTC day.

OBSERVED

		Storm	No Storm	Total
FORECAST	Storm	57	93	150
	No Storm	149	2623	2772
	Total	206	2716	2922

Table 1. Contingency table of geomagnetic storm/no storm forecasts for conditions observed at Fredericksburg, Virginia, from 1987 through 1994, one day in advance.

The "hit rate," that is, the percent of the time that a storm was predicted and one occurred <u>and</u> that no storm was predicted and none occurred, is 92%. However, only 28% of the storms that occurred were predicted, and 62% of the storm predictions were false alarms. Statistics are available for whole years from 1988 through 1999. For this period, Figure 4a shows the number of storm days that occurred, and the number that were correctly forecast one day in advance. Figure 4b shows the number of storms that were forecast, and the number of those that were correct. For four of the twelve years, no observed storms were correctly forecast. For three of the 12 years, all predicted storms were false alarms, including the most recent year, 1999, for which both ACE and SOHO/LASCO coronagraph data were available. Generally, more storms are observed than are forecast. Unfortunately, it often happens that a storm is forecast, but the timing is wrong; that is, the storm occurs one day earlier or later than expected. This situation produces a miss <u>and</u> a false alarm in the statistics. Because success is elusive, the message is clear that there is much yet to be learned about the diagnostic value of solar observations for geomagnetic activity. (Solar wind data from L1 are useful for short-term, accurate *warnings*, but not for geomagnetic storm *watches* that are produced a day or more in advance.)

Doggett [1995] showed an analysis for multiple-day forecasts (accuracy decreased with lead-time for the years examined), and some "skill score" results that compare forecast accuracy with simple forecasting schemes such as climatology, persistence or recurrence (see also *Joselyn* [1995]). Usually, but not always, on an annual basis forecaster skill is better than the simple schemes, especially for the years of high activity during cycle maximum.

Part VI. Geomagnetic activity probabilities

When a geomagnetic disturbance is expected late in a UTC day, the forecaster needs a way to predict an appropriate A index and still indicate the possibility of a disturbance. Also, because high (auroral zone) latitudes are usually relatively more disturbed than the sub-auroral, middle-latitudes of Fredericksburg and the AP network of geomagnetic observatories, separate high and middle-latitude forecasts are needed. In this section, the probability of isolated conditions appropriate to the activity categories of active, minor, major, and severe are expressed. [There are two other categories of activity: quiet and unsettled; typical geomagnetic conditions are quiet.] The probabilities are given for each of the next three days so that expected trends in activity can be communicated. A forecast verifies if a K index appropriate to the identifier occurs at any time during that day (e.g., a K of 5 is typical of minor storm conditions and 6 is typical of major storm conditions). For example, in Fig. 2, the probability forecasts for active, minor storm and major/severe storm conditions for middle latitudes for 8 June 2000 are given as 50/25/20, which adds to 95%. Therefore, the forecaster is virtually certain that a K index of 4 or greater will occur, and a K of 5 or larger is expected at the 45% level (25 + 20). But the forecaster expected the storm to peak on the following day, June 9, as shown by a 65% probability of storm conditions for middle latitudes (40 + 25), and a 90% probability of storm conditions at high latitudes (60 + 30). The trend in the forecast values shows that the storm was expected to subside on the third day. This trend can also be seen in the predicted indices for AFR and AP, which were 40 and 45 for the 8th, 60 and 75 for the 9th, and 25 and 40 for the 10th.

434

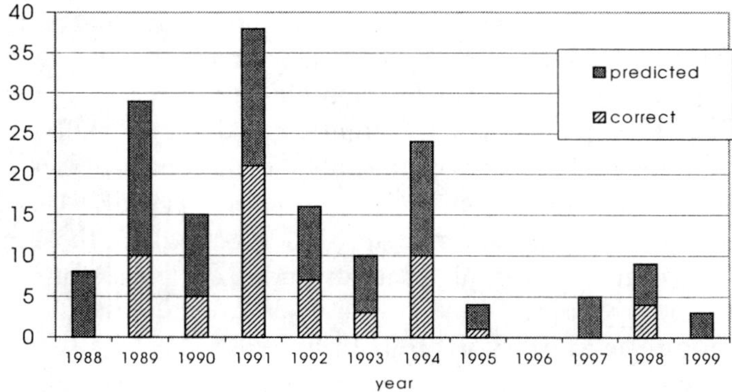

Figure 4a. Number of magnetic storms observed and predicted for Fredericksburg, Virginia, 1988-1999.   227 storms occurred but only 61 were correctly predicted.

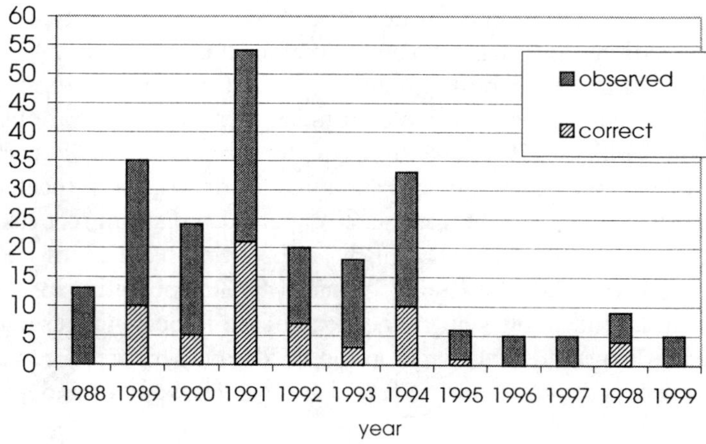

Figure 4b. Geomagnetic storms predicted for Fredericksburg, Virginia from 1988-1999. 161 storms were predicted but only 61 storms occurred. The other predictions were false alarms.

What actually happened? In fact, the storm did arrive on June 8th (a Storm Sudden Commencement was reported at 0909 UTC) but it was more intense than predicted and it subsided quickly. The AFR for the 8th was 34 (a minor storm and therefore a forecast "hit" as defined in Table 1) but on the 9th it was only 04 (indicative of quiet conditions) resulting in a false alarm. The estimated AP index for 8 June based on data available in real-time was 53; the official A_P, which was not available until the end of the month, was 64. Both of these values are consistent with major storm conditions, somewhat greater than the minor storm that was predicted. The observed K indices on the 8th at Fredericksburg [2445 6444] are consistent with a major storm because of the K of 6. Thus, the forecaster gets partial credit for the forecast because conditions were active or greater, but a higher probability of major storm conditions would have verified better (see *Doggett* [1995] for details of the ranked probability score verification algorithm used for probability forecasts). At College, Alaska (a high-latitude station), the observed K's were [2367 7764], consistent with a severe storm. Again, the probability forecast implied lower activity than was observed but the 65% forecast of storm conditions did correctly warn of the storm.

4.   SUMMARY

The NOAA Space Environment Center in Boulder, Colorado, operates an around-the-clock service center, Space Weather Operations, jointly with the US Air Force 55th Space Weather Squadron. The SWO strives to assimilate and analyze appropriate solar and geophysical data and to accurately describe and forecast the space environment, with special consideration of requirements to support those agencies and industries of strategic importance to US national security and economic health. To achieve this goal, SWO seeks out and appreciates international partners who share critical data and expertise, and encourages a growing vendor community that is able to use profitably SWO data and products to meet specific societal needs. The Space Environment Center conducts scientific research to expand and improve capabilities and draws from the larger body of worldwide scientific study and physical models. The SEC takes seriously a mandate to educate its constituency and the general public regarding space weather.

436

5.   ACKNOWLEDGEMENTS

The author is grateful to the staff of the NOAA Space Environment Center for support and useful information. In particular, Kent Doggett provided verification information and Joe Kunches, lead forecaster at NOAA Space Weather Operations reviewed this material and offered many helpful suggestions.

6.   REFERENCES

Arge, C.N., and Pizzo, V.J. (1999) Historical verification of the Wang-Sheeley Model: Further improvements in basic technique, in S.Habbal, R. Esser, J.V. Hollweg, and P.A. Isenberg, (eds.), *Solar Wind 9: Proc. 9th International Solar Wind Conference*, American Institute of Physics, Woodbury, N.Y., pp. 569-572.

Doggett, K.A. (1995) Verification of NOAA Space Environment Laboratory Forecasts: 1 January-31 December, 1994, NOAA Tech. Memo ERL SEL-90, Boulder, Colorado, 1995.

Garcia, H.A., Greer, M.S., and Viereck, R.A. (1999) Predicting solar energetic particle events from flare temperatures, in *Proc. 9th European Meeting on Solar Physics*, ESA-SP-448, pp. 983-986.

Joselyn, JoAnn (1995) Geomagnetic activity forecasting: The state of the art, *Rev. Geophys.* **33**, pp. 383-401.

Menvielle, M., and Berthelier, A. (1991) The K-derived planetary indices: Description and availability, *Rev. Geophys.*, **29**, pp. 415-435. (Correction, 1992, *Rev. Geophys.*, **30**, p. 91).

Smerd, S.F. (1969) Solar radio emission during the International Geophysical Year: 6. Some statistical results, *Ann. I.G.Y.*, **34**, pp. 331-357.

Chapter 18

Space Weather: An Air Force Research Laboratory Perspective

Gregory P. Ginet
Space Vehicles Directorate, Air Force Research Laboratory
AFRL/VSBX, 29 Randolph Rd., Hanscom AFB, MA 01731, USA

Abstract   From the energetic induced particle degradation of spacecraft microelectronics to the scintillation of trans-ionospheric communication links, the impacts of space weather have become increasingly recognized within the civilian and military space, communications, navigation, and power distribution communities. Several government agencies in the United States have been working together under the auspices of the National Space Weather Program (NSWP) to guide research and development in order to meet the space weather requirements of today's space-dependant society. The Air Force Research Laboratory (AFRL) has long been developing technologies to specify, forecast, and mitigate the hazards of the space environment with specific emphasis on the impacts to military systems. We present an overview of military space weather needs in the context of the National Security Space Architects Space Weather Architecture and the NSWP. The AFRL research program is then described and several recent accomplishments discussed.

Keywords   Space weather, solar, interplanetary, magnetosphere, ionosphere, radiation belts, scintillation, atmospheric drag.

1.   INTRODUCTION

The title and content of this paper differ to some degree from what was presented by the author at the NATO Institute on 29 June 2000. A general description of the United States' multi-agency National Space Weather Program was given as part of the presentation at the institute but will not be given here. Readers are referred to the recently released National Space

I.A. Daglis (ed.), Space Storms and Space Weather Hazards, 437–457.

438

for Meteorology, 2000) for a comprehensive exposition of the program. In this paper we focus on the other component of the author's presentation, namely, a description of the military relevance of space weather and the efforts of the Air Force Research Laboratory (AFRL) and its collaborators to perform the research and development needed to meet current and future requirements.

The AFRL Space Weather (SWx) Center of Excellence was formed in March 2000 by combining several branches of the Space Vehicles Directorate concerned with various aspects of space environment. The SWx Center executes a coherent basic research to operational prototype program in space weather aimed at meeting military space weather technology needs as efficiently and effectively as possible. The Air Force Office of Scientific Research (AFOSR), responsible for funding Air Force basic research, works together with the SWx Center to fund and execute laboratory tasks and provide grants to other government agencies, universities, and industry. However, the solar-terrestrial system is not small and the time scales important for space weather range from seconds (solar flares) to decades (the solar cycle). It is impossible for a single government agency or even a single nation to affordably cover the territory necessary for adequate specification and forecast, in terms of both data coverage and intellectual horsepower required to build field models. Crucial to meeting military space weather requirements, or anyone else's requirements, are the many joint programs and collaborations between US government agencies and other nations (e.g. Britain, European Space Agency, Australia, Germany, Canada, and Japan).

In the remainder of this paper we discuss the military relevance of space weather and then outline AFRL's current research directions, selected recent accomplishments, and future plans.

2. MILITARY RELEVANCE

Space weather impacts many military mission areas. Working closely with the National Oceanic and Atmospheric Administration's Space Environment Center (NOAA/SEC) the USAF 55th Space Weather Squadron and the Air Force Weather Agency provide a variety of real-time alerts, warnings, and post-event analysis customized for military users. The space weather phenomena of most interest are:

<u>Ionospheric Scintillation</u>. Turbulence in the Earth's ionosphere generates electron density irregularities in the active equatorial and polar regions. These variations scintillate radio signals, much like tropospheric density fluctuations scintillate or 'twinkle' star light, and can cause amplitude and phase fluctuations in ground-to-space links. Satellite UHF communications

and GPS navigation signals can be severely affected, even knocked out completely, for several hours.

Global Electron Density Profiles. Changes in the profiles of ionospheric electron density around the globe as a result of solar activity and geomagnetic storms cause variations in the total electron content (TEC) along ground-to-satellite links. Variations in the TEC alter signal propagation times with negative impacts on the accuracy of single frequency GPS receivers and geo-location systems. Dynamic electron density profiles also affect high frequency (HF) radio communications, which relies on the natural wave-guide formed between the Earth's surface and the bottom side of the ionosphere to propagate signals over long distances, and can generate clutter for space surveillance radars.

Space Radiation. Energetic particles trapped in the Earths' radiation belts, hot plasma injected towards Earth from the magnetospheric tail, solar energetic particles and cosmic rays penetrating into near-Earth space generate problems for satellite systems, astronauts, and high-flying airplane crews. Hazards include single event upsets in microelectronics, spacecraft frame and internal charging and discharging, and total dose degradation of both hardware and human tissue.

Neutral Density. Intense solar photon events and geomagnetic storms can cause substantial heating of the upper regions of the Earth's neutral atmosphere raising the average density at a given altitude and increasing the drag on satellites. Orbit prediction, collision avoidance and satellite tracking missions become more difficult as a result of an increased and often rapidly changing drag force.

The four primary areas listed above are the ones with a direct impact on military operations. However, the space environment is a strongly coupled system and any requirements to forecast global electron density profiles, for example, necessitate the specification and forecast of behavior 'upstream', i.e. in the magnetosphere, interplanetary medium, and the Sun itself. Solar and interplanetary observations and forecast models are critical to most forecasts of near-Earth effects. Of great national concern, but considered a civilian responsibility, is the stability of the national power grid which is susceptible to transformer outages arising from ground currents induced by geomagnetic storms.

To comprehensively understand and address the nation's current and future space weather needs the National Security Space Architect (NSSA) recently undertook a Space Weather Architecture Study (National Security Space Architect, 1998) involving all the government agencies with a stake in space weather, to include the Department of Defense (DoD), the Department of Commerce (i.e., NOAA), the National Science Foundation, the National Aeronautics and Space Administration (NASA), and the Federal Aviation Administration Motivated by the National Space Weather Program's Strategic Plan, the study emphasized DoD concerns but with full realization that effective and efficient space weather services are very much a joint military, civilian, and international effort. An architecture vector was developed to guide the acquisition, deployment, and operations of space weather systems that will meet the perceived requirements in the 2010-2025 timeframe. Many AFRL concepts and prototype systems, for example the Compact Environment Anomaly Sensor (CEASE) and the Scintillation Network Decision Aide (SCINDA), are part of the architecture. The NSSA also supported the need for increased space weather importance awareness out in the field and for a robust space weather research and development program. Consequent to the architecture's approval by high-ranking DoD officials in May 1999 a Transition Plan has developed and DoD is working with the NSWP to track the implementation.

As the maximum of the 11-year sunspot cycle continues to be felt in 2000-2003 the interest in space weather from the users has been growing, despite the fact that the current cycle (cycle 23) does not appear as active as previous cycles have been. For example, there has been a strong military interest in scintillation effects on communications in the equatorial regions and two new SCINDA sensors have been installed at the request of operational commands. On 14 July 2000 there was an extremely strong solar proton event (the 'Bastille Day' event) that disrupted instruments on several scientific satellites, rendered useless one Japanese scientific satellite, and probably degraded solar cells on many more DoD and civilian systems. As space weather awareness increases and space weather training and products become part of the DoD acquisition, planning, and operations infrastructure, databases will be built to statistically capture space weather effects and further refine the specification, forecast, and mitigation requirements.

3.   CURRENT RESEARCH DIRECTIONS

Space weather research and development at AFRL occurs in four main areas:

<u>Space and ground-based environment sensors</u>. Data is fundamental to any space weather specification or forecast capability. With the vast regions of space and time to cover, the relatively high costs of space instrumentation, and the hard realities of putting ground sensors in often remote or politically unstable situations, it is not surprising that the space weather field is data starved. Low-cost, miniaturized, high resolution and high dynamic range

starved. Low-cost, miniaturized, high resolution and high dynamic range sensors with as much autonomous operational capability as possible are needed for both ground and space platforms.

<u>Real-time specification and climatological model development.</u> Data is usually sparse and does not represent the exact quantity relevant to system impacts. Empirical and physics based models are needed that can assimilate different data sources, interpolate, and transform outputs to produce real-time 'snapshots' of practical value. Users in the acquisition and modelling & simulation (M&S) communities need statistical and climatological models that can often be constructed from appropriate sequences of historical snapshots.

<u>Forecast model development.</u> Given data-derived initial and boundary conditions, forecast models predict future states of the environment by using physics-based or empirically derived algorithms. Forecasting is perhaps the most challenging R&D area because a) the physics of the subject phenomena usually has to be understood at a level deeper than that needed for specification and b) the difficult problem of coupling different phenomena across space weather domains (e.g. ionosphere-magnetosphere-interplanetary) increases in proportion to the lead time required.

<u>Passive and active techniques for effects mitigation and exploitation.</u> Impacts on the technologically sophisticated systems effected by the space environment are often complex and require considerable analysis to quantify. Such quantification is necessary in order to determine what key space weather parameters need to be specified and forecast. Active and passive techniques that can mitigate hazards, such as design tools, hardened components, and charge control systems, are most desirable since they lead to 'all weather' capabilities.

Work in these areas is performed and synthesized into products along the specific paths outlined below. Selected recent accomplishments are presented in more detail in Section 4.

To meet ionospheric scintillation specification requirements the core of the Center's program in the last several years has been the development of the Scintillation Network Decision Aide, a set of ground based sensors deployed in the equatorial regions to monitor scintillation effects on satellite communication links (Groves et al., 1997). SCINDA assimilates the measurements into a regional scintillation model that provides real-time specification and short-term forecasts valid up to ~1/2 hour. The creation of SCINDA has been made possible by extensive basic research in the development and evolution of equatorial irregularities carried out in the last 20 years. Current efforts include extensive ground measurement campaigns using all-sky cameras, specialized multi-band transmitters and receivers, and radars targeted at quantifying scintillation effects on GPS navigation links and determining the source and evolution of scintillation in the complex polar regions.

Military operations in the equatorial regions demand several hour forecasts of scintillation effects. Scintillation near the equator arises from the plasma turbulence associated with density depletion regions, or 'plumes', in the nighttime F-region near the sunset terminator. Non-linear numerical plasma dynamic codes and theoretical analysis techniques are being developed to study turbulence and determine the exact relationship between global ionospheric and thermospheric drivers on one end of spatial spectrum and the resultant scintillation effects on the other. Of special importance is the subtle problem of scintillation triggers: what exactly induces the high day-to-day variability? A comprehensive analysis of satellite data taken in the upper ionosphere and magnetosphere during the last solar maximum has established a relationship between specific phases of geomagnetic storms and the onset of the scintillation-generating ionospheric instabilities. Models being developed for scintillation forecasting will ultimately become part of the Communications/Navigation Outage Forecast System (C/NOFS), described in the Future Research Directions section.

Specification of global electron density profiles is essential in order to construct global maps of arbitrary point-to-point TEC variations and the transmission properties of the 'Earth-Ionosphere' waveguide. Over the years AFRL has developed and deployed a series of sensors that are essential for electron density specification and used by both the operational and scientific communities. Ground sensors include the Digital Ionospheric Sounding System measuring bottomside electron density profiles at over 17 locations worldwide and the Ionospheric Monitoring System measuring TEC at four operationally relevant sites. Efforts are underway to validate sensor performance, improve the accuracy, reduce size and cost, and improve the autonomous operating capabilities. Space sensors include the SSIES plasma drift meter, the SSJ4/5 keV particle spectrometer, and the SSM magnetometer flown on the DMSP satellites. Building the Digital Ion Drift Meter (DIDM), a much smaller, lighter, and improved dynamic range version of the SSIES, has been a major focus of AFRL experimenters and they have recently been rewarded with a successful on-orbit test.

The primary thrusts of the global electron density modelling work have been on improving the data assimilation aspects and validating the performance of the Parameterized Real-time Ionospheric Specification Model (PRISM), a model currently operational within the DoD (Daniell and Brown, 1995). Basic PRISM climatology has been found to be equivalent to other major ionospheric models. Inclusion of real-time data from ionosondes, TEC sensors, and satellites has been shown to provide

density profiles determined from UV measurements, which will be available from the SSUSI and SSULI instruments soon to be flown on DMSP. Together with the Office of Naval Research (ONR) AFOSR is sponsoring a Multi-Disciplinary University Research Initiative (MURI) on Ionospheric Data Assimilation. Two groups, headquartered at the University of Southern California and Utah State University, respectively, are developing new models based on Kalman filtering and variational analysis techniques.

In the electron density forecasting arena, it has been recognized that accuracy and lead-time improvements to existing physics-based ionospheric forecast models, such as the Coupled Ionosphere-Thermosphere Forecast Model, will only come with a better initial specification and a better knowledge of magnetosphere-ionosphere coupling (Borer et al., 1995). Extensive analysis of data from the CRRES, DMSP and NASA International Solar-Terrestrial Physics Program satellites (e.g. ACE, WIND, and FAST) is being carried out to understand and model the effects of field-aligned currents and voltages on ionospheric properties, the triggering of storm and substorm disturbances, and the propagation of electric field disturbances through the interplanetary–magnetospheric–ionospheric system.

Quasi-empirical techniques for specifying and forecasting the neutral density effects on satellite drag developed by AFRL have undergone extensive validation in the last two years. The Modified Atmospheric Density Model revolutionized orbit tracking by applying intelligent feedback from empirical drag data for satellites in stable orbits to correct the overall atomospheric density estimate error (Marcos et al., 1998). AFRL has shown that determination of ballistic coefficients for a set of operationally relevant satellites was improved, on average, by 200% and computational variability was greatly dampened. The orbit prediction error of LEO satellites, persistently at 50% for the last 30 years, was consistently reduced to below 5%. Complementing the empirical approach, AFRL works toward an improved physical understanding of neutral density variations through improved measurement campaigns, such as the proposed Atmospheric Density Specification mass spectrometer experiment, and theoretical model development. Physics-based models are ultimately needed to forecast short-term orbit variations that occur during large geomagnetic storms.

Space Radiation to system designers, planners, and operators means the spacecraft damaging particle environment ranging from keV aurora to GeV cosmic rays. The AFRL Compact Environment Anomaly Sensor, a 4 inch x 4 inch x 4 inch "black box" with a mass of about 1 kilogram and power consumption of about 1 watt, was built to detect a broad range of particle hazards, is now undergoing its first space test and performing admirably. Designed primarily as an engineering tool to warn spacecraft operators of hazardous conditions and to reduce anomaly resolution time, a thorough

hazardous conditions and to reduce anomaly resolution time, a thorough validation and calibration of CEASE is proving that the data can be used to construct scientifically sound radiation environment models for certain dose and spectral ranges. Climatological model development is moving beyond the typical 'maximum, minimum and average' radiation models into the more complex problem of specifying probabilities of occurrence and duration of specified dose or flux levels for long duration missions. Models being developed will be easily 'upgradable', i.e. able to assimilate new databases, such as those that will become available in the near future when CEASE is launched on the joint US-European STRV-1/c satellite and on a military operations satellite, in order to quickly provide users with improved statistics and wider temporal and orbital coverage. Advances in high spectral resolution particle detection instrumentation continue with the completion of the Relativistic Electron and Energetic Proton Experiment (REEPER) suite of high-energy electron and proton sensors. When flown, REEPER will provide detailed spectral coverage of 1-30 MeV electrons and 10-400 MeV protons moving well beyond the capabilities of the previous USAF-NASA radiation belt mission (the CRRES satellite) at a fraction of the weight and cost.

The behavior of MeV electrons in the outer radiation belt is an active research area due to their detrimental impact on spacecraft via the process of deep dielectric charging. Work has concentrated on the development of models to account for adiabatic particle behavior in dynamic magnetic fields, time-dependent electric diffusion coefficients, DC electric fields, high-speed solar wind stream drivers and electron response inside geosynchronous orbit, and cyclotron-resonant wave-particle interactions. These component models are being integrated into a MeV electron transport code which will grow in capability from the slow (weeks) to the fast (minutes) time scales as acceleration processes are better understood.

Space weather begins at the Sun. To forecast events in the ionosphere and magnetosphere, whether they be short lead time forecasts for X-ray flare generated HF blackouts or long-lead time forecasts for high speed solar wind stream generated MeV electrons, a substantial degree of 'front-end' solar and interplanetary observation and modelling is required. Ground based solar observation techniques and solar surface active region modelling are the thrust of the AFRL operating location at National Solar Observatory (NSO) in Sunspot, NM. Considerably cheaper than space observations, with facilities lasting for decades rather than a typical satellite lifetime of years, ground based solar telescopes now offer superior resolution as a result of recent breakthroughs in adaptive optics. Long-term observations of iron line emissions from the coronagraph at Sac Peak have led to a new algorithm for predicting time of occurrence and intensity of solar maximum. This algorithm is undergoing validation this solar cycle. The AFRL-NSO group is also building the Improved Solar Observing Optical Network (ISOON) to replace the current world-wide SOON system as the backbone real-time solar optical data source used by USAF and NOAA forecasters. Offering improved image quality, spectral range, spatial resolution, and temporal cadence, ISOON is designed to be a remotely controlled, nearly autonomous system with reduced complexity, manning requirements, and operating costs. Models of solar surface velocity and magnetic field distributions constructed from white light and hydrogen-alpha images, such as those that will be available from ISOON, are being constructed and used to understand the build up and release of the magnetic energy ultimately responsible for solar disruptive events.

Once an explosive solar event occurs, the resultant energetic particles, plasmas, and electromagnetic fields traverse the inhomogeneous interplanetary medium before impacting the Earth's environs. Tracking the evolution of coronal mass ejections (CMEs), i.e. clouds of plasma and magnetic fields that drive dangerous geomagnetic storm-inducing shocks, all the way from the Sun source to Earth is the objective of the Solar Mass Ejection Imager (SMEI) being built by AFRL and collaborators at the University of Birmingham, England, and the University of California, San Diego. Comprised of three white-light cameras, SMEI will obtain an all sky map of the heliosphere once every 90 minutes after it is launched into a Sun-synchronous polar orbit in December 2001. Sequences of these maps will be used to observe and forecast CME speeds and transit times, improving the long lead time (1-3 days) prediction accuracy of geomagnetic storms. The impact of SMEI on long lead-time forecasts of geomagnetic storms should be comparable to that of terrestrial weather satellites for tropospheric weather forecasting. Extensive validation of the currently operational Interplanetary Shock Propagation Model over the last solar cycle has revealed deficiencies in arrival time and strength prediction that render the model essentially useless. Deficiencies in the models are of a basic physics nature: a lack of understanding of how CMEs originate and expand outwards providing 'initial conditions', no accounting for the effects of an inhomogeneous background solar wind, and insufficient knowledge of the shock-magnetosphere coupling process. The AFOSR Solar-Terrestrial Interactions laboratory task at the Center is tackling these problems through a combination of 'data mining' the SOHO, Yokoh, and ACE satellite databases, empirically determining the relevant cause-and-effect processes, and ultimately kinetic theoretic and magnetohydrodynamic modelling to improve the algorithms. Solar flares and particularly CMEs generate spacecraft and tissue damaging solar energetic particles by processes not yet understood. Researchers at the Center are defining the solar and

are developing an empirical model of solar energetic particle events to alleviate the well-known problems in current operational forecast models.

Rounding out the Center of Excellence are vigorous programs in spacecraft charging, meteor characteristics and effects, and space chemistry. Together with the NASA Space Environment Effects (SEE) program, AFRL is sponsoring the development of the NASA-Air Force Spacecraft Charging Analysis Program 2000 (NASCAP 2K) to replace the often-used NASCAP satellite design tool. Scientists from AFRL are contributing theoretical models of the dependence of geosychronous satellite charging levels on electron temperature, solid-state physics analysis determining the breakdown thresholds for dielectric materials in space, and laboratory measurements of cross sections for ion contamination reactions. In close collaboration with international experts, AFRL has developed a model of meteor storm intensity, time of occurrence and duration, to be incorporated into the AF-GEOSpace suite of space environment models (Hilmer et al, 1999). A detailed study of the effects of the deposition of metals from meteors on the night-time electron density profiles in the lower ionosphere yielded the surprising result that meteor storms increase ionization levels by over an order of magnitude for periods of hours with the consequence of increasing radio wave absorption and global HF communication capability. Necessary for this study was the accurate determination of the relevant high-energy ion charge-exchange cross sections by the AFOSR-sponsored in-house Space Chemistry laboratory task. Measurements and analyses carried out in the Hyperthermal Ion Beam Facility at AFRL are essential in determining the efficiency and damage caused by 'ion drive' thrusters. Successful quantitative remote imaging of magnetospheric plasmas employing energetic neutral atom sensors, such as being attempted by the NASA IMAGE satellite, depends critically on charge-exchange cross sections being measured by AFRL.

4. RECENT ACCOMPLISHMENTS

4.1 Operationalized space environment network display

Years of research leading to physics-based models and the space environment data-streams needed to drive them had their payoff in March 2000 when initial operating capability of the Operationalized Space Environment Network Display (Op-SEND) was achieved at 55th SWXS. For the first time, users are able to access web-based, user-friendly graphical products giving nowcast and short-term forecasts of direct operational impacts to communication, navigation, and surveillance systems. The creation of Op-SEND was also the first product of the USAF Rapid Prototyping Center. Linking together real-time sensor data input, models, and product generators Op-SEND provides the following user-impact maps:

<u>UHF SATCOM Scintillation Outage Map</u>. Utilizes real-time SCINDA sensor data and the SCINDA model to generate maps indicating low, medium, and extreme probabilities of scintillation outages from different theaters to user-specified geostationary communications satellites. The map algorithm relies on the WBMOD climatological scintillation model for regions where no real-time data is available

<u>HF Illumination Map</u>. Utilizes real-time Digital Ionospheric Sounding System (DISS) bottom-side electron density profile data and standard geomagnetic indexes to drive the Parameterized Real-time Ionospheric Specification Model (PRISM) and obtain global electron density maps. In several geographical theaters the propagation of HF radio waves is then computed using advanced statistical ray-tracing methods from several transmitter locations and for several frequencies. Users are then able to select maps indicating the distribution of HF power from the transmitter best matching their communications systems.

<u>Auroral Clutter Boundary Map</u>. Combines near real-time measurements of the equatorward boundary of the auroral oval from the SSJ4 sensor on the DMSP satellites with statistical auroral precipitation models and phenomenological radar scattering models to produce maps indicating regions where there is a high probability of radar clutter. Maps are customized for every operational space surveillance radar.

<u>GPS Single-Frequency Error Map</u>. Utilizes real-time total electron content (TEC) data from the AFRL Ionospheric Measuring System and JPL networks together with standard geomagnetic and solar indices to drive PRISM and obtain global electron density maps. For a specific time and for each point on the globe the optimal GPS configuration is determined and the single-frequency range error due to variations in the TEC is computed using the PRISM output. Users can choose maps characterizing several operational theaters. With the elimination of GPS Selective Availability the error induced by variations in ionospheric TEC becomes the largest error source for both civilian and military-single frequency systems. Figure 1 shows a typical error map.

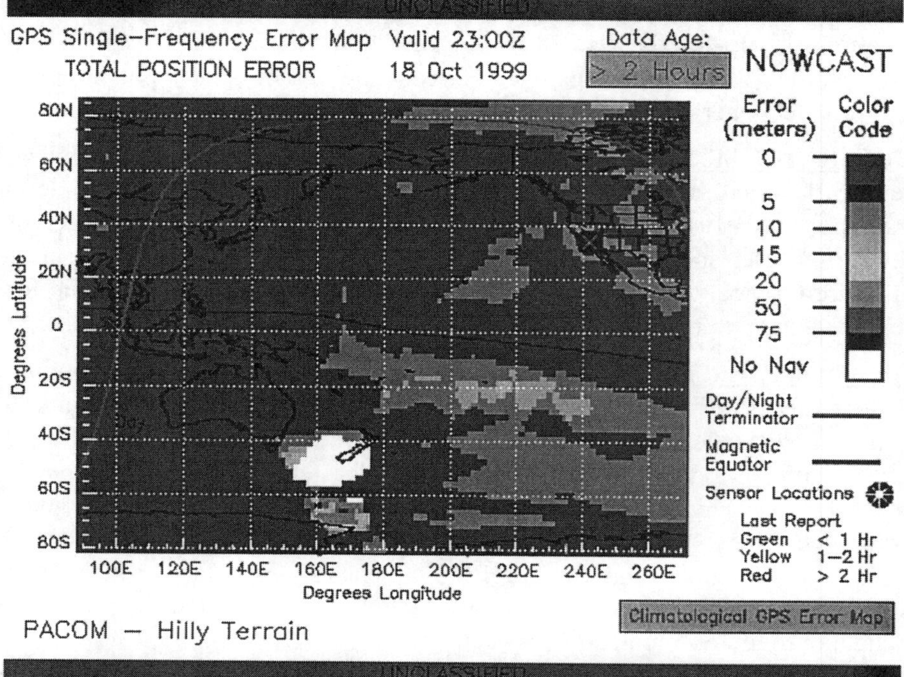

Figure 1. Op-SEND GPS Single-frequency Accuracy product showing estimates of position error at different locations.

4.2   Solar adaptive optics

The resolution of ground-based solar telescopes is seriously degraded by atmospheric turbulence, impeding our ability to resolve the fundamental physical processes causing solar variability at the spatial scales at which they occur. Together, the National Solar Observatory and AFRL have developed a solar adaptive optics (AO) system that is now becoming available to users of the Dunn Solar Telescope (DST) at NSO/Sac Peak (Rimmele et al., 1999). Using a novel correlating Shack-Hartmann wavefront sensor coupled to a deformable mirror, the first-ever AO enhanced solar spectroscopic observations (including velocity dopplergrams and magnetograms) were obtained at the DST during FY00. Marking a major advance in the quality and stability of AO technology, the resolution of the new solar measurements exceeds anything available elsewhere (Figure 2), even from space (for example, from the NASA SOHO spacecraft). The solar AO system was also used very successfully at the German solar telescope on the Canary Islands, demonstrating that its design is sufficiently flexible and robust to enable use at telescopes other than the DST. Solar AO will enhance the performance, productivity and lifetime of our existing ground-based solar telescopes (such as the DST), and represents critical enabling technology for the next-generation Advance Technology Solar Telescope being pursued by the NSO. Furthermore, solar AO has the potential to meet DoD solar event forecasting needs at a fraction of the cost of currently envisioned space-based systems.

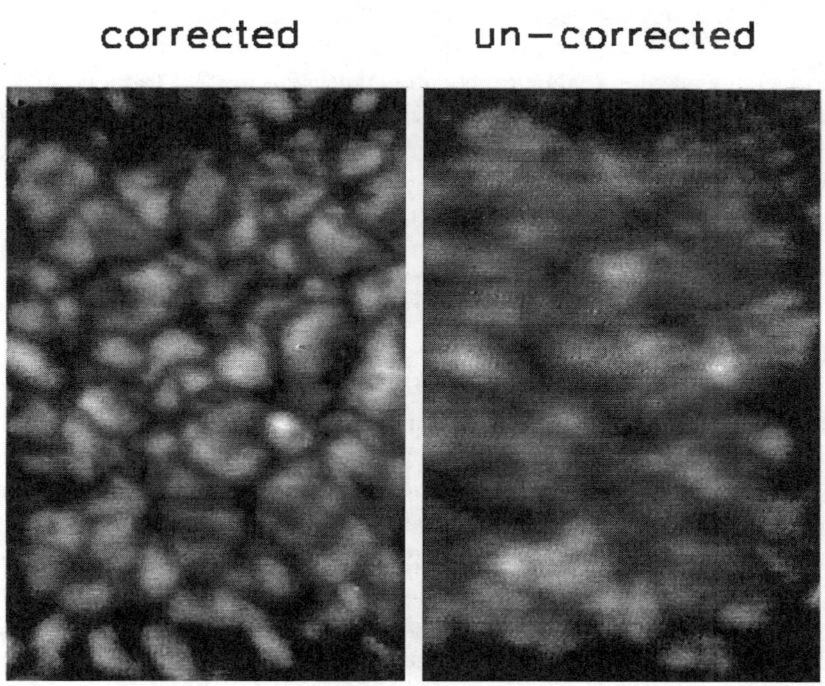

Figure 2. Comparison of solar granulation image taken with and without solar adaptive optics.

4.3　　Geomagnetic storms as scintillation triggers

Plasma density irregularities (or 'plumes') at low magnetic latitudes are the major driver of equatorial ionospheric scintillation which affects communications and navigation systems. Until recently the common wisdom was that plasma plumes were suppressed during periods of high geomagnetic activity. Recent detailed investigations of the effects of penetration electric fields on ionospheric plasma during magnetic fields have, however, constrained this understanding. In two magnetic storms investigated with measurements from the CRRES satellite, equatorial plasma plume activity was detected by DMSP satellites in the initial and main phases (Burke et al., 1998). Further investigation of approximately 15000 orbits of DMSP satellites during the solar maximum years of 1990 and 1991, in which 11 storms with strong magnetospheric ring currents (Dst < -150 nT) occurred, indicated that plasma plume formation is quite common in the initial and main phases of storms and that it is suppressed for several days after the recovery phase begins (Huang et al., 2001). These results show that the interplanetary electric field and magnetospheric compressions both provide electric fields at the magnetic equator that add to the Rayleigh-Taylor instability of the post sunset ionosphere. Both external sources must therefore be added to effects of the thermospheric dynamo in the component of the scintillation forecast model that predicts electric field strengths in the bottomside ionosphere near the dusk terminator.

4.4   Space test of the digital ion driftmeter

On 15 July 2000 the Digital Ion Drift Meter (DIDM) was launched aboard the Challenging Minisatellite Payload (CHAMP) developed by the GeoForschungsZentrum, Potsdam, Germany. CHAMP was launched from the Plesetzk cosmodrome, Russia with a COSMOS-3M rocket precisely into the planned 421 km x 475 km elliptical polar orbit. All components and functions of DIDM are performing nominally except for one of the two redundant sensors. The malfunctioning sensor passes all diagnostic tests and the problems have tentatively been attributed to a blockage of the aperture. Since the sensors are identical except for aperture size, the loss of sensor "B" will have only a modest effect on DIDM's dynamic range and sample rate. DIDM is an advanced ion velocity, temperature, and density sensor utilizing miniaturized state-of-the-art detector components and on-board digital signal processing. Data from DIDM is crucial for determining plasma densities and DC electric fields in the upper ionosphere, both key parameters needed by ionospheric hazard specification and forecast models used to meet C3I operational requirements. Figure 3 shows an artists rendition of the CHAMP satellite with the DIDM sensor indicated by the white arrow. Shown in the figure are some initial measurements by DIDM giving the ion velocity distribution as the spacecraft major axis orientation changed from approximately 10 degrees off the ram direction into the ram direction. Scientists at AFRL are currently calibrating DIDM and validating performance against other in-situ and radar scattering measurements. The unique sensing capability, small footprint, and low power consumption of DIDM make it an ideal candidate to replace the DMSP SSIES drift meter sensor on future operational missions, such as the National Polar Orbiting Satellite System (NPOESS). DIDM will next fly on the AFRL Communications/Navigation Outage Forecast System (C/NOFS) satellite scheduled for launch in 2003.

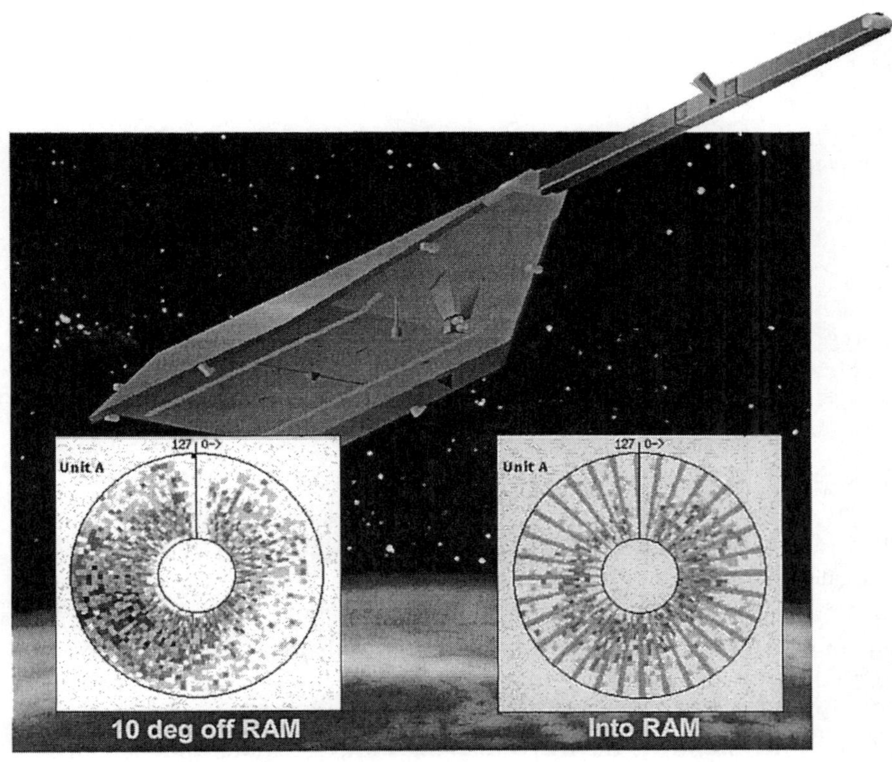

Figure 3. DIDM on the CHAMP satellite with preliminary ion velocity distribution measurements 10 deg off RAM and looking into the RAM direction.

4.5 Space test of the compact environment anomaly sensor

On 7 June 2000 the first Compact Environment Anomaly Sensor (CEASE) was launched aboard the Space Test Program's Tri-Service Experiment 5 (TSX-5) spacecraft into a 410 x 1740 km, 68° inclination elliptical orbit. The instrument was turned on 24 hours after launch and has been operating nominally ever since. CEASE is a highly miniaturized suite of sensors and particle detectors designed foremost to provide real-time alerts to operators of hazardous radiation conditions on board satellites (Dichter et al. 1998).

When operated in the higher telemetry science mode, data from CEASE provides operators and engineers a clear picture of the space radiation environment tremendously useful when resolving satellite anomalies. The science mode data stream can also be utilized to construct empirical models for improved system design and for scientific studies of space particle dynamics. Analysis of early on-orbit data, calibration against laboratory data, and validation of overall instrument performance is now underway at AFRL. Figure 4 shows a plot of the daily averages of the count rate of one of the CEASE dosimeter channels sensitive to 30-70 MeV protons as a function of time and magnetic L-shell (approximately the distance from the center of the Earth at the magnetic equator) from initial operation (9 Jun 00) to 1 Sep 00. It was fortuitous that shortly after CEASE was launched a series of solar energetic particle events occurred and the instrument was able to measure these hazardous particle populations when the orbit reached sufficiently high magnetic latitudes. The intense 'Bastille Day' event on 14 Jul 00 is readily apparent in Figure 4. Future CEASE flights include the Space Test Research Vehicle 1c (STRV-1c) mission, scheduled for launch in October 2000 and a ride on a military operational satellite, scheduled for launch in 2001.

4.6 Meteor effects in the night-time ionosphere

When meteoroids enter the Earth's ionosphere (beginning at 120 km) they are heated by collisions with atmospheric species. As the density of the atmosphere increases, the heating increases until the metals begin to evaporate from the meteoroids. Because the metals have low ionization potentials, they undergo two types of reactions that affect the ionosphere: (1) charge exchange with the ambient ions (principally O_2^+ and NO^+), and (2) ionization in the hyperthermal collisions with the atmospheric species. During the past year cross sections have been calculated for ionization by hyperthermal collisions, and charge exchange reactions involving meteor metals have also been measured. Alongside with this laboratory work, a dynamic ionosphere that incorporates the vaporization dynamics of meteoroids has been developed (McNeil et al., 2001). These achievements have made it possible to predict the effect of meteoroids on the ionosphere under quiescent conditions, during a meteor shower, and during a meteor storm of the magnitude of 1966 (150,000 meteors per hour). The latter case shows enhanced ionization that overwhelms the normal ionosphere (Figure 5). Furthermore because the Leonid meteoroids collide with the Earth at high velocities (about 72 km/s) the ionization occurs at higher altitudes, so that the atomic ions that are generated are not easily neutralized by collision with electrons. The net effect of this is to increase the total electron content by, perhaps, two orders of magnitude for periods as long as a day. This has major consequences for communication and navigation operations.

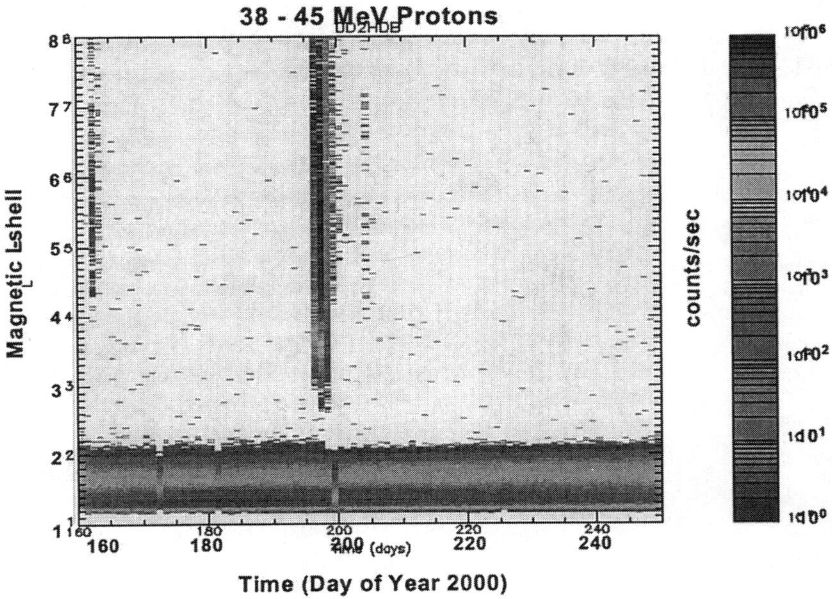

Figure 4. Preliminary results from the CEASE instrument on the TSX-5 satellite showing the 'Bastille Day' solar proton event.

5. FUTURE RESEARCH DIRECTIONS

To a large degree the future directions of the Space Weather Center of Excellence will be extrapolated from the current efforts. The development of space- and ground-based environment sensors, real-time specification and climatological models, forecast models, and passive and active techniques for effects mitigation will continue to be the pillars upon which the Center's research program will be built. There will be a natural evolution consistent with increased capabilities. As sensors proliferate in both space and on the ground more effort will be given to developing sophisticated data assimilation techniques. As real-time specification and the fundamental understanding of space weather processes improves, truly coupled models across the space weather domains can be developed to get long lead times for quantitative forecasts. Research priorities could change as implementation of the NSSA architecture under the NSWP increases space weather awareness and 'fleshes out' more requirements. At the current time the highest priority on the military space weather requirements list is the specification and forecasting of scintillation effects on satellite communications and navigation in the Earth's equatorial regions. To meet these requirements a major thrust of AFRL space weather research over the next several years will be the Communications/Navigation Outage Forecast System (C/NOFS).

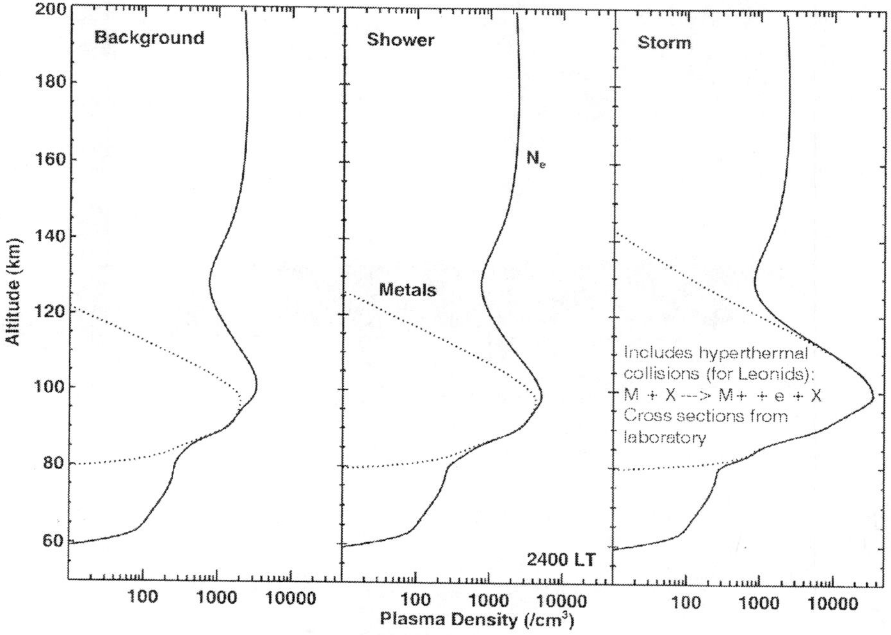

Figure 5. Modelled profiles of nighttime ionospheric electron density showing the large effect of meteor showers.

The mission of C/NOFS is to demonstrate new short-term forecast capabilities and to establish a scientific basis for long-term forecasts. To do this, C/NOFS will be composed of three core elements:

<u>C/NOFS satellite</u>. As the primary element, the C/NOFS satellites will be in an equatorial, 400 x 600 km orbit and provide in-situ measurements of seven key parameters pertinent to scintillation formation and propagation ranging from DC electric fields to electron density fluctuations. The satellite will also carry radio beacons for remote sensing of scintillation fluctuations.

<u>Ground sensor system</u>. Composed of small, simple scintillation receivers, the ground system will be an expanded version of the current SCINDA network and will augment the satellite by providing real-time specification in many theaters.

<u>Models and user products</u>. Both ground and satellite data will be assimilated to produce a global specification and forecast algorithms then applied to predict system-specific communication outages at least six hours in advance. C/NOFS maintains a research emphasis because the processes underlying the onset of scintillation are not well enough understood yet to produce forecast algorithms.

Scheduled for launch in mid 2003, the C/NOFS mission will be accomplished in three phases. In the survey mode, which lasts for the first 1-9 months after launch, key scintillation onset parameters will be identified and forecasting models refined. The forecast mode will occur 6-12 months after launch and real-time downlink of satellite data will be used to demonstrate scintillation event prediction. Finally, the prototype operational demo will last 12-36 months after launch and provide communication and navigation outage forecasts to the theaters in real-time. With luck, the USAF will begin operations of the first-ever global scintillation forecasting system in FY05.

The trend in space weather forecast models is towards multi-grid, multi-timescale computer codes covering entire domains and, in some cases, coupling across domains. Much like their tropospheric counterparts these codes will require significant computational horsepower. To meet the research community needs for space-weather-dedicated high performance computers, several government agencies have partnered to create the Community Coordinated Modeling Center (CCMC) with front-end workstations at NASA/Goddard in Greenbelt, MD connected via high-speed connections to parallel-processing supercomputers at the Air Force Weather Agency in Omaha, NB. Undergoing preliminary testing and software infrastructure construction, the CCMC will soon be open to the space weather community to run large codes mature enough to be able to predict operationally relevant events using currently available data streams. AFRL scientists look forward to using CCMC resources as large-scale forecasting codes for scintillation and space radiation effects are developed over the next several years.

6.   SUMMARY

Space is not a vacuum. Rather, it is a rapidly changing plasma, energetic particle and electromagnetic field environment posing a number of hazards to spacecraft, communications, surveillance, and navigation systems. To mitigate detrimental space weather effects, scientists and engineers at AFRL and AFOSR have built a comprehensive Sun-to-neutral atmosphere, basic research-to-operations program that addresses DoD needs in the areas of mission planning, operations, and system design. Critical to the success of National Space Weather Program, to include the USAF effort, will be the numerous collaborations and partnerships between government agencies and other nations. Space weather is truly a global system and cooperation among all interested parties is essential if specification and forecast goals are ever to be achieved.

Ultimately, every commander and satellite operator seeks an 'all weather' system where the environment will have no impact. This is not likely to ever happen is space since the design constraints and costs become too severe, and the benefits of new technological capabilities, even with environmentally induced weaknesses, is too great. Real-time space weather specification and forecasting is and will be critical to obtaining optimal performance of the rapidly growing space infrastructure.

7.   ACKNOWLEDGEMENTS

The author wishes to thank P. Bellaire, O. de la Beaujardiere, E. Cliver, K. Groves, J. Heckscher, R. Radick, W. Burke, D. Brautigam, W. Turnbull, K. Ray, D. Cooke, W. Borer, G. Bishop, E. Murad, D. Neidig, and R. Altrock for input to this manuscript and a critical reading of the preliminary drafts.

8.   REFERENCES

Burke, W. J., Maynard, N. C., Hagan, M. P., Wolf, R. A., Wilson, G. R., Gentile, L. C., Gussenhoven, M. S., Huang, C. Y., Garner, T. W., Rich, F. J. (1998), Electrodynamics of the Inner Magnetosphere Observed in the Dusk Sector by CRRES and DMSP During the Magnetic Storm of June 4-6, 1991, *J. Geophys. Res.*, **103**, 29399.

Borer, W. S., Anderson, D. N., Fuller-Rowell, T. J., Schunk, R. W., Sojka, J. J. (1995) Coupled Modeling of Equatorial Ionosphere and Thermosphere, *EOS*, **76**, F442.

Daniell, Jr., R. E., Brown, L. D. (1995)PRISM: A Parameterized Real-Time Ionospheric Specification Model, *AFRL Technical Report*, PL-TR-2061.

Dichter, B. K., McGarity, J. O., Oberhardt, M. R., Jordanov, V. T., Sperry, D. J., Huber, A. C., Pantazis, J. A., Mullen, E. G., Ginet, G. P., and Gussenhoven, M. S. (1998) Compact Environment Anomaly Sensor (CEASE): A Novel Spacecraft Instrument for in Situ Measurements of Environmental Conditions, *IEEE Trans. Nucl. Sci.*, **45**, 2800.

Groves, K. M., Basu, S., Weber, E. J., Smitham, M., Kuenzler, H., Valladares, C. E., Ssheehan, R., MacKenzie, E., Secan, J. A., Ning, P., McNeil, W. J., Moonan, D. W., Kendra, M. J. (1997) Equatorial Scintillation and Systems Support, *Radio Science*, **32**, 2047.

Hilmer, R. V., Editor, AF-GEOSpace User's Manual, Version 1.4 and 1.4P (1999) *AFRL Technical Report*, AFRL-VS-TR-1999-1551.

Huang, C. Y., Burke, W. J., Machuzak, J. S., Gentile, L. C., Sultan, P. J. (2001) DMSP Observations of Equatorial Plasma Bubbles in the Topside Ionosphere Near Solar Maximum, *J. Geophys. Res.*, **106**, *In Press.*

Marcos, F. A., Kendra, M. J., Griffin, J. M., Bass, J. N., Larson, D. R., Liu, J. J. (1998) Precision Low Earth Orbit Determination Using Atmospheric Density Calculation, *J. Astronaut. Sci.*, **46**, 395.

McNeil, W. J., Dressler, R. A., Murad, E. (2001), Impact of a Major Meteor Storm on Earth's Ionosphere: A Modeling Study, *J. Geophys. Res.*, **106**, *In Press.*

National Security Space Architect, Space Weather Architecture Study Report (1999) available at http://www.acq.osd.mil/nssa/majoreff/swx/swx.htm.

Office of the Federal Coordinator for Meteorology, National Space Weather Program Implementation Plan, Publication FCM-P31-2000 (2000), available at http://www.ofcm.gov/HOMEPAGE/text/pubs_linx.htm.

Rimmele, T. R., Dunn, R. B., Richards, K. and Radick, R. R. (1999) Solar Adaptive Optics at NSO, in *Resolution Solar Physics: Theory, Observations and Techniques*, Astr. Soc. Pacific Conf. Series **183**: High eds. T. R. Rimmele, K. S. Balasubramaniam, and R. R. Radick, p. 222.

Chapter 19

ESA Space Weather Activities

Eamonn J. Daly
*Space Environments and Effects Analysis Section, European Space Agency, ESTEC
2200 AG Noordwijk, The Netherlands*

Abstract   The European Space Agency (ESA) is undertaking a space weather initiative
in which preparatory studies are being performed and developments are being
made to pave the way for a possible future ESA space weather programme and
a possible European space weather service. This initiative is based on long-
standing activities in analysis of space environments and their effects on
European space programmes and on a successful solar terrestrial physics
programme over many years and many missions. This chapter describes these
activities, discusses space weather effects, outlines the goals of the present
phase of ESA's initiative and discusses future directions.

Keywords   ESA, space weather, space environments and effects.

1.   INTRODUCTION

The natural space environment represents a considerable hazard for
spacecraft and the European Space Agency (ESA) has for many years taken
measures to ensure that its spacecraft are able to survive and operate in it. As
spacecraft and their payloads have become more sophisticated, so their
susceptibilities to effects induced by the environment have increased.
Consequently ESA has, like other organizations, increased its efforts in
analysing these problems and in developing the means to understand and
anticipate the environment and to avoid the effects. This environment and its
complex behavior are also subjects of intensive scientific investigation
within solar-terrestrial physics. Space weather encompasses a broad range of
phenomena (solar, interplanetary, geomagnetic, ionospheric, atmospheric
and solid earth) impacting space and terrestrial technologies. It is a subject
which is of relevance to developers of technological systems but one which
relies on characterization and understanding of the solar-terrestrial system.

I.A. Daglis (ed.), Space Storms and Space Weather Hazards, 459–473.
© 2001 *Kluwer Academic Publishers. Printed in the Netherlands.*

A few high-profile space weather events have drawn attention to the effects. For example, the hazards to spacecraft from electrostatic charging (Baker et al, 1996, Fredrickson, 1996, Wrenn, 1995) and to ground-based power networks from induced current surges (Kappenman and Albertson, 1990, Lanzerotti, 1979). But these are just the "tip of the iceberg" of more numerous, less well-publicized problems and implications. There is consequently a growing appreciation that as society becomes more reliant on space-based systems for services such as communications and navigation, the disruptions to these services from space weather has become a serious issue. Apart from disruption to commercial space activities, scientific missions can be seriously affected because of their use of highly advanced technologies. Recent examples which will be further discussed are the effects on the Chandra and XMM (X-Ray Multi-Mirror) Newton X-ray astrophysics missions. The analyses of potential problems on these missions, and continuing evaluation in-flight, are good examples of the needs for accessible space weather resources including databases, "predictive" models and near-real-time data. Furthermore, over the next few years manned spaceflight will undergo a considerable expansion with the exploitation of the international space station. Space weather, in the form of enhancements to energetic particle radiation, is of crucial importance for manned space activities. Space radiation also penetrates the upper atmosphere where crew and electronic systems on aircraft can be affected. Finally, ground-based systems such as power distribution networks, pipelines and ground-to-ground radio communications can also be seriously affected.

At the time of writing, we are near the maximum of the current cycle, cycle 23. The increased chances of solar flares, coronal mass ejections and solar energetic particle events have led to added interest in space weather both from the space community and from the general public. This interest is supported by an array of excellent solar-terrestrial science missions such as the joint ESA-NASA Solar and Heliospheric Observatory SOHO (Huber and Wilson, 2000). However, it is important to recognize that the effects of Space Weather are present throughout the solar cycle and that some important aspects can be more severe away from solar maximum. For example electron flux levels in the Earth's outer radiation belt are generally higher during the decaying phase of the solar cycle.

The space weather discipline draws from both the space environments and effects domain and from the solar-terrestrial physics domain. ESA and European research agencies have strengths in both areas. There is considerable interest in Europe to investigate the marriage of the technological and scientific capabilities to address perceived user needs for space weather products and services. Whereas co-ordinated Space Weather activities are well established in the US, Europe has yet to undertake a coordinated program in this area. Past ESA workshops and studies identified the needs as well as possible European approaches to the subject (ESA, 1998, Koskinen et al., 1999). An important step towards a co-ordinated European Space Weather program has recently been taken with the initiation of broadly based studies in the context of the ESA General Studies Programme. Major parallel studies are laying the groundwork for a possible operational European space weather service. These studies will be discussed further later in this chapter.

2. SPACE WEATHER PROBLEMS FOR ESA PROGRAMMES

2.1 General

During the development of spacecraft, the expected environment needs to be carefully considered. The development process includes analyses of possible problems from the space environment and the implementation of appropriate measures to avoid or cope with effects of concern. Analyses make use of information on the environment in the form of models and tools which have developed over the years to cope with an evolving set of problems. Some of the effects on space systems are summarized in Table 1.

In this section, the major environmental effects will be outlined and their connection with space weather described.

2.2 Radiation Effects

As in other space agencies, ESA's concerns with space weather effects probably began with concerns over radiation effects on spacecraft systems. This radiation environment is due to sporadic solar particle events, energetic protons and electrons trapped in radiation belts and cosmic rays (e.g. Daly, 1989). The effects of these in damaging solar cells, electronic components and inducing upsets in large-scale-integrated electronics had long been taken into account in spacecraft development. In the last decade there has been a general increase in radiation-related problems and new types of problems have arisen.

While the preoccupation in the 1960's and 1970's was very much with the damage caused over the lifetime of a spacecraft to solar cells and components, in later years a number of other effects have arisen. The changes in states of logic elements in integrated circuit induced by the charge trail left by passage of a single energetic ion, known as single event

462

upsets (SEU), have become a major concern since effects on the spacecraft controlling functions could have devastating consequences. Single event rates can often be coped with when they occur in non-critical memories such as data stores and can often be corrected by special circuitry. Dramatic increases in the SEU rates often occur during solar energetic particle events as shown in Figure 1 from the SOHO mass memory unit. The rate rose by a factor of 6 at the peak of the November 1997 event and by a factor of 300 at the peak of the July 2000 event.

Environment	Effects
High Energy Radiation:	
Cosmic Rays	Upsets in electronics;
	Long-term hazards to crew;
	Interference with sensors;
Solar Energetic Particle Events	Radiation damage of various kinds;
	Upsets in electronics;
	Serious prompt hazards to crew;
	Massive interference with sensors;
Radiation Belts	Radiation damage of various kinds;
	Upsets in space electronics;
	Hazards to astronauts;
	Considerable interference with sensors;
	Electrostatic charging and discharges
Near-Earth Plasma Populations:	
Geomagnetic (sub-) storms	Electrostatic charging and discharges;
Ionospheric Effects	Communications disruption;
	Navigation services disruption
Others:	
Atmosphere	Increased drag on spacecraft and debris;
	Attitude perturbation
Meteoroids	Spacecraft damage

Table 1. The various space-weather environments and their effects.

Another source of interference is radiation background. Imaging detectors such as, charge-coupled devices (CCD's) are used in a variety of space applications including as imaging elements in space telescopes, in Earth observation systems or in star trackers of attitude control systems. Particles impacting a detector can give rise to signals which appear as "noise" on the image, sometimes completely overwhelming the image. Examples from the ESA-NASA SOHO spacecraft are shown in Figure 2. These images were taken during the July 2000 solar proton event and show heavy contamination of the image from particle hits on the detector. Such contamination is a feature of many space-borne detectors.

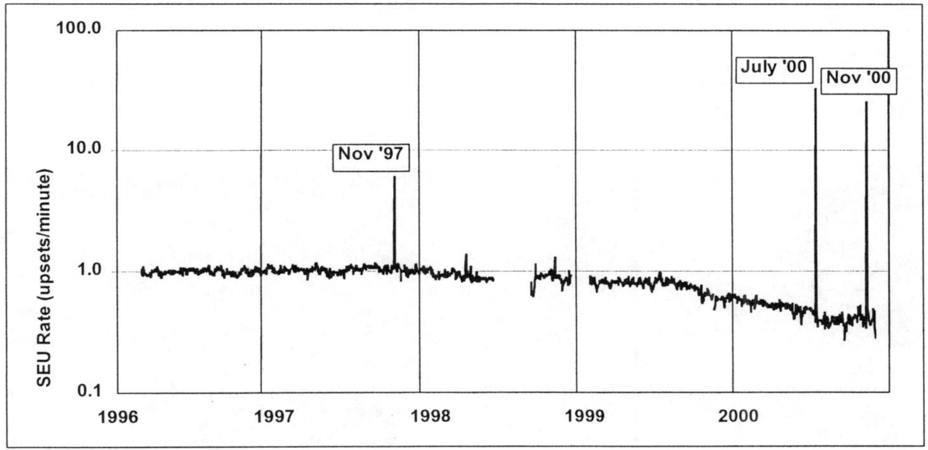

Figure 1. SOHO mass-memory single-event upsets. The spikes in November 1997 and July 2000 are due to solar particle events

ESA's Infrared Space Observatory and Hipparcos spacecraft also experienced background effects and it is also a feature of the XMM-Newton detectors. In many cases it can be removed with image processing software but if heavy contamination is present during space weather events, the data are lost. It is becoming more common to use star trackers as part of the attitude control systems of spacecraft. These help orient the spacecraft by recognizing sets of star patterns. If the image contains a lot of bright features induced by radiation, the system can become confused and several examples are known where this has occurred and led to loss of attitude. Clearly, the image background during solar energetic particle events will be very much higher than normal.

2.3   Electrostatic Surface and Internal Charging

The importance of space weather to space systems increased in the 1980's as a result of several cases of operational anomalies on geostationary communications and meteorology spacecraft. The anomalies were attributed to high-level electrostatic charging of surfaces which led to discharges and electromagnetic-induced disruption of spacecraft systems. The charging events were associated with surges of hot plasma flowing into the parts of the magnetosphere around the geostationary altitude (about 36000km) during geomagnetic "sub-storms". The affected missions include the Marecs marine communications test satellites, the ECS series of communications test satellites and the pre-operational satellites in the Meteosat meteorological satellite series. In investigating the Meteosat-1 anomalies, a decision was taken to put a plasma environment monitor on the second spacecraft in the series. The analyses of these data and their correlations with anomalies led to the conclusion that the anomalies were not due to high level surface charging.

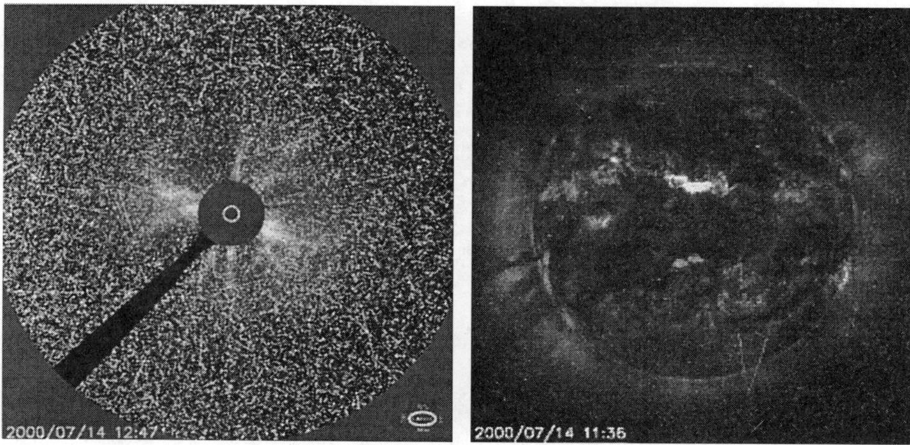

Figure 2. Images taken by the LASCO (left) and EIT (right) telescopes on the joint ESA-NASA SOHO mission during the July 2000 solar energetic particle event showing severe effects on the detector from radiation background.

About this time, it was noted by Baker et al. (1987) that anomalies on US spacecraft correlated with energetic (~MeV) electrons, implying that penetrating electrons could induce charging and discharging within spacecraft by collecting in dielectric materials or ungrounded metallic parts. Since it was a likely source of the Meteosat anomalies, Meteosat-3 contained a detector to monitor these higher energy electrons. The data showed very clear correlations with anomalies (Rodgers et al., 1998). Figure 3 shows a superposed epoch analysis of the >2MeV electron fluxes, measured in this case by a detector on the GOES geostationary satellite, for all anomalies of a particular type. This shows the average environment preceding the anomalies. The clear increase in energetic electron flux is highly indicative of internal charging as a source. Similar behavior was also reported by Wrenn (1995) for a classified UK defense satellite during the 1990's. Furthermore the energetic electron flux before the failures of the primary and back-up processors on the Equator-S mission strongly suggest that internal charging led to this total satellite loss. The environment measured by a detector on GOES-8 is shown in Figure 4 where it can be seen that preceding the failures the energetic electron fluxes were high as a result of injection events ("storms"). Equator-S was in an eccentric equatorial orbit crossing the radiation belts while GOES is in geostationary orbit. Equator-S was therefore probably exposed to a more severe environment than GOES measured. These European examples are in addition to several cases reported in recent years in the US.

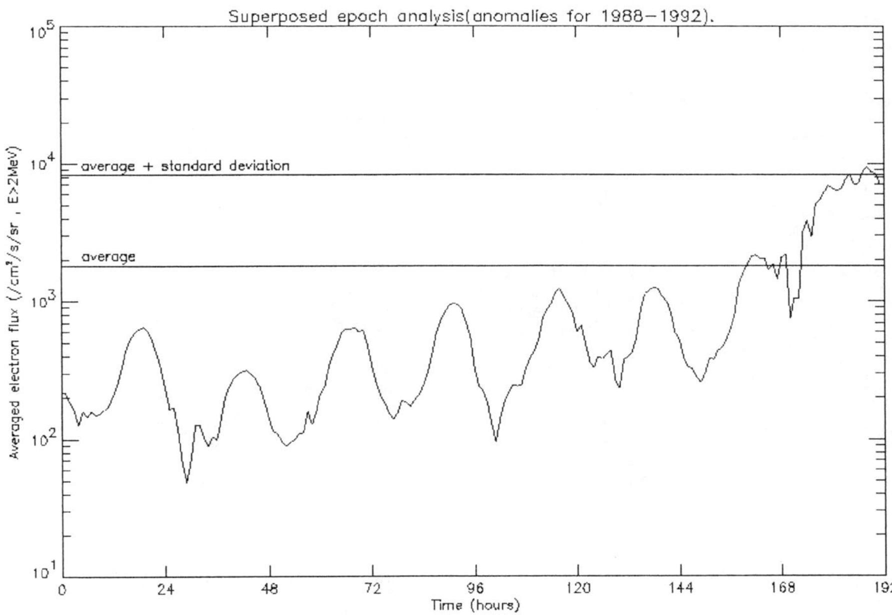

Figure 3. The average >2MeV electron environment preceding a particular type of anomaly on Meteosat-4. On average, the flux of energetic electrons increases by orders of magnitude before an anomaly

2.4 The Role of Models

To ensure that spacecraft will operate correctly in the presence of these effects, it is necessary during the development process to use models of the environments and effects for analyses, and to undertake appropriate testing. Models are intended to address the needs of the space system developer and for efficiency and usability reasons often simplify the physics involved in the phenomena. Even when physical understanding or information is incomplete, the threat still needs to be countered with some quantitative method, albeit of limited validity. It is nevertheless a long-term objective for this community to have models available which are both physically accurate and responsive to the users needs. A good example is in the area of radiation environments and effects where for many years developers have used the "standard" AP-8 (Sawyer and Vette, 1976) and AE-8 (Vette, 1991) models of the radiation belts. These models are known to be weak and do not represent the dynamic ("space weather") behavior of the electron belt. Nevertheless, in the absence of anything better, they have continued to be used. Developments have recently given hope but there still remains a usability problem.

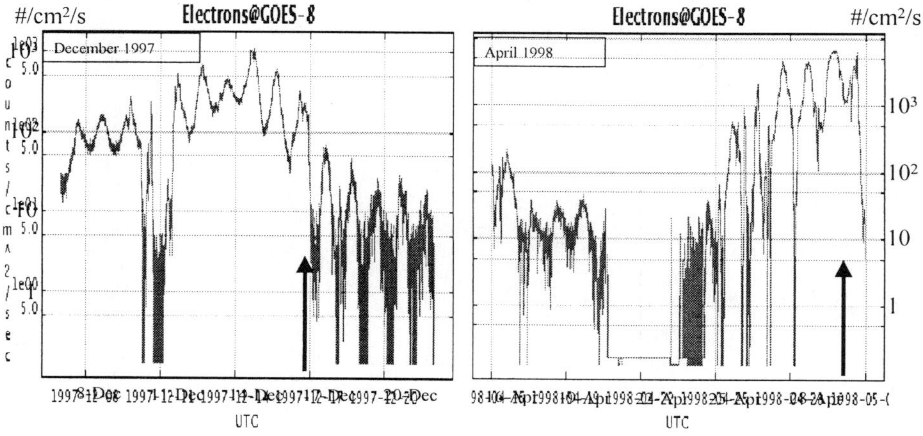

Figure 4. The environment in geostationary orbit as measured by GOES-8 detectors for the periods around the Equator-S primary and backup processor failures, indicated by the arrows

ESA's Space Environments and Effects Analysis Section has responsibility for supporting the development of ESA missions. The service it provides includes assessments of elements in Table 1. In parallel with this support function, it is responsible for the initiation and execution of technology R&D as part of a space environments and effects technical domain of ESA's Technology Research Programme (ESA, 1999). This R&D has led to developments of tools and models, as well as R&D for longer-term application. In doing this R&D and support work, the section is closely in touch with the user needs for space weather data for space system applications. Current R&D activities include (ESA, 1999):

- The Space Environment Information System (Spenvis) (Heynderickx, 1998). This is an internet/intranet-based system containing a wide range of models, tools and data concerning many aspects of space environments and their effects on space systems. It is targeted at the space systems developer who needs rapid reliable access to authoritative (often standard) methods. The system also contains link to the European ECSS engineering standard on space environment (ECSS, 2000).

- Modelling of the Earth's radiation belts where various high altitude and low-altitude data sets are studied to validate or improve models of the radiation belts. The activity also includes detailed comparison between a physical model, Salammbô, and spacecraft data of energetic electron belt dynamics
- Development of data-based analysis of space environments ("SEDAT") (CLRC, 2000) where existing spacecraft data sets are interrogated by standard and user-defined methods to derive custom "models".
- Development of engineering tools for assessment of the hazard from charging of materials inside spacecraft by energetic electrons ("internal charging") (Sørensen et al., 1999). This research also investigated the way to specify the hazard for design and the associated test methods.
- Participation with the high-energy physics community in a world-wide effort to produce a next generation of object-oriented tool-kit for simulations of particle interactions with matter ("Monte-Carlo codes"), Geant4 (Apostolakis, 2000). This effort was initiated by CERN, the European center for nuclear research. ESA's activity has resulted in space-specific features for Geant4 (Truscott et al., 2000).
- Developments of space environment monitors and the analysis and exploitation of data from them (Bühler, 1998, Desorgher at al., 1999, Daly et al., 1999).
- Analysis of electrostatic charging behavior of spacecraft in polar orbits and analyses of the correlations between anomalies and environmental parameters (Andersson et al., 1998). These studies also included research on tools for anomaly predictions (Wu et al., 1998).
- Research on AI methods in spacecraft anomaly analysis and prediction - the SAAPS (Spacecraft Anomaly Spacecraft Anomaly Analysis and Prediction System) (Wintoft, 1999).

Many other important activities have been undertaken including activities related to Martian environments, micro-particle impacts and contamination (ESA, 1999).

3.   SPACE WEATHER AND SPACE ENVIRONMENT SUPPORT TO PROJECT DEVELOPMENT

As mentioned, a key task is to support space systems development. Virtually all ESA spacecraft are supported, starting early in the process with the mission concept definition. A good case history is the XMM-Newton X-ray astronomy mission.

X-ray astronomy in space relies on the focussing of X-ray photons by low-angle scattering from shaped "shells". In most cases the "optics" consist of two sets of nested concentric shells with shapes near to sections of cones. Two grazing-incidence scatters result in focussing of the X-rays on the shell axis. ESA's XMM-Newton mission has three mirror modules of outer diameter 70 cm, each consisting of 58 nested shells which focus the X-rays onto CCD detectors some 7 m from the mirrors. XMM is in a highly eccentric orbit of apogee 114000km, perigee 7000km and inclination 39°. In this orbit it is subjected to fluxes of electrons and ions of various energies from magnetospheric and heliospheric sources.

Recently, an intensive investigation was undertaken to study potential problems to detector operation from medium-energy (100's of keV's) protons (Nartallo et al., 2000). CCD detectors are known to be radiation sensitive and much attention is given to shielding them against radiation penetrating spacecraft structural materials. However, it was found that protons of energies in the range of hundreds of keV to a few MeV could scatter at low angles through the mirror shells. These protons, because of their low energy can produce a high non-ionizing dose in unshielded CCDs and therefore pose a potential threat. Historical data on the interplanetary and magnetospheric low-energy proton environments were interrogated to determine the magnitude of the threat. Complex modelling of particle propagation through spacecraft systems was undertaken with the Geant4 Monte-Carlo toolkit. The datasets were used to establish details of observing time expected to be lost in protecting the CCDs from sporadic particle flux enhancements by closing protective shields. Several interesting points regarding space weather can be made as a result of this analysis:

- Crucial data sets used for the analysis of mission-critical engineering problems were produced by science missions (IMP, SOHO, ACE, Equator-S, ISEE) which could never foresee such applications;
- XMM-Newton has an on-board radiation monitor, to which there was resistance early in the project preparation. It is now an important resource on the spacecraft;
- Spacecraft operators are keenly interested in the state of the space weather and would certainly make use of predictions of sporadic particle enhancements should they be available.

All this effort was in addition to several space environment related analyses carried out in the course of the definition and development of the XMM-Newton mission over the preceding 10 years. In such a process, early analyses of the environments of orbit options were undertaken, followed by detailed analyses related to the final orbit and the radiation doses and particle fluxes to be anticipated for electronic components and detectors. Further analyses included assessments of the electrostatic charging hazards, analysis of the potential problems from micro-meteoroids (punctures to telescope tube and hazards to fuel tanks) and detailed analysis of radiation background sources (Dyer et al., 1995, Hilgers et al., 1998). XMM-Newton was launched in December 1999 and is operating well.

4.   ESA'S SCIENCE ACTIVITIES

ESA's science program is related to space weather in two ways. Space weather effects on science missions are an increasing concern, while on the other hand science missions can contribute crucially to space weather research. As space science missions become more complex and demanding, the need to design tolerance to space weather effects into scientific payloads as well as spacecraft systems becomes more important. Examples include sensitivity to radiation, leading to increased backgrounds and even detector damage, as well as the complete failure of key components. As mentioned above, these issues were of concern to the recently launched XMM-Newton mission (Nartallo et al., 2000).

An important spin-off of scientific missions can be to show what is possible for future service-oriented ventures. For example, the joint ESA-NASA SOHO mission is a key member of the fleet of spacecraft studying the Sun and its effects on the interplanetary environment. It is also highly useful as a resource for providing Space Weather warnings. ESA's scientific studies related to Space Weather phenomena were further enhanced with the launch of the Cluster II satellites.

As part of the competitive process for selection of future science missions, ESA recently studied future medium-sized missions. Among these were the STORMS and Solar Orbiter proposals, both of which could contribute to the world-wide space weather effort. STORMS was proposed as a set of 3 spacecraft in eccentric near-equatorial earth orbits. With apogee at about 8 Earth radii, the spacecraft pass through the radiation belts and the ring current regions. As the name suggests, the principal motivation for the mission was to study the physics of geomagnetic storms and the inner magnetosphere's responses to them. The spacecraft would carry particle and fields instruments and energetic neutral atom imagers. Solar Orbiter was proposed to orbit the sun as close as 40 solar radii (0.19 AU) and to carry out detailed solar remote sensing. Its orbit would also take it to helio-latitudes of about 33°. For part of the time the orbit would be quasi-co-rotational. Spectroscopy and imaging would be performed at high spatial and temporal resolution, along with in-situ sampling of particles and fields. Both proposals were highly rated and eventually Solar Orbiter selected for implementation. Launch is presently planned for 2009.

5. EFFORTS TOWARDS A EUROPEAN SPACE WEATHER PROGRAMME

Recognizing that there is a growing need for space weather related data for ESA programs, and also that there were issues related to the impact of space weather on non-space technologies which could be important for Europe, ESA took steps to analyze the subject in detail. While not the first ESA activity, a workshop held in 1998 (ESA, 1998) was an important event which brought together the user, science and technology communities to explore the possible ways forward. It was clear that user needs were growing. At the same time, the maturity achieved in solar terrestrial physics, allied to technological advances (in-orbit monitoring, ground-based computing power, etc.) meant that it was certainly feasible to deliver products for users in the short term and contemplate considerable improvements to them over the medium and long terms. These improvements would imply developments in the systems deployed in space and on the ground for space weather monitoring and in the science, simulation, modelling and delivery aspects of the ground-based activities. As a result ESA approved the execution of parallel wide-ranging studies. The two studies (ESA, 2000) are led by Alcatel Space and Rutherford Appleton Laboratory. In each consortium there is a strong blend of technology, science and applications. The top-level goals of the studies are to:

- investigate the needs for and the benefits of an ESA or other European space weather program
- establish the detailed data supply requirements by detailed consideration of the quantification of effects and intermediate tools;
- perform detailed analysis of potential program contents:
      a detailed definition of the space-segment
      a detailed definition and proto-typing of the service-segment
- perform an analysis of collaborative and organizational structures which need to be implemented by ESA and member states
- provide inputs and advice for preparation of a program proposal, including project implementation plan, cost estimate and risk analysis.

In association with these activities, ESA has also established *a Space Weather Working Team* consisting of European experts in various space weather and user domains, to oversee the activities and advise ESA on future activities.

While there is considerable interest in space weather in Europe, initiating any major new ESA space weather activity requires the agreement of national delegations to ESA's decision-making committees. Such a commitment can only be made after the needs for such expansion and the demonstration of its benefits are clearly established. The more scientific aspects will probably be the responsibility of ESA's science program where proposals are subject to the well-established peer review selection process. While technological research and developments will continue into space environments and effects, any large expansion of these activities for ground- and space-based space weather infrastructures is conditional upon high-level approval. The above studies and associated activities are crucial in establishing the justification.

In ESA member states many important activities related to space weather are being undertaken as part of national programs (ESA, 1998). These include activities addressing military needs. The interests of ESA's various member-states also differ. For example, Scandinavian and other nations at high latitude are keenly interested in effects on power systems, pipelines and other ground systems from auroral electrojet induced ground-level currents.

It will also be important to consider how any eventual service will be implemented in Europe. ESA's role is as an initiator and developer of technologies. The provision of fully operational end-user services should be provided by other organization in a way analogous to satellite communications or meteorology.

6.   CONCLUSIONS

The wide-ranging activities of ESA in the space weather and space environment domains have been summarized and recent important examples of space weather concerns given. In particular, the space weather effects on XMM and efforts to analyze these effects and other space environmental hazards illustrated the depth and breadth of the work that is typically necessary in this domain when preparing a complex space mission.

We have highlighted the important scientific and technological contributions that ESA in particular and Europe in general have made. We emphasize that while there is considerable interest in Europe in expanding space weather activities toward a fully-fledged program, this will be as a result of clear demonstration of real needs and benefits. These complex issues are being addressed by on-going studies.

7. ACKNOWLEDGEMENTS

I am very grateful to Alain Hilgers of ESA/ESTEC for his considerable efforts related to the ESA space weather initiatives described here. The data used in Figure 1 were provided courtesy of Helmut Schweitzer of the SOHO project team. The images on Figure 2 are courtesy of the SOHO/LASCO and SOHO/EIT consortia of the mission. SOHO is a project of international co-operation between ESA and NASA. The data used in Figure 4 were obtained from the NOAA-NGDC SPIDR service.

8. REFERENCES

Andersson L., L. Eliasson, P. Wintoft, Prediction of times with increased risk of internal charging on spacecraft, in Proceedings of the ESA Workshop on Space Weather" ESA-WPP-155, November 1998.

Apostolakis J., Geant4 status and results, CHEP-2000 Conference, to appear in proceedings, 2000

Baker, D. N., J. H. Allen, R. D. Belian, J. B. Blake, S. G. Kanekal, B. Klecker, R. P. Lepping, X. Li, R. A. Mewaldt, K. Ogilvie, T. Onsager, G.D. Reeves, G. Rostoker, H.J. Singer, H. E. Spence, N. Turner, An assessment of space environmental conditions during the recent Anik E1 spacecraft operational failure", ISTP Newsletter, Vol. 6, No. 2. June, 1996. Retrievable from http://www-istp.gsfc.nasa.gov/istp/newsletter.html

Bühler P., Desorgher L., Zehnder A. and Daly, E., Observation of the Radiation-Belts with REM, in: Proceedings of the ESA Workshop on Space Weather, WPP-155, p.333, ESA, 1998.

CLRC Rutherford Appleton Laboratory, SEDAT Project homepage http://www.wdc.rl.ac.uk/sedat/, 2000

Daly E.J., Radiation Environment Evaluation for ESA Projects, in High Energy Radiation Background in Space, eds. A.C Rester & J.I. Trombka AIP Conference Proceedings 186, NewYork 1989 483-499

Daly E.J., P. Bühler and M. Kruglanski, Observations of the outer radiation belt with REM and comparisons with models, IEEE Trans. Nucl. Sci NS-46, 6, December 1999.

Desorgher L, P Bühler, A Zehnder, E. Daly and L Adams, Modeling of the outer electron belt during magnetic storms, Radiation Measurements 30, 5 pp 559-567, 1999

Dyer C. and Truscott P., The analysis of XMM instrument background induced by the radiation environment in the XMM orbit", DERA report DRA/CIS(CIS2)/CR95032/1.0, final report on work performed for ESTEC Contract No 10932/94/NL/RE, December 1995; http://www.estec.esa.nl/wmwww/WMA/XMM/XMM.html

ECSS (European Co-operation on Space Standards), Space Environment, ed. E. Daly, ECSS-E-10-04, ECSS Secretariat, ESTEC, Noordwijk The Netherlands, 2000.

ESA, Proceedings of the Workshop On Space Weather, ESTEC, Noordwijk, The Netherlands, ESA WP-155, 1998

ESA, Space environments and effects, in: Special Issue, Preparing for the Future Vol. 9, No. 3, December 1999.

ESA Space Weather Studies www Site: http://www.estec.esa.nl/wmwww/spweather/spweathstudies.htm , ESA, 2000.

Fredrickson A.R., Upsets related to spacecraft charging, IEEE Trans. Nucl. Sci. NS-43, 2, 426, 1996.

Heynderickx, D., Quaghebeur, B., Fontaine, B., Glover, A., Carey, W.C. & Daly, E.J., New features of ESA's space environment information system (Spenvis), in: Proc. ESA Workshop on Space Weather, ESTEC, Noordwijk, The Netherlands, ESA WPP-155, pp. 245-248, ESA 1998.

Hilgers A., P. Gondoin, P. Nieminen and H. Evans, Prediction of plasma sheet electron effects on X-ray mirror missions, Proc. ESA Workshop On Space Weather, ESTEC, Noordwijk, The Netherlands, ESA WP-155 p.293, 1998.

Huber M.C.E. and Wilson A., eds., ESA's report to the 33rd COSPAR meeting", pp.29-33, ESA- SP-1241, 2000.

Kappenman J.G. and Albertson V.D., Bracing for the geomagnetic storms", IEEE Spectrum, , p5, March 1990.

Lanzerotti, L. J., Geomagnetic influences on man-made systems, J. Atm. Terr. Phys., 41, 787-796, 1979.

Nartallo R., E. Daly, H.D. Evans, A. Hilgers, P. Nieminen, J. Sørensen, F. Lei, P.R. Truscott, S. Giani, J. Apostolakis, L. Urban and S. Magni, Modelling the Interaction of the Radiation Environment with the XMM-Newton and Chandra X-ray Telescopes and its Effects on the On-board CCD detectors, Proceedings of the Space Radiation Environment Workshop, DERA Space Department, Farnborough, UK, in Press 2000.

Sawyer D.M. and Vette J.I., AP8 trapped proton environment for solar maximum and solar minimum', NSSDC 76-06, 1976.

Sørensen J., D. J. Rodgers, K. A. Ryden, P.M. Latham, G. L. Wrenn, L. Levy, G. Panabiere, ESA's tools for internal charging., in: Proceedings of the Conference on Radiation and its Effects on Components and Systems (RADECS), 1999.

Truscott P., F. Lei, C. Ferguson, R. Gurriaran, P. Nieminen, E. Daly, J. Apostolakis, S. Giani, M.G. Pia, L. Urban, M. Maire, Development of a spacecraft radiation shielding and effects toolkit based on Geant4, CHEP-2000 Proc., 2000.

Vette J.I. The AE-8 trapped electron model, NSSDC/WDC-A-R&S 91-24, 1991

Wintoft P., Development of AI methods in spacecraft anomaly predictions, IRF Web site http://eos.irfl.lu.se/saaps/ 1998.

Wrenn G.L. Conclusive evidence for internal dielectric charging anomalies on geosynchronous communications spacecraft, J. Spacecraft and Rockets 32, pp 514-520, May-June 1995

Wu, J.G, Lundstedt, H., Eliasson, L., Hilgers, A., Analysis and real-time prediction of environmentally induced spacecraft anomalies, in Proceedings of the ESA Workshop on Space Weather ESA-WPP-155, November 1999.

476

T

U

V

W

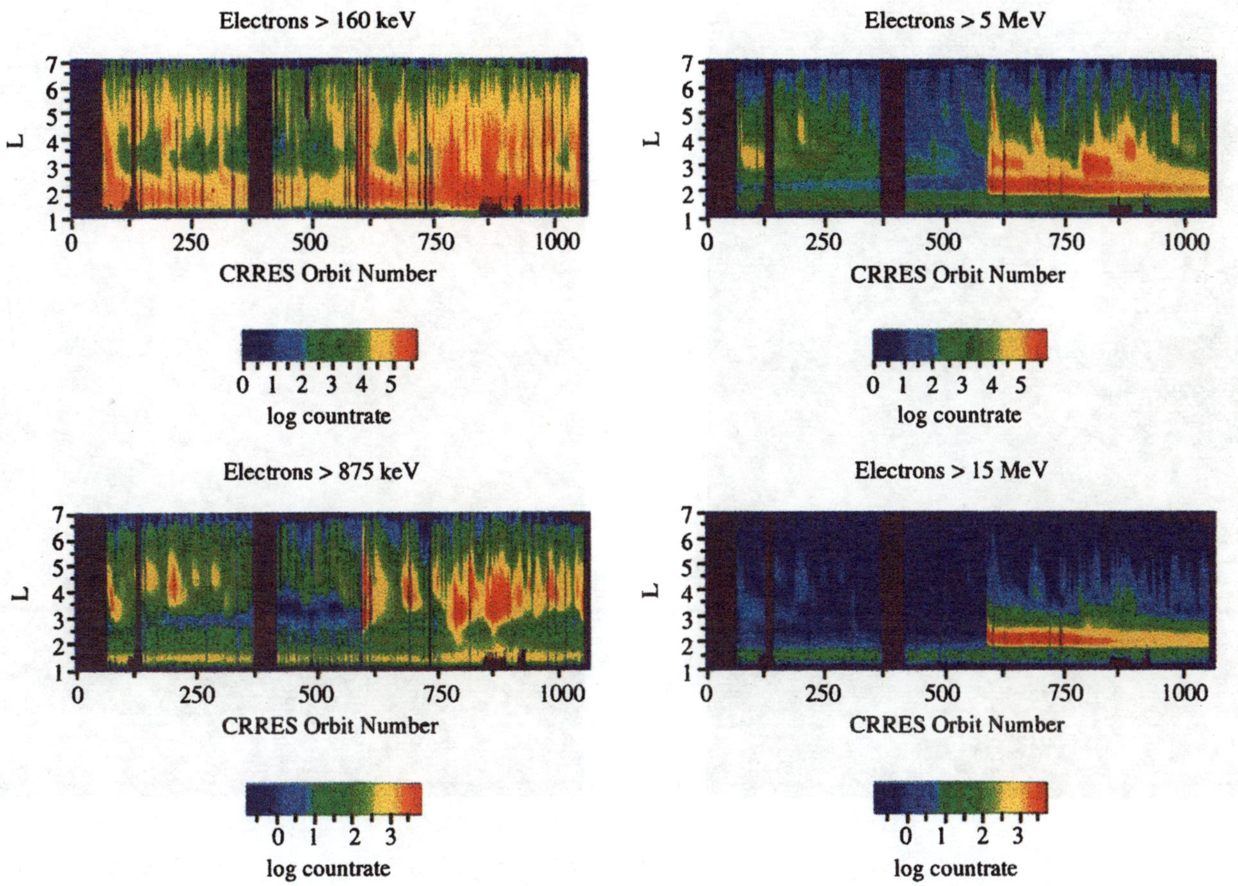

Figure 3 from p. 95

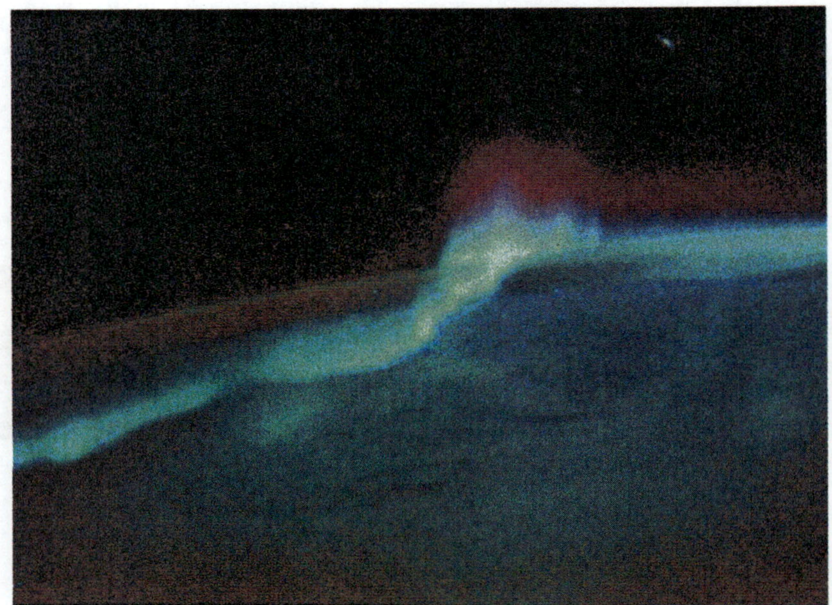

Figure 8 from p. 112

Figure 9 from p. 112

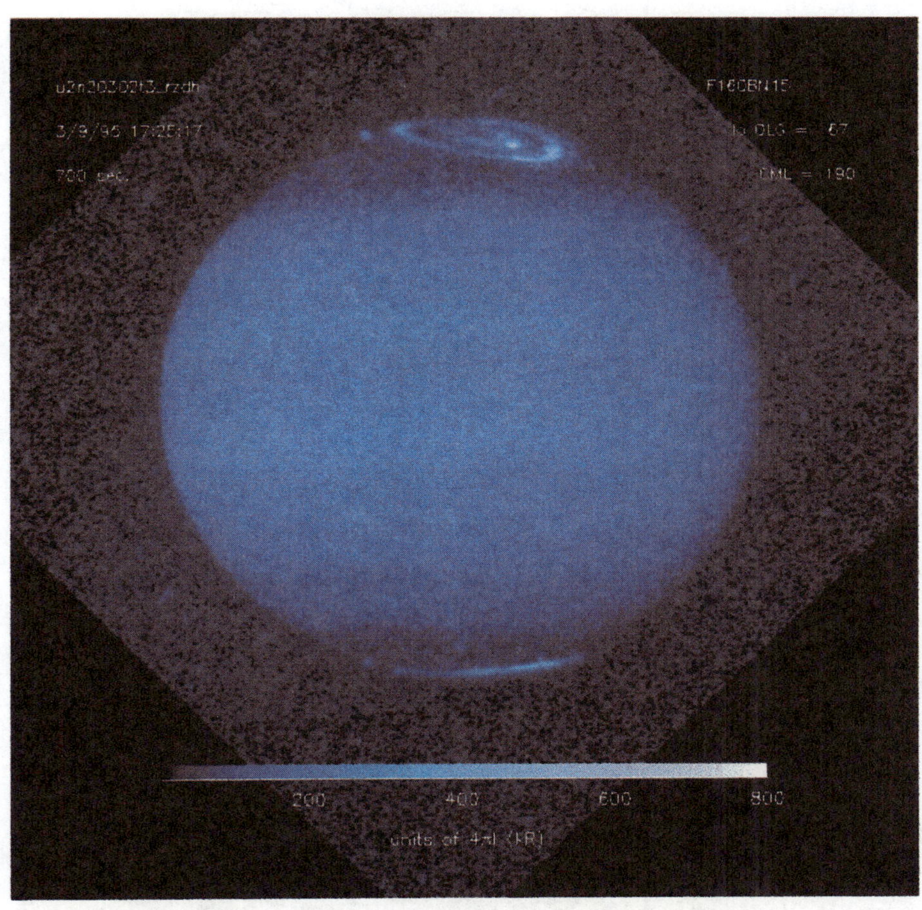

Figure 23 from p. 127

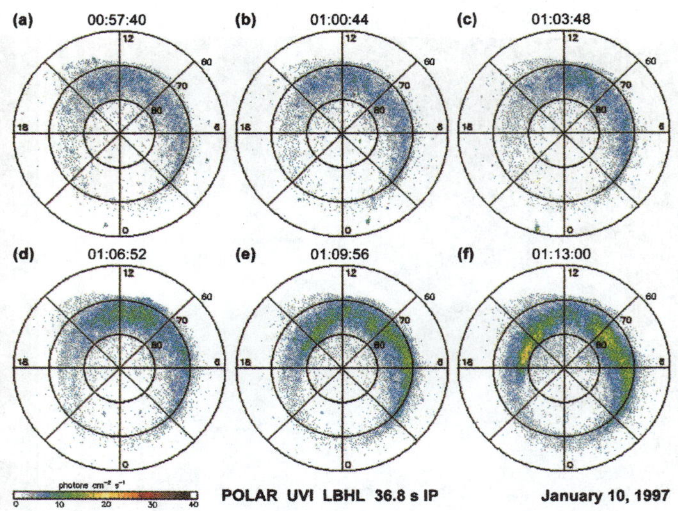

Figure 21 from p. 124

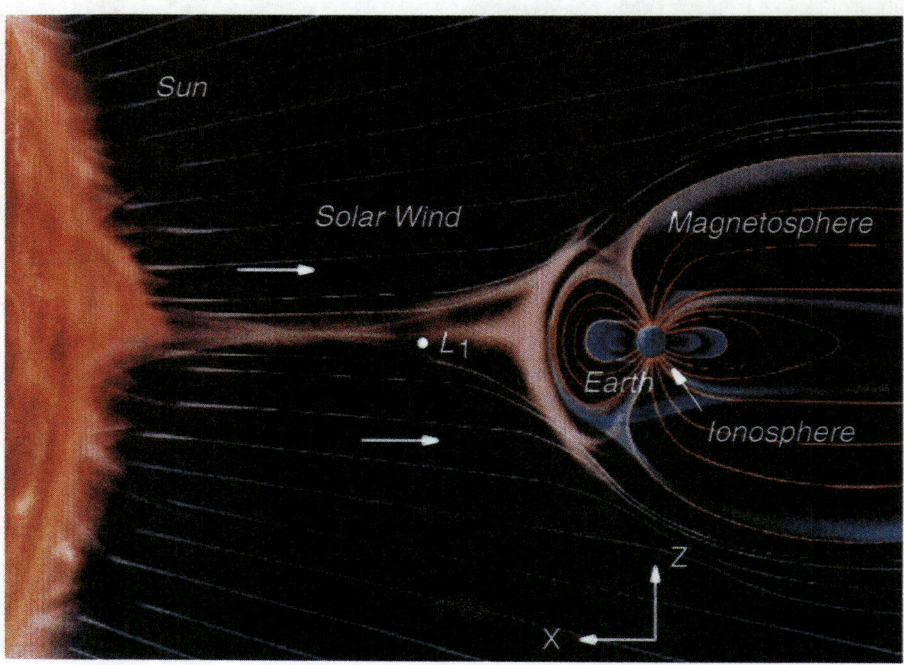

Figure 1 from p. 159